NEURAL MECHANISMS AND BIOLOGICAL SIGNIFICANCE OF GROOMING BEHAVIOR

ANNALS OF THE NEW YORK ACADEMY OF SCIENCES
Volume 525

NEURAL MECHANISMS AND BIOLOGICAL SIGNIFICANCE OF GROOMING BEHAVIOR

Edited by Deborah L. Colbern and Willem H. Gispen

The New York Academy of Sciences
New York, New York
1988

Cover (soft cover only): Rats engaged in various grooming behaviors. Shown, clockwise from upper right, are body grooming, head washing (including snout, eyes, and ears), and hindpaw licking. (Drawing by D. K. Donker, University of Utrecht.)

Library of Congress Card Number: 88-15265

CCP
Printed in the United States of America
ISBN 0-89766-441-8 (cloth)
ISBN 0-89766-446-9 (paper)
ISSN 0077-8923

ANNALS OF THE NEW YORK ACADEMY OF SCIENCES

Volume 525
May 10, 1988

NEURAL MECHANISMS AND BIOLOGICAL SIGNIFICANCE OF GROOMING BEHAVIOR[a]

Editors and Conference Organizers
DEBORAH L. COLBERN and WILLEM H. GISPEN

CONTENTS

[a] This volume is the result of a conference entitled Neural Mechanisms and Biological Significance of Grooming Behavior, which was held by the New York Academy of Sciences on October 15–17, 1986, in New York, NY.

Financial assistance was received from:

- ABBOTT LABORATORIES
- BAYER/AG MILES
- FIDIA RESEARCH LABORATORIES
- NATIONAL SCIENCE FOUNDATION
- RUDOLF MAGNUS INSTITUTE FOR PHARMACOLOGY STUDY FUND
- SANDOZ, INC.
- SCHERING-PLOUGH CORP.
- SEARLE RESEARCH AND DEVELOPMENT/DIVISION OF G. D. SEARLE & CO.
- UNIVERSITY OF UTRECHT

Preface

DEBORAH L. COLBERN

Department of Physiology and Biophysics
University of Illinois at Chicago
College of Medicine
Chicago, Illinois 60680

WILLEM H. GISPEN

Division of Molecular Neurobiology
Rudolf Magnus Institute for Pharmacology
and
Institute of Molecular Biology and Medical Biotechnology
3584 CH Utrecht, the Netherlands

In virtually every animal species, a substantial portion of waking behavior is devoted to grooming activity. Research indicates that grooming may serve a variety of adaptive functions in addition to maintenance of the external body surface and the removal of parasites. For example, animals groom when placed in unfamiliar surroundings, in conflict situations, and in various social circumstances. Injection of certain peptides into the central nervous system can induce a naturally occurring sequence of grooming behaviors, or particular components of the grooming repertoire. The biological significance of grooming in these contexts is not yet known, although several different functions have been proposed. In addition to care of the skin, pelage, or feathers, these behaviors may be involved in dearousal, stress reduction, social communication, thermoregulation, pain relief, self-stimulation, and the prevention of sexually transmitted diseases. In serving such functions, grooming behaviors apparently play a vital role in the overall homeostatic integrity of the animal.

Within the last 12 years, there has been a considerable increase in the number of publications on grooming behavior and its neurochemical and neuroanatomical substrates. The rare opportunity to investigate such intricate relationships among brain, body, and behavior is undoubtedly responsible for the excitement emerging in this field. Scientists conducting this research are located throughout the world and represent many disciplines, including anthropology, anatomy, biology, biochemistry, endocrinology, ethology, pharmacology, physiology, psychiatry, psychology, neurochemistry, veterinary science, and zoology. The quality and depth of information these scientists have produced is extraordinary. Because of the diversity of their technical and theoretical approaches, however, there has never been a context in which to share this wealth of information. This first international conference on the neural mechanisms and biological significance of grooming was organized to bring together the major contributors to this field of research, most of whom had never before had the opportunity to meet.

The conference focused primarily on self-grooming by rodents, birds, cats, dogs, rabbits, baboons, macaques, and impalas. Scientific contributions arising from this research are likely to be far-reaching, since grooming behaviors and the neuropeptides that induce them are found in diverse species, including humans. In fact, there are many indications that grooming research may be of clinical relevance as well. For example, mechanisms of peptide- and environmentally induced grooming may provide insight into the adaptive value of repetitive behaviors exhibited in some psychiatric disorders, or into the simple "nervous" behaviors associated with everyday anxiety. Indeed, certain antipsychotic, anxiolytic, and antidepressant drugs are effective in reducing grooming responses in animals. Furthermore, peptides that reliably induce the scratching component of the grooming repertoire may help delineate the systems involved in various forms of pruritus. Investigations of the neural systems underlying grooming behaviors could contribute to the development of more efficacious psychotherapeutic and antipruritic agents; perhaps certain afflictions could be prevented entirely.

This volume of the *Annals* is the first collection of research papers on the neurobiological mechanisms underlying grooming behaviors. It chronicles a unique scientific event: not just the first meeting on this topic, but a remarkable multidisciplinary forum in which biochemists, pharmacologists, ethologists and experimental psychologists effectively communicated their data and insights. As organizers of the conference, we wish to thank the participants for their excellent and enthusiastic contributions and commend them for their pioneering work.

The Development of Grooming and Its Expression in Adult Animals[a]

BENJAMIN D. SACHS

*Department of Psychology
University of Connecticut
Storrs, Connecticut 06268*

INTRODUCTION

The first section of this volume is devoted to ethological and comparative approaches to grooming. I do not wish to add to the now tiresome argument on what these approaches mean or ought to mean, but I suppose they might at least be associated with the four types of questions about behavior identified by Tinbergen[1] as needing answers: (1) What is its evolutionary history? (2) What is its functional (adaptive) significance? (3) What is its developmental history? (4) What mechanisms regulate its occurrence? The first two types of questions are commonly classified as dealing with ultimate causation, the second two with proximate causation. The theme of this volume embraces all four questions, but space limitations necessitate giving short shrift to ultimate causation, and even in addressing proximate causation, I will barely scratch the surface.

EVOLUTIONARY HISTORY OF GROOMING

Grooming or its functional equivalents must be evolutionarily ancient because it is very nearly universally represented in animal taxa. In more familiar animal forms, including mammals and birds and many insects, grooming usually takes the form of moving the extremities over the body and of mouthing the body and the extremities, although sandbathing is another common form of grooming in birds[2,3] and many mammals.[4,5] In species like fish that have neither extremities nor the flexibility to mouth portions of their body, the body may be moved over rocks or branches or sand, accomplishing many of the same functions as sandbathing in birds and mammals. Undoubtedly other animal forms have other patterns that serve the functions of grooming, but can not readily be recognized as such. (This problem is analogous to that raised in considerations of the phylogeny of play. Claims that animals in taxa "below" a certain level—usually birds or fish—do not play are severely handicapped by the likelihood that we could not recognize play among, say, worms or bivalves even if it did occur.)

Across and within phyletic groups, the particular movements of grooming behavior have been reliable enough to generate hypotheses about phylogeny.[5-8] It will be recalled that an early mainstay of Lorenz's argument[9] for the evolution of avian behavior patterns from their quadripedal ancestors was their preening behavior: that is, most

[a] This research was supported in part by USPHS research grant no. HD–08933.

birds lower their ipsilateral wing while scratching. Wing-lowering was considered to serve no function in this context (the in-place wing is well out of the leg's way), but rather to be a vestige of the quadripedal tendency to extend the ipsilateral forelimb in order to maintain balance during the temporary tripedal stance taken while scratching the head. Presumably, wing-lowering may be considered to be a more primitive trait than keeping the wing in place. This potential for grooming patterns to serve phylogenetic interpretation has been exploited most fully by Farish.[6] He recorded the grooming of 115 hymenopteran species and found the patterns sufficiently distinctive and reliable that he was able to clarify the taxonomy and probable phylogeny of the subfamilies in this major insect order. Notwithstanding the diversity of patterning in grooming that permits its use in phylogeny, some broad rules of grooming cut across very different taxa, as we shall see shortly.

FUNCTIONS OF GROOMING

The phylogenetic uses of grooming patterns arise in part from their homogeneity within species and their heterogeneity among species. Diverse too are the contexts in which grooming occurs, due in part to the many functions that grooming can serve for the individual animal. Grooming is often the primary means for caring for the outer surface of the body, thereby ridding it of detritus and parasites.[3] (In this respect grooming may represent the outer-body functional equivalent of various internal homeostatic processes,[10] including reactions of the immune system and regurgitation or diarrhea, which act to maintain the integrity of the body's interior.) However, cleaning or body maintenance is only one of many functions that may be served by many acts that we classify as grooming, or that may be served by identical grooming acts in different contexts. For example, the sequence of paw lick, face wipe, and body lick so characteristic of rodents and some other mammalian taxa has been claimed, and in most cases demonstrated, to serve the following functions in addition to cleaning (not necessarily all in the same species): counterirritation, thermoregulation,[11,12] social signaling (including establishing or maintaining social status and spreading of pheromones),[5,13,14] increasing arousal, decreasing arousal,[15-18] and self-stimulation. The last term especially may require clarification, as well as differentiation from the arousal functions of grooming.

By self-stimulatory function I mean the type of phenomenon analyzed by Roth and Rosenblatt[19-21] and by Celia Moore and her colleagues.[22-26] Roth and Rosenblatt demonstrated about 20 years ago that over the course of pregnancy rats devote an increasing proportion of their grooming time to their ventral surface, including both the nipple lines and the anogenital region. If the rats were fitted with collars that prevented them from licking their ventral regions, then mammary development was inhibited. The inference was that the preparation for nursing that occurs during pregnancy is due in part to the self-stimulation provided by the female.

An elegant body of research from the laboratory of Celia Moore has continued this epigenetic tradition. Moore has found two important sexual dimorphisms in grooming in rats. First, maternal rats spend more time licking the anogenital region of male pups than female pups, a difference that affects the pups' potential for the later display of masculine sexual behavior.[22,25] This effect is of less concern to us here only because my emphasis is on autogrooming, and maternal grooming falls outside this realm. The other dimorphism, of greater relevance in the present context, is in the genital autogrooming of peripubertal and adult rats.[23,24,26] Prepubertal males aged 33 to 41 days spend more time in anogenital autogrooming than do females.[19] If males are prevented by collars from engaging in this autogrooming of the body from day 27 to 48, then the development of sexual maturity, as reflected in the weight of the sex accessories, is delayed significantly.[26] In short, just as female rats promote their readiness for lactation by ventral grooming, so do male rats promote their readiness for reproduction by self-stimulatory genital grooming.

In both these examples, the self-stimulatory (as opposed to body maintenance) function of grooming was uncovered in part by attending to state-dependent changes from the normal temporal patterning or sequencing of grooming. Fentress[27-34] and others (e.g., Dawkins,[35] Richmond and Sachs[36]) have demonstrated how attention to these intraindividual variations in grooming can help identify the diverse functions that grooming may serve, as well as the different mechanisms that may regulate grooming.

In discussing the self-stimulatory roles of grooming, I have already made part of the transition from functional questions to developmental questions. It is now time to make that transition formally, while noting in passing that research addressed at one of the classic ethological questions often, sometimes inevitably, also addresses one or more of the others.

THE DEVELOPMENT OF GROOMING AND ITS EXPRESSION IN ADULTS

Development of Grooming

Until the last decade or so, good quantitative descriptions of the development of grooming in any species were lacking. Our own research on rats was not intended just to fill that gap, but to develop data with which to compare our observations of grooming in adult rats, and thereby to check preliminary findings[37] indicating that the sequential organization ("ethogeny") of grooming in adults paralleled the sequence in which the elements of grooming emerged during ontogeny.

Gail Richmond and I[36] studied the development of grooming in six litters of rats observed in their nests and home cages after the mother had been removed. The results are summarized in FIGURE 1, which shows the rank order in which grooming of various body areas first was observed, and the mean age at first occurrence. Thus, the first grooming act to appear was almost invariably forepaw movement toward the snout (mean age at first appearance, 1.1 days), and the last acts to appear were anogenital grooming (18.4 days) and tail grooming (20.4 days). Note that there is a general cephalocaudal progression of grooming by the forepaws and the mouth. However, the participation of the hindpaws in grooming seemed to follow an independent sequence. It appeared first, before the emergence of mouthing of the body, in the form of head scratching on day 6, and on day 11 scratching and mouthing of the hindpaws were combined. (This is not to say that the precise form of the movements did not also change with age after their initial appearance. See Golani and Fentress[38] for evidence to the contrary.)

We inferred that there are two developmental schedules: the one for mouthing and forepaw grooming has a rather clear cephalocaudal progression, and the other for hindpaw scratching has a separate, and as yet unclear, progression of development. Viewed in hierarchical terms (more on this momentarily), it appears that there are at least two major subbranches in the grooming system, a conclusion supported by analyses of grooming in adult rats, to which we now turn.

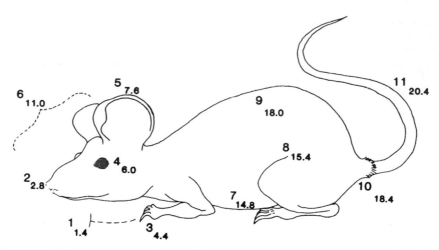

FIGURE 1. The development of grooming in rats. *Large numbers* represent the rank order in which body areas were first groomed, and *small numbers* indicate the mean age in days at which grooming of those body areas first appeared. (Adapted from Richmond and Sachs.[36])

Grooming in Adults

Our description[36] of grooming in adult rats was based on video recordings of grooming emitted by five male hooded rats in their home cages. (An earlier analysis[37] based on direct visual recording of four males and four females yielded the same basic pattern of results.) The 30 bouts selected for analysis met several criteria, including the occurrence of one or more instances of head grooming and licking or biting at

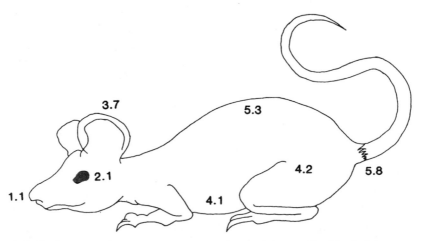

FIGURE 2. The sequential organization of grooming in adult rats. *Numbers* depict mean rank order in which body areas were groomed during a bout of grooming. (Adapted from Richmond & Sachs.[36])

TABLE 1. Sequence of Infant Development and Adult Grooming in *Meriones unguiculatus*[a]

Measure	Grooming Sequence					
	Mouth-Nose	Face	Ears	Flank	Ventrum	Tail
Age at appearance of infant components (mean days of age)	11.1	14.8	19.5	19.9	26.8	33.5
Mean rank order of components in adults	1.0	2.3	3.0	3.1	3.4	4.1

[a] Adapted from the *Journal of Comparative Psychology* (Thiessen *et al.*[14]).

least one area of the body posterior to the head. For complete details of the procedure, see Richmond and Sachs.[36]

In adult rats there is a clear cephalocaudal progression of grooming bouts (FIG. 2). Grooming most commonly starts with nose wipes by the forepaws (mean rank order = 1.1), and most often ends with anogenital grooming (mean rank order = 5.8). The sequence of acts is by no means random between beginning and end either. Analysis of the transition probabilities confirms a general cephalocaudal progression over the head (nose before eyes before ears), and over the body (back or side before hip before belly before anogenital region). However, scratching of the head by the hindleg can appear at any stage of the sequence, suggesting again that scratching belongs to a different branch of the hierarchy than does the rest of the grooming sequence.

The cephalocaudal progression of grooming in adults has now been described by numerous investigators in diverse species, including cat,[10] mouse,[27] parakeet,[39] wallaby,[40] squirrels of several species,[7,41] cricket,[42] kestrel,[43] and giant rat,[44] and this sequence has been found to recapitulate the developmental sequence in those studies where a relation has been sought. Among the clearest data are those based on observations of gerbils by Thiessen, Pendergrass, and Young[45] (TABLE 1). They found a correlation of +1.0 between the rank order of appearance of six components in an adult grooming bout and the order in which these acts developed. A subsequent calculation of this statistic for our data on rats[36] yielded a similar value of +0.87 for seven components. Reliable as the parallel between these sequential patterns in adults and juveniles may be, its interpretation is unresolved. Richmond and Sachs[36] rejected hypotheses based on cephalocaudal neural development and on increased body flexibility, and concluded tentatively that the sequential development of grooming in juveniles and the sequential patterning of grooming in adults reflect, respectively, "the maturation and activation of functional units or programs in the central nervous system" (p. 92). Testing this hypothesis will not be easy.

Hierarchical Organization of Grooming

I have already alluded to a hierarchical organization of grooming, and I would like to return to this concept now. Dawkins[35,46] and Fentress[27,29,32-34] have explored the utility of the construct of hierarchies in general, and as applied to grooming in particular. Dawkins[35] especially has offered hierarchical organization as a "candidate principle" in understanding behavior, though by no means the only such principle. For example, Dawkins and Dawkins[46] concluded that grooming in blowflies was adequately accounted for by expectations from postural facilitation, without recourse to explanations in terms of hierarchical organization. (Simply put, the postural facilitation hypothesis states that certain movements are grouped or combined because doing so is posturally convenient and therefore energetically cost-effective. Conversely, other movements are not combined because their simultaneous performance is inefficient. For example, in an extreme case, grooming with forelegs and hindlegs at the same time would cause the animal to fall down.) In contrast, LeFebvre and his coworkers[7,39,42,43] have consistently analyzed and interpreted their data from diverse species in terms of hierarchical organization. For example, they have used cluster analyses to demonstrate that in the budgerigar,[33] as in other species they have examined, there are major sub-hierarchies (subdirectories?, subroutines?) which tend to be expressed in a cephalocaudal sequence. Furthermore, the analysis of grooming sequences in the budgerigar was more consistent with a hierarchical explanation than with an alternative hypothesis based on postural facilitation.

The sequence of grooming acts in rodents, and the apparent independence of some subroutines of grooming, for example, scratching, from others have also suggested a hierarchical organization to grooming. This segregation of scratching from other forms of grooming was also clearly visible in a study by Hansen and Drake af Hagelsrum.[47] In the presence of female rats, male rats exposed to a variety of treatments that prevent copulation (i.e., recent castration, medial preoptic lesions, unreceptive females) displayed substantial increases in "displacement activities," including drinking and hindlimb scratching, but not other aspects of grooming. Several other studies (see, for example, refs. 48–50) using pharmacological treatments also have dissociated effects on scratching and other forms of grooming.

In the study of Hansen and Drake af Hagelsrum,[47] scratching and drinking were classified as displacement activities because of the thwarted-behavior context in which these behaviors occurred. Others may object to this term, preferring to identify them as "boundary shift" phenomena[29,33,51] or something else, and I prefer to remain agnostic on this question of nomenclature. (The boundaries that shift are those between external, stimulus-sensitive control and internal, stimulus-insensitive control. The shifts depend on such factors as the internal state of the animal, the intensity of stimulation, and the duration of the animal's involvement in an ongoing activity.[33,34]) However, the data stand on their own and clearly confirm that in some contexts scratching is different from — and may be organized and regulated quite independently of — other types of grooming. One implication of these results (with apologies to Gertrude Stein) is that grooming is not grooming is not grooming. Investigators who lump together diverse grooming acts to obtain a total grooming score do so at the peril of overlooking important distinctions. More difficult to assess in the trade-off between "lumping" and "splitting" categories of grooming is the value of such heroic descriptive methods as the Eshkol-Wachmann movement notation system.[38]

Further evidence for a separation of grooming elements within the major branches of a hierarchy, and a hint of how transitions between hierarchical elements may be signaled, also emerged from the study by Richmond and Sachs.[36] We made detailed video analyses of head-grooming sequences, involving repeated forepaw wipes directed over nose, eyes, and ears, interspersed with licking of the forepaws. Examination of the duration of each stroke revealed that, as illustrated in FIGURE 3, the last stroke of a sequence of strokes over the nose, the eye, or the ear was always significantly longer than the preceding strokes, which did not differ significantly from each other in duration. This increase in stroke duration was not affected by the number of preceding strokes in the sequence, nor by the target to which the next stroke was directed. That is, the change in stroke duration was predictive of a change in target, but not of the locus of the new target.

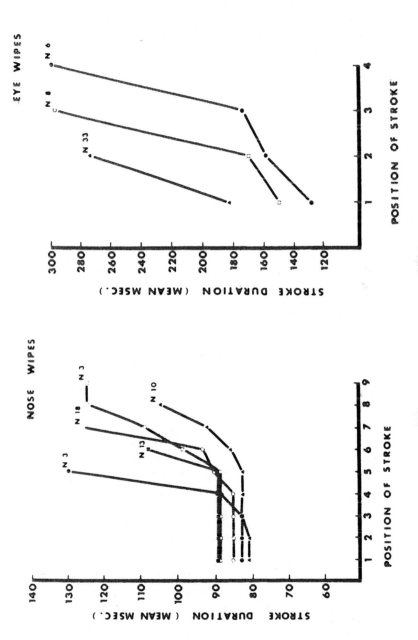

FIGURE 3. Duration of forearm strokes as a function of ordinal position in chains of nose wipes or eye wipes. (From Richmond and Sachs.[36] Reprinted by permission from *Behaviour*.)

These data are consistent with a number of alternative explanations involving competition between behavioral tendencies, perhaps between different branches of the grooming hierarchy. Fentress[29] noted that "rapid movements during grooming can serve to isolate the 'grooming system' from both dependence upon and sensitivity to interruption by sensory factors" (p. 142). Here, however, we see that within the "grooming system" the decreased speed of one target movement (hierarchical element?) may indicate its interruptibility by another target movement. The data are also reminiscent of the criteria used by Dawkins[35] (p. 24; see also ref. 52) for "decisions," *i.e.*, events which themselves "could not easily be predicted, but from which future events can be predicted." In the data on face wiping in rats one can not predict from the penultimate stroke that the last stroke is to follow, but from the last stroke one can predict that a target shift is imminent. I do not want to belabor this particular interpretation. Rather, the point these data help to reinforce is that grooming is a highly dynamic activity, one that has a distinct grammar not only of sequencing, but of impending sequence changes.

The metaphor of syntactic grammar has often before been applied to the sequential organization of grooming and other behaviors (see, for example, refs. 35, 53, and 54). The data just reviewed are reminiscent of two other linguistic metaphors. The first is intonation patterns, such as the rising inflection at the end of questions or the falling one at the end of sentences. Increased stroke duration may represent such a transition marker. Indeed, Klatt[55] has reviewed data from English and other languages indicating that slowing down is a reliable marker for ends of words, phrases, sentences, and conceptual units, and he raised the provocative question of whether there is "a natural tendency to slow down at the end of all motor sequences or planning units" (p. 1212).

The second linguistic metaphor, not incompatible with the first, is the rules of conversational turn-taking. In each society there are cues that indicate when a speaker has finished a turn or is soliciting a turn from another speaker, as well as cues indicating when another person wants to interrupt with a conversational turn. Perhaps increased stroke duration in grooming reflects a cue from the current operation that it will yield the behavioral floor to the next activity. Whether that cue originates entirely from the ongoing activity or is prompted by a signal from an impending activity can not yet be assessed.

Structure and Function of Genital Grooming

I turn now to another context in which it seems to be true that there are varieties of grooming. As I showed earlier, genital grooming in rats is usually the last event in a grooming sequence. However, during copulation genital grooming is usually the first, and often the only, event in a grooming bout. Most male rodents invariably groom their genitalia after each intromission as well as after ejaculation. (For rats the average duration of this activity is about 5–10 s.) Male rats also engage in genital grooming after about 80% of mounts without intromission: approximately 60% of mounts preceding the first intromission and 90% thereafter (unpublished data). This genital grooming occurs so immediately upon dismount that the ventroflexion to permit genital grooming begins even as the hips withdraw after the last thrust of mounting or intromission.

The suggestion that genital grooming in this copulatory context is not part of the normal grooming pattern, or at least not governed by the normal rules of grooming, emerges not just from its violation of the normal cephalocaudal sequence, but also from two other bits of evidence (B. Sachs, D. Bitran, and A. Molloy, unpublished data). First, the duration of genital grooming is independent of whether or not the preceding

TABLE 2. Duration of Genital Grooming after Copulatory Events

	Preceding Event		
	Mount	Intromission	Ejaculation
Duration (s)			
Mean	6.8	7.8	11.8[a]
SEM	1.4	0.5	1.2

NOTE: Data based on 8 males. Only events that ended mount bouts[60] were included: 71% of mounts, 90% of intromissions, 100% of ejaculations.
[a] Significantly different from mounts and intromissions, $p < 0.02$.

mount included penile insertion, and hence appears to be independent of the amount of stimulation received by the penis (TABLE 2). Second, application of topical anesthetics to the glans, which reduces considerably the incidence of mounts with intromission, has little or no effect on the probability or duration of genital grooming after mounts and intromission patterns (TABLE 3), again reflecting substantial independence of the act from its sensory causes and consequences.

Another basis for the suggestion that genital grooming in the context of copulation is a distinct entity is the demonstration by Fentress[29,33,34] that the probability of a switch from an ongoing behavioral system (e.g., grooming) to another (e.g., feeding) decreases as a function of the length and intensity of the animal's involvement with the first activity. Hence, one would not expect an animal strongly involved in mating to begin either feeding or grooming. This expectation is only partly confirmed by an examination of the mating-feeding-grooming priorities of male rats.[56,57] That is, males deprived of food for several days tend to copulate to ejaculation before starting to eat food scattered throughout the mating chamber, despite the intervals of 30–60 s between intromissions during which they might take time to eat. Nonetheless, these food-deprived, copulation-preoccupied rats groom their genitalia after each intromission, suggesting again that genital grooming in this context is a part of the copulatory system and not of the grooming system. (The prevention of postcopulatory genital grooming does not alter any parameter of copulation,[58] and its function may only now have been discovered.[59])

Another question raised by genital grooming in the context of copulation is more problematic. Assuming that copulatory activity reflects a relatively high state of systemic arousal, both the boundary-shift hypothesis and the high-arousal hypothesis[29,33]

TABLE 3. Probability and Duration of Genital Grooming after Copulatory Events following Application of Topical Anesthetic to the Glans Penis of Rats

	Mount		Intromission	
Treatment with:	Oil	Cetacaine	Oil	Cetacaine
Number of events	21	108	66	66
Probability of grooming	0.67	0.62	1.0	1.0
Grooming duration (s)				
Mean	4.26	4.82	6.27	7.46
SEM	0.52	0.87	0.46	1.14
t		0.62		0.92
df		5		10

NOTE: Only mounts and intromissions that ended mount bouts[60] were included in the analysis. Statistics based on t-tests for correlated means. Cetacaine (Cetylite Industries) comprises 14% benzocaine, 2% tetracaine hydrochloride, 2% butyl aminobenzoate, and inert ingredients.

would predict less attention to peripheral stimulation in the context of copulation. The immediate genital grooming after mounts and intromissions suggests high attention to genital stimulation. However, the previously described automaticity of genital grooming in this context, the lack of effect of topical anesthetics, and the lack of a difference in the duration of genital grooming after mounts and intromissions suggests a more centrally programmed ("self-organized," in the sense used by Fentress) activity than at first it may appear to be.

In any event, the male rat's attention to peripheral stimuli during a copulatory episode is at the very least stimulus-dependent and time-dependent. Before ejaculation, the male is highly sensitive to cues from the female and to mildly painful flank shock, both of which potentiate copulation. After ejaculation, during the so-called absolute refractory period, the male rat is virtually unresponsive to the female and to the same intensity of flank shock (reviewed in ref. 60), but he is highly sensitive to cues from intruding males, which he is likely to attack intensely.[61,62]

Lest the preceding section convey the impression that I consider afference irrelevant to genital grooming or other aspects of grooming, I shall now present some evidence on the role of afference, even in some surprising contexts. Fentress[27,29,30,32-34] has written extensively on the context dependency of the stimulus regulation of grooming, and his own studies (see for example, refs. 28 and 63) and others[24,51] amply support this dependency. Among the strongest evidence for stimulus independence is the classic study by Fentress of mice that had their forelimbs amputated neonatally.[28] These mice nonetheless displayed grooming behavior that, except for the absence of the forelimbs, appeared quite normal, even to the point of the animals licking the air or the substrate at those times when the forepaws would have been in front of the mouth, and closing an eye when the ipsilateral forearm would have been moving over the eye. Other evidence (ref. 63, Sachs and Bitran, unpublished) also supports the view that, in many contexts, grooming proceeds quite normally without normal afference or reafference.

The other side of that coin is that grooming in many circumstances is afference sensitive or afference dependent. One of the most ignored and striking testaments to stimulus control of grooming is the mutilation that often results from autophagia after denervation such as dorsal rhizotomy[64] or transection of the pudendal nerve or spinal cord. Whether the excessive chewing results from numbness or from "phantom" pain or itching is unknown, but the *absence* of pain from chewing is clearly the proximate cause of this mutilation. This may be the best place to make explicit what has been only implicit, namely the fundamental need to resolve whether treatments that alter grooming frequency or duration are doing so by acting more or less directly on a grooming pattern generator, or whether they are simply increasing the animal's itches, or whether they are mimicking a postprandial condition or other motivational state.[65] As thorny as this question may be, many results may be uninterpretable if it is not addressed.

The other example of stimulus-dependent grooming that I want to discuss deals again with genital grooming and is appropriate to the themes of this volume because some pharmacological treatments that increase grooming have also been reported to increase penile reflexes outside the context of copulation.[16,17,48,66,67] In order to observe erections ex copula we restrain the males in supine position with the head and anterior torso enclosed in a loosely fitting cylinder and the posterior torso and legs restrained manually or with paper ("masking") tape (FIG. 4). In addition, the penile sheath is retracted and maintained that way to keep the glans penis exposed, but no phasic stimulation is applied to the penis. When erections are displayed, they usually start after a latency period of 5–10 min, and occur in clusters of 3–10 brief erections of the penile corpora (spongiosum and cavernosum). The time between erections within

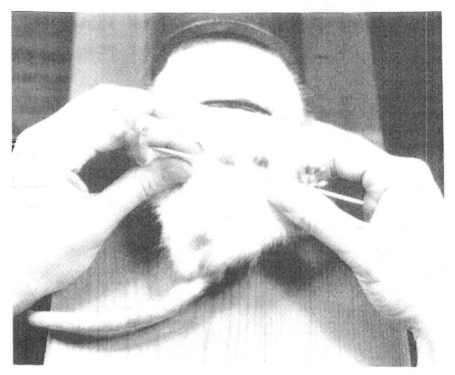

FIGURE 4. Test situation for the evocation of penile erections in rats.

a cluster is rarely more than 2–3 s. Pauses longer than 15 s define an intercluster interval, but intervals of 1–2 min are most common.[68]

Restraint is normally considered a stressor for animals,[16,17] and it is for our rats at first: during the first minutes of the their first supine test they commonly struggle, urinate, defecate, and vocalize at 22 kHz. During such activities the males show little grooming and no penile reflexes. However, often within the first 20-min test, and usually by the second or third test, the signs of stress decline, the males groom themselves, and they begin to display penile erections. This co-occurrence of grooming and erection is reliable. FIGURE 5, based on unpublished data generously sent by John T. Clark, shows the results of three samples of intact rats during their first supine test and without any other treatment. Of the 50 males, 33 males displayed erections in their first test, and 28 of these exhibited facial grooming. (The diameter of the cylinder prevented other forms of grooming.) None of the nonresponders showed any grooming. It appears that the conditions that promote grooming during restraint also promote the appearance of penile erection.

In my laboratory we have begun to look more closely at the relation between grooming and penile reflexes in this ex copula context, particularly comparing grooming that occurs during clusters of erections with grooming displayed during the middle of the intervals between clusters.[69] We use cylinders loose enough for the rat to flex and lick his anterior ventrum, and assume that most of the ventral licking would have been genital licking if not for the restraint. In this test situation, grooming occurs with greater probability during erection clusters than between them. Licking of the paws

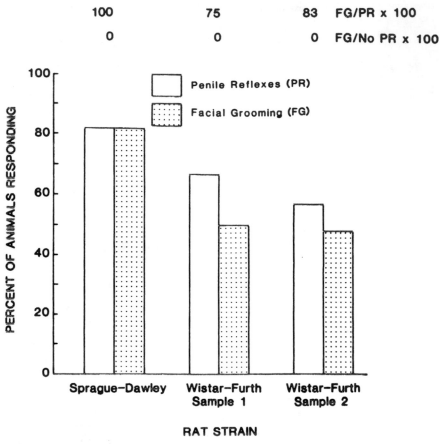

FIGURE 5. Concordance of occurrence of grooming and penile reflexes in three samples of rats receiving their first test for penile reflexes. Relation of *open columns* (penile reflexes) to *stippled columns* (facial grooming) is extended at top of graph. The first line gives the percentages of males with erections that also groomed. The second line gives the percentages of males that did not have erections but that did groom. (Based on unpublished data from J. T. Clark.)

is the most prevalent activity, and it increases reliably during clusters of erections, but ventral grooming, which occurs infrequently during intercluster intervals, has its rate trebled during clusters of erections. These data suggest to us that the onset of grooming outside the context of copulation is provoked by afference resulting from penile erection, either from the genitalia, or from the smooth and striated muscles that cause penile erection, or from the activity of the motoneurons that activate the penile responses.

Supporting evidence for this view comes from two sources. First, preliminary electromyographic recordings from the striated penile muscles of freely moving rats reveal a reliable pattern of EMG activity preceding the onset of bouts of grooming directed immediately toward the genitalia (TABLE 4). Additional support for the view that erections may provoke genital grooming comes from the study of Moore and Rodgers[26] alluded to earlier. Their peripubertal males showed a significant increase in genital

TABLE 4. Relation of Grooming Bouts to Clusters of EMG Bursts Recorded from Bulbospongiosus Muscle (Preliminary Data; $n = 2$ Rats)

	Occurrences of Grooming Relative to EMG Burst Clusters		
	Before	After	Without
Genital grooming	3	31[a]	7
Nongenital grooming	5	1	18

[a] Includes five cases in which grooming and EMG burst appeared to start simultaneously.

grooming, but not in other grooming, beginning at about the age that males start to show spontaneous erections.[70] The sex difference in total genital grooming was due to a difference in number of genital-grooming bouts started; there was not a difference in the mean duration of genital-grooming bouts that occurred. These results may mean that the increased genital grooming displayed by males was due to the increased frequency of spontaneous erections and the stimuli arising from these erections. An implication of these data for research on grooming is that any procedure that increases grooming in general, and especially any procedure that affects genital grooming, may do so indirectly via its effects on penile reflexes.

SUMMARY AND CONCLUSIONS

We have seen that grooming is a class of heterogeneous activities, widely represented in animal taxa, yet sufficiently homogeneous within some phyletic groups to generate and test phylogenetic hypotheses. In the life of the grooming animal, the functions served by these activities are also diverse. Similar acts of grooming may serve different functions in different species or in different contexts. Sometimes these different functions can be discovered by careful attention to variations in the spatiotemporal patterning or sequencing of grooming elements.

In several species a general cephalocaudal progression has been noted during both the ontogeny of grooming and during its expression in adults. During early development and in adulthood, the components or functional units of grooming appear to be hierarchically organized. Scratching with the hindpaw, for example, appears in rodents to be separate from the hierarchical branches in which one finds licking and face wiping. At least some transitions between functional units can be predicted from changes in the temporal patterning of one grooming unit (*e.g.,* eye wipe) just prior to the onset of another unit (*e.g.,* ear wipe).

Analyses of genital and other types of grooming during two forms of sexual activity (copulation and the display of penile erections ex copula) were used to demonstrate once again that the stimulus regulation of grooming is context dependent.

Among the implications of this review for the physiological study of grooming are the following:

1. Careful attention should be given to the spatiotemporal patterning of grooming, in order to reduce errors of "lumping" and "splitting" in classifying grooming acts, and also to detect alterations in the patterning when they occur. Such changes in patterning may be assumed to reflect changes in the physiological state of the animal.
2. Grooming acts that appear formally similar in different contexts may operate under rather different physiological systems.

3. Experimental manipulations may affect grooming directly, that is, by generating grooming efference without changing afference, or indirectly, by altering the "motivational state" or by creating stimulation that potentiates grooming, as from itching or from reafference due to provoking other behavior patterns.

ACKNOWLEDGMENTS

I am indebted to D. Bitran, G. Holmes, R. Leipheimer, D. McQuade, L. Miller, A. Molloy, and G. Morali for assistance in the collection and analysis of previously unpublished data, to D. McQuade and G. Boudle for assistance with the figures, and to D. Bitran and J. Sachs for valuable comments on earlier drafts of the manuscript.

REFERENCES

1. TINBERGEN, N. 1951. The Study of Instinct. Oxford Univ. Press. Oxford.
2. BORCHELT, P. L. 1977. Development of dustbathing components in bobwhite and Japanese quail. Dev. Psychobiol. 10: 97–103.
3. BORCHELT, P. L. 1980. Care of the body surface (COBS). In Comparative Psychology: An Evolutionary Analysis of Animal Behavior, M. R. Denny, Ed.: 363–384. John Wiley & Sons. New York.
4. BORCHELT, P. L, J. G. GRISWOLD & R. S. BRANCHEK. 1976. An analysis of sandbathing and grooming in the kangaroo rat (Dipodomys merriami). Anim. Behav. 24: 347–353.
5. EISENBERG, J. F. 1963. A comparative study of sandbathing behavior in Heteromyid rodents. Behaviour 22: 16–23.
6. FARISH, D. J. 1972. The evolutionary implications of qualitative variation in the grooming behaviour of Hymenoptera (Insecta). Anim. Behav. 20: 662–676.
7. FERRON, J. & L. LEFEBVRE. 1982. Comparative organization of grooming sequences in adult and young sciurid rodents. Behaviour 81: 110–127.
8. THELEN, E. & D. J. FARISH. 1977. Analysis of grooming behaviour of wild and mutant strains of Bracon hebetor (Braconidae-Hymenoptera). Behaviour 62: 70–102.
9. LORENZ, K. Z. 1958. The evolution of behavior. Sci. Am. Dec.: 34–42.
10. SWENSON, R. M. & W. RANDALL. 1977. Grooming behavior in cats with pontile lesions. J. Comp. Physiol. Psychol. 91: 213–236.
11. ROBERTS, W. W., E. H. BERGQUIST & T. C. L. ROBINSON. 1969. Thermoregulatory grooming and sleep-like relaxation induced by local warming of preoptic area and anterior hypothalamus in opossum. J. Comp. Physiol. Psychol. 87: 182–188.
12. TANAKA, H., K. KANOSUE, T. NAKAYAMA & Z. SHEN. 1986. Grooming, body extension, and vasomotor responses induced by hypothalamic warming at different body temperatures. Physiol. Behav. 38: 145–151.
13. HARRIMAN, A. E. & D. D. THIESSEN. 1985. Harderian letdown in male Mongolian gerbils (Meriones unguiculatus) contributes to proceptive behavior. Horm. Behav. 19: 213–219.
14. THIESSEN, D., M. PENDERGRASS & R. K. YOUNG. 1983. Development and expression of autogrooming in the Mongolian gerbil, Meriones unguiculatus. J. Comp. Psychol. 97: 187–190.
15. DELIUS, J. D. 1970. Irrelevant behaviour, information processing and arousal homeostasis. Psychol. Forsch. 33: 165–188.
16. GISPEN, W. H. & R. L. ISAACSON. 1981. ACTH-induced excessive grooming in the rat. Pharmacol. Ther. 12: 209–246.
17. GISPEN, W. H. & R. L. ISAACSON. 1986. Excessive grooming in response to ACTH. In Neuropeptides and Behavior, Vol. 1. CNS Effects of ACTH, MSH, and Opioid Peptides. International Encyclopedia of Pharmacology and Therapeutics, Section 117. D. de Wied, W. H. Gispen & Tj. B. van Wimersma Greidanus, Eds.: 273–312. Pergamon. Oxford and New York.

18. JOLLES, J., J. ROMPA-BARENDREGT & W. H. GISPEN. 1979. ACTH-induced excessive grooming in the rat: The influence of environmental and motivational factors. Horm. Behav. **12:** 60-72.

19. ROTH, L. L. & J. S. ROSENBLATT. 1966. Mammary glands of pregnant rats: Development stimulated by licking. Science **151:** 1403-1404.

20. ROTH, L. L. & J. S. ROSENBLATT. 1967. Changes in self-licking during pregnancy in the rat. J. Comp. Physiol. Psychol. **63:** 397-400.

21. ROTH, L. L. & J. S. ROSENBLATT. 1968. Self-licking and mammary development during pregnancy in the rat. J. Endocrinol. **42:** 363-378.

22. MOORE, C. L. 1983. Maternal contributions to the development of masculine sexual behavior in laboratory rats. Dev. Psychobiol. **17:** 347-356.

23. MOORE, C. L. 1986. A hormonal basis for sex differences in the self-grooming of rats. Horm. Behav. **20:** 155-165.

24. MOORE, C. L. 1986. Sex differences in self-grooming of rats: Effects of gonadal hormones and context. Physiol. Behav. **36:** 451-455.

25. MOORE, C. L. & G. A. MORELLI. 1979. Mother rats interact differently with male and female offspring. J. Comp. Physiol. Psychol. **93:** 677-684.

26. MOORE, C. L. & S. A. ROGERS. 1984. Contribution of self-grooming to onset of puberty in male rats. Dev. Psychobiol. **17:** 243-253.

27. FENTRESS, J. C. 1972. Development and patterning of movement sequences in inbred mice. *In* The Biology of Behavior. J. A. Kiger, Jr., Ed.: 83-131. Oregon State Univ. Press. Corvallis, Oregon.

28. FENTRESS, J. C. 1973. Development of grooming in mice with amputated forelimbs. Science **179:** 704-705.

29. FENTRESS, J. C. 1976. Dynamic boundaries of patterned behaviour: Interaction and self-organization. *In* Growing Points in Ethology. P. P. G. Bateson & R. A. Hinde, Eds.: 135-169. Cambridge Univ. Press. Cambridge.

30. FENTRESS, J. C. 1977. The tonic hypothesis and the patterning of behavior. Ann. N. Y. Acad. Sci. **290:** 370-395.

31. FENTRESS, J. C. 1977. Opening remarks: Constructing the potentialities of phenotype. Ann. N. Y. Acad. Sci. **290:** 220-225.

32. FENTRESS, J. C. 1978. Conflict and context in sexual behaviour. *In* Biological Determinants of Sexual Behaviour. J. B. Hutchison, Ed.: 579-614. John Wiley & Sons. New York.

33. FENTRESS, J. C. 1983. Ethological models of hierarchy and patterning of species-specific behavior. *In* Handbook of Behavioral Neurobiology, Vol. 6. Motivation. E. Satinoff & P. Teitelbaum, Eds.: 185-234. Plenum. New York.

34. FENTRESS, J. C. 1987. Expressive contexts, fine structure, and central mediation of rodent grooming. This volume.

35. DAWKINS, R. 1976. Hierarchical organisation: a candidate principle for ethology. *In* Growing Points in Ethology. P. P. G. Bateson & R. A. Hinde, Eds.: 7-54. Cambridge Univ. Press. Cambridge.

36. RICHMOND, G. & B. D. SACHS. 1980. Grooming in Norway rats: The development and adult expression of a complex motor pattern. Behaviour **75:** 82-96.

37. SACHS, B. D. 1977. Ethogeny recapitulates ontogeny: The grooming sequence in adult rats parallels the ontogeny of grooming. Presented at Northeast Regional Meeting, Animal Behavior Society, St. John's, Newfoundland.

38. GOLANI, I. & J. C. FENTRESS. 1985. Early ontogeny of face grooming in mice. Dev. Psychobiol. **18:** 529-544.

39. LEFEBVRE, L. 1982. The organization of grooming in budgerigars. Behav. Processes **7:** 93-106.

40. RUSSELL, E. M. & D. C. GILES. 1974. The effects of young in the pouch on pouch cleaning in the tammar wallaby, *Macropus eugenii* Desmarest (Marsupialia). Behaviour **51:** 19-37.

41. HORWICH, R. H. 1972. The Ontogeny of Social Behavior in the Gray Squirrel (*Sciurus carolinensis*). Paul Parey. Berlin.

42. LEFEBVRE, L. 1981. Grooming in crickets: Timing and hierarchical organization. Anim. Behav. **29:** 973-984.

43. LEFEBVRE, L. & R. JOLY. 1982. Organization rules and timing in kestrel grooming. Anim. Behav. **30:** 1020-1028.

44. EWER, R. F. 1967. The behaviour of the African giant rat (*Cricetomys gambianus* Waterhouse). Z. Tierpsychol. **24:** 6–79.
45. THIESSEN, D. 1983. The thermoenergetics of communication and social interaction among Mongolian gerbils. *In* Symbiosis in Parent-Young Interactions. L. Rosenblum & H. Moltz, Eds.: 113–144. Plenum. New York.
46. DAWKINS, R. & M. DAWKINS. 1976. Hierarchical organization and postural facilitation: rules for grooming in flies. Anim. Behav. **24:** 739–755.
47. HANSEN, S. & L. J. K. DRAKE AF HAGELSRUM. 1984. Emergence of displacement activities in the male rat following thwarting of sexual behavior. Behav. Neurosci. **98:** 868–883.
48. SPRUIJT, B. M., V. HOGLAND, W. H. GISPEN & B. J. MEYERSON. 1985. Effects of ACTH on male rat behavior in an exploratory, copulatory and socio-sexual approach test. Psychoneuroendocrinology **10:** 431–438.
49. SPRUIJT, B. & W. H. GISPEN. 1983. ACTH and grooming behaviour in the rat. *In* Hormones and Behaviour in Higher Vertebrates. J. Balthazart, E. Pröve & R. Gilles, Eds.: 118–136. Springer-Verlag. Berlin.
50. VAN WIMERSMA GREIDANUS, TJ. B., D. K. DONKER, R. WALHOF, J. C. A. VAN GRAFHORST, N. DE VRIES, S. J. VAN SCHAIK, C. MAIGRET, B. M. SPRUIJT & D. L. COLBERN. 1985. The effects of neurotensin, naloxone, and haloperidol on elements of excessive grooming behavior induced by bombesin. Peptides **6:** 1179–1183.
51. COHEN, J. A. & E. O. PRICE. 1979. Grooming in the Norway rat: Displacement activity or "boundary-shift"? Behav. Neur. Biol. **26:** 177–188.
52. DAWKINS, R. & M. DAWKINS. 1973. Decisions and the uncertainty of behaviour. Behaviour **45:** 83–103.
53. FENTRESS, J. C. & F. P. STILWELL. 1973. Grammar of a movement sequence in inbred mice. Nature **244:** 52–53.
54. VOWLES, D. M. 1970. Neuroethology, evolution, and grammar. *In* Development and Evolution of Behavior: Essays in Memory of T. C. Schneirla. L. R. Aronson, E. Tobach, D. S. Lehrman & J. S. Rosenblatt, Eds.: 194–215. W. H. Freeman. San Francisco.
55. KLATT, D. H. 1976. Linguistic uses of segmental duration in English: Acoustic and perceptual evidence. J. Acoust. Soc. Am. **59:** 1208–1221.
56. SACHS, B. D. & E. D. MARSAN. 1972. Male rats prefer sex to food after six days of food deprivation. Psychon. Sci. **28:** 47–49.
57. BROWN, R. E. & D. J. MCFARLAND. 1979. Interaction of hunger and sexual motivation in the male rat: A time-sharing approach. Anim. Behav. **27:** 887–896.
58. HART, B. L. & C. M. HAUGEN. 1971. Prevention of genital grooming in mating behaviour of male rats (*Rattus norvegicus*). Anim. Behav. **19:** 230–232.
59. HART, B. L., E. K. KORINEK & P. L. BRENNAN. 1987. Postcopulatory grooming in male rats prevents sexually transmitted diseases. This volume.
60. SACHS, B. D. & R. J. BARFIELD. 1976. Functional analysis of masculine copulatory behavior in the rat. Adv. Study Behav. **7:** 92–154.
61. THOR, D. H. & K. J. FLANNELLY. 1979. Copulation and intermale aggression in rats. J. Comp. Physiol. Psychol. **93:** 223–228.
62. FLANNELLY, K. J., R. J. BLANCHARD, M. Y. MURAOKA & L. FLANNELLY. 1982. Copulation increases offensive attack in male rats. Physiol. Behav. **29:** 381–385.
63. BERRIDGE, K. & J. C. FENTRESS. 1986. Contextual control of trigeminal sensorimotor function. J. Neurosci. **6:** 325–330.
64. DENNIS, S. G. & R. MELZACK. 1979. Self-mutilation after dorsal rhizotomy in rats: Effects of prior pain and pattern of root lesions. Exp. Neurol. **65:** 412–421.
65. BEAGLEY, W. K. 1976. Grooming in the rat as an aftereffect of lateral hypothalamic stimulation. J. Comp. Physiol. Psychol. **90:** 790–798.
66. OGAWA, S., S. KUDO, Y. KITSUNA & S. FUKUCHI. 1980. Increase in oxytocin secretion at ejaculation in male. Clin. Endocrinology **13:** 95–97.
67. DRAGO, F., C. A. PEDERSEN, J. D. CALDWELL & A. J. PRANGE, JR. 1986. Oxytocin potently enhances novelty-induced grooming behavior in the rat. Brain Res. **368:** 287–295.
68. SACHS, B. D. 1983. Potency and fertility: Hormonal and mechanical causes and effects of penile actions in rats. *In* Hormones and Behaviour in Higher Vertebrates. J. Balthazart, E. Pröve & R. Gilles, Eds.: 86–110. Springer-Verlag. Berlin.

69. SACHS, B. D. & A. G. MOLLOY. 1987. Selective association of grooming acts and penile reflexes in rats. This volume.
70. SACHS, B. D. & R. L. MEISEL. 1980. Pubertal development of penile reflexes and copulation in male rats. Psychoneuroendocrinology 4: 287–296.

Expressive Contexts, Fine Structure, and Central Mediation of Rodent Grooming[a]

JOHN C. FENTRESS

Departments of Psychology and Biology
Dalhousie University
Halifax, Nova Scotia, Canada B3H 4J1

INTRODUCTION

The behavioral analysis of rodent grooming has clarified a number of important issues, among them:

1. How are individual actions defined, and how are these actions linked together, and to their broader expressive surrounds?
2. What are the precise rules of sequential ordering among movement components that allow us to define a behavioral "syntax"?
3. How might these rules of sequential structure reflect not only broadly defined behavioral events (*e.g.*, those concerned with "motivation"), but also the operation of specific sensory and motor mechanisms?
4. What can developmental analyses tell us about these events and mechanisms?
5. Can analyses of rodent grooming, as a model system, provide useful insights into mammalian behavior and neural function more generally, including both theoretical and clinical properties?

It is my contention in this paper that studies of rodent grooming have indeed provided a powerful model system relevant to each of these questions. The three fundamental issues of (1) expressive contexts in which grooming occurs, (2) the fine structure of grooming, and (3) the central mediation of grooming, will be taken as illustrations.

EXPRESSIVE CONTEXTS

Rodent grooming, like all forms of mammalian behavior, occurs within definable contexts. These contexts can range from abstracted statements about "motivational state" to concrete observations of the sequential ordering of specific motor events. Whatever level of analysis one prefers to employ, it is critical to recognize that *any* given expressive event occurs within the context of other such events. No aspect of behavioral expression is an island unto itself.

The relevance of this combination of contextual and component concerns for those interested in mechanisms is that the events underlying different aspects of behavioral expression (*i.e.*, grooming versus something else) are both separable and yet linked to

[a] The research summarized here was supported in part by Canadian MRC and NSERC grants to the author.

18

varying degrees to one another. Abstracted neurobehavioral events may have differing degrees of independence from one another, and the dependencies that do occur may be either synergistic or antagonistic, depending upon the specific circumstances. In addition, there are sequencing rules of behavioral expression that, by definition, can be evaluated only through careful analyses of preceding and subsequent expressions. There are the issues of *behavioral context*. Behavioral context, as thus regarded, is an essential supplement to the search for mechanisms that might otherwise be restricted to artificially isolated events.

As illustration, rodent grooming often occurs between protracted quiet and active states (*e.g.*, sleep and exploration). It frequently serves as a "transitional element" in this sense, as if the animal grooms before it settles clearly into something else. One is reminded here of the scratching before smoking, etc., actions shown by humans when they are in otherwise behaviorally uncertain situations (such as whether to stay in the dentist's office as their appointment approaches!). Darwin[1] commented insightfully upon such behavioral expressions, and in more recent years ethologists have discussed them under the general rubric of "displacement activities."[2,3]

It is useful to recall why concepts such as "displacement activity," as studied by ethologists, came into vogue. Many displacement activities, rodent grooming included, occur in surprising contexts. Why, for example, should a wild vole groom after it has been presented with a model predator?[4,5] The original displacement concepts presumed an accumulation of behavioral "energy" that was, somehow, re-expressed ("displaced") to a new channel.[6] The channel selected was most often a behavior that was frequently expressed anyway, and basically "simple" in its organizational structure.[2,6] The original term is a classic case of confounding between descriptive statements and interpretation (*i.e.*, descriptively surprising = "energy displacement").[2]

More sophisticated models soon followed. These represented attempts to disentangle descriptive and causal statements, and also to provide a more biologically relevant causal framework. The "disinhibition" hypothesis thus came into vogue.[7] The basic premise behind this hypothesis is that actions are mutually incompatible. Thus, if one blocks an action that normally blocks another, then the latter may be "disinhibited."[8,9] There are a number of neurobiological supports for such a proposition.[10]

However, the proposition as normally stated does not allow for the possibility that other factors, or even the same factors under different conditions, may activate (facilitate or arouse) a broader range of behavioral expressions than assumed by most of the earlier ethological and even neurobiological models. Of critical importance to the evaluation of these various models is the careful dissection of (1) qualitative, (2) quantitative, and (3) temporal aspects of the variables concerned.[11] For example, it is now well established that a number of moderately intense events can facilitate rather than block concurrent behavioral sets relevant to well-formed or "habitual" actions, whereas more intense occurrences of the same stimuli will have an inhibitory influence upon these same actions. Less well-established, more variable, and less vigorously expressed action patterns are more easily disrupted.[2,11] Also, since behaviorally relevant events typically take some period of time to achieve their full effectiveness and also decline in their effectiveness over time, "intensity" and "temporal," as well as "qualitative" considerations can often be played off against one another.[11] Qualitatively different variables, as expressed at a number of different levels, can also be expected to follow somewhat independent rules.

One lesson from these recent neurobehavioral studies is that the rules of organization of behavioral expression are fundamentally dynamic. That is, a given variable (factor or whatever) can have different influences at different times, and under different circumstances.[3,10,12] Studies of rodent grooming, in association with alternative behavioral patterns, have provided useful insights.[13-15] Many of these insights have derived from more refined analyses of grooming structure in relation to surrounding events.

FINE STRUCTURE

It is thus often critical to provide a more detailed statement of how individually defined actions are expressed in various contexts. For example, Fentress and Stilwell[16] found that individual grooming strokes in mice could be arranged in a hierarchical sequence, Golani and Fentress[17] showed that different aspects of mouse grooming changed differentially during development, and Berridge, Fentress, and Parr[14] have recently established "syntactical" rules for the fine structure of grooming behavior in laboratory rats. I shall emphasize these latter "syntactical" rules here, for they are of special interest from both descriptive and mechanistic perspectives.

There has long been a recognition of the need to provide better ways for characterizing and measuring the sequential organization of behavior.[14,18,19] By implanting rats with oral cannulae, through which different taste substances can be presented, it is possible to elicit a number of ingestive, aversive, and grooming actions, both during and following the taste presentations. These various actions follow rules of sequencing that can be defined. For example, it is often possible to separate actions perceptually (with such separations agreed upon at nearly a 100% level across observers), and to determine rules by which these actions are combined in time.[3,16]

Berridge, Fentress, and Parr[14] have recently completed a detailed study in which several thousands of spontaneous grooming and taste-elicited ingestive/aversive movements in rats were videotaped and scored with a microcomputer. Rules of behavioral sequencing were determined in three ways: (1) information analyses of linkage tightness (stereotypy) among movement components for complete sequence chains, (2) examination of specific sequential-reciprocal transitions between individual actions and action groups, and (3) visual evaluation of both loosely structured and tightly structured linear action chains.

Rhythmic tongue protrusions, downward forelimb strokes of varying amplitude across the face, paw and body licking, and flailing movements of the forepaws are representative of movements that are seen both in ingestive/aversive contexts and during spontaneous grooming. Thus both specific sequential linkages of individual movements and more broadly defined motivational contexts can be evaluated for rules of behavioral organization.

Infusions of sucrose (1.0 and 0.03 M), HCl (0.1 and 0.01 M), and quinine HCl (3×10^{-3} and 3×10^{-4} M) through chronically implanted oral cannulae in nineteen rats were provided in a 1 ml volume at a constant rate over 1 min, 5 min into a trial, in a counterbalanced order. The rats were also observed for 6 min after an infusion, and videotape records were made for all grooming bouts as well as during the entire infusion period. A tilted mirror reflected each rat's face and mouth to facilitate detailed recording of the animal's movements. Videotapes were scored at 1/10 to 1/15 actual speed, and as each action occurred its code was entered into a microcomputer. In this way we obtained a record of frequency, duration, and sequential order for each action. For basically discrete actions such as tongue protrusions, gapes, headshakes, and facewashing strokes, an entry was made for each occurrence. For more continuous actions such as paw licking, mouth movements, locomotion or passive dripping from the mouth of an infused solution, the action bouts were divided into 5-s bins (to minimize the possibility of artificially inflated "bouts" of brief actions that normally occur in long strings).

Information analyses of sequential dependencies among actions were carried out for action pairs and triplets. These analyses take into account the number of individually defined actions (H_0), their relative probabilities (H_1), and their first- and second-order transitional constraints (H_2 and H_3). Sample sizes ranged from 920 to 6798 observed transitions, with a mean sample size of 3326 transitions per stimulus condition. For both taste-elicited and grooming sequences calculation of the relative frequencies of the 12 recorded actions (H_1) reduced uncertainty to one out of two or three (versus one out of 12 for equally probable actions). Knowledge of the immediately preceding action (H_2) increased the ability to predict the subsequent action by approximately 25% to 50%, an indication of action sequential dependencies. Knowledge of preceding action *pairs* (H_3) increased the ability to predict by 5% to 10%. While there were some quantitative differences between grooming and other action patterns at this level,[15] the above figures represent the fundamental principles of nonrandom behavioral occurrences in each case.

Transition tabulations allow one to determine which particular actions are associated with one another. Two trends were immediately clear. The first is *perseveration*, in which a given action follows itself in immediate sequence. The second is *reciprocity*, in which the number of transitions between action *A* and *B* is similar in both directions. Together, these trends reflect an *alternating perseveration* rule of the type *AAAABBBBBAAAABBBB*. This rule, revealed through three-way matrices, must be combined with any tendency to progress linearly through a behavioral sequence (*ABCD*). Reciprocity is particularly striking since it is maintained even when the individual actions have quite different probabilities of expression, as measured by the transition matrix as a whole. For example, although scored mouth movements (400) occurred more than three times as frequently as lateral tongue protrusions (132) during postprandial grooming, two-way transitions between mouth movements and lateral tongue protrusions were essentially identical (102 and 101, respectively). Single lateral tongue protrusions were almost invariably associated with either immediately preceding or subsequent mouth movements, whereas mouth movements had a broader range of reciprocal associations. Together the properties of simple perseveration, paired alternation, and alternating perseveration account for approximately 75% of all transitions during grooming and taste-elicited sequences.

Linear chains of the type *ABCD* were absent from ingestive/aversive responses, but did make up one phase of grooming. During much of grooming, complex patterns of paw licking and downward strokes of more or less progressive amplitude were variably distributed and compatible with the general rule of alternating perseveration. However, during a large number of protracted grooming bouts a highly rhythmic series of paw licking led to a phase with (1) 5 to 9 rapid elliptical forepaw strokes at 6–7 Hz, followed by (2) a short series of slower (2–4 Hz) strokes of generally ascending amplitude until (3) they reached above and caudal to the pinnae, followed in turn by (4) licking of the ventrolateral body surface. This highly stereotyped chain of movements was seen often during spontaneous grooming but only infrequently during taste-elicited or aversive sequences that contained similar action components. With practice we could anticipate the occurrence of this stereotyped chain by the precise rhythmicity and spatial focus of paw licking strokes. These were in contrast to other forms of paw licking that were less regular in their timing and/or displaced to various parts of the face.

Such observations indicate that the form of movement, as well as movement classification into broader categories such as paw lick, can provide useful insights into rules of behavioral organization. Movement form itself is multidimensional, and demands analysis from complementary descriptive perspectives. Golani and Fentress[17] pursued this line of reasoning in a developmental study of grooming in mice. The basic technique was to describe grooming movements separately for (1) their kinematic structure (*e.g.*, rotational and planar movements around a single joint), (2) their trajectories in space, and (3) the resultant contact pathways between forepaws and face. Using a strobe light and cine film taken at 100 fps in a mirrored chamber that also provided support for the mice, grooming ontogeny was traced with daily records from birth through postnatal day 14.

During the first 100 hours after birth, individual kinematic components were often poorly coordinated, trajectories were variable and often poorly directed, and contact pathways between the forepaws and face were more or less fortuitous. Both single- and double-handed grooming bouts from one to ten strokes occurred, but with minimal functional cohesion. Between postnatal hours 100 and 200, kinematics became more tightly coordinated, the trajectories were both more restricted and precise, and the contact pathways were restricted to the anterior part of the face. Almost all grooming "bouts" consisted of a single stroke. Beyond 200 hours the movements became re-elaborated, with protracted bouts of clearly recognizable movement classes. During this time consistent contact pathways could be achieved through a wide range of often compensatory head and forelimb movements, shifts in body posture, etc. It is this richness in grooming behavior that offers particularly exciting prospects for the analysis of peripheral and central routes of mediation.

CENTRAL MEDIATION

Questions of central mediation in grooming can be referred both to the broad ("motivational") contexts within which grooming occurs, and to its fine structure. It is precisely this combination of opportunities that make grooming behavior such a useful model system for behavioral and neural scientists alike.

Infant and centrally damaged rodents may show not only functionally impoverished grooming,[20-22] but also initiate grooming at apparently inopportune times. For example, in each case the animals may suddenly initiate grooming if the forepaws happen to pass fortuitously near the face, such as during locomotion or even swimming.[2] Such animals also often fail to display the compensatory movements among individual body segments that together ensure functional coherence, even in situations where the animal is subjected to experimental perturbations. For example, Golani and I[17] found that by the time mice are approximately 10 days old they can achieve symmetrical contact pathways by the forepaws across the face even when the movements of certain limb segments are restricted (as when young mice have a limb jammed among littermates). They achieve this through a number of bodily adjustments. Younger and CNS-damaged animals often fail to make such compensatory movements.

A somewhat clearer picture of the control priorities in grooming can be obtained through simple manipulations. Thus, when one limb of a young adult mouse was pulled gently away from the face via a thread after grooming was initiated, its free limb sacrificed its contacts with the face to perform oscillations that were symmetrical with the perturbed limb, *in front of* the face.[17] Further, if one or more limbs were pulled from the face during grooming, a mouse was likely to terminate grooming during its variable phase, but persist in making grooming movements during the stereotyped phase.[10,13] Such observations suggest a dynamic interplay between central and peripheral control pathways.

Here I shall illustrate some of these central and peripheral factors that have been clarified through recent experiments with trigeminal deafferentation and striatal lesions in rats, conducted in collaboration with K. Berridge.[12,15,23] Questions we asked include the following: (1) To what extent are the grooming patterns that we observed the result of stimulus-mediated "reflex systems" as opposed to central control mechanisms? (2) Do the rules of central and peripheral control vary in a rule-governed manner as a function of motivational context and/or the particular phase of the movement sequence being observed? (3) What processes underlie the switching between sequencing rules? (4) Might striatal mechanisms within the rodent CNS contribute to these higher-order sequencing rules, and thus complement the operation of the caudal brainstem mechanisms known to be responsible for individual actions?[22,24]

These questions were guided in part by the conceptual realization that to assign any action solely to a reflex or endogenously generated mechanism presumes an unlikely dichotomy. Even intermediate viewpoints commonly presuppose that actions are organized in a static manner: *i.e.*, that the properties of control in one context can be generalized readily to other contexts. Our behavioral analyses indicate that grooming reflects more flexible organizational principles. These principles might be of widespread relevance to the sequencing of mammalian behavior, and as such indicate important contributions of rostral (*e.g.*, striatal) brain mechanisms.[23-26]

Since basic movement components such as rhythmic tongue protrusions, forelimb strokes across the face, and high-frequency forelimb flails are emitted in various combinations by rats, both during taste-elicited ingestion/aversion and postprandial grooming, it is possible to determine the extent to which these acts are differentially affected by peripheral and central manipulations in different expressive contexts. Berridge and Fentress[12,15] found that trigeminal deafferentation (removal of tactile somatosensation and pain from the facial region through sections of appropriate mandibular and maxillary nerve branches) produced deformations of rhythmic tongue protrusions in ingestive but not in grooming contexts, whereas alterations of forelimb movements were, conversely, restricted to grooming contexts. Further, the deformations that occurred during the variable phases of grooming were absent during the stereotyped sequence described earlier in this paper.

Movement changes during the sequentially variable phase of postprandial grooming were of a wide variety. As illustration, forepaw licking was often impoverished and poorly coordinated, individual paws occasionally became "stuck" on the face due to excessive pressure, stroke amplitudes were more homogeneously restricted to intermediate levels, otherwise normally appearing elliptical strokes were displaced laterally on the face, lateral forepaw accelerations interrupted several sequences, and forelimb flails often intruded in a disruptive manner. Yet, once the stereotyped sequence phase was initiated through the signature of tight forepaw ellipses around the mouth, this sequence phase was almost invariably completed in a manner that did not allow us to distinguish between deafferented and control animals.

Taken as a whole, these observations indicate clearly that motivational context, as well as sequential phase of grooming behavior, are important determinants of control rules (*cf.* Sachs, this volume[3]). Normal peripheral information from the facial region is important for the maintenance of normal forelimb flail probability during postprandial grooming but not ingestive/aversive contexts, and a number of movement changes that occur during the variable phases of grooming are absent during the stereotyped phases. In contrast, the proportion of midline rhythmic tongue protrusions is affected only during taste-elicited action sequences. Contextual dependencies are both important and different for different classes of action. Actions that are formed in one context through a combination of sensory and central pathways may be maintained in other contexts when these same sensory pathways are eliminated.

In our most recent studies we have begun to examine the integrative role of rostral brain mechanisms, namely the corpus striatum. Two obvious reasons for this are that

(1) the striatum is known to modulate the effectiveness of peripheral sensory events,[27-29] and (2) other investigators have suggested a role of striatal processes in the mediation of higher-order sequencing properties in behavior.[24] Disorders of the corpus striatum in man can also lead to a number of severe pathologies, such as Huntington's chorea and Parkinsonism.[30,31] The biology of grooming behavior may thus provide valuable information of "syntactical" rules of sequentially ordered movements of broad adaptive significance.[14] The most interesting prediction is that disruption of striatal circuitry will interfere with the normal execution of otherwise well-established and regularly occurring movement sequences, more or less independently of the individual action components from which these sequences are composed. The stereotyped phases of grooming, shown to be immune from otherwise dramatic effects of trigeminal deafferentation (above), provide a useful model system.

In the study by Berridge and Fentress,[23] 50 male rats were anesthetized and given (1) bilateral and stereotactically located injections of kainic acid in the anterior striatum, (2) bilateral kainic acid injections in the posterior corpus striatum (including globus pallidus), (3) vehicle solution controls, (4) trigeminal lesions, and (5) no pretest treatment other than implantation of oral cannulae as in all other groups. During the second and third weeks after surgery the rats were sprayed lightly with a water mist to induce grooming, and placed within the mirrored recording apparatus described earlier.

Kainic acid lesions of both the anterior and posterior corpus striatum produced a significant disruption of chaining completion during the stereotyped phases of grooming behavior. This disruption of movement chaining could not be accounted for by reduction in the number of stereotyped grooming sequences that were *initiated*. Complementary analyses also precluded explanation of the chaining disruptions solely in terms of postural disorders; *i.e.*, they reflect a disruption of centrally mediated sequencing rules. Further, the lesioned rats emitted all action components of normal grooming.

One interesting possibility is that activation of striatal circuitry serves to focus well-established central pathways involved in the sequential ordering of grooming movements with respect to both sensory and competing central events. There is indeed much ethological evidence to support such a contention. The basic hypothesis is that strongly activated circuits become increasingly "self-organized," and that their independence from potentially competing sensory and central "surrounds" is thereby increased.[2,10,11,20] A point that is evident without possible refutation from these studies is that both dynamic and relational (contextual) properties of behavioral organization in mammalian species deserve further attention at a number of analysis levels from a number of complementary perspectives. Rodent grooming, as witnessed by the present conference, may provide an especially useful model system to do just that.

CONCLUSIONS

Rodent grooming is obviously a rich source of behavioral and biological information. In order to benefit fully from the utility of grooming as a model system, it will be necessary to combine careful analyses of the expressive contexts within which grooming occurs with the analysis of grooming fine structure, as well as with a systematic dissection of the dynamic interplay among specified peripheral and central determinants. Two other properties of rodent grooming provide special incentives for further analysis. First, grooming represents a behavioral pattern that is rich yet rule-given in its structure, and that can be approached at a number of complementary analysis levels.

Secondly, the time frames relevant to the analysis of grooming range from the moment to moment expressions of individual movement properties, through more broadly considered changes of motivational state, to issues of development and even evolution.

The linkage of such apparently disparate levels and time frames of analysis is obviously a critical goal in both the behavioral and neurobiological disciplines. We have seen in this conference how cooperative investigations of rodent grooming behavior can help lead us toward a better understanding of these levels and time frames of analysis that in other areas are all too often pursued independently. Finally, I suspect that rodent grooming may hold tractable clues to a number of issues of obvious import to our own human lives, such as the deeper understanding of what constrains syntactical rules of behavioral sequencing, the dynamic interplay among various central and peripheral events in normal and abnormal behavioral temporal processing, the value of developmental dissections of adult integrative events, and even (one would hope) the clearer elucidation of central nervous system processes that may contribute to the understanding and alleviation of human health disorders.

ACKNOWLEDGMENTS

I thank the New York Academy of Sciences and conference organizers D. L. Colbern and W. H. Gispen for the opportunity to participate in this conference.

REFERENCES

1. DARWIN, C. 1872. The Expression of the Emotions in Man and the Animals. Murray. London.
2. FENTRESS, J. C. 1983. Ethological models of hierarchy and patterning of species-specific behavior. In Handbook of Behavioral Neurobiology, Vol. 6. E. Satinoff & P. Teitelbaum, Eds.: 185–234. Plenum. New York.
3. SACHS, B. D. 1987. The development of grooming and its expression in adult animals. This volume.
4. FENTRESS, J. C. 1968. Interrupted ongoing behaviour in voles (Microtus agrestis and Clethrionomys britannicus): I. Response as a function of preceding activity and the context of an apparently "irrelevant" motor pattern. Anim. Behav. 16: 135–153.
5. FENTRESS, J. C. 1968. Interrupted ongoing behaviour in voles (Microtus agrestis and Clethrionomys britannicus): II. Extended analysis of intervening motivational variables underlying fleeing and grooming activities. Anim. Behav. 16: 154–167.
6. TINBERGEN, N. 1951. The Study of Instinct. Clarendon Press. Oxford.
7. IERSEL, J. J. A. & A. C. A. VAN BOL. 1958. Preening of two tern species. Behaviour 13: 1–88.
8. ROWELL, C. H. F. 1961. Displacement grooming in the chaffinch. Anim. Behav. 9: 38–63.
9. SEVENSTER, P. 1961. A causal analysis of a displacement activity (fanning in Gasterosteus aculeatus L.). Behaviour 9: (Suppl.) 1–170.
10. FENTRESS, J. C. 1986. Ethology and the neural sciences. In Relevance of Models and Theories in Ethology. R. Campan & R. Zayan, Eds.: Privat. Toulouse. 77–107.
11. FENTRESS, J. C. 1976. Dynamic boundaries of patterned behavior: Interaction and self-organization. In Growing Points in Ethology. P. P. G. Bateson & R. A. Hinde, Eds.: 135–169. Cambridge Univ. Press. Cambridge.
12. BERRIDGE, K. C. & J. C. FENTRESS. 1986. Contextual control of trigeminal sensorimotor function. J. Neurosci. 6: 325–330.
13. FENTRESS, J. C. 1984. The development of coordination. J. Motor Behav. 16: 99–134.
14. BERRIDGE, K. C., J. C. FENTRESS & H. PARR. 1987. Natural syntax rules control action sequence of rats. Behav. Brain Res. 23: 59–68.
15. BERRIDGE, K. C. & J. C. FENTRESS. 1987. Deafferentation does not disrupt natural rules of action syntax. Behav. Brain Res. 23: 69–76.

16. FENTRESS, J. C. & F. P. STILWELL. 1973. Grammar of a movement sequence in inbred mice. Nature **244:** 52–53.
17. GOLANI, I. & J. C. FENTRESS. 1985. Early ontogeny of face grooming in mice. Dev. Psychobiology **18(6):** 529–544.
18. LASHLEY, K. S. 1951. The problem of serial order in behavior. *In* Cerebral Mechanisms in Behavior. L. A. Jeffress, Ed.: 112–136. John Wiley & Sons. New York.
19. FENTRESS, J. C. & P. J. McLEOD. 1986. Motor patterns in development. *In* Handbook of Behavioral Neurobiology, Vol. 8. Developmental Psychobiology and Developmental Neurobiology. E. M. Blass, Ed.: 35–97. Plenum. New York.
20. FENTRESS, J. C. 1977. The tonic hypothesis and the patterning of behavior. Ann. N. Y. Acad. Sci. **290:** 370–395.
21. RICHMOND, G. & B. D. SACHS. 1980. Grooming in Norway rats: the development and adult expression of a complex motor pattern. Behaviour **75:** 82–96.
22. KOLB, B. & I. Q. WHISHAW. 1981. Decortication of rats in infancy or adulthood produced comparable functional losses on learned and species-typical behaviors. J. Comp. Physiol. Psychol. **95:** 468–483.
23. BERRIDGE, K. C. & J. C. FENTRESS. 1987. Impairment of natural grooming sequence chains after striatopallidal lesions. Psychobiology. Submitted.
24. COOLS, A. R. 1985. Brain and behavior: Hierarchy of feedback systems and control of input. *In* Perspectives in Ethology. P. P. G. Bateson & P. M. Klopfer, Eds.:109–168. Plenum. New York.
25. DIVAC, I. 1983. Two levels of functional heterogeneity of the neostriatum. Neuroscience **10(4):** 1151–1155.
26. DUNNETT, S. B. & S. D. IVERSEN. 1982. Sensorimotor impairments following localized kainic acid and 6-hydroxydopamine lesions of the neostriatum. Brain Res. **248:** 121–127.
27. LIDSKY, T. I., C. MANETTO & J. S. SCHNEIDER. 1985. Consideration of sensory factors involved in motor functions of the basal ganglia. Brain Res. Rev. **9:** 133–146.
28. ROLLS, E. T., S. J. THORPE, S. MADDISON, A. ROPER-HALL, A. PUERTO & D. PERRET. 1979. Activity of neurones in the neostriatum and related structures in the alert animal. *In* The Neostriatum. I. Divac & R. G. E. Oberg, Eds. Pergamon. New York.
29. SCHNEIDER, J. S. 1984. Basal ganglia role in behavior: Importance of sensory gating and its relevance to psychiatry. Biol. Psychiatry **19(12):** 1693–1710.
30. FIBIGER, H. C. 1978. Kainic acid lesions of the striatum: A pharmacological and behavioral model of Huntington's disease. *In* Kainic Acid as a Tool in Neurobiology. E. G. McGeer, Ed. Raven Press. New York.
31. MARSDEN, C. D. 1980. The enigma of the basal ganglia and movement. Trends Neurosci. **3:** 284–287.

Body Temperature and Grooming in the Mongolian Gerbil

DELBERT D. THIESSEN

Department of Psychology
University of Texas
Austin, Texas 78712

When I was a graduate student at the University of California at Berkeley, self-grooming among a number of species was believed to be primarily a displacement act resulting from the conflict of opposing motivations.[1] It certainly does occur frequently in social contexts where approach and avoidance behaviors are simultaneously evoked. But that is only the tip of the theoretical iceberg. We now know that self-grooming in rodents can have a variety of functions, and is not merely a marker for motivational conflict. It may limit ectoparasitism[2]; it is a primary means by which rodents spread saliva for evaporative cooling[3]; and it also contributes to the general conditioning of the pelage.[4] Moreover, as the papers in this volume demonstrate, self-grooming can serve as an index of neurological activity and as a model for studying developmental processes, complex behavioral patterns, and drug effects. A behavior that once appeared to be trivial and of little interest has finally achieved a status as a significant and challenging process that could have implications for individual survival and reproduction.[5,6] This conference is testimonial to our growing concern.

Our own work has concentrated on two facets of self-grooming in the Mongolian gerbil *Meriones unguiculatus*, chemocommunication and the use of Harderian materials for thermoregulation. Basically what we have found is that Harderian material released during a self-groom is an attraction chemosignal used during social interactions, and that the material, when spread on the pelage, insulates the animal against environmental changes in ambient temperature. The theoretical importance of this work is that it has led us to the conclusion that chemocommunication and thermoregulation may be intimately related.[7,8] It appears to us that the efficiency of sensory and motor mechanisms underlying communication, as well as the successful operation of cognitive processes, are directly related to rather small variations in body temperature. Grooming can regulate these variations.

This paper concentrates on the relationships between self-grooming, the release and spread of Harderian material and saliva, and an animal's thermoregulatory capacity. I will attempt to make the points that (1) grooming frequency is temperature dependent, (2) grooming is associated with the release of both Harderian material and saliva, (3) Harderian lipids and pigments which are spread on the pelage insulate the animal against cold and wetness and alter the reflectance of radiant energy from the pelage, (4) the amounts of saliva and Harderian material released and spread over the hair are associated with ambient and body temperature changes, and (5) animals control the amounts of pelage lipids by altering rates of self-grooming and sandbathing. The general picture to emerge is that self-grooming is a necessary and dynamic interface between the environment and homeostatic regulation. The use of Harderian materials and saliva can essentially free the animal from disturbing temperature perturbations in its environment.

THE HARDERIAN SYSTEM IN PROFILE

The Harderian system is an essential part of self-grooming. The gland itself is located in the occular cavity of most vertebrate species. In the Mongolian gerbil, at least, the gland releases its secretions through the lacrimal canal and out the external nares of the nose. The exit of the material is associated with multiple eyeblinks that occur during a self-groom.[9] The position of the gland relative to the lacrimal canal, and the area initially covered with Harderian material during a facial groom, are shown for *Meriones unguiculatus* in FIGURE 1.

The gland appears sporadically among mammalian species and does not seem to be associated with a particular ecology or lifestyle.[10,11] It does often appear in species that are furred or feathered, and is most evident in species with nictitating membranes. A list of mammalian orders and species where the gland has been noted is given in TABLE 1. In our lab we have observed the groom-related release of Harderian material in *Peromyscus, Norvegicus, Mesocricetus, Meriones unguiculatus, M. tristrami, M. shawi,* and *M. libycus.* The gland apparently does not appear in bats, dogs, cats, horses, or primates, although my impression is that comparative observations have not been extensive or thorough.

In the Mongolian gerbil the paired gland weight is approximately 182 mg in the male and 169 mg in the female. Corrected for body weight differences, the sexes do not differ in Harderian weights. Indeed, none of our work involving hormonal manipulations and behavioral observations suggests any sexual dimorphism. The morphology and histochemistry of the gland have recently been described.[12,13] The major pigment in *Meriones unguiculatus* is presumably protoporphyrin,[14,10] and the primary fatty acids are derivatives of 2,3-alkanediol diacyl esters.[15,16] The proportional amounts of

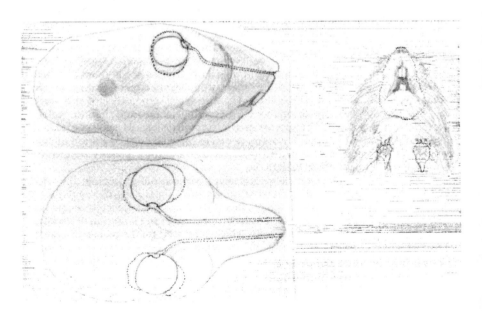

FIGURE 1. Position of the Harderian gland in the orbit of the eye, showing the connecting lacrimal path to the external nares and the region of facial spread for Harderian material.

TABLE 1. Taxonomic Representation of Harderian Gland[a]

Marsupials	opossums, *Didelphis, Caluromys,* and *Marmosa*
Insectivores	hedgehog *Erinaceus*
	big-clawed shrew *Sorex*
	short-tailed shrew Blarina
Edentata	nine-banded armadillo Dasypus
Lagomorphs	rabbit *Oryctolagus* and *Sylvilsgus*
	Pika *Ochotona*
Rodents	Mouse *Mus*
	rat *Tattus* and *Norvegicus*
	hamster *Mesocricetus* and *Cricetalus*
	gerbil *Meriones* (sps.)
	guinea pig *Cavia*
	squirrels *Tamiasciurus, Sciurus* (sps.), and *Citellus*
	chipmunk *Tamias*
	deer mouse *Peromyscus*
Cetaceans	Mysticetes *B. alaenoptera* (sps.), *Megaptera, Balaena,* and Odontocetes *Delphinnus, Phocaena, Delphinapterus,* and *Hyperodon* Dolphin *Platanista* and *Pontoporia*
Pinnipeds	walrus *Odobenus*
	seal *Phoca*
Sirenians	dugong *Dugong*
Artiodactyls	pig *Sus*
	deer *Cervus* and *Dama*

these lipids are seen in TABLE 2. Protoporphyrin does not seem to be an active chemosignal,[17,7] but the fatty acids have yet to be tested for their biological significance. Both the pigment and the fatty acids could contribute to variations in the coat color and the insulative quality of the pelage.

The protoporphyrin is extremely important as a marker for the release of the Harderian material by the gerbil and several other species with this form of porphyrin. It fluoresces dark red when stimulated by long-wave ultraviolet light, making it possible to determine exactly when the material exits from the nose. This technique has allowed us to quantify the relative amounts of Harderian substances released by rating the intensity of the fluorescence on the face immediately following a groom. The fluorescent quality of the material has a half-life of about 2 minutes.

SELF-GROOMING AND HARDERIAN FUNCTIONS

What follows is a brief description of some of our published and unpublished findings

TABLE 2. Major Lipids (Derived from 2,3-Alkanediol Diacyl Esters) of the Harderian Gland of *Meriones unguiculatus*[16]

Capric	3.0%
Lauriac	16.3%
Myristic	16.5%
Stearic	5.1%
Arachedic	16.4%
Behenic	9.8%

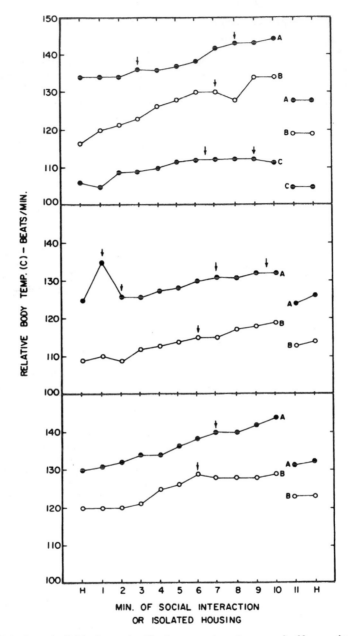

FIGURE 2. Seven individual records of body temperature change and self-grooming frequency (*arrows*) during paired male-male social interactions; *H* = home cage.

for the Mongolian gerbil, *Meriones unguiculatus*. The general laboratory conditions and testing procedures have been published previously.[8,18] All measurements were done on adult male gerbils. Each experimental and control group is composed of from 10 to 12 animals. Mini Mitter (Sun River, OR) biotelemetry units were used to evaluate body temperature in freely moving animals. Lipid levels on the pelage were obtained by shaving animals and extracting the lipids with an organic solvent. I will refer to specific testing conditions only when it is necessary to provide clarification.

The Basic Characteristics of Self-Grooming

The self-groom in the gerbil is a fairly stereotyped sequence of behaviors that results in the release and spread of Harderian material and saliva. Adult gerbils self-groom their body parts in the order mouth-nose, face, ears, flank, ventrum, and tail. This is the identical order in which the groom matures during the first 33 days of life, suggesting that the ontogenetic sequence specifies the neural pattern of the adult groom.[19] Harderian material is first released around 15 days of age, at a time in which the eyes open, hair growth is complete, and the groom has nearly all of it components. The adult continues to groom periodically throughout its life, always showing an association between the fixed action pattern of grooming and the release and spread of Harderian material and saliva.

Grooms are provoked by a number of environmental events, including exposure to novelty, wheel running, chewing food pellets, and social interactions.[20] Elevations in body temperature, as measured by the number of biotelemetry transmitter pulses per minute, are especially effective in triggering self-grooms, as evidenced by the data in FIGURE 2. In this case body temperature was recorded for animals living in isolation and when they were paired with conspecifics. Individual temperature curves are presented, with self-grooms indicated with arrows. Grooms rarely appear when the animals are isolated in their home cages at ambient temperatures. During social interactions, however, the frequency of grooming is increased considerably, and every groom for all seven animals is followed either by a stabilization or a fall in body temperature. This latter could be caused by the evaporative cooling of saliva or the momentary depression of general activity associated with the act of grooming. Presumably all of the activities leading to grooming are thermogenic and result in the spread of saliva and Harderian substances. When body temperature increases between two and three degrees, the animal switches from the common form of self-grooming associated with eye blinking to one in which copious amounts of saliva are spread over the ventrum in a more or less continuous fashion.[21] Paradoxically, in cold environments body temperature increases (compensatory thermogenesis) are potent stimulants of grooming (see below).

There is an intimate relationship between Harderian release and saliva production and release. We have worked out a method of measuring the relative amounts of both substances immediately following a groom. The quantity of Harderian material is measured by blotting the nostrils with a preweighed piece of filter paper at the point of greatest release (correlated with eye blinking). Saliva production is measured in an analogous way, by placing a cotton pledget in the mouth for one minute beginning at a corresponding time. TABLE 3 gives data for animals who were restrained and measured following the eye blink of the groom, or who were allowed to complete the groom (nonrestrained). Measures were taken on separate occasions for animals at times 0 minute, 1 minute, 2 minutes, and 3 minutes. Harderian material decreases steadily over

TABLE 3. Temporal Variation in Saliva and Harderian Secretion (mg) following an Autogroom for Nonrestrained and Restrained *Meriones unguiculatus*

Minutes Postgroom	Nonrestrained		Restrained	
	Saliva Secretion	Harderian Secretion	Saliva Secretion	Harderian Secretion
0	42.4 ± 3.29	4.8 ± 0.39	43.2 ± 3.94	6.2 ± 1.60
1	32.6 ± 2.41	3.3 ± 0.80	26.8 ± 1.79	4.10 ± 0.62
2	25.7 ± 1.60	2.9 ± 0.43	28.1 ± 4.1	4.30 ± 0.70
3	33.3 ± 2.66	2.4 ± 0.45	41.3 ± 5.2	2.50 ± 0.37

minute and then begins to increase again toward its high level. The results are comparable in restrained and nonrestrained animals, suggesting that it is not merely the spread of the material with the forepaws that accounts for the temporal variations. The material breaks down or evaporates in a relatively short period of time. One cannot compare the Harderian and saliva measures directly, as the former is a measure of material expressed during that second of release, while the latter is a measure extending over 1 minute. In any case, it is evident that a groom is associated with a copious release of both Harderian material and saliva, and that the time course for the recovery may be different for the two. I will refer to data in the next section which shows that the amounts of Harderian material and saliva vary according to the ambient temperature and take on quite different patterns.

The Thermoregulatory Link between Grooming and Harderian Excretion

Many rodents either passively or actively accumulate lipids on their pelage.[22] The lipids originate from skin sebaceous glands, or in the case of the Mongolian gerbil, from either the sebaceous or Harderian glands. Both *Meriones* and *Dipodomys*[2] have the ability to regulate pelage lipids according to environmental conditions. *Dipodomys* and *Meriones* remove excess pelage lipids while living on sand, and *Meriones* adds lipids by self-grooming. FIGURE 3 shows data for *M. unguiculatus* to indicate that (1) Harderianectomy (HX) results in a reduced level of pelage lipids within 8 to 16 days, (2) shampooing lipids from the pelage is followed by a recovery of lipids within 16 days, and the rate of recovery is faster in cool conditions than in ambient conditions, (3) animals living on sand decrease their pelage lipids within 16 days, whereas those animals living on wire mesh, and thus unable to sandbathe, accumulate excess lipids within that same time frame, and (4) shampooing lipids from the pelage of Harderianectomized gerbils is followed by only a minor increase in pelage lipids (presumably sebaceous in nature).

These results are an obvious demonstration of the dynamic changes that occur in animals subjected to various environmental conditions. Animals are dependent upon their Harderian gland for pelage lipids, and hardly at all on sebaceous glands. They will remove pelage lipids through sand living if levels are high, and accumulate these lipids if prevented from sandbathing. It is worthwhile also to note that gerbils apply more lipids on the pelage when the environmental temperature is lowered. These variations are apparently an important part of the homeostatic devices used to counter environmental threats.

The obvious question is, Do these variations in pelage lipids, rates of grooming,

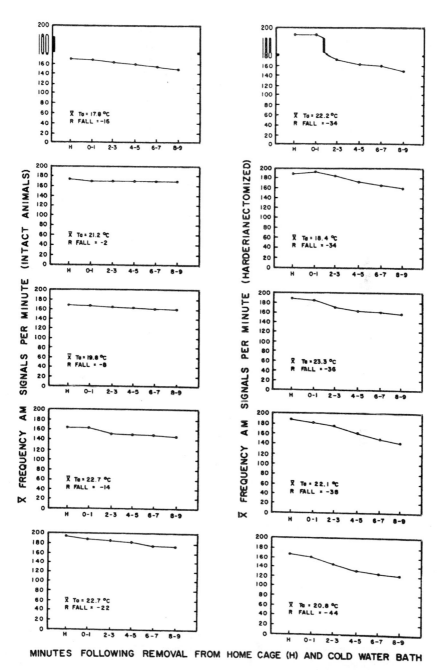

MINUTES FOLLOWING REMOVAL FROM HOME CAGE (H) AND COLD WATER BATH

FIGURE 3. Individual body temperature variations in intact and Harderianectomized gerbils following a cold-water dip.

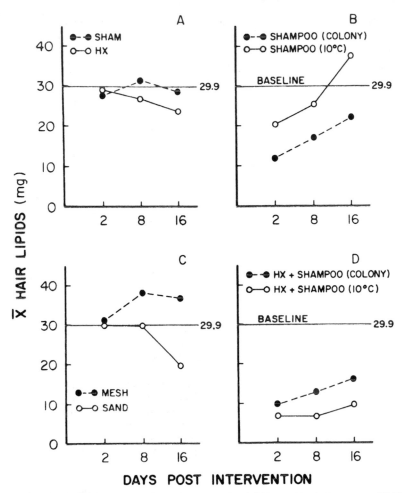

FIGURE 4. Variation in pelage lipids among animals of different Harderian status. **(A)** Effects of Harderianectomy (*HX*) on pelage lipids. **(B)** Reacquisition of pelage lipids among intact animals as the result of self-grooming in ambient and cold conditions. **(C)** Accumulation of excess pelage lipids on animals prevented from sandbathing, and the loss of pelage lipids from animals living on sand. **(D)** Change of sebaceous hair lipids among Harderianectomized gerbils living at ambient and cold conditions. The average baseline lipids among intact animals living in ambient temperatures is shown by the horizontal line at 29.9 mg.

to a 1-second dip in ice water and then compared their body temperature changes to those of nonoperated controls.[18] The results are seen in FIGURE 4. Body temperature dropped more rapidly over a 9-minute period in those animals without harderian material on their pelage. We found that shampooing resulted in the same rapid drop in body temperature, and that the drop in temperature in Harderianectomized animals was attenuated if lipids were artificially replaced on the pelage prior to the test. Ap-

TABLE 4. Effects of Variations in Harderian-Pelage Status on Pelage Reflectance and Body Temperature Increases under Radiant Energy for *Meriones unguiculatus*[23]

Conditions	Solar Reflectance (%)	Temperature Increase (%)
Harderianectomied	22.12 ± 3.38	3.30 ± 0.40
Control	19.56 ± 3.10	5.84 ± 0.49
Shampooed	16.98 ± 2.95	1.34 ± 0.49
Control	14.63 ± 2.59	2.68 ± 0.31
Sand Exposed	24.74 ± 2.90	3.34 ± 0.56
Control	18.70 ± 2.12	4.84 ± 0.38

pelage and increases the rate of radiant absorption.[23] When we measured the percent reflectance of body hair with a Bausch and Lomb Spectronic 20 spectrophotometer equipped with a reflectance attachment, we found a consistent relationship between the Harderian status of the animal and the percent reflectance. Moreover, when these same animals are exposed to radiant energy from a tungsten lamp for 5 minutes, body temperature changes were directly related to the Harderian status of the animal. These data are seen in TABLE 4. Specifically, animals with reduced levels of Harderian material on their pelage, because of Harderianectomy, shampooing, or sand living, reflect more radiant energy in the visual range and show reduced levels of body temperature increases. In other words, a lightened pelage, however, obtained, leads to more reflectance of light and a protection against temperature increases. The ability to regulate Harderian lipids and pigments on the pelage could allow an animal to better control its body temperature during the daylight hours. This could be especially important for the Mongolian gerbil who spends a large share of its daytime activity in the winter on the surface of the ground.[24]

Pendergrass[25] and Pendergrass and Thiessen[26] have subsequently found that gerbils living in the cold (5°C) accumulate more pelage lipids and increase radiant energy absorption, whereas gerbils living in a warm environment (34°C) reduce pelage lipids and decrease radiant energy absorption. These changes are compatible with attempts to optimize pelage insulation and radiant absorption under cold and warm conditions respectively.

Under varying ambient temperatures gerbils release differing amounts of Harderian material and saliva during a groom (see TABLE 5). Harderian release is exaggerated for every self-groom in cold environments and decreased in warm environments. Conversely, saliva secretion is reduced in cold environments and increased in warm environemnts. If Harderian material is used for insulation and to increase radiant absorption, and if saliva is used for evaporative cooling, then these changes in release rate are expected—animals in a cold environment benefit from applying Harderian lipids and pigments to the pelage, and animals in a warm environment benefit from

TABLE 5. Effects of Ambient Temperature on Saliva and Harderian Secretion and Latency to Groom for *Meriones unguiculatus*

Ambient Temperature	Saliva Secretion	Harderian Secretion	Latency to Groom (sc)
7°C	52.4 ± 8.0	5.8 ± 0.47	458 ± 68.1
22°C	49.2 ± 5.4	5.0 ± 0.66	392 ± 42.1
34°C	100.2 ± 10.6	4.8 ± 0.51	635 ± 84.5

low amounts of Harderian material on the pelage and higher amounts of saliva secretion for evaporative cooling. We also have some recent data to indicate that the frequency of grooming is higher among animals exposed to cold environments, and is lower among animals exposed to warm environments. Apparently gerbils can adjust the amounts of Harderian material and saliva by altering both the amounts released during any particular groom and the frequency with which these products are discharged.

The Mongolian gerbil has one other major way of adjusting its pelage lipids, and that is by way of sandbathing. I have repeatedly pointed out that animals living on sand are able to remove excessive amounts of pelage lipids. They apparently do this by passive exposure to the sand and by actively engaging in sandbathing. We reasoned, therefore, that if sandbathing is a mechanism to remove hair lipids, then the frequency of sandbathing should also be temperature dependent.[26] From an adaptive perspective, animals should sandbathe more frequently when ambient temperatures are high, and less frequently when temperatures are low. That is, in the first case they should remove insulating lipids and in the second case they should preserve that insulative protection. FIGURE 5 shows data adding confirmation to that hypothesis. What we do not know at this point is how much lipid can be removed by sandbathing bouts. Our impression is that it may take several days for significant amounts of lipids to be removed, just as it does following Harderianectomy. Evidently, the manipulation of Harderian lipids and pigments on the pelage is geared toward long-term adaptations to environmental temperature variations. It takes days, or even weeks, for an animal to set optimal levels of Harderian spread. Saliva secretion, to the extent that it is used for evaporative cooling, is a short-term adaptation to sudden increases in body temperature. In combination, Harderian and saliva spread can help an animal stabilize its body temperature over long and short periods of time.

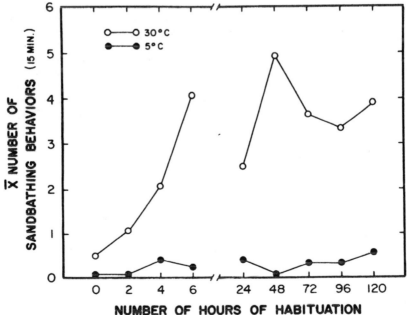

FIGURE 5. Frequency of sandbathing bouts over a 120-h period at 5°C and 30°C.

Grooming as an Ecological Adaptation

Self-grooming in *Meriones unguiculatus* serves several purposes, only some of which we have explored in any detail. Our hope is to show how grooming, Harderian spread, saliva spread, sandbathing, and thermoregulation are integrated. Our basic model suggests that Harderian spread mediates long-term changes in the insulation and radiant absorption qualities of the pelage. Self-grooming and thermoregulation are closely associated. We have found that cold ambient conditions result in thermogenesis and the triggering of grooms.[21] An insulative coat is thereby accumulated which can protect the animal from cold and dampness. Moreover, Harderian lipids and pigments on the pelage can increase radiant absorption and thus elevate body temperature. Under cold conditions the animal reduces sandbathing and hence the removal of Harderian substances from the pelage. On the other hand, when ambient temperatures are high, the animal reduces Harderian spread, thus decreasing the insulative and absorption characteristics of the pelage. Beyond that, the animal removes much of the Harderian material from the body by sandbathing, and exaggerates the loss of body heat through saliva spread and evaporative cooling. As is expected, the amount of Harderian material released during a self-groom is increased when the animal is exposed to a cold environment and is decreased upon exposure to a hot environment. The amount of saliva produced and spread for evaporative cooling is increased when the animal is exposed to a hot environment and reduced upon exposure to a cold environment.

The intricacy of the system for actively controlling body temperature suggests its importance for the Mongolian gerbil. This gerbil normally lives in an arid desert region in Mongolia and Northeast China.[27] Day-night temperatures can vary as much as seasonal temperatures, and the animals are subjected to annual monsoons. The mechanisms I have discussed can help the gerbil cope with this wide range of ecological conditions. Obviously the gerbil has other physiological and behavioral mechanisms for controlling its body temperature. It can, for instance, modify its day-night activity cycle by altering the times spent above ground, thus optimizing its energy budget under different seasonal conditions.[24] It can expose itself to increased radiant energy by becoming diurnal in the winter, and it can reduce its exposure to high radiant levels by becoming nocturnal in the summer. Thus, by modifying its circadian rhythm, in combination with variations in grooming and sandbathing, the gerbil can stabilize its body temperature while remaining active throughout the year.

Other rodent species might be expected to rely on similar mechanisms for adjusting body temperature, especially those who are to some extent diurnal and who live in an arid environment. We have seen a similar release and spread of Harderian substances in other species of gerbils and in deer mice, hamsters, house mice, and rats (see TABLE 1 and the earlier discussion), but we have no evidence that identical processes are used for thermal conservation. A comparative attack on this problem is certainly needed. Whether nonrodent species use Harderian material in a comparable way is entirely an open question. A large number of mammalian species apparently do not have the gland at all or do not self-groom in similar ways. An important step would be to find out if species differ in their use of lipids or other mechanisms for thermoregulation. For example, animals without Harderian glands may depend more on skin sebaceous lipids for insulative protection, and they may also use sweat glands for evaporative cooling.[22] It has also been suggested that animals with large lacrimal glands tend to have small or absent Harderian glands.[10] This could be a clue to help explain adaptive differences. Finally, ecological variations and body size differences might determine the relative importance of the Harderian gland and other thermoregulatory devices. Clearly there is a pressing need for comparative information.

The emerging picture for the Mongolian gerbil is complex. Yet studies to date have given us the impression that self-grooming is a significant behavior that allows an animal to cope with a range of environmental variations. It is important to note, I think, that self-grooming is an essential link to physiological conservation. Grooms are used as a compensatory behavior to maintain a more constant body temperature. They are not manipulative, in the same sense as behaviors for nest building, burrowing, or locomotion. They do not alter the environment. Rather, self-grooms stabilize metabolic events and increase the efficiency of other processes, such that manipulative behaviors can continue uninterrupted. To serve this end effectively, grooms need to be energy efficient, modifiable within a changing environment, and properly integrated with other behaviors. Self-grooming in *Meriones unguiculatus* meets all of these criteria.

SUMMARY

The Mongolian gerbil, *Meriones unguiculatus*, like many other rodents, releases a complex mixture of pigments and lipids from the Harderian gland during a self-groom. The material exits from the external nares of the nose, is mixed with saliva, and spread widely over the pelage. Cold temperatures, especially, are effective in initiating grooming. A self-groom is associated with an increaase in body temperature (compensatory thermogenesis in the cold). In addition to acting as a chemosignal, the Harderian material serves two major homeostatic functions: (1) the lipids on the pelage act to insulate the animal against cold and wetness, and (2) the lipids and pigments darken the pelage and increase radiant absorption. Body temperature is thus maintained at a higher level than would otherwise be the case. The amount of Harderian material found on the pelage varies with Harderianectomy, sandbathing, and ambient temperatures. Animals prevented from sandbathing accumulate excess lipids on the pelage, and cold temperatures facilitate the acquisition of lipids on the pelage. Under hot temperatures the grooming of Harderian substances is repressed and the frequency of sandbathing is increased. Thus pelage lipids are reduced in two ways. The amount of Harderian material released during an autogroom is inversely related to the ambient temperature, whereas the amount of saliva used for evaporative cooling is positively related to ambient temperature. The net effect is that pelage lipids are increased and maintained during cold conditions, and are reduced during hot conditions. In hot environments the gerbil switches from the spread of Harderian material for insulation to the spread of saliva for evaporative cooling. The gerbil optimizes its body temperature by varying the frequency of grooming and sandbathing, and by altering the amount of Harderian material and saliva released. Other species living in arid environments may use similar mechanisms to stabilize body temperature. Self-grooming is a critical behavior for meeting thermal needs, and is complexly integrated with related processes.

REFERENCES

1. LORENZ, K. 1965. Evolution and Modification of Behavior. Univ. of Chicago Press. Chicago.
2. BORCHELT, P. L., J. G. GRISWOLD & R. S. BRANCHEK. 1976. An analysis of sandbathing and grooming in the kangaroo rat (*Dipodomys merriami*). Anim. Behav. **24**: 347–353.
3. HAINSWORTH, F. R. & E. M. STRICKER. 1970. Salivary cooling by rats in the heat. Physiology and Behavioral Temperature Regulation. J. D. Hardy, P. Gagge & J. Stolwijk, Eds.: 611–626. Thomas. Springfield, IL.
4. PATENAUDE, F. & J. BOVET. 1983. Self-grooming and social grooming in North American beaver (*Castor canadensis*). Can. J. Zool. **62**: 1872–1878.

5. MOORE, C. & S. ROGERS. 1984. Contribution of self-grooming to onset of puberty in male rats. Dev. Psychobiol. **17(3):** 243–253.
6. WETTERBERG, L., E. GELLER & A. YAWILER. 1969. Harderian gland: An extraretinal photoreceptor influencing the pineal gland in neonatal rats? Science **167:** 884–886.
7. THIESSEN, D. 1977. Thermoenergetics and evolution of pheromone communications. Prog. Psychobiol. Physiol. Psychol. **7:** 91–191.
8. THIESSEN, D. 1983. Thermal constraints and influences on communication. Chemical Signals in Vertebrates 3. D. Muller-Schwarze & R. M. Silverstein, Eds. Plenum. New York.
9. THIESSEN, D., A. CLANCY & M. GOODWIN. 1976. Harderian gland pheromone in the Mongolian gerbil *Meriones unguiculatus*. J. Chem. Ecol. **2(2):** 231–238.
10. KENNEDY, G. Y. 1970. Harderoporphyrin: A new porphyrin from the Harderian glands of the rat. Comp. Biochem. Physiol. **36:** 21–36.
11. SAKAI, T. 1981. The mammalian Harderian gland: Morphology, biochemistry, function and phylogeny. Arch. Histol. Jap. **44(4):** 299–333.
12. JOHNSTON, H. S., J. MCGADEY, G. G. THOMPSON, M. R. MOORE & A. P. PAYNE. 1983. The Harderian gland, its secretory duct and porphyrin content in the Mongolian gerbil (*Meriones unguiculatus*). J. Anat. **137(3):** 615–630.
13. SAKAI, T. & T. YOHRO. 1981. A histological study of the Harderian gland of Mongolian gerbils, *Meriones meridianus*. Anat. Rec. **200:** 259–270.
14. CARRIERE, R. 1985. Ultrastructural visualization of intracellular porphyrin in the rat Harderian gland. Anat. Rec. **213:** 496–504.
15. OTSUKA, H., O. OTSURU, T. KASAMA, A. KAWAGUCHI, K. YAMASAKI & Y. SEYAMA. 1986. Stereochemistry of 2,3-alkanediols obtained from the Harderian gland of Mongolian gerbil (*Meriones unguiculatus*). J. Biochem. **99:** 1339–1344.
16. OTSURU, O., K. OTSUKA, T. KASAMA, Y. SEYAMA, T. SAKAI & T. YOHRO. 1983. The characterization of 2,3-alkanediol diacyl esters obtained from the Harderian glands of Mongolian gerbil (*Meriones unguiculatus*.) J. Biochem. **94:** 2049–2054.
17. PAYNE, A. P. 1979. The attractiveness of Harderian gland smears to sexually naive and experienced male golden hamsters. Anim. Behav. **27:** 897–904.
18. THIESSEN, D. & E. M. W. KITTRELL. 1980. The Harderian gland and thermoregulation in the gerbil (*Meriones unguiculatus*). Physiol. Behav. **24:** 417–424.
19. THIESSEN, D., M. PENDERGRASS & R. K. YOUNG. 1983. Development and expression of autogrooming in the Mongolian gerbil (*Meriones unguiculatus*). J. Comp. Psychol. **97(3):** 187–190.
20. THIESSEN, D., M. GRAHAM, J. PERKINS & S. MARCKS. 1977. Temperature regulation and social grooming in the Mongolian gerbil (*Meriones unguiculatus*). Behav. Biol. **19:** 279–288.
21. PENDERGRASS, M. L. & D. THIESSEN. 1981. Body temperature and autogrooming in the Mongolian gerbil (*Meriones unguiculatus*). Behav. Neural Biol. **33:** 524–528.
22. SOKOLOV, V. E. 1982. Mammal Skin. University of California Press. Berkeley, Calif.
23. THIESSEN, D., M. PENDERGRASS & A. E. HARRIMAN. 1982. The thermoenergetics of coat colour maintenance by the Mongolian gerbil (*Meriones unguiculatus*). J. Therm. Biol. **7:** 51–56.
24. RANDALL, J. A. & D. THIESSEN. 1980. Seasonal activity and thermoregulation in *Meriones unguiculatus*: A gerbil's choice. Behav. Ecol. Sociobiol. **7:** 267–272.
25. PENDERGRASS, M. L. 1985. Bioenergetic effects of hair lipids in the Mongolian gerbil, *Meriones unguiculatus*. Ph.D. thesis, University of Texas, Austin, Tex.
26. PENDERGRASS, M. L. & D. THIESSEN. 1983. Sandbathing is thermoregulatory in the Mongolian gerbil, *Meriones unguiculatus*. Behav. Neural Biol. **37:** 125–133.
27. THIESSEN, D. & P. YAHR. 1977. The Gerbil in Behavioral Investigations. Univ. of Texas Press. Austin, Tex.

Preening and Associated Comfort Behavior in Birds[a]

JUAN D. DELIUS

Experimentelle Tierpsychologie
Psychologisches Institut
Ruhr-Universität
D 4630 Bochum, Federal Republic of Germany

INTRODUCTION

The integument of birds has a considerably more complex structure than that of mammals.[1] In most avian species the plumage also has an important extra function compared with the pelage of mammals: flight. Accordingly preening, as a behavior that conditions the bird's integument, can be expected to play a more important role within the activity repertoire of birds than grooming does in that of mammals. Indeed preening is generally a very elaborate behavior that takes up a considerable proportion of the time budget of birds. For example, van Rhijn[2] has impressively documented the complexity of the preening behavior of gulls, and I have found that during the breeding season and at daytime 15% of their time was taken up by this activity.[3] During moonlit nights (unpublished observations) the preening took up almost as much of the gulls' time (12%). Pigeons interrupt sleeping in complete darkness virtually only to preen[4]; gulls can be expected to do likewise.

Although preening is such a frequent behavior among birds, our knowledge about it is still sparse and disconnected. To give this brief review the wisp of a plot, it takes the form of a somewhat personal narrative. Besides preening proper (drawing feathers through the beak), a number of other comfort patterns that are more or less closely related to preening, either motivationally or functionally, will be mentioned. The reader should realize that there are several other behaviors of this kind that, though obviously important to birds, I will not refer to. As a striking, rather exotic example, one can mention anting, or bathing in a heap of ants, an activity many bird species seem to love!

OCCURRENCE AND STRUCTURE OF PREENING

Compared with many other behavioral responses, the timing of preening is usually not critically dependent on specific environmental events. By and large, as a preventive body-surface maintenance activity, it does not have to be done at any particular time as long as it is done sometime and at intervals that are not too far apart. It would

[a] Most of the research done in our laboratory was supported by the Deutsche Forschungsgemeinschaft through its Sonderforschungsbereich 114. Pierre Deviche (now at the Dept. of Zoology, Oregon State University, Corvallis), who contributed much to the research reported, held Humboldt and NATO fellowships while in Bochum.

seem to be a kind of spare-time activity that is performed whenever there is nothing more important to do. Gulls in the wild, subject as they are to all kinds of selective pressures, spend less time preening than well-provided-for, aviary caged gulls (15% versus 26% of their time, $p < 0.01$; own observations). In contexts that make little demand on the animals, and in an undisturbed environment, rhythmicities with periods on the order of several tens of minutes may be predominant in determining the occurrence of preening.[5] Over and above these short cycles, a diurnal cycle is often very apparent[6] (but compare ref. 3). It is not certain whether it is directly due to a central circadian rhythm, or secondary to the dark-light cycle. It is well documented that periodic external stimuli such as light pulses can entrain the short-term cyclic occurrence of preening.[7]

Some preening-related behavior can, however, become strongly stimulus dependent in particular situations. I have observed skylarks to respond forthwith with sunbathing behavior during prolonged cloudy-foggy weather spells whenever the sun managed to break through and to immediately begin with rainbathing behavior during extended hot-dry weather spells whenever it began to drizzle.[6] Deprivation of the opportunity to dustbathe leads to a cumulating increase in the probability that chickens will show this behavior as soon as dust is made available.[8] There are also, undoubtedly, events that require immediate preening attention by birds. This might be the case with the more offensive activities of integument parasites,[9] disturbances due to feather loss and growth (molting seems to intensify preening in pigeons), and disturbances due to wind gusts or the behavioral activities of the bird itself or those of other birds. Particularly short duration bouts of preening are thought to be primarily elicited by such acute cutaneous, and sometimes visual, stimuli.[10]

This evidence indicates that local somesthetic stimuli may, on the one hand, trigger and direct preening. On the other hand, they are probably not essential for preening to occur and to be structured. Lefebvre and Joly[11] (see also ref. 5), for example, argue on the basis of a detailed analysis of extensive observational data that the longer preening bouts of kestrels are not initiated by any particular stimulus. They also conclude that

TABLE 1. Local and Overall Preening of Dorsally Denervated and Sham-Operated Control Pigeons Unstimulated and Stimulated with Itching Powder Applications to the Dorsal Skin

	Dorsal Denervated ($n = 3$)		Sham Operated ($n = 3$)	
	Before	After	Before	After
Unstimulated				
% of periods with preening directed at dorsum	28%	24%	19%	21%
Average number of 15-s periods in which preening took place (max. score = 120)	35	37	28	45
Stimulated with itching powder				
% of periods with preening directed at dorsum	39%[a]	25%	40%[a]	47%[a]
Average number of 15-s periods in which preening took place (max. score = 120)	68[a]	32	71[a]	83[a]

Based on 3 preoperative and 3 postoperative 30-min sessions on 6 birds. All figures are rounded to the nearest whole number.

[a] Significantly different from equivalent (unsuperscripted) averages ($p < 0.05$ or less).

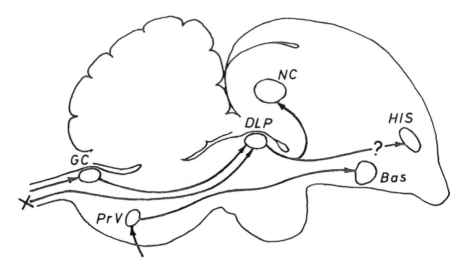

FIGURE 1. Scheme of the somatosensory projections to the forebrain of the pigeon: *DLP*, nucleus dorsolateralis posterior; *GC*, nucleus gracilis et cuneatus; *NC*, neostriatum caudale; *HIS*, hyperstriatum intercalatus superior; *PrV*, nucleus sensorius principalis nervi trigemini; *Bas*, nucleus basalis prosencephali (updated from Delius and Bennetto[15]).

the organization of these bouts, in terms of the body areas sequentially preened, is determined by a hierarchical, stochastic control system (a noisy "grammar") that is largely independent of specific external stimuli.

Our own unpublished experimental data (partly obtained in collaboration with K. Benetto) on pigeons that had a large patch of the dorsal skin surgically denervated by removing a small length of six to eight pairs of dorsal cutaneous nerves and control animals that were sham operated support this view. The pigeons continued to preen spontaneously the dorsum as much, but not significantly more, than they had done before surgery, not differing in this from the controls. Itching powder (cowhage) applied to the dorsal skin of the sham-operated pigeons markedly augmented the proportion of preening movements directed at this region but also increased somewhat the preening directed at other body areas. The dorsally denervated pigeons did not show a special response to the dorsal application of itching powder, preening this region about as much as they did normally, when not specially stimulated, confirming that they had been effectively denervated (TABLE 1). Thus local stimulation is not essential to organize the normal sequence of preening movements, although it can certainly initiate and direct preening. The evidence suggests that bird preening might not be much different from mammalian grooming in this respect.[12] Incidentally, in view of the context the comparisons with mammals will be kept to a minimum; readers are referred to the relevant contributions in this volume.

Contrary to former opinions, the furnishing of the avian skin with sensory receptors is at least comparable to that of mammals.[13] Some arrangements are quite specialized for detecting plumage derangement.[14] There is thus a substrate that can mediate preening triggered by cutaneous stimuli. An effort was made to localize the somesthetic projection areas in the avian brain in the hope that they may somehow be involved in this control. Using the evoked potential technique, three such areas were located in the forebrain of pigeons, two in the telencephalon, and one in the thalamus. A further, purely trigeminal, telencephalic projection area was already known[13,15] (FIG. 1). Exploratory experiments involving lesions of the two telencephalic areas (the thalamic and trigeminal projections were not investigated), however, did not reveal any modification of preening. Neither did explicit electrical stimulation of these areas through implanted electrodes elicit any excess preening (see also below). It thus would seem that the somesthetic sensory system links up with the preening motor system at a lower level of the neuraxis than the telencephalon (but see below).

FIGURE 2. Response of a gull to a slight, brief disturbance. After an alarm response (neck stretch, *1*) it stares down (*2*), preens (*3*), stretches one wing and leg (*4*), yawns (*5*), and after moving away a few steps squats down and becomes drowsy (*6*). Between these responses it repeatedly eyed the source of disturbance.

DISPLACEMENT PREENING

In addition to normally occurring preening, some other comfort patterns make occasional appearances in contexts that are anything but relaxed. This has been described by ethologists as displacement behavior.[16,17] In situations that can be calculated to be somewhat stressful (involving motivational conflict), birds (and indeed also other animals including humans) often show brief bouts of comfort behavior, and some other equally irrelevant responses, embedded among behaviors more directly related to resolving the situation inducing the stress. FIGURE 2 shows a sequence of behavior that demonstrates the point. The photographer waggled a finger out of a slit of a hide placed in the middle of a gull breeding colony when the animals were at ease. Birds nearby immediately responded with an attentive neck elongation, wide-open eyes, and a few alarm calls; some actually flew away. Then the finger was withdrawn again and before relaxing most of the remaining birds showed a peculiar sequence of responses including a brief bout of preening. The bird photographed finally moved a few steps away and squatted down, soon showing signs of drowsiness (see also ref. 18). On a more global scale, a similar but more intensive delayed preening response to a stressing event was observed when surrounding birds were alarmed by brief and energetic arm waving from a tower emplaced in the middle of the gull colony. A high level of behavioral arousal was so induced. The gradual return to normal that followed was marked by a burst of preening and indeed other comfort responses not recorded in detail[3] (FIG. 3).

Experiments involving electrical brain stimulation through chronically implanted electrodes in awake, unrestrained (but captive) gulls showed that certain brain loci pro-

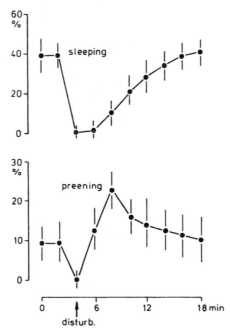

FIGURE 3. Time course of preening and sleeping responses of gulls following a brief, strong disturbance. Means (± SD) based on 10 sessions, sampling of >100 individuals (after Delius[3]).

duced a behavior syndrome bearing a close similarity to the response pattern described above.[19] Upon brief stimulation of such a locus, two or more of the following behavioral responses were elicited: staring down, pecking, preening, relaxation, yawning, squatting, drowsiness, and sleeping (compare with FIG. 1; pecking sometimes follows staring down, and sleeping frequently ensues after drowsiness). These responses were shown upon stimulation, most often in the sequence listed with latencies that varied between a few seconds to a few minutes (FIG. 4a). The anatomical location of the stimulation sites that yielded this preening cum relaxation syndrome are illustrated in FIGURE 4b. Most of them appear to cluster along the lateral wall of the anterior section of the forebrain ventricle, including the nucleus accumbens, the medial neostriatum intermedium, the lateral neostriatum intermedium, the posterior hippocampal complex, and perhaps the posterior ventral hypothalamus. It may be worthwhile to point out also the two loci near the nucleus basalis, since this structure will be mentioned again later on.

A key fact was that the activation of these brain sites tended to elicit drowsiness and sometimes even sleep, besides the other responses. This can be related to the frequent mention of sleep behavior, along with comfort and a few other behaviors, in the displacement activity literature.[20] Together with some other evidence these findings led to the formulation of a dearousal hypothesis of displacement activities. It was meant to compete with a then-current disinhibition theory of the same phenomenon. According to this latter theory there is a mutual inhibition between motives activated by the stress situation that leads to the disinhibition of a background motive yielding irrelevant displacement responses.[18,21]

Stressful situations of the kind that elicit displacement activities were thought, according to the dearousal hypothesis, to involve an excessive information processing activity in the brain (having to do with the operation of deciding whether and what corrective behavioral action is to be taken in response to the stressing situation). This was conceived as leading to a greater activation of neural networks than could be coped with effectively. Such an information overload, or overarousal, would trigger corrective action such that mechanisms designed to dampen the information processing rate

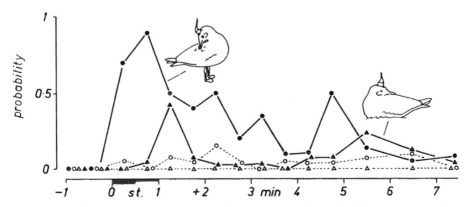

FIGURE 4a. Brain-stimulation-elicited preening and dearousal in gulls. Time course of probability of occurrence of preening and sleeping per 30-s periods. Averages for two lateral neostriatum intermedium sites, 12 trials each. 50-μA stimulation trains lasted 30 or 60 s (*st.*). *Open symbols* and *dashed lines* are results of control dummy trials.

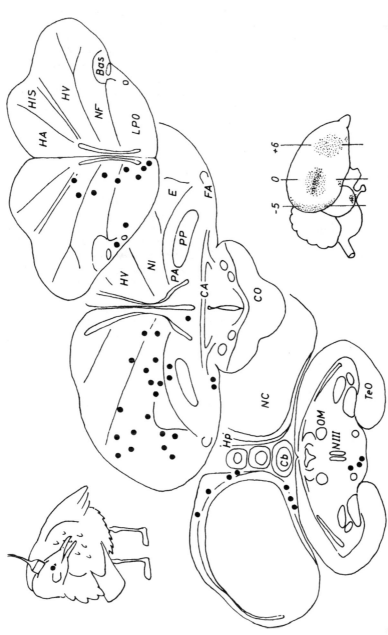

FIGURE 4b. Locations of brain sites in gulls that yielded a dearousal syndrome including preening upon electrical stimulation. Electrode tips were identified histologically. For convenience all sites have been plotted on the left brain half and on only three frontal sections. The actual anterior-posterior distribution of the sites is indicated by the stippling in the brain sideview insert. Besides the 37 sites reported in Delius,[19] 10 additional comparable sites studied afterwards have been included. *Bas*, Nucleus basalis prosencephali; *Hp*, hippocampus; *HV*, hyperstriatum ventrale; *LPO*, lobus paraolfactorius; *NI*, neostriatum intermedium; *NF*, neostriatum frontale.

for the purposes of neural rest or regeneration (drowsiness, sleep) would act to bring the activation back into the range where efficient (adaptive) information processing was again possible. The concept is one of neural arousal homeostasis[20] (see also ref. 22). Preening and other behavior that is neurally coupled to sleep (see results of the stimulation experiments mentioned above) would be incidentally coactivated or perhaps even contribute to the dearousing process. Weak, repetitive cutaneous stimulation as might arise during preening, for example, has been said to act in such a manner.

Some efforts were made to support the arousal homeostasis hypothesis with further kinds of evidence, now using pigeons. The essential element of the hypothesis is the assumption that there is an arousal-regulating negative feedback loop in operation. The most direct analytical approach suggested by elementary systems theory considerations is to attempt to cut open the loop. Based on some of the brain stimulation results (FIG. 4b), it was surmised that the hippocampal formation could be the critical link of such a circuit. In a first experiment pigeons with bilateral hippocampal lesions, when exposed to a mildly stressing situation, a novel cage, showed a significantly lesser preening and a significantly greater behavioral arousal increment than control sham-operated pigeons, even though the preening activity of the hippocampal birds in their home cage environment did not differ from that of the controls. The result accorded nicely with the hypothesis, but in later experiments they could not be replicated. These experiments rather revealed a considerable inter- and intraindividual variation of the preening evoked by the stressing stimulus. The handling associated with the procedure was probably stressing in itself and also caused plumage disturbances that directly induced preening (see above), masking any differences that may nevertheless have been due to the different surgical treatments. If hippocampal lesions have an effect on stress-induced preening, it is more subtle than our methods could reveal. In any case even the demonstration of a weak effect would be insufficient to support the specific hippocampus-mediated version of the dearousal hypothesis. The negative results do not, however, disprove the more general idea, as structures other than the hippocampus may mediate the dearousal feedback postulated.

More generally, we have found it difficult to devise a laboratory test that reliably yields a quantifiable displacement response in pigeons, even though casual observations of feral pigeons leave little doubt about the existence of the phenomenon, including preening and other comfort behaviors, in this species. It may be that operant conditioning techniques could provide a solution.[23]

A regularly occurring, brief, preening-like behavior of pigeons has been labeled "displacement preening."[24] It is often incomplete in that the animal's bill does not touch the scapular feathers at which it is directed.[25] In the opinion of a pigeon sexual behavior expert (M. Abs, personal communication), the movement is part and parcel of their courtship ceremony, that is, a highly ritualized display[17] that no longer has the character of a proper displacement activity.

Another approach attempted to substantiate further the assumption that preening is truly associated with neural dearousal. As an index we employed the electroencephalogram (EEG) recorded from chronic depth electrodes variously placed in the forebrain of freely moving pigeons. Birds show EEG changes, from the desynchronized pattern to the synchronized pattern, as they proceed from a very awake state to an asleep state, much as do mammals[26] (so-called rapid eye movement, desynchronized EEG sleep is rare in birds). Preening was indeed found to be accompanied by slow wave, high-amplitude activity. With the technique employed, however, it was possible that subtle cable-movement artifacts were interfering with the recording. Pecking, also involving rapid head movements, for example, also correlated with the appearance of slow waves in the EEG. Still, that would in itself not necessarily be incompatible with the dearousal hypothesis, as pecking also occurs as a displacement activity in birds. More recent pigeon EEG recordings, however, made during experiments with a different intention but with a technique that excludes cable artifacts, confirm the original observations that slow waves are often associated with preening and pecking. The objection now is that these newer recordings may suffer from another kind of movement artifact, namely those due to a mechanical displacement of brain tissue itself. This issue is as yet unresolved.

CORTICOTROPIN PREENING?

It is well known that stress activates the adrenocortical system. Could stress-induced behavior be an effect due to the corresponding hormonal processes? Inspired by the report of Gessa *et al.*[27] that intracerebroventricular (i.c.v.) administration of adrenocorticotropic hormone (ACTH) elicited yawning and stretching behavior in cats and other mammals and the fact that these responses have been reported as displacement activities in the relevant species, we examined the effect of the same treatment in birds. The result was that doses of 1 or more international units (IU) of ACTH injected into the lateral forebrain ventricles of pigeons through chronically implanted cannulae yielded yawning and headshaking with short latencies[28] (FIG. 5). Even though we specifically looked for an effect on preening we could not demonstrate a quantifiable one. A more detailed study[29] confirmed but also extended the above results. The frequency of wing-flapping and a behavioral arousal index increased, and the frequency of eye closing and of single wing-stretches decreased with i.c.v. ACTH. Preening was again unaffected by the hormone administration, even though the dosage was varied. No short-term changes of behavior could be observed in pigeons when ACTH was administered intravenously. Initially this was seen as indicating that the i.c.v. effect was purely a pharmacological, rather than a physiological effect. It is now clear, however, that ACTH and other peptides related to it function also as ventricular liquormones and as synaptic modulators.

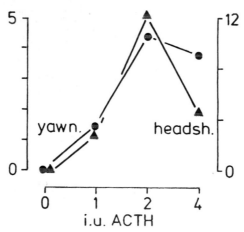

FIGURE 5. Yawning and headshaking elicited by i.c.v. administration of ACTH in pigeons. *Abcissae:* Mean number of responses shown during 50 min after the injections. Each datum is a mean of nine trials with four pigeons. (From Delius *et al.*[28]).

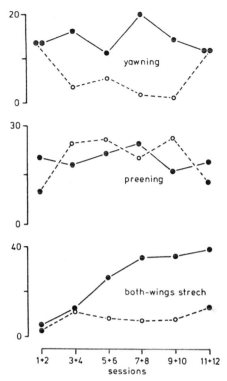

FIGURE 6. Is there sensitization or tolerance with i.c.v. ACTH-induced behavior in pigeons? Before sessions 1, 2, 11, and 12, all pigeons received 2 IU ACTH (●). Before remaining sessions experimental pigeons (*solid lines*, $n = 4$) also received ACTH but control pigeons (*dashed lines*, $n = 3$) received saline injections (O). *Abcissae*: number of 60-s periods in which the relevant behavior occurred (max. 60). Comparison between mean scores of sessions 1 and 2 and scores of sessions 11 and 12 reveals a significant sensitization effect for both-wings stretch ($p < 0.05$), but no effects for yawning and preening (see also text; after Neveling[32]).

It was then suspected that some of the behavioral responses to ACTH might be affected by sensitization or tolerance effects induced by repeated injections. Such effects are known to occur with many behaviorally active drugs (see for example refs. 30 and 31). A study directed by Pierre Deviche in our laboratory was designed to detect these effects.[32] For preening, as well as other comfort patterns, including yawning, headshaking and wingflapping (see above), neither a sensitization nor a tolerance effect was found. In the case of the both-wing-stretch pattern, however, there was sensitization (FIG. 6). This behavior can thus also be elicited by i.c.v. ACTH, but only after a preceding course of several daily hormone injections.

Another study[45] supervised by Deviche investigated whether the ACTH syndrome was perhaps due to the activation of opiate receptors. ACTH administered by the i.c.v. route (20 μg/4 μl) elicited significantly increased yawning, headshaking, bodyshaking, wingflapping, and decreased preening. Bodyshaking had not been described to be part of the i.c.v. ACTH syndrome before. Naloxone, an opiate antagonist given either i.c.v. (20 μg/4 μl) or i.m. (2 mg/0.5 ml) 10 min before the ACTH did not modify the response to the hormone, except perhaps in the case of preening where the inhibition caused by ACTH appeared to be slightly reduced. Naloxone given by itself did not have any effect on the occurrence of any of the behaviors mentioned. Furthermore, i.c.v. injections of up to 5 µg β-endorphin and up to 10 µg Met-enkephalin had no effect on the comfort behavior of the pigeons.

In a further study under Deviche's direction, von Uslar[33] found that $ACTH_{1-39}$ and $ACTH_{1-24}$ is superior to [D-Phe-7]$ACTH_{4-10}$ in eliciting comfort response in pigeons. The two long chain peptides were also found to augment the occurrence of mandibulation (small jaw and tongue movements), a behavior that had not been scored previously. Preening, on the other hand, was depressed by them. Only in releasing head and body shaking was [D-Phe-7]$ACTH_{4-10}$ just as effective as the longer peptides. Simple $ACTH_{4-10}$ was found to be completely ineffective. All peptides were given at a dose of 20 µg per pigeon. Incidentally, this dose (as well as those mentioned previously) are some 10 times higher than those needed in mammals (pigeons weigh about 500 g). We have no explanation except that birds generally seem to require higher dosages of pharmacological agents. Their higher metabolic turnover rate may have something to do with it.

In spite of the fact that in the studies mentioned there was no increase, and sometimes even a decrease, in preening (but see ref. 34 for different results in chickens), the impression persisted that this behavior was somehow affected in an interesting way by i.c.v. ACTH even though this was not reflected in the data recorded. Data to support this impression was finally obtained by von Uslar.[33] Her scoring scheme for preening was more detailed than that used in our previous studies. She recorded as separate categories nonpreening and the preening of eight different body regions (breast, belly, shoulder, wing outside, flank-wing inside, back, tail, and head rubbing or scratching) with a time resolution of 0.5 seconds. Based on observations of six ventricle cannulated pigeons and using a balanced intrasubject treatment design, she established that within a period of 45 min after an ACTH injection (20 µg) the pigeons preened on average only for 27.8 s, whereas after control saline injections the same pigeons preened on average for 194.5 ($p < 0.01$). The difference would not have emerged as significant had our usual simpler system of check-scoring been employed.

Defining a bout as a sequence of preening acts interrupted by not more than 30 s of nonpreening (a criterion that we had found useful in previous studies; see also ref. 5), she also showed that the number of preening bouts produced by the pigeons while ACTH injected did not differ from that shown while treated with saline (6.1 and 6.5 bouts respectively, $p \gg 0.05$). Consequently, however, the durations of the

TABLE 2. Composition of Preening of Pigeons after i.c.v. Injections of ACTH and Saline[a]

	ACTH	Saline
Belly	6.5	9.8
Breast	12.2	7.2
Shoulder	13.7	8.4
Wing (outside)	29.9	27.2
Wing (inside)	5.6	13.0
Back	20.3	25.2
Tail	7.7	6.0
Head	3.5	2.6

[a] Composition of preening is indicated as percentages of total preening time directed at body areas, based on averages derived from 45-min sessions after injections, 2 sessions per condition, $n = 6$ pigeons (from von Uslar[33]).

bouts during the ACTH treatment were considerably briefer than under the control situation (4.6 and 29.9 s, $p < 0.01$). She also demonstrated that under ACTH and saline the composition of preening in terms of the different body areas preened does not differ. Indeed the relative times spent on preening the various areas is equivalent (TABLE 2, $p \gg 0.05$). What happens is that when treated with ACTH, the animals preen each of the body regions for less time than when treated with saline (on average 1.6 s versus 4.6 s per region, $p < 0.05$). This agrees well with the difference between normal and displacement preening reported by Duncan and Wood-Gush[35] for chickens. Based on cinematographic recordings, these authors established that displacement preening was more hurried than normal preening, in agreement with frequent but unsupported statements in the displacement behavior literature.

Still, stress also induces an increase in preening (see FIG. 3). Thus the ACTH syndrome in pigeons does not succeed fully in matching their displacement behavior. There is also a striking disparity with mammals where the stress hormone does activate the equivalent of preening, grooming, in addition to yawning and stretching.[36,31]

DOPAMINERGIC PREENING

Recent pharmacological experiments on pecking in pigeons (a theme not unrelated to displacement activities, as commented earlier) have yielded some results that are

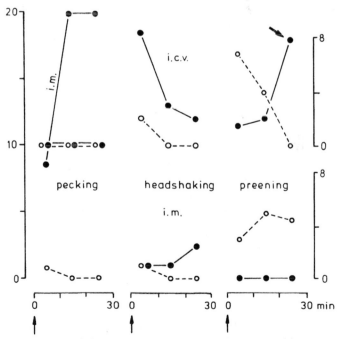

FIGURE 7. Response elicited in pigeons by apomorphine (●) and saline (○). *Below*, i.m. (1.5 mg) injections; *above*, i.c.v. (60 µg) injections. *Abcissae*, number of 30-s periods per 10 min with the relevant behavior (max. 20); arrows indicate injections. For pecking the scale of the i.m. apomorphine response extends into the top of the figure. Note delayed, significant ($p < 0.01$) increase of preening after i.c.v. injections (modified from Deviche[39]).

relevant in the present context. It is well known that apomorphine, a potent dopamine (DA) agonist, elicits persistent fits of pecking in pigeons when injected intramuscularly (i.m.) at doses above 0.75 mg per pigeon.[37,30] The occurrence of preening is at the same time strongly inhibited. Lower doses of i.m. apomorphine (0.05 mg), however, may facilitate preening.[38] Apomorphine given i.c.v. (60 μg) does not, remarkably, elicit any pecking at all but with some delay significantly enhances preening (FIG. 7) Headshaking, incidentally, is elicited by both i.m. and i.c.v. apomorphine injections.[39] This suggests that similarly to pecking, preening can be driven by dopaminoceptive neural elements although it is easily overshadowed by more potently elicited responses.

In an attempt to identify the central nervous structure where apomorphine actually elicits pecking, Lindenblatt[40] injected intracerebrally (i.c.) microquantities (20 μg in 1 μl solvent) of its L-isomer into various brain structures. She employed cannulae that were acutely inserted through previously stereotactically implanted guide tubes. The control treatment consisted of injections of the inactive D-apomorphine isomer.

Although other brain structures may also be relevant, preliminary results already imply that the main effort so far is concentrated on the avian nucleus basalis area in the frontolateral telencephalon (see FIG. 1 and FIG. 4b for its general location). It is now certain that apomorphine stereospecifically induces pecking at this site.[41] Remarkably, however, preening is also activated by the L-apomorphine injection at most of the basalis sites. Of nine histologically verified basalis injection sites, 7 (77%) yielded more preening with L- than with D-apomorphine injections, whereas in the surrounding brain tissue, of 18 sites only 6 yielded more preening (30%) with L-apomorphine. FIGURE 8 gives an example of the time course of the effect and also locates the relevant positive injection sites as well as some of the negative sites.

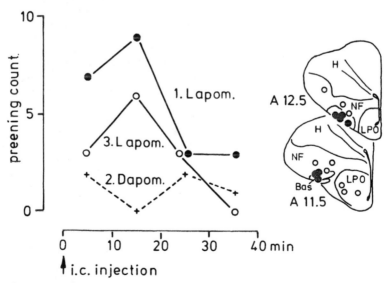

FIGURE 8. *Left,* Preening after intracerebral injections of L-apomorphine (first and third injection) and D-apomorphine (second injection) into the nucleus basalis of the pigeon. *Abcissa,* Number of 30-s periods with preening. *Right,* Sites where L-apomorphine elicited more preening than D-apomorphine (●) and some of the sites where no such effect was found (O). Data from Lindenblatt[40] and personal communication).

This evidence of an involvement of dopaminergic mechanisms in avian preening is interesting because a similar situation seems to apply to grooming in mammals even though the particulars seem to be in dispute.[31,42] At this point it is premature to speculate about the possible links between stress, ACTH, and DA in birds because of lack of data. That the nucleus basalis may be concerned with the control of preening is very remarkable. Until now this exclusively avian structure was thought to be a substrate dedicated to the sensorimotor coordination of feeding, or more precisely concerned with the pecking and grasping of food.[30] The various sensory afferences (auditory, vestibular, olfactory) it receives besides the somesthetic illustrated in FIGURE 1 certainly qualify it for that role.[43,44] On the other hand they may make it just as well suited to control the subtle head and bill movements that constitute preening. Grasping grains or feathers are after all not very different operations. A link of this kind between pecking and preening may help to explain the peculiar association of these patterns in displacement contexts. However that may be, the dopamine connection promises to be a profitable research avenue to the central mechanisms of avian preening.

ACKNOWLEDGMENTS

I thank Dagmar Hagenkötter for patient help in preparing, and Julia Delius for editing, this manuscript. G. Keim assisted with the artwork.

REFERENCES

1. LUCAS, A. M. & P. R. STETTENHEIM. 1972. Avian Anatomy, Integument. U.S. Department of Agriculture. U.S. Government Printing Office. Washington, D.C.
2. VAN RHIJN, J. G. 1977. The patterning of preening and other comfort behaviour in a herring gull. Behaviour 63: 71–109.
3. DELIUS, J. D. 1970. The effect of daytime, tides and other factors on some activities of lesser black-backed gulls, *Larus fuscus*. Rev. Comp. Anim. 4: 3–11.
4. TRARADRI, V. 1966. Sleep in pigeon. Arch. Ital. Biol. 104: 516–521.
5. LEFEBVRE, L. 1982. The organization of grooming in budgerigars. Behav. Proc. 7: 93–106.
6. DELIUS, J. D. 1969. A stochastic analysis of the maintenance behaviour of skylarks. Behaviour 33: 137–178.
7. SLATER, P. J. B. & A. M. WOOD. 1977. Does activation influence short-term changes in zebra finch behaviour? Anim. Behav. 25: 736–746.
8. VESTERGAARD, K. 1982. Dust-bathing in the domestic fowl — Diurnal rhythm and dust deprivation. Appl. Anim. Ethol. 8: 487–495.
9. BROWN, N. S. 1974. The effect of louse infestation, wet feathers, and relative humidity on the grooming behavior of the domestic chicken. Poultry Sci. 53: 1717–1719.
10. SLATER, P. J. B. 1974. Bouts and gaps in the behaviour of zebra finches, with special reference to preening. Rev. Comp. Anim. 8: 47–61.
11. LEFEBVRE, L. & R. JOLY. 1982. Organization rules and timing in kestrel grooming. Anim. Behav. 30: 1020–1028.
12. FENTRESS, J. C. 1972. Development and patterning of movement sequences in inbred mice. *In* The Biology of Behaviour. J. A. Krieger, Ed. Oregon Press. Eugene, Oregon.
13. NECKER, R. 1983. Somatosensory system. *In* Physiology and Behaviour of the Pigeon. M. Abs, Ed.: 171–219. Academic. London.
14. NECKER, R. 1985. Observations on the function of a slowly-adapting mechanoreceptor associated with filoplumes in the feathered skin of pigeons. J. Comp. Physiol. A 156: 391–394.
15. DELIUS, J. D. & K. BENNETTO. 1972. Cutaneous sensory projections to the avian forebrain. Brain. Res. 37: 205–221.

16. TINBERGEN, N. 1951. The Study of Instinct. Oxford Univ. Press. Oxford.
17. HINDE, R. A. 1970. Animal Behaviour: A Synthesis of Ethology and Comparative Psychology. 2d edit. McGraw-Hill. New York.
18. VAN IERSEL, J. J. A. & A. C. A. BOL. 1958. Preening in two tern species. A study of displacement activities. Behaviour 13: 1-88.
19. DELIUS, J. D. 1967. Displacement activities and arousal. Nature 214: 1259-1260.
20. DELIUS, J. D. 1970. Irrelevant behaviour, information processing and arousal homeostasis. Psychol. Forsch. 33: 165-188.
21. ROWELL, C. H. F. 1961. Displacement grooming in the chaffinch. Anim. Behav. 9: 38-63.
22. HOLLAND, H. C. 1976. Displacement activity as a form of abnormal behaviour in animals. In Obsessional States. H. R. Beech, Ed.: 161-173. Methuen. London.
23. LYON, D. O. & L. TURNER. 1972. Adjunctive attack and displacement preening in the pigeon as a function of the ratio requirement for reinforcement. Psychol. Rec. 22: 509-514.
24. FABRICIUS, E. & A. M. JANSSON. 1963. Laboratory observations on the reproductive behaviour of the pigeon (Columba livia) during the pre-incubatory phase of the breeding cycle. Anim. Behav. 11: 534-547.
25. SPITERI, N. J. 1975. Social, especially agonistic behaviour in the pigeon. Masters thesis, Durham, England.
26. BOLTON, T. B. 1976. Nervous system. In Avian Physiology. 2d edit. P. D. Sturkie, Ed: 1-28. Springer. Berlin.
27. GESSA, G. L., M. PISANO, L. VARGIU, F. CRABAI & W. FERRARI. 1967. Stretching and yawning movements after intracerebral injection of ACTH. Rev. Can. Biol. 26: 229-236.
28. DELIUS, J. D., B. CRAIG & C. CHAUDOIR. 1976. Adrenocorticotropic hormone, glucose and displacement activities in pigeons. Z. Tierpsychol. 40: 183-193.
29. DEVICHE, P. & J. D. DELIUS. 1981. Short-term modulation of domestic pigeon (Columba livia L.) behaviour induced by intraventricular administration of ACTH. Z. Tierpsychol. 55: 335-342.
30. DELIUS, J. D. 1985. The peck of the pigeon: Free for all. In Behaviour Analysis and Contemporary Psychology. C. F. Lowe, M. Richelle, D. E. Blackman & C. M. Bradshaw, Eds.: 53-86. Lawrence Erlbaum Associates. Hillsdale, N. J.
31. GISPEN, W. H. & R. L. ISAACSON. 1981. ACTH-induced excessive grooming in the rat. Pharmacol. Ther. 12: 209-246.
32. NEVELING, U. 1981. Verhaltensänderungen bei Tauben induziert durch wiederholte intraventrikuläre Gaben von Adrenocorticotropem Hormon. Dipl. thesis, Bochum.
33. VON USLAR, A. 1983. Die Wirkung von Corticotropin und verwandten Peptiden auf das Verhalten der Taube. Dipl. thesis, Bochum.
34. WILLIAMS, N. S. & D. L. SCAMPOLI. 1984. Handling, ACTH, ACTH 1-24, and naloxone effects on preening behavior in domestic chickens. Pharmacol. Biochem. Behav. 20: 681-682.
35. DUNCAN, I. J. H. & D. G. M. WOOD-GUSH. 1972. An analysis of displacement preening in domestic fowl. Anim. Behav. 20: 68-71.
36. DUNN, A. J., E. J. GREEN & R. L. ISAACSON. 1979. Intracerebral adrenocorticotropic hormone mediates novelty-induced grooming in the rat. Science 203: 281-283.
37. BRUNELLI, M., F. MAGNI, G. MORRUZI & D. MUSUMECI. 1975. Apomorphine pecking in the pigeon. Arch. Ital. Biol. 113: 303-325.
38. DEVICHE, P. 1985. Behavioral response to apomorphine and its interaction with opiates in domestic pigeon. Pharmacol. Biochem. Behav. 22: 209-214.
39. DEVICHE, P. 1983. Stereotyped behavior affected by peripheral and intracerebroventricular apomorphine administration in pigeons. Pharmacol. Biochem. Behav. 18: 323-326.
40. LINDENBLATT, U. 1986. Die dopaminerge Auslösung des Pickverhaltens bei Tauben. Ph.D. thesis, Bochum.
41. LINDENBLATT, U. & J. D. DELIUS. 1986. Nucleus basalis prosencephali, a substrate for apomorphine elicited pecking in the pigeon. Submitted.
42. DUNN, A. J., J. E. ALPERT & S. D. IVERSEN. 1984. Dopamine denervation of frontal cortex or nucleus accumbens does not affect ACTH-induced grooming behaviour. Behav. Brain Res. 12: 307-315.

43. SCHALL, U. & J. D. DELIUS. 1986. Sensory inputs to the nucleus basalis prosencephali, a feeding-pecking centre in the pigeon. J. Comp. Physiol. A. **159:** 33–41.
44. SCHALL, U., O. GÜNTÜRKÜN & J. D. DELIUS. 1986. Sensory projections to the nucleus basalis prosencephali of the pigeon. Cell Tissue Res. **245:** 539–546.
45. SCHUBERT, B. 1982. Corticotropin induziertes Verhalten bei Tauben, Einfluss von Naloxon. Dipl. thesis, Bochum.

Biological Influences on Grooming in Nonhuman Primates[a]

ROBERT W. GOY, GARY KRAEMER, AND DAVID GOLDFOOT

Wisconsin Regional Primate Research Center
University of Wisconsin
Madison, Wisconsin 53715

Two conspicuous factors that are repeatedly seen to influence grooming frequencies are sex and species. Differences between the sexes are not constant across all species or even within a species under varying conditions. The sex difference that characterizes one species may be reversed in other species. One factor that may account for the reversal is the difference between male-resident and female-resident species.[1] In the former (like the chimpanzee) males groom more than females,[2] whereas in the latter (like the baboon) females groom more than males.[3]

The pervasive nature of sex differences in grooming suggests that gonadal hormones might play a role in the expression of the behavior, even though the influence of the hormones might change with evolution and speciation. This report focuses on the rhesus macaque. It is a widely studied species, and a reasonable representative of Old World, semiarboreal monkeys. For the past 15 years, we have studied the development of behavior in small groups of laboratory-reared rhesus. These groups typically consist of five animals (range: 4 to 6), and they usually include both males and females. During the first year of life they live with their mothers in a communal pen, and at about 1 year of age they are weaned. After weaning, they live together continuously except for 3 months of each year when they are housed individually to permit standardized testing procedures. In short, the subjects on which we are reporting are not socially deprived, and they develop strong interpeer bonds early in life that are maintained throughout juvenile development and into early adolescence.

Our data on grooming presented here are limited to those standardized observations carried out when the animals were 24 to 27 months of age and repeated when they were 36 to 39 months old. These two periods of observation are referred to respectively as run 3 and run 4. These periods were selected because most subjects (both males and females) have not shown conspicuous signs of endocrine puberty at 24 months, and nearly all of both sexes have manifest indications of endocrine puberty by 36 months. The data on grooming are collected using the focal animal technique, by means of which each member of a peer group is observed for 5 minutes daily while interacting with its peers. Observations are carried out for 50 successive days (5 days per week) and the frequency of occurrence, but not the duration, of grooming is recorded. A record is also made of the particular partner for each grooming episode. Since the sizes of groups vary, and hence the number of actual partners available, results are calculated as the average number of grooming episodes per available partner per 50 days of observation. All of the statistical analyses were carried out on the grooming

[a] This work was supported by NIH grant no. RR00167 to the Wisconsin Regional Primate Research Center (WRPRC) and no. MH40748. This is WRPRC Publication No. 27-014.

data following square-root transformation to achieve greater homogeneity of variances. Tables and figures, however, present nontransformed means and standard errors.

Sex differences in the frequency of grooming by males and females during runs 3 and 4, that is, at 2 and 3 years of age, are illustrated in FIGURE 1. These data were collected on 68 normal males and 85 females from more than 30 different rearing groups. As shown in FIGURE 1, females groomed more than males during both run 3 and run 4 ($F_{1,151} = 4.23$, $p = 0.039$) and grooming by both sexes increased reliably from run 3 to run 4 ($F_{1,151} = 55.11$, $p < 0.001$).

Although frequency of grooming was influenced by the sex of the groomer, it was not reliably influenced by the sex of the partner (FIG. 2, $F_{1,151} = 0.69$, $p > 0.58$). Failure of the sex of the partner to influence the frequency of grooming by male and female groomers was apparent even during run 4, when endocrine puberty was manifest. In other words, at a time when increased interest in oppositely sexed partners might have been hypothesized, one might have expected grooming to be biased toward a heterosexual preference. In fact, however, pubertal status did bias the grooming behavior of the participants (TABLE 1) but the bias was only conspicuous in those groups

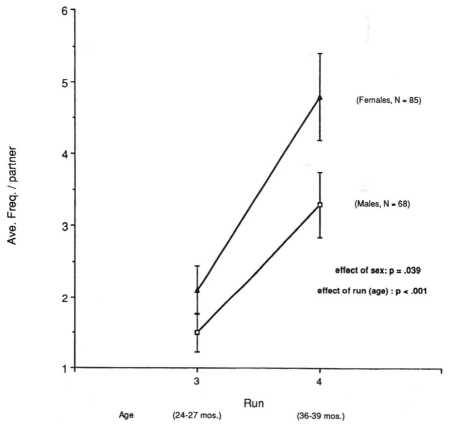

FIGURE 1. Grooming by male (□) and female (△) rhesus before (run 3) and after (run 4) the commencement of puberty.

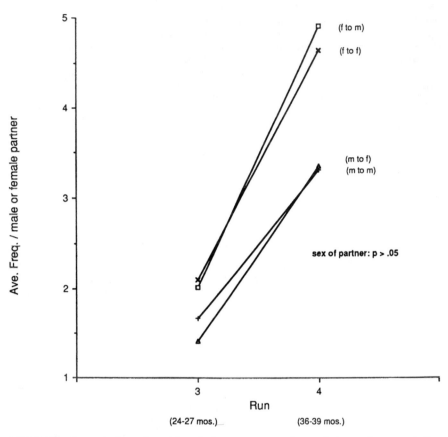

FIGURE 2. Grooming by male and female rhesus to same and oppositely sexed partners before (run 3) and after (run 4) the commencement of puberty.

TABLE 1. Influence of Pubertal Sexual Activity on Frequency of Grooming: Average Frequency over 50 Days

		Active Partner			Inactive Partner			
	Subjects (n)	Partners (n)	m	SD	Partners (n)	m	SD	Totals
Males that display intromission	14	13 females	10.8	13.4	14 females	2.8	3.8	13.6
Males that do not display intromission	14	13 females	2.7	4.4	14 females	2.4	4.2	5.1
Females that receive intromissions	13	14 males	9.3	12.3	14 males	6.2	9.8	15.5
Females that do not receive intromissions	14	14 males	5.2	6.7	14 males	4.6	4.5	9.8

in which complete sexual behavior was displayed. Twelve of the groups contained a total of 28 males and 27 females. Within each group the sexually ready male(s) displayed intromissions, and generally this intromissive behavior was shown exclusively or predominantly with only one of the two or three available female peers. As the means shown in TABLE 1 illustrate, the sexually active male(s) groomed his (their) sexual partner(s) three times as often as he (they) groomed the female peer with whom no intromissions were displayed ($F_{1,26} = 8.56$, $p = 0.004$). This high frequency of grooming by the intromissive males was not reciprocated, and although the female sexual partner groomed the intromissive male more often than the nonintromissive male, the difference was not statistically significant ($p > 0.50$). In contrast, the sexually inactive males and females distributed their lower frequencies of grooming almost equally between active and inactive partners. It is important to note as well that males displaying intromissions did not show significantly more grooms towards females overall ($m = 6.8$) than did nonintromissive males ($m = 2.6$, $F_{1,26} = 3.18$, $p > 0.08$), although there was a trend in that direction. These results suggest that directed preferential grooming may be accompanied by a reduction in grooming behavior towards that female with whom the male does not display intromissions.

Results presented so far might be interpreted to mean that, despite special influences of pubertal sexual activity, age alone acts to increase the frequency of grooming by all animals living in heterosexual groups. This could be true if gonadal hormones contributed little or nothing to the increase in grooming observed from run 3 to run 4. However, concentrations of gonadal hormones in blood are increasing during this developmental stage, and it should be reemphasized that grooming increases in sexually inactive animals as well as in the sexually active. The possibility that estrogens facilitate the expression of grooming is consistent with the increase in grooming from run 3 to run 4 and also with the sex difference in grooming, since males would have lower levels of available estrogen than females at both ages. The hypothesis that estrogen regulates both the attributed age and sex effects cannot be entirely discounted. If it is true, however, then they are the opposite of the actions of estrogen reported by Michael and his colleagues[4] for changes in grooming during the ovarian cycle of the rhesus. In those studies, females, on average, groomed less at midcycle when estrogens were at their highest concentrations in the blood than they did at other times in the cycle. Moreover, the estrogen facilitation hypothesis is inconsistent with findings for seven females spayed prior to 1 year of age and studied in our laboratory in a manner similar to that used for the intact females. These spayed females had intact males available for partners, and they groomed them more often than sexually inactive intact females groomed their male partners. In addition, they showed an increase in grooming from run 3 (8.1 ± 2.9) to run 4 (10.6 ± 4.6), as did intact females.

Because of the existing evidence that seems to rule out facilitatory actions of estrogen, it is reasonable to interpret the present results in an alternative endocrinological framework. The alternative framework assumes that age (maturation) alone does account for the increase in grooming seen in both sexes from run 3 to run 4. In addition, the framework permits the hypothesis that the sex difference in grooming is affected by an inhibitory influence of androgens.

We are able at this time to evaluate to a limited extent this hypothesis about inhibitory influences of androgen. First of all, we know from extensive past work that androgens can affect the sexually dimorphic social behavior of rhesus monkeys. Insofar as we have been able to ascertain from these past studies, sexually dimorphic behavioral traits in rhesus, as in other species, can only be influenced by androgen acting during early developmental stages such as prior to birth, during late stages such as puberty or adulthood, or by virtue of its special actions during both early and late developmental stages.[5] By comparing intact males at puberty with males castrated during the postnatal period, we can ascertain whether or not actions of endogenous androgens at early developmental stages alone are sufficient to establish the male-typical frequency of grooming peers. Furthermore, by including in these comparisons females who have been somatically and psychologically virilized by early androgen exposure, we can ascertain whether or not the male genotype per se contributes to the behavioral sex difference in grooming. For these comparisons an analysis has been made of the grooming behavior of 257 rhesus monkeys distributed among 55 different peer groups. The 257 individuals include 99 intact males, 108 females, 32 pseudohermaphroditic females, and 18 males castrated within 90 days after birth. Analysis was limited to the data collected during run 4 primarily to ensure that most of the intact males would have endogenous pubertal androgens, but also because grooming activity is more frequent at that age in both sexes. In addition, the analysis ignored sex of partner as a factor.

The four sex classes (intact males, intact females, pseudohermaphroditic females, and castrated males) differed statistically (FIG. 3) in frequency of grooming peers ($F_{3,245} = 5.14$, $p = 0.002$). Intact males showed reliably fewer grooms of peers than any of the other three sex classes. Furthermore, there were no statistically reliable differences between castrated males, females, and female pseudohermaphrodites. The

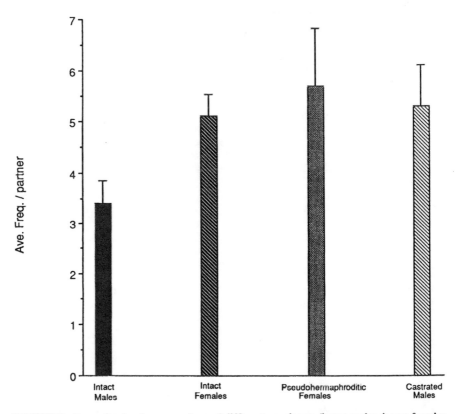

FIGURE 3. Grooming by rhesus monkeys of different sex classes (intact males, intact females, pseudohermaphroditic females, and castrated males) during run 4.

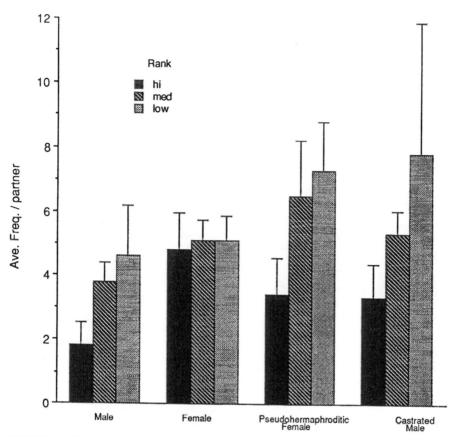

FIGURE 4. Effect of social rank on grooming by rhesus monkeys of different sex classes.

grooming performance of our castrated males resembled that reported by Loy *et al.*[6] In that study castrated males groomed more often than intacts, and the difference was more pronounced with male than with female partners.

These results can be interpreted to mean that androgens at early developmental stages alone, regardless of genotype, are insufficient to establish the male-typical frequency of grooming. The corollary inference is that either androgens during puberty are sufficient to inhibit grooming, or that actions of androgen during both early and late developmental stages are required. Our present data do not permit a decision between these latter two possibilities. However, as a result of the futile comparison between castrated males and females, we can conclude again that ovarian estrogens are not essential to the female-typical frequency of grooming peers.

As with most behavioral systems that we have studied in rhesus, even though the expression of grooming is hormonally regulated, it can be expected to be strongly modified by social factors as well. Social status, based on the extensive body of findings in the primate literature relating it to grooming,[3] may be such a factor, and it might account for the differences among sex classes described above. In order to evaluate this possibility, every monkey's status within its own peer group was ranked as either high, medium, or low, and a more comprehensive analysis of variance was carried out. Although dominance rank reliably influenced frequency of grooming ($F_{2,245} = 4.27$, $p < 0.015$), intact males at each dominance rank were consistently lower in frequency of grooming than females, pseudohermaphroditic females, or castrated males at corresponding ranks (FIG. 4). Furthermore, there was no statistically reliable interaction between sex class and dominance rank ($F_{6,245} = 0.38$, $p > 0.89$). Accordingly, the hypothesis that the sex difference in grooming results from inhibitory actions of androgen remains tenable.

In our examination of social status thus far we have ignored possible effects of the sex of the partner, and have demonstrated that the relationship between status and grooming was inverse. That is to say, highest ranked (alpha) animals generally showed lower frequencies of grooming than lowest ranked (omega) animals when the frequencies were averaged across partners of both sexes. However, this inverse relationship between status and grooming frequency does not exist for grooming partners of the opposite sex. When alpha males ($n = 22$) were compared to omega males ($n = 7$) for frequency of grooming female peers during run 4, average frequencies were 2.6 and 1.5 respectively ($F_{1,27} = 0.4$, $p > 0.50$). Correspondingly, alpha females ($n = 11$) groomed male partners 5.2 times on average, whereas omega females ($n = 21$) groomed male partners only 3.8 times ($F_{1,30} = 0.71$, $p > 0.50$). Thus, in neither sex did the lowest ranked animals groom oppositely sexed partners more often than the highest ranked animals. These rankings, alpha and omega, are based on status relative to the group as a whole, but it is possible to look at the status of males relative only to each other, and to do the same for females. Readers can think of such estimates of status as "intrasexual rankings" and we refer to them by the terms dominant and subordinate as distinct from alpha, beta, etc. When this is done, a somewhat different picture of how status affects the grooming of oppositely sexed peers emerges. For this analysis, we considered only those peer groups that contained at least two males and two females. We identified the dominant and subordinate males and females in each group by the usual measures of agonistic behavior. We then analyzed for influences of sex and status of groomer as well as status of the oppositely sexed partner. In an overall analysis, sex and status of the groomer were not statistically significant determinants of the frequency of grooming, but they interacted significantly ($F_{1,115} = 5.27$, $p = 0.022$) such that dominant males groomed least frequently (mean = 2.9) and dominant females groomed most frequently (mean = 5.9). Subordinate males and females on average groomed 4.6 and 3.8 times. Moreover, there was a significant influence of the status of the oppositely sexed partner, and dominant partners were groomed more often than subordinate partners ($F_{1,115} = 7.34$, $p = 0.008$) regardless of their sex or the sex or status of the groomer.

It is worth reexamining these grooming performances of the males with oppositely sexed partners in the context of the pubertal sexual experience of intromission. This reexamination of pubertal sexual experience involves all males, and unlike our earlier analysis (TABLE 1) is not restricted to males that distributed all or most of their intromissions to only one female. For this purpose, we classified all of the males into two categories: those that displayed intromissions during run 4 ($n = 28$) and those that did not ($n = 36$). This classification did not produce a statistically significant main effect on grooming performance ($F_{1,60} = 0.40$, $p > 0.50$). However, it interacted significantly with the social status of the female partners such that males displaying intromission groomed dominant females more often than they groomed subordinate females (7.5 ± 2.3 vs. 1.2 ± 0.5), whereas males that did not display intromissions groomed dominant and subordinate females 3.8 ± 1.0 and 3.1 ± 0.7 times respec-

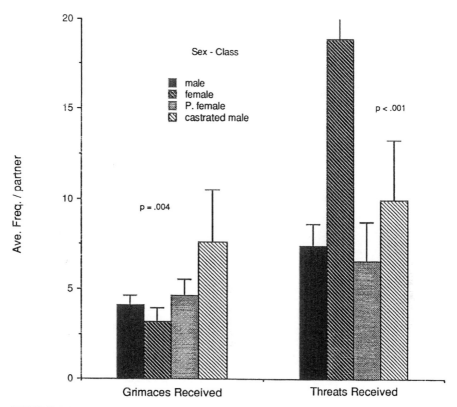

FIGURE 5. Effect of sex class on the frequency of grimaces and threats received during run 4.

tively ($F_{1,60}$ = 7.59, p = 0.008). Thus the experience of intromission biases the grooming performance of males toward females that have high intrasexual status. This suggests that once males have developed erotic capability, they utilize their grooming primarily to facilitate sexual access to females of high social rank. It should be pointed out that the females receiving the intromissions were most often females with high social status. When corresponding classifications were made for females, there was no significant effect of the experience of intromission, and no interaction of this experience with female status or status of the male partner. There was a marginally significant interaction ($F_{1,51}$ = 3.36, p = 0.069) of female status with male partner status, suggesting that females of high rank select males of high rank as their preferred partners for grooming regardless of whether or not these males display intromission with them.

The difference between the sexes with regard to frequency of grooming and type of preferred partner suggests that the sexes can utilize grooming for different purposes. Whereas pubertal males use grooming to consolidate sexual relationships, females seem to use grooming to consolidate social relationships consistent with their social status.

The possibility that still other psychological or behavioral traits might differ among the sex classes and contribute to the sex difference in grooming performance was ex-

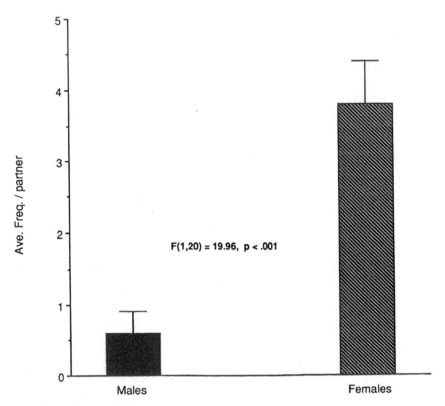

FIGURE 6. Grooming by male and female rhesus reared and housed in all-male and all-female (isosexual) groups.

amined. The traits examined for such influences included (1) the amount of grooming received from partners, (2) the number of grimace gestures received from partners, (3) the number of threats given to partners, and (4) the number of threats received from partners. In separate analyses each of these traits was reliably related to rank (all F's > 12.2, all p's < 0.001). Only two of the traits, grimaces received ($p < 0.004$) and threats received ($p < 0.001$), were reliably related to sex class (FIG. 5). These differences among the sex classes were not consistent, however. In the case of grimaces received, mean scores for castrated males exceeded those of all other groups. In the case of threats received, mean scores for females exceeded those of all other groups. Moreover, the pattern of frequencies distributed over the four sex classes did not correspond to the pattern found for frequency of grooming by these sex classes. Finally, correlations between frequency of grooming and frequency of grimaces received on the one hand or threats received on the other hand were trivial ($r = 0.015$ and -0.062, respectively, not significant). Thus, these behavioral traits clearly did not account for any of the variation found for grooming.

We have studied other groups of monkeys that provide yet another possibility for evaluating how environmental factors might influence the frequencies of grooming that are typical of each sex. A few years ago we showed that rearing males in all-male

groups and females in all-female groups caused an enhancement (known to be reversible) of specific female-typical traits in the males (presenting) and certain male-typical
traits (mounting) in the females.[7] We referred to this rearing system as the isosexual
condition and, in all, 15 males distributed equally among three groups, and 15 females
also distributed among three groups, were studied. Our earlier report did not include
data on grooming behavior, and we present them here.

The difference between the sexes (reared and tested only with same-sexed peers)
in grooming overall (FIG. 6) was statistically significant ($F_{1,20}$ = 19.96, $p < 0.001$).
There was no overall effect of rank (FIG. 7) or of run (FIG. 8); however, there was a
significant interaction between run and sex (FIG. 9) and grooming frequency increased
in females from run 3 to run 4, whereas the frequency for males decreased.

Isosexual rearing and testing did not measurably affect the grooming of females.
During run 4, isosexual females groomed their partners 5.2 ± 0.8 times on average,
whereas heterosexually reared and tested females groomed their female partners 5.1
± 0.9 times ($F_{1,102}$ = 0.039, $p > 0.90$). In contrast, during the same run, isosexual
males groomed their partners 0.25 ± 0.1 times, whereas heterosexually reared males
groomed their male partners 3.5 ± 0.6 times ($F_{1,103}$ = 12.38, $p < 0.015$). This differen-

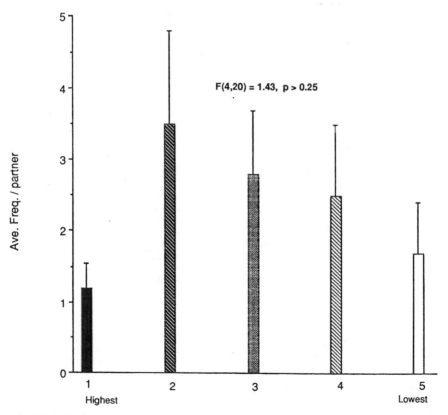

FIGURE 7. Lack of effect of social rank on isosexual grooming by male and female rhesus
combined.

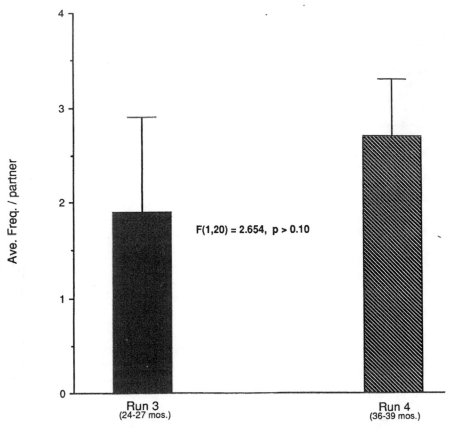

FIGURE 8. Lack of effect of puberty (run) on isosexual grooming by male and female rhesus combined.

tial effect of isosexual conditions on the males and females suggests that males are stimulated to groom same-sexed partners when females are members of the same social group. When females are not present, as in isosexual groups of males, then the hypothesized inhibitory effects of pubertal androgen on grooming are very pronounced.

It would be a valuable aid to a final decision regarding this hypothesis of androgenic inhibition if isosexual groups of castrated males had been studied. In order to support the hypothesis, such males ought not to show a decline in grooming from run 3 to run 4, and their frequencies of grooming same-sexed peers ought to be reliably higher than those of intact males. However, such groups were not available.

The biological basis of social grooming in rhesus monkeys encompasses a wide-ranging set of factors that include maturational, hormonal, psychological, and social variables. In addition, specific experiences with a partner can markedly influence grooming behavior. Identification of these variables in the present study is entirely consistent with Yerkes's earliest surmises about the phylogeny and ontogeny of grooming[8] and its primary function as a social service, albeit noticeably modulated by self-interest in lower primates. The interaction of all of these variables regulates

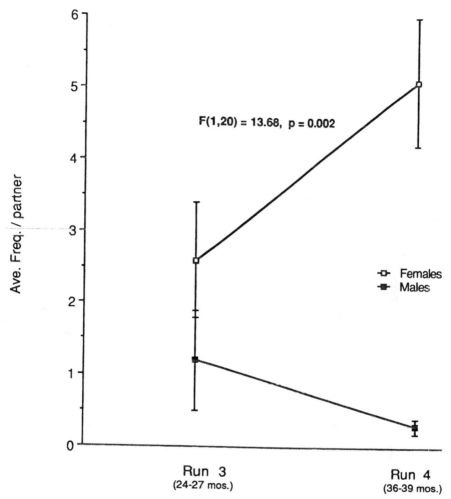

FIGURE 9. Interaction of sex and puberty (run) on isosexual grooming by male and female rhesus.

the frequency with which an individual grooms a specific partner in a specific environmental context. Of these variables, the most important in accounting for sex differences in rhesus social grooming seems to be the hormonal variable. More specifically, the sex difference arises as a result of pubertal androgens acting to inhibit the expression of grooming in males.

It is not likely that this inhibitory effect is a direct and simple consequence of androgenic actions. In fact, the term "androgenic inhibition" could be quite misleading if it is interpreted to mean a direct action of androgen on some special set of androgen-sensitive neuronal elements involved in the mediation of grooming. Based on our analysis of special experiential events, it seems more likely that the less frequent expression of grooming by pubertal males (compared with females) is a by-product of the effects of androgen on their sexual capabilities. In short, the androgenic activation of sexual behavior brings about a specialization in the uses of grooming by males and a corresponding reduction in the number of partners toward which grooming is displayed. Sexually active males, for the most part, restrict and concentrate their grooming activities on their preferred sexual partners or on dominant females with which they are attempting to establish consort relationships. Even pubertal males that do not yet show sexual behavior groom dominant females preferentially. In brief, pubertal androgens induce eroticism in males, and in association with this phenomenon, affiliative behaviors such as grooming undergo a corresponding eroticization of function and their performance becomes more restricted to interactions with the erotic object (partner). Accordingly, when data for frequency of grooming for an individual are based on the average frequency across all available partners, then the average for a male restricting his grooming to one partner will likely be lower than the average for a female that grooms all available partners. Since pubertal males more evidently restrict their grooming behavior than females, it is not unreasonable to account, at least in part, for the overall sex difference in grooming by this mechanism. The sex difference is further magnified by the failure of females to restrict grooming to their erotic partners. Indeed, among females, there is no evidence to suggest that grooming undergoes erogenous channelization.

Restriction of grooming to one or a few partners cannot, however, be the only variable contributing to the sex difference, since the frequency of grooming is also lower in males that have not established an erotic relationship with a partner than it is in females. Furthermore, the reader is again reminded that males with only male partners available (isosexual males) show the lowest frequency of grooming of all of the combinations studied. These caveats reinforce the need for a multivariate approach to grooming behavior in primates.

ACKNOWLEDGMENT

The authors express their appreciation to Edith Chan for typing the manuscript.

REFERENCES

1. HARCOURT, A. H. & K. J. STEWART. 1983. Interactions, relationships and social structure: The great apes. *In* Primate Social Relationships: An Integrated Approach. R. A. Hinde, Ed.: 307–314. Sinauer Associates. Sunderland, MA.
2. SIMPSON, M. J. A. 1973. The social grooming of male chimpanzees. *In* Comparative Ecology and Behaviour of Primates. R. P. Michael & J. H. Crook, Eds. Academic. London.
3. SEYFARTH, R. M. 1977. A model of social grooming among adult female baboons. Anim. Behav. **24:** 917–938.
4. MICHAEL, R. P., J. HERBERT & J. WELEGALLA. 1966. Ovarian hormones and grooming in the rhesus monkey (*Macaca mulatta*) under laboratory conditions. J. Endocrinol. **36:** 263–279.
5. GOY, R. W. & B. S. MCEWEN, Eds. 1980. Sexual Differentiation of the Brain. MIT Press. Cambridge, MA.
6. LOY, J. D., K. LOY, G. KEIFER & C. CONAWAY. 1984. The behavior of gonadectomized rhesus monkeys. *In* Contributions to Primatology, Vol. 20. F. S. Szalay, Ed.: 1–114. Karger. Basel.
7. GOLDFOOT, D. A., K. WALLEN, D. A. NEFF, M. C. MCBRAIR & R. W. GOY. 1984. Social influences on the display of sexually dimorphic behavior in rhesus monkeys: Isosexual rearing. Arch. Sex. Behav. **13:** 395–412.
8. YERKES, R. M. 1933. Genetic aspects of grooming, a socially important primate behavior pattern. J. Soc. Psychol. **4:** 3–25.

Comparison of Adaptive Responses in Familiar and Novel Environments: Modulatory Factors

SANDRA E. FILE,[a] PETER S. MABBUTT,
AND JACQUELINE H. WALKER

MRC Neuropharmacology Research Group
Department of Pharmacology
The School of Pharmacy
University of London, London WC1N 1AX, United Kingdom

The links among anxiety, stress, and grooming are not clearly established. Yule[1] found increased grooming when rats were placed in unfamiliar cages. Thompson and Higgins[2] found more grooming in shocked versus unshocked rats, and Cohen and Price[3] found that rats exposed to the sound of rat screams showed more grooming. These findings might suggest that increased grooming occurs in response to an increase in fear or arousal, but Doyle and Yule[4] found no correlation between grooming and the incidence of other indications of fear or anxiety, such as urination, defecation, and freezing. Gispen and Isaacson[5] suggested that in fact increased grooming occurred on termination of, or habituation to, a stressful situation. In support of this suggestion, Roth and Katz[6] found that rats placed in an open field after an hour-long exposure to loud noise and bright light showed higher levels of grooming than their unstressed controls. They found a significant change with repeated testing over days, but unfortunately the direction of the effect was not specified.

The purpose of experiment 1 was to study the influence on grooming of factors that are known to modulate anxiety. In experiment 1 the time spent grooming and the type of grooming (face wash, front paw lick, body wash, genital lick, or hind limb scratch) was studied in the home cage and in the test conditions used in two animal tests of anxiety. In the social interaction test of anxiety, rats are placed in a test arena that is either familiar or unfamiliar to them and is lit by high or low light.[7,8] In the elevated plus-maze, anxiety is generated by open elevated maze arms.[9] Thus the effects on grooming of manipulating environmental familiarity, light level, and elevation were studied. Both of these tests generate behavioral changes that indicate increased anxiety and both raise plasma corticosterone concentrations. Strain and sex differences have been found in the social interaction and elevated maze tests,[9-11] and therefore comparisons were made between male Wistar and male and female hooded Lister rats.

The relationship between anxiety and grooming can also be explored by examining the effects of drugs. ACTH and CRF have both been reported to increase grooming,[12-15] and these peptides also have an anxiogenic effect (shown by a specific reduction in social interaction) in the social interaction test.[16-18] However, Roth and Katz[6] found that although adrenalectomy increased grooming in the open field, hypophysectomy

---

[a] SEF is a Wellcome Trust Senior Lecturer.

was without effect. Opioid antagonists have an anxiogenic effect in the social interaction test,[19] but Roth and Katz[6] found that naltrexone *reduced* grooming of previously stressed and of unstressed rats placed in the open field. The anxiolytic drug ethanol also increased grooming in mice,[20] and so the relationship between drug-induced changes in anxiety and grooming may not be a simple one. In experiment 2, the effects of three anxiogenic drugs were examined on grooming in the home cage, in the social interaction test, and in the elevated maze. The three drugs were chosen because each exerts its anxiogenic effects by acting at a different receptor (see ref. 21 for review). Pentylenetetrazole acts at the chloride ionophore on the GABA-benzodiazepine receptor complex, FG 7142 acts at the benzodiazepine binding site, and yohimbine acts at α_2 adrenoceptors.

METHODS

Animals

Male Wistar rats (home bred from Bantin and Kingman stock), male hooded Lister rats (Olac Ltd., Bicester) and female hooded Lister rats (home bred from Olac stock) were housed singly with food and water freely available. They were housed in an 11-h light : 13-h dark cycle with lights on at 0600 h. The cages were opaque plastic, 35 × 24 × 17 cm, with wire grid floor and lids. The animals were housed in low light (30 scotopic lux).

Apparatus

The arena normally used for the social interaction test was a wooden box, 60 × 60 × 35 cm, with a solid floor. A camera was mounted vertically above the arena and self-grooming was scored from a monitor in an adjacent room. The observer, who was blind to the drug treatment of the rats, recorded the duration and incidence of each type of grooming. All sessions were recorded on videotape to permit rescoring. For the high light test conditions the illuminance on the floor of the box was 300 scotopic lux and for the low light test conditions it was 30 scotopic lux.

The elevated open arm was wooden, 50 × 10 cm, and was raised 50 cm above the floor. The light level on the arm was 185 lux.

Drugs

Pentylenetetrazole (Sigma) and yohimbine (Sigma) were dissolved in water, and FG 7142 (Ferrosan) was suspended in water with a drop of Tween-20, to various concentrations to give a constant injection volume of 2 ml/kg. Control animals received equal volume injections of water or the water-Tween vehicle. All injections were intraperitoneal.

Procedure

Experiment 1

Each group of rats was randomly allocated ($n = 10$ per group) among the test conditions: home cage; elevated arm; low light, familiar arena; low light, unfamiliar arena; high light, familiar arena; high light, unfamiliar arena. In order to avoid any circadian changes all testing was conducted between 0630 and 1200 h. Each test period lasted 10 min and the frequency and duration of each type of grooming was recorded separately in four 2.5-min time bins. The following categories of grooming were scored: face wash; front paw lick; body wash; genital lick; hind limb scratch.

Home Cage Test. In this test the rats were left in their home cages, but these were placed so that the observer could clearly see the rat. The observer sat 64 cm from the home cage and observed from above. Each rat was observed for 10 min starting immediately after the cage was moved.

Elevated Arm. Each rat was placed on the elevated open arm for 10 min. The observer sat 64 cm away from the arm.

Wooden Test Arena. Rats to be tested in the familiar arenas received the two 10-min familiarization sessions on the 2 days preceding the test, in the appropriate light level. On the test day, each rat was placed in the center of the arena and observed for 10 min.

Experiment 2

The same three test conditions were used as in experiment 1. Within each condition male hooded rats were allocated randomly to control, pentylenetetrazole (15 or 30 mg/kg), FG 7142 (2.5 or 5 mg/kg) or yohimbine (1.25 or 2.5 mg/kg). There were 10 rats in each condition. Injections were given 30 min before observations started for the yohimbine and FG 7142 groups and 5 min before for the pentylenetetrazole groups.

RESULTS

It can be seen in FIGURE 1 that, whereas there were no differences among the different rat groups in the time they spent grooming in the home cage, there were both strain and sex differences in the increase in grooming in the elevated arm. The male Wistars spent significantly longer grooming in the elevated arm than did the male hooded rats (test $\times$ strain $F[1,36] = 5.48$, $p < 0.03$), and the female hooded rats also spent longer grooming than their male counterparts (test $\times$ sex $F[1,36] = 5.07$, $p < 0.04$). FIGURE 1 also shows the distribution of grooming over the 10-min time period. There was a significant effect of time for the two male strains ($F[3,108] = 5.0$, $p < 0.005$) and a significant time $\times$ strain $\times$ test interaction ($F[3,108] = 2.82$, $p < 0.05$), because the male hooded rats showed the steepest rise at the end of the home cage test, whereas the Wistars showed more change over time in the elevated maze. The female hooded rats showed the same change over time as their male counterparts, with the steepest rise being in the last 2.5 min of the home cage test (see FIG. 1).

FIGURE 2 shows the time spent grooming in each of the four types of wooden test arena. There was a significant effect of light level ($F[1,35] = 9.41$, $p < 0.005$), with less grooming in the high light test conditions. The familiarity of the test arena was without effect on the time spent grooming ($F[1,35] = 1.01$), nor were any of the interactions with strain, sex, or light significant. There was a significant effect of the time spent in the test situation for the two male strains ($F[3,105] = 11.16$, $p < 0.0001$). There were also significant interactions for time $\times$ strain ($F[3,105] = 2.85$, $p < 0.05$) and time $\times$ strain $\times$ familiarity $\times$ light ($F[3,105] = 3.57$, $p < 0.02$); this was because the

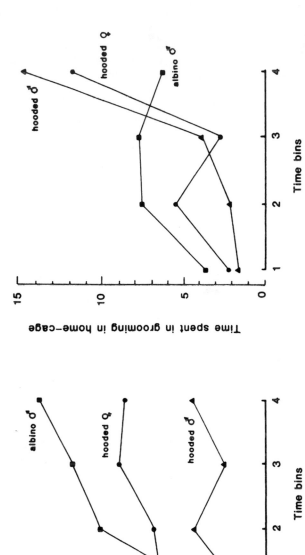

FIGURE 1. Mean time(s) spent self-grooming in an elevated arm (*left*) and home-cage (*right*) for male hooded rats, female hooded rats, and male Wistar rats. The scores are shown for each successive 2.5-min time bin.

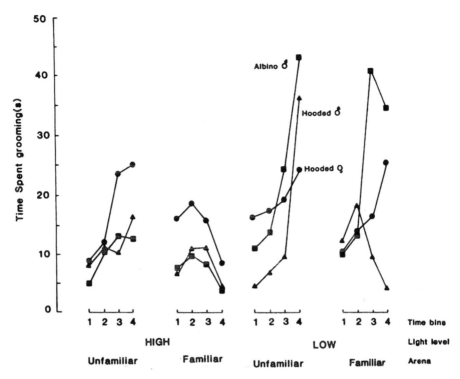

FIGURE 2. Mean time(s) spent in self-grooming by rats placed for 10 min in an unfamiliar or familiar test arena that was lit by high or by low light. The scores are shown for each successive 2.5-min time bin and for male Wistar, male hooded Lister, and female hooded Lister rats.

Wistars showed more change with time in the low light conditions and the hoodeds showed more change in the high light unfamiliar condition (see FIG. 2). The significant familiarity × time interaction (F[3,105] = 4.97, $p < 0.005$) reflects the different time course seen in unfamiliar and familiar conditions. In the unfamiliar arena there was a steady increase over each successive time period, whereas in the familiar test arenas the peak time spent grooming was the second or third time period (see FIGURE 2). The significant time × light interaction (F[3,105] = 5.54, $p < 0.005$) reflects the much greater rise of grooming over the test period in the low light conditions (see FIG. 2). The female hooded rats showed significantly more grooming than the males (F[1,35] = 4.49, $p < 0.05$) and there was a significant familiarity × time × sex × light interaction (F[3,105] = 5.20, $p < 0.005$), because the females showed the greatest rise with time in the high light, unfamiliar condition (see FIG. 2).

FIGURE 3 shows the frequency with which each type of grooming occurred for the three groups of rats in all the test conditions. It is clear from this figure that the most common forms of grooming in all the conditions were face wash and forepaw lick.

FIGURE 4 shows the time spent grooming in the home cage and in the elevated arm for male hooded rats injected with vehicle or with pentylenetetrazole or FG 7142 5 or 30 min prior to observation. As can be seen, pentylenetetrazole increased the time spent grooming in both the home cage and the elevated arm (F[2,54] = 13.38, $p <$

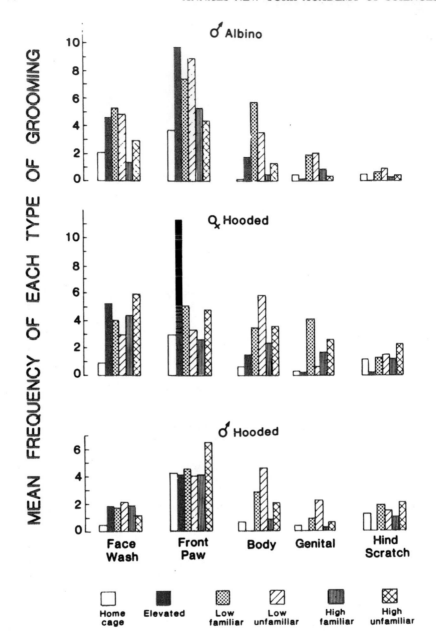

FIGURE 3. Mean frequencies of face wash, front paw wash, body wash, genital licking, and hind limb scratching for male Wistar (*top panel*), female hooded (*center panel*), and male hooded (*bottom panel*) rats observed for 10 minutes.

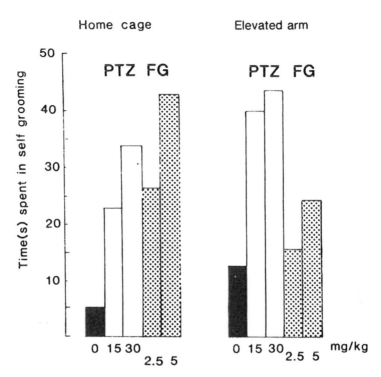

FIGURE 4. Mean time(s) spent in self-grooming in the home cage (*left*) and in the elevated arm (*right*) for rats injected 5 min before with pentylenetetrazole (PTZ, 15 or 30 mg/kg) or 30 min before with FG 7142 (FG, 2.5 or 5 mg/kg).

0.0001). There was a significant effect of time ($F[3,162] = 3.88, p < 0.02$) and a time × pentylenetetrazole interaction ($F[6,162] = 2.35, p < 0.05$), because the higher dose group showed a particularly marked increase with time in both the home cage and the elevated arm. FG 7142 also caused a significant increase in the time spent grooming ($F[2,54] = 5.63, p < 0.01$), and there was a significant time × FG × test condition interaction ($F[6,162] = 2.20, p < 0.05$), because the increase in grooming in the rats treated with FG 7142 occurred particularly in the last time period of the home cage test. There was no overall effect of yohimbine on the time spent grooming, however there was a test × yohimbine interaction of borderline significance, because yohimbine increased the time spent grooming only in the home cage ($F[2,53] = 2.39, p = 0.10$). Although the increase in time spent grooming was not significant, there was a significant increase in the frequencies of front paw licking and body washing in the home cage after yohimbine ($p < 0.05$, see FIGURE 5). Pentylenetetrazole and FG 7142 also selectively increased these forms of grooming in the home cage ($p < 0.05$), leaving unchanged the incidence of other types of grooming.

Pentylenetetrazole (30 mg/kg) significantly decreased the time spent grooming in the social interaction test ($F[2,53] = 3.69, p < 0.05$), but neither FG 7142 nor yohimbine had a significant effect (in both cases $F < 1.0$). There was no significant change in the frequency of any type of grooming with administration of anxiogenics in this test.

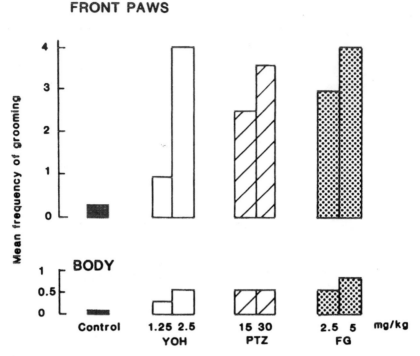

FIGURE 5. Mean frequency of front paw and body washing during a 10 min observation period in the home cage after (left to right): vehicle injection (5 or 30 min i.p. before—scores combined), yohimbine (YOH, 1.25 or 2.5 mg/kg i.p. 30 min before), pentylenetetrazole (PTZ, 15 or 30 mg/kg i.p. 5 min before) or FG 7142 (FG, 2.5 or 5 mg/kg i.p. 30 min before).

DISCUSSION

TABLE 1 summarizes the effects of various subject and environmental parameters on anxiety, as measured in the social interaction and elevated plus-maze tests, and on the time spent in self-grooming. Female hooded and male albino rats showed more self-grooming than male hooded rats. However, although female rats are *less* anxious than their male counterparts in the elevated plus-maze,[10] no differences in anxiety were found between male Wistar and hooded Lister rats.[9] The three environmental factors and three drugs were all chosen because they increase behavioral indexes of anxiety. However, as can be seen in TABLE 1, they produced increases, decreases, and no change in self-grooming.

If self-grooming increases in response to the termination of, or habituation to, anxiogenic stimuli, then one would predict increased grooming over time in the test situation. This was found in the home cage, the elevated arm, and in the low and high light test arenas, whether they were familiar or unfamiliar to the rats. As can be seen in FIGURE 2, the time course was different for familiar and unfamiliar test arenas, with the former showing an inverted-U curve (most clearly seen in the male Wistar rats).

TABLE 1. Summary of Effects of Sex, Strain, Light, Familiarity, Elevation and Three Anxiogenic Drugs on Self-Grooming and on Behavioral Indexes of Anxiety

	Grooming	Anxiety
Sex: females	↑	↓
Strain: Wistar	↑	=
Light: high	↓	↑
Familiarity: unfamiliar	=	↑
Elevation	↑	↑
PTZ	↑	↑
FG 7142	↑	↑
Yohimbine	=	↑

↑, Increase.
↓, Decrease.
=, No change.

This suggests that with increased time in the situation and hence habituation to stress, self-grooming eventually decreases. It also suggests that the habituation occurs both within and between test sessions. Thus, although the overall time spent grooming in the familiar and unfamiliar arenas did not differ, the distribution of grooming over time was different.

If grooming is elicited in response to habituation to stress, it is extremely difficult to make any overall predictions about the effects on grooming of any particular anxiogenic stimulus. The overall level of grooming will be the joint result of the level of stress generated, the rate of habituation to the stressful stimuli, and the duration of the test period. Thus there was greater self-grooming in the elevated arm than in the home cage, because the former was more stressful and sufficient within-session habituation to the stress occurred for the rise in grooming to be manifest. The difference in distribution over time of grooming in the high and low light test arenas suggests that the male rats were habituating more slowly to the high light. If the test session had been prolonged to, say 1 hour, then one would predict a gradual rise in self-grooming in the high light, and a decrease in the low light. However, what is interesting is that the within-session decrease in grooming occurs (at least for the male Wistar rats) in the high light, familiar condition without high levels of grooming ever being reached. This suggests between-session habituation may have different effects on grooming than within-session habituation. In general, the changes in grooming as a result of our environmental manipulations are consistent with self-grooming being a response to habituation to anxiogenic stimuli. With relatively low levels of stress the rat can habituate to the stress within the 10-min test session. Thus for stimuli within this range there is more grooming as the stress is increased. Anxiogenic conditions enhance self-grooming because there is within-session habituation to the stimuli. With higher levels of stress the overall increase in self-grooming is less because of slower within-session habituation.

Can this hypothesis account for our sex and strain differences? We would have to predict that female hooded and male Wistar rats would show more rapid habituation to anxiogenic or stressful stimuli than male hooded rats. Unfortunately there are no data yet on this question. On the basis of habituation to stress, what can be made of the data on self-grooming in response to anxiogenic drugs? First it must be assumed that the anxiogenic effects of the drug and the test session are additive. PTZ was injected only 5 minutes before the test began, and therefore there was little opportunity for the rats to habituate to the anxiogenic effects of the drug before the test. The increased grooming in PTZ-injected rats, compared with controls in the home cage and elevated arm, suggests that considerable within-session habituation occurred to the anxiogenic stimuli generated by the drug. The significant decrease in grooming seen with the combination of the higher dose of PTZ and the social interaction test arena could be explained if the combined stress of PTZ and the test arenas was sufficient to slow habituation. Both yohimbine and FG 7142 were injected 30 min before testing began. The failure of yohimbine to change self-grooming could be accounted for if complete habituation had occurred in this time to the anxiogenic effects of the drug. From previous experiments we know that it is still possible to detect anxiogenic effects of yohimbine 30 min after injection,[16] but it is possible that self-grooming is triggered by a higher level of stress than is needed to reduce social interaction. The results with FG 7142 would suggest that habituation to its anxiogenic effects was not complete by the time the tests began. Thus, the combination of FG 7142 and the wooden test arenas was stressful enough to decrease the rate of habituation, whereas habituation to the drug could still occur in the elevated arm and the home cage.

In conclusion, the results of these experiments suggest that there is no simple relationship between anxiogenic conditions and the incidence of self-grooming. The results can be reasonably well explained if self-grooming is a response to habituation to anxiogenic stimuli. Given equal rates of habituation, within a given time period the more anxiogenic stimuli will generate more self-grooming. But, at least above a certain level of stress, there seems to be an interaction between the level of anxiety and the rate of habituation. Studies of the time course of grooming over a prolonged period are necessary to verify this hypothesis. Additionally, further studies are needed to clarify the relationship between between-session habituation and grooming.

REFERENCES

1. YULE, E. P. 1957. The open field test. Proc. S. Afr. Psychol. Assoc.
2. THOMPSON, W. R. & W. H. HIGGINS. 1958. Emotion and organised behavior. Can. J. Psychol. **12:** 61–68.
3. COHEN, J. A. & E. O. PRICE. 1979. Grooming in the Norway rat: Displacement activity or "boundary shift"? Behav. Neural Biol. **26:** 177–188.
4. DOYLE, G. & E. P. YULE. 1959. Grooming activities and freezing behaviour in relation to emotionality in albino rats. Anim. Behav. **7:** 18–22.
5. GISPEN, W. H. & R. L. ISAACSON. 1981. ACTH-induced grooming in the rat. Pharmacol. Ther. **12:** 209–224.
6. ROTH, K. A. & R. J. KATZ. 1979. Stress, behavioral arousal, and open field activity—a reexamination of emotionality in the rat. Neurosci. Biobehav. Rev. **3:** 247–263.
7. FILE, S.E. 1980. The use of social interaction as a method for detecting anxiolytic activity of chlordiazepoxide-like drugs. J. Neurosci. Methods **2:** 219–238.
8. FILE, S. E. 1985. Animal models for predicting clinical efficacy of anxiolytic drugs: Social behaviour. Neuropsychobiology **13:** 55–62.
9. PELLOW, S., P. CHOPIN, S. E. FILE & M. BRILEY. 1985. The validation of open:closed arm entries in an elevated plus-maze as a measure of anxiety in the rat. J. Neurosci. Methods **14:** 149–167.
10. FILE, S. E. & S. V. VELLUCCI. 1979. Behavioural and biochemical measures of stress in hooded rats obtained from different sources. Physiol. Behav. **22:** 31–35.
11. JOHNSTON, A. L., S. E. FILE, F. FARABOLLINI & C. A. WILSON. 1986. The advantages of a female brain: Decreased anxiety and increased activity. Horm. Behav. Submitted.
12. DUNN, A. J., E. J. GREEN & R. L. ISAACSON. 1979. Intracerebral adrenocorticotropic hormone mediates novelty-induced grooming in the rat. Science **203:** 281–283.
13. GISPEN, W. H., V. M. WIEGANT, H. M. GREVEN & D. DE WIED. 1975. The induction of

excessive grooming in the rat by intraventricular application of peptides derived from ACTH: Structure-activity studies. Life Sci. **17:** 645–652.

14. MORLEY, J. & A. S. LEVINE. 1982. Corticotropin-releasing factor, grooming and ingestive behavior. Life Sci. **31:** 1459–1464.

15. VELDHUIS, H. D. & D. DE WIED. 1984. Differential behavioral actions of corticotropin-releasing factor (CRF). Pharmacol. Biochem. Behav. **21:** 707–713.

16. DUNN, A. J. & S. E. FILE. 1986. Corticotropin-releasing factor displays an anxiogenic action in the social interaction test. Horm. Behav. Submitted.

17. FILE, S. E. & A. CLARKE. 1980. Intraventricular ACTH reduces social interaction in male rats. Pharmacol. Biochem. Behav. **12:** 855–859.

18. FILE, S. E. & S. V. VELLUCCI. 1978. Studies on the role of ACTH and of 5-HT in anxiety using an animal model. J. Pharm. Pharmacol. **30:** 105–110.

19. FILE, S. E. 1980. Naloxone reduces social and exploratory activity in the rat. Psychopharmacology **71:** 41–44.

20. ALLAN, A. M. & R. L. ISAACSON. 1985. Ethanol-induced grooming in mice selectively bred for differential sensitivity to ethanol. Behav. Neural Biol. **44:** 386–392.

21. PELLOW, S. & S. E. FILE. 1984. Multiple sites of action for anxiogenic drugs: Behavioural, electrophysiological and biochemical correlations. Psychopharmacology **83:** 304–315.

Effect of Perinatal Exposure to Therapeutic Doses of Chlorimipramine on Grooming Behavior in the Adult Rat[a]

E. L. RODRÍGUEZ ECHANDÍA, M. R. FÓSCOLO, AND A. GONZALEZ

Laboratorio de Investigaciones Cerebrales
National University of Cuyo
5500 Mendoza, Argentina

Exposure of laboratory animals to several psychotropic drugs during a vulnerable period of brain development can produce behavioral impairment in later life. Transient or permanent alterations have been reported with perinatal treatment with amphetamines, alcohol, barbiturates, neuroleptics, antidepressants, and other drugs. Tricyclic antidepressants (TAD) are extensively used in the treatment of depression, and therefore identification of their eventual deleterious effects upon development of brain mechanisms controlling behavior is of great interest.

DELETERIOUS EFFECTS OF PRENATAL EXPOSURE TO TAD

Species differences in susceptibility to perinatal TAD treatment have been reported. The doses producing toxic reactions in the rabbit are without effect in the pregnant mouse.[1] Very high doses of TAD (150 mg/kg) are tolerated by the pregnant mouse without apparent embryotoxic or teratogenic effects, whereas doses of 15 mg per kg per day produce fetal toxicity in the rat[1] and cause fetal and neonatal mortality.[2] Daily subcutaneous injections of 5–15 mg per kg per day of TAD before and during pregnancy fail to produce fetal defects in rats, but retard maturation, reflex ontogeny, and exploratory behavior in the surviving offspring.[2] Such effects might be mediated in part by maternal toxicity, however. It has been reported that chronic exposure of adult rats to large doses of TAD reduces motor activity[3-5] as well as food and water intake, causing decrease in body weight[6,7] and impairment of body weight increase during pregnancy.[8]

Though the metabolism of TAD in rats was early shown to be similar to man[9,10] in general, doses of TAD used in rat experiments are several times higher than doses used in humans. Therapeutic doses of TAD in the treatment of depression are usually 1–3 mg per kg per day. According to some reports, chronic ingestion of TAD by pregnant women results in increase in the incidence of birth defects,[11] but this has not been

[a] This work was supported by CONICET grant no. 3–053500/85.

confirmed by other authors.[12] When pregnant rats are submitted to chronic ingestion of therapeutic doses of the tertiary amine tricyclic, chlorimipramine (CI), the treatment is apparently devoid of maternal intoxication and fetal or neonatal mortality, does not impair physical maturation of offspring, and does not alter maternal behavior.[13]

BEHAVIORAL EFFECT OF PRENATAL AND NEONATAL TAD EXPOSURE

The prenatal treatment with the secondary amine tricyclic, imipramine, was shown to impair social and environmental interactions in infant and adult rats.[14,15] In addition, it has been reported that the imipramine-exposed offspring fail to show the increase in visual cortical depth as a result of enriched rearing conditions reported by Rosenzweig and colleagues.[16] This effect was correlated with the behavioral unresponsiveness of the imipramine offspring in an enriched environment and was considered, therefore, as a subtle teratogenic effect of prenatal exposure to the drug.[14,15] More recently, File and Tucker[17] reported that daily i.p. injections of CI (3 and 10 mg per kg per day) to pregnant rats from day 8 to day 21 of gestation cause the adolescent offspring to display reduced anxiety in their social interaction test of anxiety.[18] The authors also found that these CI offspring habituate to familiar environments and stimuli more rapidly than control animals.[17,19] With a different experimental approach (therapeutic oral doses of CI from day 15 prior to mating until delivery), we have reported[13] that CI offspring behave normally in open field tests performed at day 21 of age. However, they showed definite behavioral abnormalities in other test situations. When tested at 30–40 days of age in a familiar environment, the CI-exposed rats showed increased grooming and digging responses. In emergence tests they emerged into and explored the novel environment less than controls. In social behavior tests the group-reared CI offspring showed scarce interactions with an unfamiliar partner.[13] These responses were interpreted as expressions of increased emotionality. Some of them disappeared when rats were retested at adulthood (130 days) but others remained unchanged.[13]

Our results conform to those reported by Coyle and Singer,[14,15] but as far as prenatal CI exposure effects on emotionality of offspring are concerned, they conflict with the results reported by File and Tucker.[17,19] Such differences might be due to the duration of the treatment. In our experiments dams were deliberately exposed to CI before conception to reproduce a situation occurring frequently in human females submitted to antidepressant treatment. This was also the case in the experiments of Singer and Coyle.[14,15] In this regard, a chronic CI treatment ending immediately before mating has been reported to impair postnatal development in the offspring.[8] Furthermore, our prenatal treatment (also that of Coyle and Singer[14,15]) comprised both organogenesis and fetal periods. In the experiments of File and Tucker[17,19] the treatment was circumscribed to the fetal period of pregnancy only.

Some reports show that TAD exposure during lactation can also cause behavioral alterations in later life. Daily i.p. injections of CI (30 mg per kg per day) from 1 to 3 postnatal weeks was shown to impair rapid eye movement sleep and would cause lifelong anomalies in emotionality and sexual behavior.[20] These findings were interpreted as indicating that CI-exposed offspring were more anxious at adulthood than normal rats. However, in other reports[21] it is shown that a similar CI treatment can impair development of body weight but fails to cause an increase in anxiety in social interaction tests performed at the third month of age. In addition, the behavioral responding of these animals treated neonatally with high doses of CI was normal in open field or hole board tests and in tests of sexual behavior or aggressive behavior.[21]

The early postnatal treatment with lower doses of CI also seems not to cause behavioral effects in later life.[13,21]

We have reported that the behavioral effects of the prenatal CI treatment were potentiated when the treatment continued throughout lactation.[13] Recently, we have analyzed (unpublished results) the behavioral responses of prenatal and prenatal plus neonatal CI-exposed offspring when submitted to a hole board trial for the first time. It was found that the 2-month-old offspring of dams ingesting 3 mg per kg per day CI from day 5 prior to mating until parturition and cross-fostered at birth gave scores for locomotion, head dipping, grooming, immobility duration and defecation that were similar to those for control offspring. The only differences found were that the CI-exposed rats entered the central squares of the field (1.0 × 1.0 m divided in 20.0 × 20.0 cm squares, with 16 holes spaced 20.0 cm apart from one another) and explored the central holes less than controls. The prenatal CI-exposed offspring that continued with their own CI-ingesting mothers up to weaning displayed, in addition, less time spent head dipping and greater frequency of grooming than control rats (FIG. 1). These results also suggest that the behavioral effects of the prenatal CI treatment is potentiated by the neonatal CI exposure. Changes in maternal care are unlikely to have been responsible for such potentiating effects. Maternal behavior of CI-consuming dams during nursing was normal and such was the case also for the body weight of their pups at weaning. It is speculated, therefore, that when brain development was altered by CI the neonatal treatment may impair compensatory processes occurring in early postnatal life.

EFFECT OF PRENATAL AND NEONATAL TAD EXPOSURE ON RESPONSES TO STRESS

It is known that some stresses can induce behavioral depression in rats[22-24] that has been ascribed to the degree of fear experienced. We have tested the hole board response to a short restraint stress (5 min) in male offspring of rats ingesting therapeutic doses of CI prior to mating until parturition or weaning (unpublished results). It was found that behavioral responsiveness to this stress was greater in the CI-exposed offspring at 50–60 days of age. In the rats submitted to prenatal CI exposure the restraint stress caused stronger inhibition of locomotion and greater immobility duration than in the controls (FIG. 1). Consistent with our previous report,[13] when the CI treatment of dams continued throughout the lactation period, the effects on stress response of offspring were potentiated. FIGURE 1 shows that in the rats submitted to prenatal plus neonatal CI exposure not only the scores of locomotion and immobility duration but also the latency to start locomotion and the time spent head dipping reached significant differences when compared with controls. These results suggest that young rats exposed to therapeutic doses of CI from conception up to parturition or weaning are more vulnerable to a mild restraint stress than normal rats.

EFFECT OF PRENATAL AND NEONATAL TAD EXPOSURE ON GROOMING BEHAVIOR

In accordance with Gispen and Isaacson,[25] grooming behavior is defined here as bouts of heterogeneous constituents comprising face washing, body licking, paw licking, head and body shaking, scratching, and genital licking. Some TAD, such as amitryptyline,

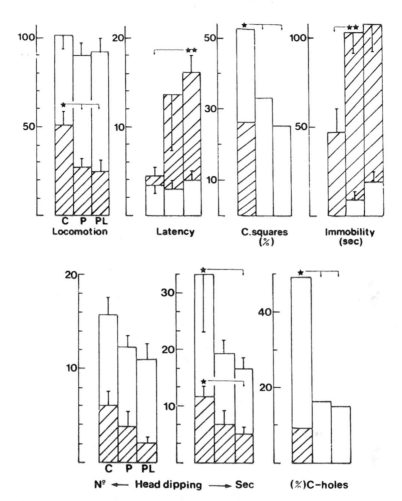

FIGURE 1. Hole board scores (means ± SEM for 5 min) of controls (*C*), *n* = 13; prenatal CI treatment (*P*), *n* = 12; and prenatal plus lactation exposure (*P*), *n* = 13. *Open columns:* baseline scores; *hatched columns:* post restraint stress scores; * $p < 0.05$; ** $p < 0.01$. When means are expressed in proportions the chi-square test was used. Other comparisons: two-way analysis of variance and Scheffe's test for multiple comparisons.

and other types of antidepressant drugs are known to reduce the excessive grooming in adult rats.[26] Little attention has been paid, however, to the effects of prenatal and neonatal TAD exposure on grooming behavior in later life.

Coyle and Singer[14,15] reported that the male offspring exposed prenatally to imipramine (5 mg per kg per day) and reared in an enriched environment showed normal grooming responses in home cage tests performed both in infancy and adulthood. In other experiments,[13] the infant male offspring of dams ingesting 3 mg per kg per day

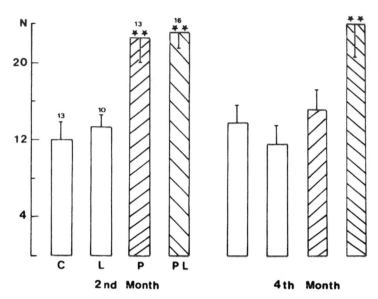

FIGURE 2. Familiar environment test. Frequency of grooming bouts (mean ± SEM for 10 min) in second and fourth months of age: *C*, controls; *L*, CI exposure through lactation; *P*, prenatal exposure; *PL*, prenatal plus lactation exposure. *Numbers* represent number of rats; **$p < 0.02$ (two-way ANOVA and Scheffe's test for multiple comparisons). (Redrawn after Rodríguez Echandía and Broitman[13] and including scratching activity).

CI prior to mating until parturition displayed excessive grooming when placed singly in a familiar cage containing clean wooden shavings (FIG. 2). It is interesting that when these CI offspring were assigned to a social behavior test at adolescence they showed less social interactions than normal but displayed greater frequency of self-grooming than their control partners (FIG. 3). These excessive grooming responses were just a

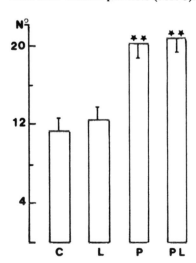

FIGURE 3. Social behavior test. Frequency of self-grooming (mean ± SEM for 10 min) in second-month tests. For number of animals and explanation see FIGURE 2; **$p < 0.01$ (ANOVA and Duncan's multiple range test).

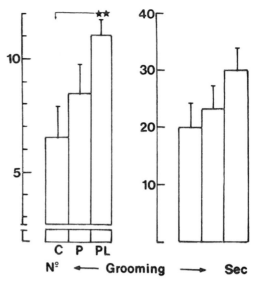

FIGURE 4. Hole board grooming (frequency and time spent) of male rats at two months of age (means ± SEM for 5 min; $n = 13$ per column). For explanation see FIGURE 2; **$p < 0.01$ (ANOVA and Duncan's multiple range test).

transient effect of the prenatal CI treatment, since grooming was normal when animals were retested at adulthood.[13] The rats exposed to CI through the mother's milk only displayed normal grooming responses at any age. The prenatal plus lactation CI treatment produced permanent effects on grooming behavior. These animals showed excessive grooming responses at both infancy and adulthood[13] (FIG. 2). In recent experiments (unpublished results) we found that 2-month-old male offspring of dams ingesting CI prior to mating until delivery and cross-fostered at birth displayed normal grooming responses when tested in the hole board for the first time. However, the offspring that continued to be exposed to CI through the mother's milk up to weaning showed excessive grooming (FIG. 4). These results are also consistent with the view that when prenatal CI treatment is complemented with CI exposure throughout lactation the alteration in grooming behavior becomes permanent.

Grooming behavior in rodents has been interpreted to represent maintenance of the fur. However, grooming displayed in an unfamiliar environment has also been interpreted as a reflection of fear[27] or a reaction to a stress situation.[28] Excessive grooming in rats can be induced by water immersion,[25,29] exposure to cold,[36] handling and novelty,[31–33] i.p. injections of physiological saline,[34] a short restraint stress (discussed herein) and other relatively mild stressors. As shown above, the prenatal CI exposure, especially the prenatal plus neonatal exposure, would increase vulnerability to a mild stress in rats. If such were the case, a heightened stress-induced release of ACTH and prolactin (PRL) might be expected in these animals. Both ACTH and PRL have been shown to cause excessive grooming in rats,[25,35] and this effect would be mediated by dopamine and opioid transmission in the striatum.[36] The dopamine receptor antagonist, haloperidol, and the opiate receptor blocker, naloxone, antagonize both PRL- and ACTH-induced excessive grooming.[34,40] On the other hand, both ACTH and PRL have been shown to affect noradrenaline, serotonin, and GABA systems in the brain.[41–43] We have reported that the excessive grooming in response to a mild stressor is not prevented by relatively high doses (20 mg/kg) of the α antagonists, phentolamine, or the β antagonist, propranolol.[34] However, the α_2 receptor agonist, clonidine, was shown to be effective in reducing the ACTH-induced excessive grooming.[25] It appears therefore, that noradrenaline systems in the brain may contribute to the regulation of grooming responses. The serotonergic systems would also participate. It has been reported that two serotonin antagonists, methysergide and pizotifene, are effective in blocking the appearance of excessive grooming in response to a mild stress.[34] This effect would not be due to depression of general behavior because the serotonin antagonists did not affect locomotion.[34]

The GABAergic system might be involved also in the regulation of grooming behavior, perhaps by its inhibitory effects on monoaminergic neurons. It is known that gamma-vinyl-GABA (G-V-G) can facilitate GABAergic transmission by an irreversible inhibition of GABA-transaminase, which produces a long-lasting increase of endogenous GABA in the brain areas containing GABAergic terminals or perikarya.[44] It is of interest that grooming and other behaviors can be inhibited by injection of G-V-G into the locus ceruleus region in rats.[45] Such effect would be mediated by noradrenaline neurons since it was not apparent when G-V-G was applied in rats recovered from selective destruction of locus ceruleus-noradrenaline neurons by local injection of 6-OHDA. Further work is necessary to elucidate the hormonal and neurotransmitter systems involved in the excessive grooming caused by the prenatal and the prenatal plus neonatal treatment with CI. Chronic antidepressant therapy produces alterations of uptake, storage, synthesis, and release of monoaminergic transmitters and causes adaptive changes in receptor functioning. These adaptations occur both presynaptically (down regulation of α_2 adrenoceptors and of dopaminergic presynaptic receptors) and postsynaptically (decreased β adrenoceptor and 5-HT$_2$ receptor binding, see ref. 46). Chronic antidepressant treatment was shown to induce also an increase in [^{3}H]GABA "B" binding.[47] It is conceivable, therefore, that some of these effects on neurotransmitter systems might be responsible for the behavioral alterations caused by the perinatal TAD exposure.

REFERENCES

1. HARPER, K. H., A. K. PALMER & R. E. DAVIS. 1965. Effects of imipramine upon the pregnancy of laboratory animals. Arzneimittel-Forsch. **15:** 1218-1221.
2. COYLE, I. R. 1975. Changes in the developing behavior following prenatal administration of imipramine. Pharmacol. Biochem. Behav. **3:** 799-807.
3. HERR, F., J. STEWART & M. P. GJAREST. 1961. Tranquilizers and antidepressants. A pharmacological comparison. Arch. Int. Pharmacodyn. Ther. **134:** 328-342.
4. HOROVITZ, Z. P., A. R. FURGIUELE, J. P. HIGH & J. C. BURKE. 1964. Comparative activities of substituted dibenzothiazepine of imipramine and of desipramine. Arch. Int. Pharmacodyn. Ther. **151:** 180-191.
5. FURGIUELE, A. R., M. H. AUMENT & Z. P. HOROVITZ. 1964. Acute and chronic effects of imipramine and desipramine in normal rats and in rats with lesioned amygdalae. Arch. Int. Pharmacodyn. Ther. **151:** 170-179.
6. STENGER, E. G., I. AEPPLI & I. FRATA. 1965. Teratogenic effect of N-(dimethyl aminopropyl) iminodibenzyl-HCl (Tofranil) in animals. Arzneimittel-Forsch. **15:** 1222-1224.
7. ZABIC, J. E., R. M. LEVINE, J. H. SPAULING & R. P. MAICKEL. 1977. Interactions of tricyclic antidepressant drugs with deprivation-induced fluid consumption by rats. Neuropharmacology **16:** 267-271.
8. BROITMAN, S. T. & A. O. DONOSO. 1978. Effects of chronic imipramine and clomipramine oral administration on maternal behavior and litter development. Psychopharmacology **56:** 93-101.
9. BICKEL, M. H. & H. J. WEDER. 1968. The total fate of a drug: Kinetics of distribution, excretion and formation of 14 metabolites in rats treated with imipramine. Arch. Int. Pharmacodyn. Ther. **173:** 433-463.

10. CRAMMER, J. L., B. SCOTT, H. WOODS & B. ROLFE. 1968. Metabolism of ^{14}C-imipramine. I. Excretion in the rat and man. Psychopharmacologia **12:** 263–277.
11. HENDRICKX, A. G. 1975. Teratologic evaluation of imipramine in pregnancy. Brit. Med. J. **11:** 745.
12. KUENSSBERG, E. V. & J. D. E. KNOX. 1972. Imipramine in pregnancy. Brit. Med. J. **2:** 292.
13. RODRIGUEZ ECHANDIA, E. L. & S. T. BROITMAN. 1983. Effect of prenatal and postnatal exposure to therapeutic doses of chlorimipramine on emotionality in the rat. Psychopharmacology. **79:** 236–241.
14. COYLE, I. R. & G. SINGER. 1975. The interaction of post-weaning housing conditions and prenatal drug effects on behaviour. Psychopharmacologia **41:** 237–244.
15. COYLE I. R. & G. SINGER. 1975. The interactive effects of prenatal imipramine exposure and postnatal rearing conditions on behaviour and histology. Psychopharmacologia **44:** 253–256.
16. ROSENZWEIG, M. R., E. L. BENNETT & M. C. DIAMOND. 1972. Brain changes in response to experience. Sci. Am. **226:** 22–30.
17. FILE, S. E. & J. C. TUCKER. 1983. Prenatal treatment with clomipramine has an anxiolytic profile in the adolescent rat. Physiol. Behav. **31:** 57–61.
18. FILE, S. E. & J. R. G. HYDE. A test of anxiety that distinguishes between the actions of benzodiazepines and those of other minor tranquilizers and stimulants. Pharmacol. Biochem. Behav. **11:** 63–69.
19. FILE, S. E. & J. R. G. TUCKER. 1984. Prenatal treatment with clomipramine: Effects on the behaviour of male and female adolescent rats. Psychopharmacology **82:** 221–224.
20. MIRMIRAN, M., N. E. VAN DER POLL, M. A. CORNER, H. G. VAN OYEN & H. L. BOUR. 1981. Suppression of active sleep by chronic treatment with chlorimipramine during early postnatal development: Effects upon adult sleep and behavior in the rat. Brain Res. **204:** 129–146.
21. FILE, S. E. & J. C. TUCKER. 1983. Neonatal clomipramine treatment in the rat does not affect social, sexual and exploratory behaviors in adulthood. Neurobehav. Toxicol. Teratol. **5:** 3–8.
22. CADLAND, D. K. & M. NAGY. 1969. The open field behavior: Some comparative data. Ann. N. Y. Acad. Sci. **159:** 831–851.
23. WILLIAMS, D. I. 1972. Effects of electric shocks on exploratory behavior in the rat. Quart. J. Exp. Psychol. **24:** 544–546.
24. RODRIGUEZ ECHANDIA, E. L., S. T. BROITMAN & M. R. FOSCOLO. 1982. Chronic treatment with therapeutic doses of chlorimipramine causes hyperactivity in male rats. IRCS Med. Sci. **10:** 366–367.
25. GISPEN, W. H. & R. L. ISAACSON. 1981. ACTH-induced excessive grooming in the rat. Pharmacol. Ther. **12:** 209–246.
26. TRABER, J., H. R. KLEIN & W. H. GISPEN. 1982. Actions of antidepressant and neuroleptic drugs on ACTH- and novelty-induced behavior in the rat. Eur. J. Pharmacol. **80:** 407–414.
27. TINBERGEN, N. 1952. Derived activities. Q. Rev. Biol. **27:** 1–32.
28. GISPEN, W. H. 1982. Neuropeptides and behavior: ACTH. Scand. J. Psychol. **1:** 16–25.
29. COLBERN, D. L., R. L. ISAACSON, J. G. HANNIGAN & W. H. GISPEN. 1981. Water immersion, excessive grooming and paper shredding in the rat. Behav. Neural Biol. **32:** 428–437.
30. HANNIGAN. J. H. & R. L. ISAACSON. 1981. Conditioned excessive grooming in the rat after footshock. Effect of naloxone and situational cues. Behav. Neural Biol. **33:** 280–292.
31. BINDRA, D. & N. SPINNER. 1958. Response to different degrees of novelty. The incidence of various activities. J. Exp. Anim. Behav. **1:** 341–350.
32. COLBERN, D. L., R. L. ISAACSON, E. G. GREEN & W. H. GISPEN. 1978. Repeated intraventricular injections of $ACTH_{1-24}$: The effects of home or novel environment on excessive grooming. Behav. Biol **23:** 381–387.
33. JOLLES, J., J. ROMPA-BARENDREGT & W. H. GISPEN. 1979. ACTH-induced excessive grooming in the rat: The influence of environmental and motivational factors. Horm. Behav. **12:** 60–72.
34. RODRIGUEZ ECHANDIA, E. L., S. T. BROITMAN & M. R. FOSCOLO. 1983. Effect of serotonergic and catecholaminergic antagonists on mild-stress-induced excessive grooming in the rat. Behav. Neurosci. **97:** 1022–1024.

35. DRAGO, F., B. BOHUS, W. H. GISPEN, U. SCAPAGNINI & D. DE WIED. 1983. Prolactin-enhanced grooming behavior: Interaction with ACTH. Brain Res. **263:** 277–282.
36. DRAGO, F., B. BOHUS, P. L. CANONICO & U. SCAPAGNINI. 1981. Prolactin induces grooming in the rat: Possible involvement of nigrostriatal dopaminergic system. Pharmacol. Biochem. Behav. **15:** 61–63.
37. DRAGO, F., J. M. VAN REE, B. BOHUS & D. DE WIED. 1981. Endogenous hyperprolactinemia enhances amphetamine and apomorphine-induced stereotypy. Eur. J. Pharmacol. **72:** 249–253.
38. DRAGO, F., W. H. GISPEN & B. BOHUS. 1982. Behavioral effects of prolactin: Involvement of opioid receptors. *In* Advances in Endogenous and Exogenous Opioids. H. Takagi & E. J. Simon, Eds.: 335–337. Kodansha Press. Kyoto.
39. WIEGANT, V. M., W. H. GISPEN, L. TERENIUS & D. DE WIED. 1977. ACTH-like peptides and morphine: Interactions at the level of the nervous system. Psychoneuroendocrinology **2:** 63–69.
40. WIEGANT, V. M., J. JOLLES & W. H. GISPEN. 1978. B-endorphine grooming in the rat: Single dose tolerance. *In* Characteristics and Functions of Opioids. J. M. van Ree & L. Terenius, Eds. Elsevier/North-Holland Biomedical Press. Amsterdam.
41. HOHN, K. G. & W. O. WUTTKE. 1978. Changes in catecholamine turnover in the anterior part of the mediobasal hypothalamus and the medial preoptic area in response to hyperprolactinaemia in ovariectomized rats. Brain Res. **156:** 241–252.
42. VERSTEEG, D. G. H. 1980. Interaction of peptides related to ACTH, MSH and B-LPH with neurotransmitters in the brain. Pharmacol. Ther. **11:** 535–557.
43. MASKY, T., E. DUEKER & W. WUTTKE. 1981. Hypothalamic and limbic GABA turnover and glutamate concentration following ovariectomy and hyperprolactinaemia. Neurosci. Lett. **7:** 254.
44. CASU, M. & K. GALE. 1981. Intracerebral injection of gamma-vinyl-GABA: Method for measuring rates of GABA synthesis in specific brain regions in vivo. Life Sci. **29:** 681–688.
45. RODRIGUEZ ECHANDIA, E. L., S. T. BROITMAN, M. FOSCOLO & A. GONZALEZ. 1986. Gamma-vinyl-GABA injection in the locus coeruleus region of the rat. Effect on spontaneous behavior and stress responses. *In* GABA and Endocrine Function. G. Racagni & A. O. Donoso, Eds.: 93–102. Raven. New York.
46. SUGRUE, M. F. 1983. Chronic antidepressant therapy and associated changes in central monoaminergic receptor functioning. Pharmacol. Ther. **21:** 1–35.
47. LLOYD, K. G. & A. PILC. 1984. Chronic antidepressants and GABA synapses. Neuropharmacology **23:** 841–842.

An Ethological Analysis of Excessive Grooming in Young and Aged Rats

BERRY M. SPRUIJT, PHILIP WELBERGEN, JAN BRAKKEE, AND
WILLEM HENDRIK GISPEN

Institute of Molecular Biology and Medical Biotechnology
and
Rudolf Magnus Institute for Pharmacology
University of Utrecht
3584 CH Utrecht, The Netherlands

INTRODUCTION

Age-related behavioral changes have been studied predominantly in learning paradigms. The declines in memory and cognitive performance are among the most salient effects of aging and dementia.[1-5] The changes in cognitive performance of aging animals have been interpreted as deficits in acquisition and retention of learned skills.[6-13] However, age-related changes in cognitive functioning are not necessarily caused by deficits in mechanisms involved in the retention and acquisition of new behavioral patterns. To assess the contribution of possible other causes the changes in physical strength and motor coordination have been investigated. It appeared that perception also markedly changes with age.[14] In addition, senescent animals showed a decrease in tests measuring motor performance.[14-17] The main interest of these studies on noncognitive behavioral changes concerned their possible contribution to the impaired performances of aged subjects in learning tests. As a consequence, current concepts such as the Jacksonian principle[18] of age-related behavioral alterations is restricted to behavior displayed in problem-solving tasks. In order to extend the spectrum of behavioral changes in aging animals and to evaluate the relevance of current aging models, other behavioral systems should be studied as well. In this paper the excessive grooming behavior of the rat is used as an appropriate noncognitive behavior.

EXCESSIVE GROOMING

Grooming or maintenance behavior is a common species characteristic movement pattern with readily definable components. In the laboratory rat, grooming behavior may occupy as much as 25–40% of the awake time, depending on the housing conditions, with most of the behavior seen just prior to and after the diurnal sleep period.[19] The grooming of the rat and the cat, the preening or sandbathing of the bird, the rubbing and sweeping of the fly, all serve a role in the care of the body surface. Therefore, this class of behaviors is also designated as care of body surface.[20] Since it is a very reproducible and easily measurable behavior, it has been used extensively in experimental studies, especially peptide-induced excessive grooming. Peptide-induced excessive grooming has been applied for different purposes: (1) structure-activity studies were

89

used to identify where the crucial information is hidden in peptide-inducing peptides (for reviews see refs. 23 and 24 and other papers in this volume), and (2) local application of peptides in specific brain areas in combination with dopaminergic and GABAergic agents were applied to unravel the underlying neural circuitry, i.e., the modulating dopaminergic and GABAergic systems and the possible site of action of the peptide.[23-25]

The application of ethological methods in experimental studies yields a more sensitive tool in studying changes in behavior. In peptide-induced grooming the distinction between different grooming elements showed that different peptides induce grooming patterns with different distribution of frequencies of elements.[26,27] Even pharmacological manipulation of peptide-induced grooming with haloperidol and naloxone showed a differential suppression of, respectively, grooming and scratching.[28] An ethological method was also used to compare ACTH-induced excessive grooming with spontaneously occurring grooming in order to assign biological relevance to ACTH-induced grooming. The meaning of behavior is not only determined by the frequency and/or the durations of its components but to a great extent by the serial ordering of the different elements.

A sequential analysis of ACTH-induced excessive grooming demonstrated the similarity in the sequential structure of this behavior to the structure of saline-induced excessive grooming.[29] In the present study the combined use of ACTH-induced excessive grooming and a sequential analysis of the behavior were used to show in detail subtle differences in the sequential organization of grooming behavior between young and aged rats. In addition, on a higher level of behavioral organization the occurrence of grooming in relation to other behavioral systems such as sociosexual behavior in young and aged rats was studied. For this purpose a situation was created that elicited, apart from grooming behavior, sexual and social behaviors.

GROOMING BEHAVIOR IN YOUNG AND AGED RATS

Firstly, the grooming behavior of 10 young (3 months) and 12 old (24 months) WAG (rij) rats was observed in a setting in which almost no other behavior was induced. The experiment was carried out according to conditions described extensively elsewhere.[30] Fifteen min after the intracerebroventricular injection of ACTH (0.3 µg/0.3 µl) the animals were individually observed for 55 minutes. A complete record of the behavior displayed by each animal was made using the following behavioral elements: forepaw vibration (VI), face washing (FW), body grooming (BG), anogenital grooming (AG), tail sniffing (TS), scratching (SC), body shake (BS), and stretching and yawning (SYS). Exploratory behaviors, sleeping, and other nongrooming elements were taken together into one rest category (RE). These elements have been extensively described by Gispen and Isaacson[21] and Spruijt and Gispen.[22]

GROOMING DURING AN ENCOUNTER WITH A SEXUALLY RECEPTIVE FEMALE

Eighteen young (3 months) and 20 old (24 months) Brown Norway rats were exposed to estrous females of a similar age, according to standard procedures of Meyerson.[31] The sexual and social behaviors displayed in a 25 min encounter were registered. After the males had the opportunity to explore the cage (80 cm × 40 cm × 40 cm) the female entered the area. Ovariectomized females were treated with estradiol benzoate

(25 µg/kg) and 48 h later with progesterone (1 mg/rat) and used as stimulus objects 4 h after the injection.

Apart from the differences in sexual abilities, which will be described elsewhere, special attention was paid to the role of grooming behavior. Twenty-two behaviors, 11 for the male and 11 for the female, have been observed in this test. Since the behavior of the females is not discussed here we will mention only the behaviors of the males. They are: (1) approach towards the female, (2) mounting (clasping the flanks of the female and performing pelvic thrusts), (3) intromission (mount with a vigorous backward lunge followed by genital licking), (4) ejaculation (prolonged mount with intense clasping of the female followed by a slow dismount and subsequent genital licking), (5) exploration, (6) genital licking, (7) grooming, (8) aggression, (9) resting (lying down), (10) allogrooming, and (11) crawling over the female.

It must be emphasized that genital licking was clearly distinguished from grooming, since genital licking in this context is closely associated with sexual behaviors such as intromission, which is always followed by genital licking. The sociosexual behaviors of the male rat have been described by Meyerson and Hoglund.[31]

DATA ANALYSIS

In both experiments frequencies, durations, and in case of the sexual test, the latencies of each behavior were registered. The analysis of sequences was based on the assumption that the display of a given act is only determined by the immediate preceding act of the animal itself or in a social situation of its partner. For the observations and the analysis of the data previously described programs were used.[29] The sequential analysis was performed by comparing the frequencies of all combinations of two behavioral elements of both age classes. The combinations of elements were coded in so-called transition matrices, in which each cell represents the frequency of the concerning combination of elements. The next step was to sum the matrices per age class over all individuals. The two resulting total matrices were subjected to a chi-square followed by the calculation of the adjusted residual per cell. The adjusted residual per cell represents a parameter indicating the significance of occurrence of that combination of behavioral elements. All values exceeding ± 1.96 are supposed to have a higher frequency of occurrence than can be expected according to a random expectation.

ACTH-INDUCED EXCESSIVE GROOMING IN YOUNG AND AGED RATS

The scores of the different grooming elements for young and aged rats have been depicted in FIGURES 1A and 1B. The increase in the grooming scores and the level of ACTH-induced excessive grooming in aged rats is not different from the behavior displayed by young animals. Moreover, young and old rats show a similar distribution of elements as far as the frequencies are concerned. With respect to the durations of the defined elements only a difference is noticed in the amount of tail sniffing (FIGURE 1E). Thus, possible aged-related deficits in motor performance are not reflected in the durations and frequencies of different grooming behaviors. This similarity at first sight formed an excellent base for the sequential analysis, since apparently aged animals have the ability to display all components of grooming behavior.

In FIGURE 2 significant combinations of elements are indicated by arrows connecting the different behaviors. Each arrow can be read as either "is specifically followed by"

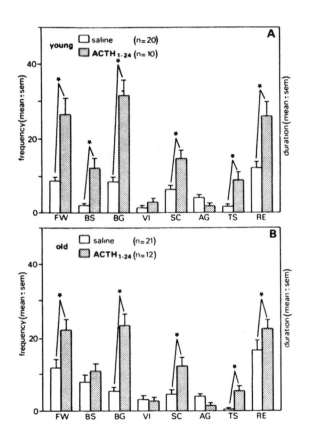

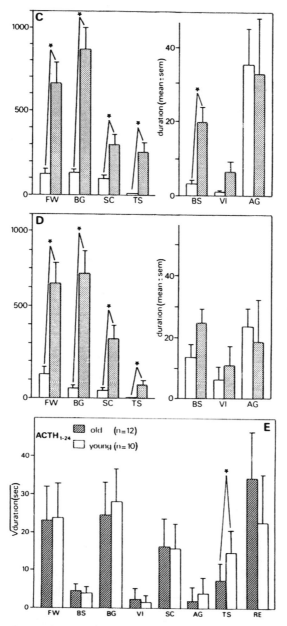

FIGURE 1. Excessive grooming in young and old rats. The frequencies and durations of face washing (*FW*), body shake (*BS*), body grooming (*BG*), vibration (*VI*), scratching (*SC*), anogenital grooming (*AG*), tail sniffing (*TS*), and other behaviors (*RE*) of young rats after saline and ACTH treatment are presented in, respectively, **A** and **C**. In **B** and **D** frequencies and durations of aged rats are shown. In **E** the durations of young versus old rats both treated with ACTH is compared; an *asterisk* indicates $p < 0.05$.

excitatory pathways

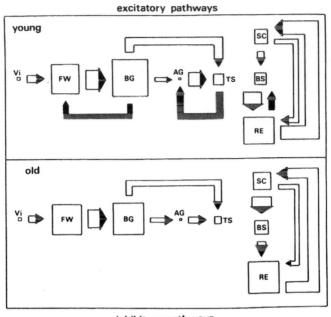

inhibitory pathways

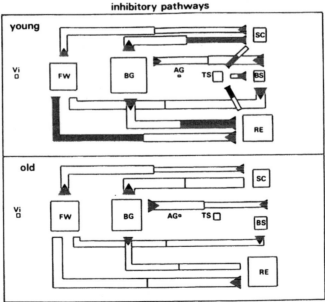

FIGURE 2. Differences in the structure of ACTH-induced grooming in young and old rats (indicated by *black arrows*). The area of each square represents the duration of the respective element (for the explanation of abbreviations see legend to FIG. 1). The width of the arrows is related to the transition frequency in excitatory pathways, in contrast to inhibitory pathways, in which arrowheads show the combination of elements not followed by each other. For further explanation see text.

(pointing outwards) or as "is certainly not followed by" (pointing inwards) representing, respectively, excitatory and inhibitory pathways. In the pathway diagrams of both the young and old animals, the main characteristic is the division into two distinct clusters of elements. One consists of the grooming elements head grooming, body grooming, anogenital grooming, and tail sniffing, and seems to be displayed in this cephalocaudal order. The order consists of scratching, body shake, and all other behavior, mostly exploration and lying down. Apparently, scratching does not fit in this cephalocaudal order of expression of acts. The special position of scratching in the expression of grooming has also been noticed in the ontogeny of grooming behavior.[32] There are more reasons to make a distinction between scratching and grooming. Lesions in the periaqueductal gray showed that grooming behavior was far more affected than scratching.[25] Pharmacological suppression of bombesin-induced scratching by naloxone showed that scratching was preferentially affected by the opioid antagonist.[28] Based on these findings it was hypothesized that there may be differences in the neural substrate of grooming and scratching.

COMPARISON BETWEEN YOUNG AND OLD

With respect to positive residuals — the excitatory pathways — the difference between young and old rats lies in the number of arrows and the difference in the width of one arrow. Young animals seem often to display the combination body grooming and face washing in both directions, while this reversed order is not present in aged animals, at least not more than can be expected according to a random model. In addition, the combination anogenital grooming and tail sniffing is displayed in the reversed order in the young animals only (see FIG. 2). The preference to perform tail sniffing after anogenital grooming is in young animals much greater than in aged animals (compare the widths of the two arrows).

Regarding the negative adjusted residuals, the inhibitory pathways of young animals again show more specific arrows (inhibitions). The combinations, which will certainly not be performed by young animals, are combinations of elements from either of the two clusters. In senescent animals, however, grooming of the body may be followed by scratching or resting, and tail sniffing may be followed by body shake. The aged animals may also shift from resting or scratching to body grooming or tail sniffing, whereas young animals always start grooming with vibration or face washing (notice the inhibitory connections to the other grooming elements). One must keep in mind that the absence of any arrow between two behavioral elements means that these behaviors may be "neighbors" in a string of elements according to a random expectation.

It is inferred from these findings that triplets of grooming elements occur more frequently in young animals than in old animals. TABLE 1 shows the relative frequency of some triplets of elements. Series of three grooming elements are more frequently seen in the young animals (see TABLE 1, a,c), whereas a combination of two elements of one cluster with one of the other cluster is seen more in the older animals (see TABLE 1,b,d,e,f). The cephalocaudal display seems more often interrupted in the older animals by behaviors such as body shake, rest, and scratching, probably resulting in a diminishment in duration of the terminating tail sniffing behavior. From these grooming patterns disrupted by other behaviors it is inferred that the nongrooming behaviors, displayed in the same test, are also more interrupted by grooming. To investigate the effect of a scattered grooming behavior on the sequential organization of nongrooming behaviors a setting was chosen in which other well-known behaviors and grooming are known to occur, namely a situation in which sociosexual behaviors will be performed.

TABLE 1. Triplets of Grooming Elements

	Young Rats	Old Rats
a. FW→BG→FW	53.8%	38.4%
b. FW→BG→SC	5.4%	14.0%
c. FW→BG→TS	13.4%	9.1%
d. BS→RE→FW	2.1%	13.5%
e. BS→RE→SC	15.6%	9.7%
f. RE→BG→RE	9.7%	15.1%

The relative frequencies of several triplets of grooming and nongrooming elements. FW: Face washing; BG: body grooming; SC: scratching; BS: body shake; TS: tail sniffing; RE: all other behavior.

The sexual behavior of rats has a known characteristic pattern and grooming bouts are seen after the sexual performance has taken place.

GROOMING BEHAVIOR IN SOCIOSEXUAL CONTEXT IN YOUNG AND AGED RATS

The sexual ability of young and old Brown Norway rats will be extensively described elsewhere. In brief, the time required to reach ejaculation is much shorter for young animals; moreover, the number of ejaculations reached in 25 min is lower in aged rats.[33] However, during the first 5 min no major differences between the two age classes could be assessed (see FIG. 3). In order to base the comparison of the sequential organization of behavior on similar distributions of frequencies, the analysis was performed on the first 5 min of the session. In the pathway diagram in FIGURE 4 the sequential structure of the male performance is depicted.

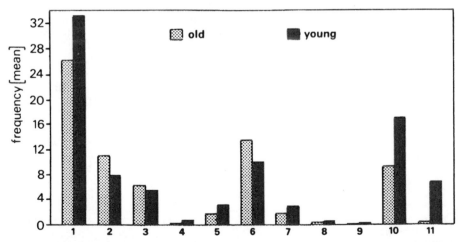

FIGURE 3. The frequencies of 11 behavioral elements measured during an encounter with an estrous female. Behaviors: *1*, approach; *2*, mounting; *3*, intromission; *5*, exploration; *6*, genital licking; *7*, self-grooming; *9*, resting, lying; *10*, grooming the female; *11*, crawling over. The only significant differences are found in exploratory behavior and crawling over.

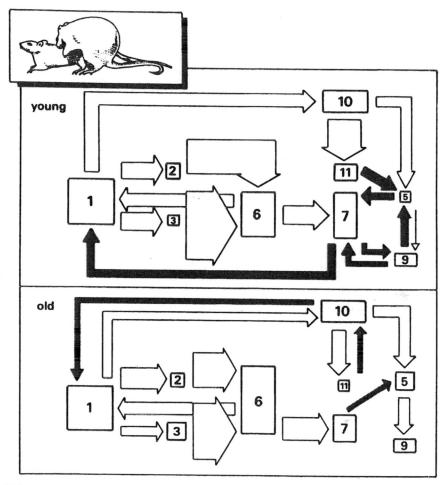

FIGURE 4. Sexual behavior of Brown Norway rats. The width of the arrows is related to the transition frequencies (see legend to FIG. 2). The area of the squares represents the duration of the respective behaviors. Differences between young and old rats are indicated by *black arrows*. See FIGURE 3 for an explanation of the numbers.

The well-known order of sexual behavior is clearly visible in both age classes: approach followed by mounting-intromission, and then genital licking and self-grooming. The black arrows, representing the differences between young and aged rats, are predominantly concentrated around the grooming behavior. Grooming behavior in young animals is specifically seen after genital licking, exploration, and resting. A lack of arrows towards grooming behavior can be noticed in old rats, whereas the frequency of grooming is similar in both age classes. This lack of arrows towards grooming (despite an equal frequency for the occurrence of grooming), and with an equal frequency for the transition from genital licking to grooming (see the width of the arrow from 6 to 7) means that the specific transitions from 5 to 7 and from 9 to 7 seen in young animals are divided over all behavioral elements in the older ones.

After self-grooming the young animals show a preference either for resting or for approaching the female. The aged rats show no specific preference at all for a certain following behavior, which means that any behavior may follow self-grooming. Thus, taken together a difference between the behavioral profiles of young and aged animals emerges. In young rats grooming has a more specific place in the profile, it is primarily performed after the sexual activity or exploratory activity, and it precedes either resting behavior or approaching the female again. In older animals it is also displayed after sexual activity, but more than in young animals it may precede or follow any other behavior.

A more random display of grooming in this sociosexual context leads to a sexual performance, which is more often interrupted by grooming behavior. Since no major differences are noticed either in the duration or in the frequency of grooming behavior, the grooming bouts must be of shorter duration.

Generally an encounter of a sexually experienced young male and a receptive female starts with copulatory behaviors and afterwards exploratory behaviors, grooming, and resting will be displayed. It can be seen in old animals that the fixed pattern of approach, mounting or intromission, and genital licking is interrupted by grooming. This phenomenon resembles the similar finding in grooming behavior, which was discussed above. In both behavioral systems a rather fixed pattern of elements was interrupted in aged individuals by a behavior to be performed at the end of a string of elements in young individuals. In addition, the interruption of grooming behavior by other behaviors — the rest category in the previous experiment — presumes the interruption of those "rest behaviors," in this case sociosexual behaviors, by grooming behavior, provided that the frequencies of the different behavioral elements are similar. In this sense the interruption of grooming is in agreement with the interruption of sociosexual behaviors by grooming.

The behavior studied in both settings demonstrates a drop in the number of specific combinations of behavioral elements in aging animals. The loss in sequential organization in grooming behavior is most clearly reflected in a decreased display of tail sniffing. Interestingly, tail sniffing is ontogenetically the last appearing element of the grooming repertoire. Such behavioral breakdown resembles the behavioral breakdown seen in learning paradigms (last learned, first forgotten). This so-called Jacksonian principle of behavioral degeneration is seemingly not confined to learning paradigms. The change in the sociosexual performance is probably the result of an interaction of many age-related deficits varying from degenerating peripheral organs, genitals, peripheral and central hormonal changes, testosterone, LH, LHRH, etc. But a loss in behavioral organization such as described for grooming behavior contributes to the above-described change in sexual performance. To summarize, in a variety of behavioral systems changes in aging animals may be characterized as Jacksonian degeneration. Ethological methods may provide appropriate tools to reveal such principles and to conceive possible consequences for other behavioral systems.

REFERENCES

1. CRAIK, F. I. M. & M. BYROL. 1982. Aging and cognitive deficits: The role of attentional responses. *In* Aging and Cognitive Processes, F. I. M. Craik & S. Trehub, Eds.: 191–211. Plenum. New York.
2. BOTHWINNICK, J. 1981. Neuropsychology of Aging. *In* Handbook of Clinical Neuropsychology. S. B. Filskov & T. J. Boll, Eds.: 135–171. Wiley. New York.
3. MOSCOVITCH, M. A. 1982. A neuropsychological approach to perception and memory in

normal and pathological aging. *In* Aging and Cognitive Processes. F. I. M. Craik & S. Trehub, Eds.: 55–78. Plenum. New York.

4. JOLLES, J. & J. HIJMAN. 1983. The neuropsychology of aging and dementia. Dev. Neurol. 7: 227–250.
5. FLICKER, CH., S. H. FERRIS, T. CROOK, R. T. BARTUS & B. REISBERG. 1985. Cognitive function in normal aging and early dementia. Adv. Appl. Neurol. Sci. 2: 2–17.
6. DOTRY, B. A. 1966. Age and avoidance conditioning in rats. J. Gerontol. 21: 287–290.
7. GOODRICK, C. L. 1972. Learning by mature young and aged Wistar albino rats as a function of test complexity. J Gerontol. 27: 353–357.
8. MEDIN, D. L., P. O'NEIL, P. SMELTZ & R. T. DAVIS. 1973. Age differences in retention of concurrent discrimination problems in monkeys. J. Gerontol. 28: 63–67.
9. OLTON, D. S. & R. J. SAMUELSON. 1976. Remembering of places passed: Spatial memory in rats. J. Exp. Psychol. 2: 97–116.
10. BARNES, C. A., L. NADEL & W. K. HONIG. 1980. Spatial memory deficits in senescent rats. Can. J. Psychol. 34: 29–39.
11. JENSEN, R. A., J. L. MARTINEZ, J. L. McGAUGH, R. B. MESSING & B. J. VASQUEZ. 1980. The psychobiology of aging. *In* The Aging Nervous System. G. J. Meletta & F. J. Pirozzlola, Eds.: 110–125. Praeger. New York.
12. WALLACE, J., E. E. KRAUTER & J. A. CAMPBELL. 1980. Animal models of declining memory in the aged rat: Short-term and spatial memory in the aged rat. J. Gerontol. 35: 355–363.
13. DEAN, R. L. & R. T. BARTUS. 1985. Animal models of geriatric cognitive dysfunctions: Evidence for an important cholinergic involvement. Adv. Appl. Neurol. Sci. 2: 269–282.
14. CAMPBELL, B. A., E. E. KRAUTER & J. E. WALLACE. 1980. Animal models of aging: Sensory-motor and cognitive function in the aged rat. *In* Psychology of Aging: Problems and Perspectives. D. G. Stein, Ed.: 201–226. Elsevier Biomedical Press. Amsterdam.
15. ALTMAN, J. & K. SUDAUSHAN. 1975. Postnatal development of locomotion in the laboratory rat. Anim. Behav. 13: 896–920.
16. MARSHALL, J. F. & N. BRUNIOS. 1979. Movement disorders of aged rats: Reversal by dopamine receptor stimulation. Science 206: 477–479.
17. GAGE, F. H., S. J. DUNNETT & A. BJÖRKLUND. 1984. Spatial learning and motor deficits in aged rats. J. Neurol. 5: 43–48.
18. CAMPBELL, B. A., C. B. SANANES & J. R. GADDY. 1985. Animal models of Jacksonian dissolution of memory in the aged. Adv. Appl. Neurol. Sci. 2: 283–291.
19. BOLLES, R. C. 1960. Grooming behavior in the rat. J. Comp. Physiol. 53: 306–310.
20. BORCHELT, P. L. 1980. Care of the body surface. *In* Comparative Psychology: An Evolutionary Analysis of Animal Behavior. R. M. Denny, Ed.: 362–384. Wiley. New York.
21. GISPEN, W. H. & R. L. ISAACSON. 1981. ACTH-induced excessive grooming in the rat. Pharmacol. Ther. 12: 209–246.
22. SPRUIJT, B. M. & W. H. GISPEN. 1983. ACTH and grooming behavior in the rat. *In* Hormones and Behavior in Higher Vertebrates. J. Balthazart, E. Pröve & R. Gilles, Eds.: 118–136. Springer-Verlag. Berlin.
23. SPRUIJT, B. M., A. R. COOLS, B. A. ELLENBROEK & W. H. GISPEN. 1986. Dopaminergic modulation of ACTH-induced grooming. Eur. J. Pharmacol. 120: 249–256.
24. SPRUIJT, B. M., A. R. COOLS, B. A. ELLENBROEK & W. H. GISPEN. 1986. The colliculus superior modulates ACTH-induced excessive grooming. Life Sci. 39: 461–470.
25. SPRUIJT, B. M., A. R. COOLS & W. H. GISPEN. 1986. The periaqueductal gray: A prerequisite for ACTH-induced excessive grooming. Behav. Brain Res. 20: 19–25.
26. ALOYO, V. J., B. M. SPRUIJT, H. ZWIERS & W. H. GISPEN. 1983. Peptide-induced excessive grooming in the rat: The role of opiate receptors. Peptides 4: 833–836.
27. VAN WIMERSMA GREIDANUS, TJ. B., D. K. DONKER, F. F. M. ZINNICQ BERGMAN, R. BEKENKAMP, C. MAIGRET & B. M. SPRUIJT. 1985. Comparison between excessive grooming induced by bombesin or by ACTH: The differential elements of grooming and development of tolerance. Peptides 6: 369–372.
28. VAN WIMERSMA GREIDANUS, TJ. B., D. K. DONKER, R. WALHOF, J. C. A. VAN GRAFHORST, N. DE VRIES, S. J. VAN SCHAIK, C. MAIGRET, B. M. SPRUIJT & D. L. COLBERN. 1985. The effects of neurotensin, naloxone and haloperidol on elements of excessive grooming behavior induced by bombesin. Peptides 6: 1179–1183.

29. Spruijt, B. M. & W. H. Gispen. 1984. Behavioral sequences as an easily quantifiable parameter in experimental studies. Physiol. Behav. **32:** 707–710.
30. Gispen, W. H., V. M. Wiegant, H. M. Greven & D. de Wied. 1975. The induction of excessive grooming in the rat by intraventricular application of peptides derived from ACTH. Structure-activity studies. Life Sci. **17:** 645–652.
31. Meyerson, B. J. & U. Hoglund. 1981. Exploratory and sociosexual behavior in the male laboratory rat: A methodological approach for the investigation of drug action. Acta Pharmacol. Toxicol. **48:** 168–180.
32. Richmond, G. & B. D. Sachs. 1980. Grooming in Norway rats: The development and adult expression of a complex motor pattern behavior. Behavior **75:** 82–95.
33. Spruijt, B. M., B. J. Meyerson, & U. Höglund. 1987. Aging and sexual behavior in the male rat. In preparation.

Analysis of Age-related Changes in Stress-induced Grooming in the Rat

Differential Behavioral Profile of Adaptation to Stress

HIDEKI KAMETANI[a]

Department of Psychology
Tokyo Metropolitan Institute of Gerontology
Tokyo-173, Japan

In the study of brain function and behavior, there is increasing interest in determining the biological significance and neural mechanisms of grooming behavior in animals.[1] One group of investigators, for example, is interested in grooming behavior as it relates to the development of motor control systems.[2,3] Since grooming behavior can be well defined and also consists of several sequential motor acts, it provides appropriate behavioral patterns for a possible model of motor control development. Ethologically, grooming behavior under a high arousal state has been described as a displacement activity that may be essential in restoration of homeostatic status.[4-6] In this context, grooming has been considered as an index of behavioral adaptation to a stressful situation. Furthermore, there is a meaningful correlation between some endogenous brain systems and grooming behavior. It has been well-established that intracerebroventricular administration of certain pro-opiomelanocortin peptide fragments elicit excessive grooming in rodents.[7,8] Opioid and dopaminergic systems are involved in the induction of this peptide-induced excessive grooming.[9,10]

Analysis of grooming provides several intriguing problems for aging research. First, grooming may provide appropriate patterns of behavior to use in investigating age-related changes in motor function. Second, aging is well characterized as the inability to maintain homeostatic status by environmental demands.[11] Consequently, behavioral adaptations of animals to a stressful environment, such as grooming, should alter as the animals grow older. In the study of the neural mechanisms of grooming behavior, significant correlations can be expected between age-related changes in peptidergic and dopaminergic systems[12,13] and behavioral changes such as grooming.

The principal question addressed in this article is whether there are qualitative or quantitative differences in grooming behavior exhibited by young and aged rats under several different observation conditions, and how such changes may relate to differences in reaction to stress and to fundamental neural mechanisms for maintaining homeostasis.

Despite the wide documentation of age-related decline of psychomotor behavior indicators such as motor reflexes, swimming performance, and exploratory behavior in an open field,[14] few experiments have been conducted to examine grooming behavior

---

[a] Present address: Gerontology Research Center, National Institute on Aging, National Institute of Health, Francis Scott Key Medical Center, Baltimore, MD 21224.

in aged animals. I will, therefore, review our previous work, which showed striking age differences in grooming activity under stressful situations in the rat.[15,16]

AGE DIFFERENCES IN NOVELTY-INDUCED GROOMING IN RATS

Placing an experimentally naive rat into a novel environment elicits excessive grooming activity.[17,18] This novelty-induced grooming is thought of as an after reaction to stress that lowers the high arousal state.[19] Our first experiment focused on age-related changes in novelty-induced grooming to clarify the pattern of behavioral adaptation pattern to a stressful novel situation in aged rats.[15]

Home Cage and Novelty-induced Grooming

In previous experiments, we have examined age differences in the grooming activity of Fischer-344 female rats under several different conditions, *i.e.,* a home cage, undisturbed situation; a home cage after brief handling treatment; and a small, novel testing chamber. Several characteristics of these grooming activities (duration, number of grooming bouts, composition of grooming elements, etc.) in young and aged rats were then analyzed.

The methods used in the experiment have been previously described.[15] Briefly, subjects were Fischer-344 virgin female rats, both young adult (6–8 months old) and aged (26–28 months old), obtained from the Tokyo Metropolitan Institute of Gerontology colony. Only those animals judged healthy by external appearance and with no gross pathology by necropsy were included. They were individually housed 1 week before the experiment in transplant home cages with food and water ad libitum. The colony room was maintained on a 12-h/12-h light-dark schedule (lights on between 0800 and 2000) and the room temperature was maintained at 23–24°C.

On the first day, baseline grooming behavior was observed in the home cage in the colony room under undisturbed conditions for 50 min in the light phase (1100–1400). Grooming elements analyzed included face washing, licking (licking the body fur, genital area, limbs, and tail), and scratching. Immediately after the observation, the rat was picked up, handled briefly (approximately 10 sec), and then returned to the cage. The grooming behavior after this handling treatment was then recorded for 50 min. On the second day, the home cage was transported from the colony room to the animal observation room. After 1 hour for the animal to adapt to the observation room, the rat was placed in a novel testing chamber, consisting of a wooden box (40 × 40 × 40 cm) painted gray. Grooming behavior, rearing, and locomotion in this novel chamber were recorded by a video camera for 50 min. This testing took place between 1100 and 1500 and was repeated for 7 consecutive days. All recorded data were analyzed by a microcomputer using ETHOL, a behavioral analysis program developed by Hendrie and Bennett.[20]

The mean total grooming duration in the 50-min observation for each age and test condition is indicated in FIG. 1. As shown in the figure, no age difference in baseline grooming activity was found between young and aged rats during daytime observation. Under this condition, there was a considerable range of individual differences with respect to the incidence of grooming, *i.e.,* some rats only slept throughout the observation period, whereas a few rats groomed for a long period. Although brief handling treatment elicited a slight increase in grooming in both young adult and aged rats, there were no differences between baseline grooming and handling-induced

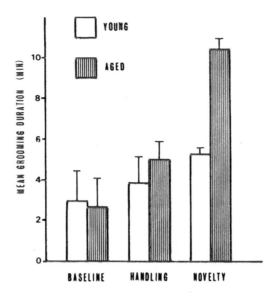

FIGURE 1. Mean grooming duration of young and aged rats (±SEM) observed for 50 min in an undisturbed home cage condition, a home cage after brief handling, and a novel testing chamber. (From Kametani *et al.* [15] Reprinted by permission from *Behavioral and Neural Biology.*)

grooming within each group or between the young and aged groups. In contrast, exposure to a novel chamber produced excessive grooming in both age groups, a result that is consistent with previous reports on novelty-induced grooming in the rat.[5,6,18] Additionally, we found that aged rats exhibited a markedly greater increase in novelty-induced grooming as compared to their younger counterparts. These results indicate that aged rats exhibit greater changes in grooming behavior after exposure to a novel situation than young adults without any differences in the amount of baseline and handling-induced grooming. More recently, Continella *et al.*[21] reported the same results using young and aged male Wistar rats. Thus, the tendency for an increase in novelty-induced grooming with advanced age seems a general phenomenon that crosses sex and strain boundaries in the rat.

For a more detailed analysis of the age difference in novelty-induced grooming in young and age rats, the number of grooming bouts, the mean length of each bout, the composition of grooming elements, and the mean duration of face washing and body licking were examined. These results are summarized in TABLE 1. Regarding the age-associated differences in these variables, aged rats showed a significantly increased number of grooming bouts. Thus, aged rats groomed themselves more frequently when placed in a novel situation. However, the mean duration of the grooming bouts did not differ between the two age groups. Moreover, the mean duration of face washing and body licking elements were also prolonged and the percentage of face washing was increased in aged rats as compared to the young rats.

The age differences in grooming response reported here may reflect different levels of stress or arousal experienced by the two age groups upon being placed in a stressful situation. Although there are discrepancies among data concerning age-related changes in emotionality, it is likely that aged rats experienced a higher level of stress compared

TABLE 1. Elements and Time Course of Novelty-induced Grooming in Young (n = 12) and Aged (n = 8) Rats ($\pm$ SEM)

Index	Day 1	Day 4	Day 7
Duration of each grooming bout(s)			
Young rats	23.6 ± 3.2	35.6 ± 3.7	27.6 ± 4.5
Aged rats	29.7 ± 2.5	51.9 ± 6.5	52.3 ± 4.0
Number of grooming bouts			
Young rats	14.5 ± 1.8	8.9 ± 1.2	10.1 ± 1.8
Aged rats	22.1 ± 2.5	15.1 ± 1.8	17.3 ± 2.2
Percentage of face washing			
Young rats	54.0 ± 2.3	42.7 ± 4.0	44.9 ± 4.1
Aged rats	60.1 ± 2.6	49.2 ± 2.6	45.1 ± 2.8
Duration of face washing element(s)			
Young rats	5.2 ± 0.6	5.2 ± 0.5	5.4 ± 0.5
Aged rats	8.1 ± 0.9	6.0 ± 0.6	5.4 ± 0.3
Duration of licking element(s)			
Young rats	5.5 ± 0.3	6.8 ± 0.6	6.0 ± 0.5
Aged rats	6.9 ± 0.6	7.5 ± 0.8	7.2 ± 0.7

to young adults in a novel situation, since their defecation score was significantly greater than young rats.

Time Course of Grooming by Repeated Exposure to Novelty

The aging animal is well characterized by its inability to adapt to and recover from stress. For example, Sapolsky *et al.* observed an elevation in basal corticosterone levels and an impaired capacity to adapt to and recover from restraint and cold stress in aged rats.[22] Also, deficits of habituation of exploratory behavior in an open field were reported by Brennan *et al.*[23]

Our observation of the time course of grooming during 1 week of daily testing is consistent with these findings. As shown in FIGURE 2, the total duration of grooming activity in young rats was constant while aged animals displayed progressively increased grooming throughout the 7 consecutive days. Specifically, grooming in young rats was slightly increased during the first half of the observation period but progressively decreased in consecutive daily trials during the second half. Overall, there was no significant change in the total grooming duration. This is consistent with the observation of Hannigan and Isaacson.[24] In aged rats, as in young animals, grooming activity during the first half of the observation period increased progressively. In contrast to young animals, however, grooming remained relatively constant during the second half of the daily trials. Thus, the total duration of grooming in aged rats increased, as shown in FIGURE 2. It has been reported that aged animals exhibit habituation deficits both within and between sessions.[23,25] The increase in grooming throughout the 7-day testing period may reflect such habituational deficits in aged rats. With repeated testing, the number of grooming bouts declined in both groups; only aged rats showed prolongation of each grooming bout. Furthermore, the percentage of face washing decreased

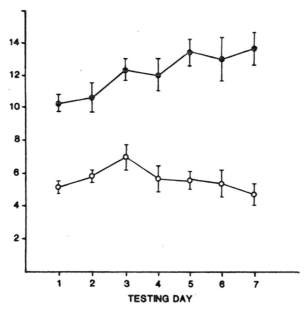

FIGURE 2. Mean grooming duration (±SEM) in a novel testing chamber for 7 consecutive days. *Open circles* indicate a young group, and *closed circles* indicate an aged group. (From Kametani et al[15] Reprinted by permission from *Behavioral and Neural Biology.*)

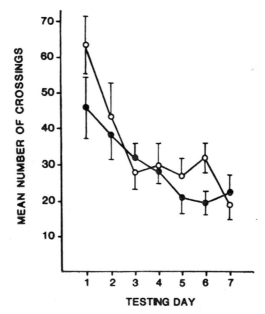

FIGURE 3. Mean number of line crossings (±SEM) in a novel testing chamber for 7 consecutive days. *Open circles* indicate a young group, and *closed circles* indicated an aged group.

in both groups. This compositional change in grooming is due to the fact that the first element of a grooming bout is usually face washing and during the initial testing days, grooming is easily interrupted at this first phase. Taken together, these findings suggest that the increase in grooming with repeated testing reflects a disability of aged rats to adapt to a novel circumstance.

Age Differences in Other Types of Behaviors

In a novel observation chamber, the rats exhibited typical behavioral patterns that reflect the novelty, *i.e.,* exploratory type behavior such as locomotion and rearing. These exploratory activities and resting periods were also recorded throughout the testing sessions. The floor of the chamber was divided by crossed lines into four equal sections with locomotion being defined as the number of lines crossed. The resting state was defined as immobility in the absence of any kind of apparent activity such as grooming, locomotion, rearing, or sniffing.

Results are shown in FIGURES 3–5. With repeated testing, rearing and locomotion decreased and resting periods increased in both age groups. An age difference was only found in rearing at day 6. Thus, there were no clear differences in exploratory behavior between young and aged rats comparable to the grooming activity differences described above. In view of these results, it might be suggested that grooming and exploratory behavior are indices of different aspects of an animal's response to

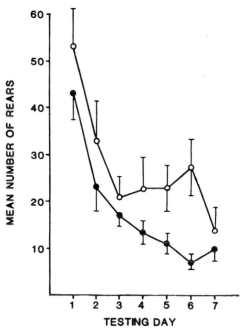

FIGURE 4. Mean number of rears (±SEM) in a novel testing chamber for 7 consecutive days. *Open circles* indicate a young group, and *closed circles* indicate an aged group.

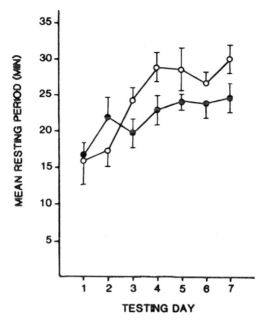

FIGURE 5. Mean duration of resting periods (±SEM) in a novel testing chamber for 7 consecutive days. *Open circles* indicate a young group, and *closed circles* indicate an aged group.

novelty and that grooming is a more sensitive measure for exploring age-related changes in behavioral adaptation to novelty stress.

AGE DIFFERENCES IN GROOMING BEHAVIOR INDUCED BY FOOT SHOCK

Other than novelty, mild stressors such as water immersion, unfamiliar sound, and exposure to a cold (5°C) environment seem capable of inducing excessive grooming in the rat.[1,26] In addition, Hannigan and Isaacson found that reexposure to an apparatus in which a brief electric shock had been previously applied can produce excessive grooming.[24] In our second experiment, the effect of foot shock was examined to confirm whether the increase in grooming of aged rats in a novel environment is a general behavioral reaction to stress in aged animals. Furthermore, the effect of administering the opiate antagonist naloxone was examined to explore the possible involvement of opiate-sensitive systems in excessive grooming in aged animals.[16]

Our experimental procedures were similar to those used by Hannigan and Isaacson.[24] In brief, eight naive female Fischer-344 young (8–10 months old) and eight aged (30–32 months old) rats were used. The housing conditions before the start of the experiment were the same as in the first experiment (described earlier). A rat was individually placed into a Plexiglas novel observation chamber (23 × 19 × 21 cm) for 50 min on each of three consecutive days to estimate the baseline novelty-induced grooming level. On the fourth day, the rats were injected with saline 10 min before testing and given four

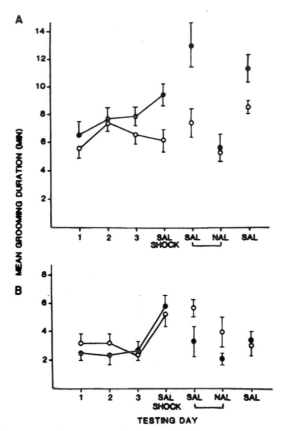

FIGURE 6. Mean (±SEM) duration of grooming over days for (**A**) first and (**B**) second contiguous 25-min observation period. Days 1–3 constitute baseline. On day 4, all animals received saline injection before the observation and footshock between periods. On day 5, animals received saline (*SAL*) or naloxone (*NAL*) injections (l mg/kg, i.p.). On day 6, all animals received saline injection. *Open circles* indicate young groups, and *closed circles* indicate aged groups. (From Kametani *et al.*[16])

scrambled foot shocks, each of 2-s duration (ISI = 13 s, 1 mA), halfway through the testing session. On the fifth day, testing was done after pretreatment with either saline or naloxone (1 mg/kg, i.p.). On the sixth day, all animals were again pretreated with saline before observation. Grooming and rearing were recorded using a video camera during these sessions and the duration of grooming and the frequency of rearing were analyzed using a microcomputer.

As shown in FIGURE 6, no significant age difference in novelty-induced grooming was found during the first three days, although the trends in the first half of the testing session were similar to those observed in the previous experiment. This result seems to be due to the fact that the aged animals used in this experiment were older than those in the first experiment. General activity level or responsiveness to novelty, per se, may decline in these very old rats. However, both young and aged rats showed ex-

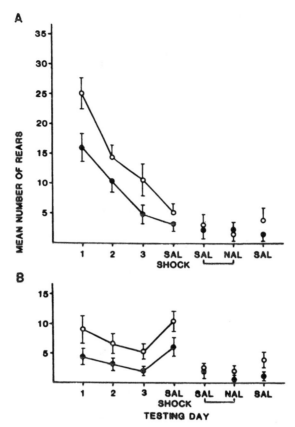

FIGURE 7. Mean (±SEM) number of rears over days for (A) first and (B) second contiguous 25-min observation period; for further details, see legend to FIGURE 6.

cessive grooming after foot shock presentation. On the day following the foot shock session, the saline-treated, aged group showed a significant increase in grooming activity during the first half of the testing session relative to the activity during the third baseline day. In contrast, the naloxone-treated, aged group showed no increase in grooming. Although young rats showed the same trends in grooming behavior as the aged rats on the fifth testing day, the changes were not significant. As shown in FIGURE 7, the frequency of rearing observed in young rats was slightly greater than in the aged group, but the frequency of rearing in both groups decreased equally during the first 3 baseline days. The effect of foot shock was observed only in the second half of the shock session. In contrast to the increased grooming behavior seen in the saline-treated rats as compared with the naloxone-treated animals, the frequency of rearing on the day following the foot shock session did not significantly differ from that on the third baseline day in either treatment group. These observations confirm the assumption that the grooming response to stress is greater in aged rats than in young rats and that opiate-sensitive systems are crucial in inducing the increase in stress-induced grooming.

IMPLICATIONS OF THE INCREASE IN STRESS-INDUCED GROOMING IN AGED RATS

Many experiments have shown motor disturbances in aged animals.[14,27-30] The rate of decline varies depending upon the motor function studied and the measuring task used. Simple movements, including postural adjustments and placing reactions, exhibit smaller changes than the performance of more complex sensorimotor tasks such as swimming in a water tank, traversal of an elevated platform, or rotorod performance.[31] Although grooming behavior also requires the coordination of several motor acts, the present results suggest less alteration in the elemental aspects of grooming with age than expected. The duration of face washing and body licking increase slightly in aged animals, which may be attributed to the slowness of each washing and licking stroke and/or differences in body size between young and aged rats. The most apparent age differences in grooming behavior under stress are quantitative in nature.

Various explanations for the mechanism of stress-induced excessive grooming have been proposed.[1] In one ethological explanation, grooming is thought of as a low priority activity that is usually inhibited by other competing behavioral tendencies. When such competing behavior is suppressed in a conflict situation or by habituation, grooming activity is easily disinhibited. According to this notion, the increase in novelty-induced grooming in aged rats can be explained by the fact that competing behavioral tendencies such as locomotion and rearing decline, albeit slightly in our experiments, with aging.

Several peripheral and central factors may be involved in the tendency toward increased stress-induced grooming with aging. Among the putative central factors involved in the induction of excessive grooming under stress, central ACTH has been implicated as a mediator of increased grooming.[1,5-7] The effect of ACTH is probably mediated in an agonistic fashion on opiate-sensitive systems, since the opiate antagonist naloxone counteracts both novelty- and ACTH-induced grooming.[9,24] Furthermore, the fact that haloperidol also reduces grooming suggests the involvement of the dopaminergic systems in the induction of stress-induced grooming.[9]

The increased stress-induced grooming in aged rats reported here seems to be due to alteration in the endogenous opioid system, since age differences can be obliterated by the administration of naloxone. Thus, one might postulate an increase in these peptidergic neural activities with aging. While a number of studies have observed reduced ACTH and β-endorphin content in several regions of the aging brain[32] and decreased opiate receptor concentration,[33] several studies have examined indirectly the influence of increasing age on opioid activity. For instance, many reports indicate that aged animals are less sensitive to painful stimuli.[33,34] In view of this apparent decline in pain sensitivity and the observation of reduced brain opioid receptor concentrations in several brain regions, a plausible explanation suggested by Simpkins is that the decline in opioid receptor number occurs secondary to hyperactivity of opioid neurons.[35] Such hypersensitivity of the opioid systems may be related to the increase in stress-induced grooming in aged rats.

An alternative explanation exists for the apparently contradictory findings that age-related increase in grooming occurs with a decrease in ACTH–β-endorphin content and opiate receptor levels. It has been argued that grooming is a response that occurs after the stress-inducing stimuli have been removed or their effect is decreased through habituation; it is likely that the central neurohormonal concomitants of stress are a waning.[36] Prolonging the duration of stimulation or of central neuropeptide activation acts to inhibit excessive grooming.[36] If older animals have less ACTH released or if the CRF-induced processes are less efficient, more grooming might be anticipated on the basis of a decrease in competition from the activating ACTH–β-endorphin systems.[15]

Thirdly, other neural systems besides ACTH–β-endorphin might be involved in stress-induced grooming. Drago *et al.* reported that i.c.v. injection of prolactin enhances novelty-induced but not home-cage grooming in rats.[37] Endogeneous hyperprolactinemia, as induced by pituitary homografts under the kidney capsule, is also accompanied by excessive grooming.[38,39] Also, plasma prolactin levels increase with aging due to a lack of dopaminergic inhibitory feedback.[40] Based on these findings, Continella *et al.* suggested that increased prolactin might explain the increase in stress-induced grooming behavior in aged rats.[21] Further experiments are necessary to determine the influential factors on increased stress-induced grooming in aged rats.

CONCLUSION

Our experiments demonstrated the following: (1) No differences were observed in grooming activities between young adult and aged rats under the nonstressful circumstance. (2) Aged rats showed significantly more grooming activities under stressful circumstances compared with young adult rats. (3) Opioid systems may be related to the increased grooming activities of aged rats under stressful circumstances. The age differences seen in stress-induced grooming support the idea that aging can be characterized as the inability to return to homeostasis after exposure to a stressful environment.

ACKNOWLEDGMENTS

The author gratefully thanks Drs. A. Sato, K. Inoue, K. Fukuzawa, and M. M. Dooley for their critical comments during the preparation of the manuscript. The author also thanks Sankyo Pharmaceutical Co., Ltd., Tokyo, Japan, for donation of the naloxone.

REFERENCES

1. GISPEN, W. H & R. L. ISAACSON. 1981. ACTH-induced excessive grooming in the rat. Pharmacol. Ther. **12:** 209–246.
2. FENTRESS, J. C. 1973. Development of grooming in mice with amputated forelimbs. Science **179:** 704–705.
3. FENTRESS, J. C. 1981. Order in ontogeny: Relational dynamics. *In* Behavioral Development. K. Immelmann, G. W. Barlow, L. Petrinovich & M. Main, Eds.: 338–371. Cambridge Univ. Press. Cambridge.
4. COHEN, J. A. & E. O. PRICE. 1979. Grooming in the Norway rat: Displacement activity or "boundary-shift"? Behav. Neural Biol. **26:** 177–188.
5. JOLLES J., J. ROMPA-BARENDREGT & W. H. GISPEN. 1979. ACTH-induced excessive grooming in the rat: The influence of environmental and motivational factors. Horm. Behav. **12:** 60–72.
6. JOLLES, J., J. ROMPA-BARENDREGT & W. H. GISPEN. 1979. Novelty and grooming behavior in the rat. Behav. Neural Biol. **25:** 563–572.
7. DUNN, A. J., E. J. GREEN & R. L. ISAACSON. 1979. Intracerebral adrenocorticotropic hormone mediates novelty-induced grooming in the rat. Science **203:** 281–283.
8. GISPEN, W. H., V. M. WIEGANT, A. F. BRADBURY, E. C. HULME, D. G. SMYTH, C. R. SNELL & D. DE WIED. 1976. Induction of excessive grooming in the rat by fragments of lipotropin. Nature **264:** 794–795.
9. GREEN, E. J., R. L. ISAACSON, A. J. DUNN & T. H. LANTHORN. 1979. Naloxone and

haloperidol reduce grooming occurring as an after effect of novelty. Behav. Neural Biol. **27**: 546–551.

10. WIEGANT, V. M., A. R. COOLS & W. H. GISPEN. 1977. ACTH-induced excessive grooming involves brain dopamine. Eur. J. Pharmacol. **41**: 343–345.

11. SELYE H. & B. TUCHWEBER. 1976. Stress in relation to aging and disease. In Hypothalamus, Pituitary and Aging. A. V. Everitt & J. A. Burgess, Eds.: 553–569. Thomas. New York.

12. DE WIED, D. & J. M. VAN REE. 1982. Neuropeptides, mental performance and aging. Life Sci. **31**: 709–719.

13. FINCH, C. E. 1973. Catecholamine metabolism in the brains of aging male mice. Brain Res. **52**: 261–276.

14. ELIAS, M. F. & P. K. ELIAS. 1977. Motivation and activity. In Handbook of Psychology of Aging. J. E. Birren & K. W. Shaie, Eds.: 357–383. Van Nostrand-Reinhold. New York.

15. KAMETANI, H., H. OSADA & K. INOUE. 1984. The increase in novelty-induced grooming in aged rats: A preliminary observation. Behav. Neural Biol. **42**: 73–80.

16. KAMETANI, H., H. OSADA & K. INOUE. 1985. Stress-induced grooming in aged rats. Jpn. J. Psychopharmacol. **5**: 169–170.

17. BINDRA, D & N. SPINNER. 1958. Responses to different degrees of novelty: The incidence of various activities. J. Exp. Anal. Behav. **1**: 341–350.

18. COLBERN, D. L., R. L. ISAACSON, E. J. GREEN & W. H. GISPEN. 1978. Repeated intraventricular injections of $ACTH_{1-24}$: The effect of home or novel environments on excessive grooming. Behav. Neural Biol. **23**: 381–387.

19. ISAACSON, R. L., J. H. HANNIGAN, J. H. BRAKKEE & W. H. GISPEN. 1983. The time course of excessive grooming after neuropeptide administration. Brain Res. Bull. **11**: 289–293.

20. HENDRIE, C. A. & S. BENNETT. 1983. A microcomputer technique for the detailed analysis of animal behavior. Physiol. Behav. **30**: 233–235.

21. CONTINELLA, G., F. DRAGO, S. AUDITORE & U. SCAPAGNINI. 1985. Quantitative alteration of grooming behavior in aged male rats. Physiol. Behav. **35**: 839–841.

22. SAPOLSKY, R. M., L. C. KREY & B. S. McEWEN. 1983. The adrenocortical stress response in the aged male rat: Impairment of recovery from stress. Exp. Gerontol. **18**: 55–64.

23. BRENNAN, M. J., A. DALLOB & E. FRIEDMAN. 1981. Involvement of hippocampal serotonergic activity in age-related changes in exploratory behavior. Neurobiol. Aging **2**: 199–203.

24. HANNIGAN, J. H. & R. L. ISAACSON. 1981. Conditioned excessive grooming in the rat after footshock: Effect of naloxone and situational cues. Behav. Neural Biol. **33**: 280–292.

25. BRENNAN, M. J. & D. QUARTERMAIN. 1980. Age-related differences in within-session habituation: The effect of stimulus complexity. Gerontologist **20**: 71.

26. COLBERN, D. L., R. L. ISAACSON & J. H. HANNIGAN. 1981. Water immersion, excessive grooming, and paper shredding in the rat. Behav. Neural Biol. **32**: 428–437.

27. GOODRICK, C. L. 1971. Variables affecting free exploration responses of male and female Wistar rats as a function of age. Dev. Psychol. **4**: 440–446.

28. KRAUTER, E. E., J. E. WALLACE & B. A. CAMPBELL. 1981. Sensory-motor function in the aging rat. Behav. Neural Biol. **31**: 367–392.

29. MACHIDA, T., K. MATSUMOTO, S. KOBAYASHI & T. NOUMURA. 1982. Age-associated changes in the open field behavior of male mice of the ICR/SLC strain. Zoological Magazine **91**: 263–271.

30. MARSHALL, J. F. 1982. Sensorimotor disturbances in the aging rodents. J. Gerontol. **37**: 548–554.

31. WALLACE, J. E., E. E. KRAUTER & B. A. CAMPBELL. 1980. Motor and reflexive behavior in the aging rat. J. Gerontol. **35**: 364–370.

32. GAMBERT, S. R., T. L. GARTHWAITE, C. H. PONTZER & T. C. HAGAN. 1980. Age-related changes in central nervous system β-endorphin and ACTH. Neuroendocrinology **3**: 252–255.

33. HESS, G. D., J. A. JOSEPH & G. S. ROTH. 1981. Effect of age on sensitivity to pain and brain opiate receptor. Neurobiol. Aging **2**: 49–55.

34. NICAK, A. 1971. Changes of sensitivity to pain in relation to postnatal development in rats. Exp. Gerontol. **6**: 111–114.

35. SIMPKINS, J. W. 1983. Changes in hypothalamic hypophysiotropic hormones and neurotrans-

mitters during aging. *In* Neuroendocrinology of Aging. J. Meites, Ed.: 41–59. Plenum. New York.

36. ALLAN A. M. & R. L. ISAACSON. 1985. Ethanol-induced grooming in mice selectively bred for differential sensitivity to ethanol. Behav. Neural Biol. **44:** 386–392.

37. DRAGO, F., P. L. CANONICO, R. BITETTI & U. SCAPAGNINI. 1980. Systemic and intraventricular prolactin induces excessive grooming. Eur. J. Pharmacol. **65:** 457–458.

38. DRAGO, F. & B. BOHUS. 1981. Hyperprolactinaemia-induced excessive grooming in the rat: Time-course and element analysis. Behav. Neural Biol. **33:** 117–122.

39. DRAGO F., B. BOHUS, R. BITTETI, U. SCAPAGNINI, J. M. VAN REE & D. DE WIED. 1986. Intracerebroventricular injection of antiprolactin serum supresses excessive grooming of pituitary homografted rats. Behav. Neural Biol. **46:** 99–105.

40. GULDELSKY, G. A., D. D. NANSEL & J. C. PORTER. 1981. Dopaminergic control of prolactin secretion in the aging male rat. Brain Res. **204:** 446–450.

Cross-Species Comparison of the ACTH-induced Behavioral Syndrome

A. BERTOLINI, ROSANNA POGGIOLI, AND A. VALERIA VERGONI

Institute of Pharmacology
University of Modena
I-41100 Modena, Italy

THE ACTH-INDUCED BEHAVIORAL SYNDROME

The first description of the behavioral syndrome induced by the intracerebroventricular (i.c.v.) injection of ACTH-MSH peptides (melanopeptides[1]) was given by Ferrari[2] in 1955, and was communicated at another New York Academy of Sciences conference 24 years ago by Ferrari, Gessa and Vargiu.[3] They reported:

> Most of the dogs given ACTH intracisternally exhibit apparently normal behavior for about half an hour. Thereafter . . . the dogs are drowsy, yawn frequently and after about one hour they start to stretch in the way they usually do when they awake from physiological sleep. The intervals between successive stretching acts become shorter and shorter until a stretching act begins immediately after the preceding one. Despite this peculiar behavior, the dogs seem to remain in contact with the environment, as they are responsive to calls and perform normal activities without fear or aggressiveness. . . . Rabbits, cats, and rats given ACTH and MSH also exhibit stretching crises qualitatively similar to those of dogs. In rabbits and rats the stretching syndrome is preceded by increased grooming activity and scratching. Rabbits yawn more frequently than dogs. In cats, intracisternal injection of ACTH caused a marked drowsiness interrupted by stretching movements. However, doses of ACTH up to 0.02 I.U./kg injected intrarachnoidally in men caused nausea and vomiting but not stretching.

In those and in the following years, the behavioral effects of melanopeptides have been increasingly and widely studied in different laboratories by injecting these peptides either intravenously, into the cerebrospinal fluid (CSF) or into discrete brain areas.

At present, we can say that ACTH and its behaviorally active fragments induce the following effects: excessive grooming,[1-5] stretching and yawning (stretching-yawning syndrome, SYS),[1-3, 6-10] penile erections and ejaculations in adult males,[6-11] increased lordosis in females,[11-15] whole-body shakes,[5] anorexia,[16,17] increased motivation and attention,[10,18,19] facilitated avoidance acquisition and retention,[19,28] delayed extinction of rewarded behavior,[19,29-31] facilitated sexual motivation,[18,19,32,33] and hyperalgesia.[34-36] Moreover, the injection of ACTH in certain well-defined areas of the central nervous system induces quite peculiar syndromes. For example, injection into the periaqueductal gray matter of naive rats results in a dose-dependent, so-called "explosive motor behavior" characterized by fearful hyperreactivity and symptoms like those of the opiate abstinence syndrome (hyperreactivity to previously neutral auditory and visual stimuli, jumping, teeth chattering, squealing on touch, etc.),[37] while unilateral injection into the locus ceruleus of rats results in postural asymmetry and locomotor impairment, the animal showing a leaning posture ipsilateral to the microinjection side, a disruption in normal locomotion, and other typical movement disorders[38] (TABLE 1). Finally, the i.c.v. injection of ACTH or MSH in morphine-dependent rats precipitates a quasi-opiate abstinence syndrome.[39]

Some behavioral effects of melanopeptides (grooming, stretching and yawning, penile erection) can be observed only after direct intracranial injection, while other effects (motivation, attention, learning, and memory) can also be obtained by systemic administration.

As already noticed by Ferrari, Gessa and Vargiu,[3] the behavioral syndrome induced by melanopeptides has different features in the different species: rabbits and rats exhibit both stretching and yawning acts and penile erections, monkeys mostly yawning, whereas in cats and dogs stretching prevails.[9] Also, quite recent studies have shown that, in the same species, the ACTH-induced behavioral syndrome has different features at different ages.[40]

We report here the results of a cross-species comparison study where the following components of the ACTH-induced behavioral syndrome were evaluated in mice, rats, and rabbits: grooming, yawning, stretching, penile erection, whole body shake, food intake. The effect of ACTH on pain sensitivity in rats and mice was also studied.

EXPERIMENTAL PROCEDURES

Animals

Adult male Wistar rats (200–240 g), Swiss albino mice (24–30 g), and rabbits of various strains (2500–3500 g) were used. They were housed in temperature- (22 ± 0.2°C) and humidity- (60%) controlled colony rooms, with a natural light-dark cycle and, except in the case of food-intake studies (see below), free access to food and water. At least 1 week prior to the behavioral test, stainless steel (rabbits) or polyethlyene (rats and mice) cannulae were stereotaxically implanted[41,42] into a brain lateral ventricle under pentobarbital anesthesia. Cannulae placements were routinely verified after autopsy, and results obtained from improperly implanted animals were discarded. In

TABLE 1. Behavioral and Behavior-affecting Symptoms Induced by ACTH-MSH Peptides

excessive grooming
stretching and yawning
penile erection and ejaculation
increased lordosis in females
facilitated sexual motivation
increased motivation and attention
facilitated avoidance acquisition and retention
delayed extinction of rewarded behavior
whole-body shakes
reduction of food intake
reduced threshold to pain and other stimuli
quasi-opiate abstinence syndrome (following injection into the periaqueductal gray matter)
postural asymmetry and locomotor impairment (following injection into the locus ceruleus)

all experiments naive animals were used, and tests were performed in the light phase of the cycle, between 0900 and 1400 hours.

Drug and Treatment

All these studies were performed using the synthetic tetracosapeptide, $ACTH_{1-24}$. It was freshly dissolved in saline immediately before treatment, and i.c.v. infused into unrestrained, conscious animals, in a volume of 3 (mice), 5 (rats), or 10 (rabbits) microliters, at the rate of 1 microliter per 10 seconds. Control animals received the same volume of saline by the same route and at the same rate of infusion.

Behavioral Testing

Apparatus

Grooming, whole-body shakes, stretching, yawning, and penile erections were scored by placing the animals, immediately after treatment, into individual glass boxes with a mirror underneath in a naturally lit sound-proof observation room. The observer was unaware of the treatment.

Grooming

The behavioral procedure was essentially the same as described by Gispen *et al.*[4,5] Animals were scored every 15 s for a period of 50 min (rats and mice) or of 110 min (rabbits), starting 10 min after i.c.v. treatment. The observer noted whether or not one of the behavioral components of the grooming act was displayed. If so, a positive grooming score was given for the time interval, so that a maximum of 200 (rats and mice) or 440 (rabbits) positive grooming scores could be obtained. In order to facilitate comparison, data were transformed into the percentages of the total number of 15-s-grooming scores possible.

Stretching, Yawning, and Penile Erections

Animals were observed continuously for 2 h starting 10 min after treatment, and each stretch, yawn, and penile erection was scored. Only those penile erections were scored that took place spontaneously, not those preceded by genital grooming. The penile erection elicited by melanopeptides is accompanied by pelvic thrusting as during copulation, and usually ends with an ejaculation; during or at the end of the penile erection episode, the animal usually bends down and sucks its penis.

Food Intake

Animals were treated after a 24-h food-deprivation period (water ad libitum), and testing was done in their individual plastic cages without wood shaving. Immediately after ACTH or saline injection, preweighed food pellets were placed in the cages. The amount of food eaten was measured every hour during the 4 hours following treatment, careful correction being made for spillage.

Whole-Body Shakes

These were scored for 1 h, starting immediately after treatment, at room temperature (22±0.2°C).

Pain Sensitivity

Pain threshold was measured by the hot plate test, with a platform temperature of 45 ± 0.2°C and a cutoff time of 60 seconds. The time elapsing until the animals licked their paws was recorded before and 10 min after ACTH or saline injection. After-treatment values were transformed into the percentage of the respective pretreatment nociception thresholds.

RESULTS

Grooming

Rats display the highest grooming response to i.c.v. injection of ACTH, and rabbits the lowest (FIG.. 1). The effect is dose-dependent up to a maximal effective dose. In our experimental conditions the maximal effective doses were 0.1 µg per animal in mice, 2 µg in rats, and 10 µg in rabbits; and their effect, expressed as a percentage

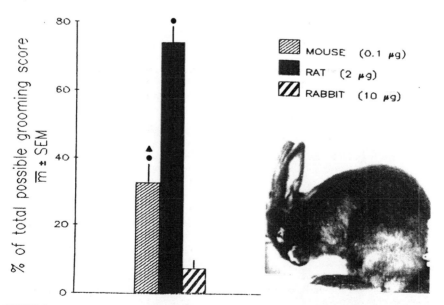

FIGURE 1. Total amount of grooming scored in mice, rats, and rabbits following the intracerebroventricular (i.c.v.) injection of the maximal active dose of $ACTH_{1-24}$. At least 10 animals per dose were observed; ● = < 0.01 vs. rabbits, and ▲ = < 0.01 vs. rats (overall ANOVA followed by multiple comparison test).

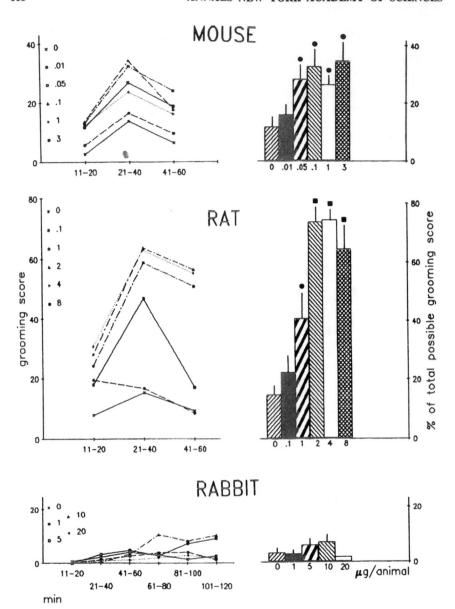

FIGURE 2. Excessive grooming induced by the i.c.v. injection of ACTH$_{1-24}$. At least 10 animals per dose were observed. *Left*, Time-course of the grooming behavior after ACTH or saline injection. *Right*, Cumulative grooming scored with the different doses used. Values reported are means ± SEM; ● = < 0.01, and ■ = < 0.001 vs. saline (overall ANOVA followed by Dunnett's test).

of the total possible grooming score, was 32.86 ± 5.22 in mice, 74.19 ± 5.09 in rats, and 7.4 ± 2.22 in rabbits. Rats also have the highest basal grooming behavior ($15.7 \pm 2.76\%$ in saline-treated animals, FIG. 2). While ACTH-induced excessive grooming in rats and mice is maximally displayed during the first hour after treatment and all components are present (face washing, body grooming, scratching, paw and tail licking, genital grooming, shakes), in rabbits it persists, albeit very sporadically, for at least 2 hours, and consists almost exclusively of face washing and paw licking, a facial grooming episode often immediately preceding a penile erection.

Yawning and Stretching

These two components of the ACTH-induced behavioral picture are usually associated and constitute a "syndrome within the syndrome", the so-called stretching and yawning syndrome (SYS).[3] In monkeys, yawning prevails, while in cats and dogs stretching is far more prevalent than yawning.[9] In rodents, both stretching and yawning are fairly well represented; however, stretching is usually three to five times more prevalent than yawning. The SYS is most marked in mice, while rats and rabbits display a response that is roughly 50% that of mice (FIGS. 3 and 4). In our experimental conditions, the maximal effective doses were 1 µg per animal in mice, 4 µg in rats, and 10 µg in rabbits. These doses caused the following acts of yawning and stretching, respectively, during the 2-h observation period: in mice, 17.33 ± 4.77 and 62.67 ± 15.08; in rats, 8.70 ± 1.74 and 33 ± 4.77; in rabbits, 7.14 ± 2.38 and 24.0 ± 6.20 (FIGS. 5 and 6).

Yawning was virtually absent in saline-treated mice and rabbits, while saline-treated rats yawned occasionally (1.42 ± 0.58 acts of yawning in 2 hours).

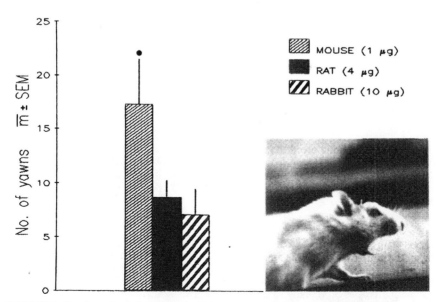

FIGURE 3. Total number of yawns scored in mice, rats, and rabbits during 2 h following the i.c.v. injection of the maximal active dose of $ACTH_{1-24}$. At least 10 animals per dose were observed; ● = < 0.05 vs. rabbits (overall ANOVA followed by multiple comparison test).

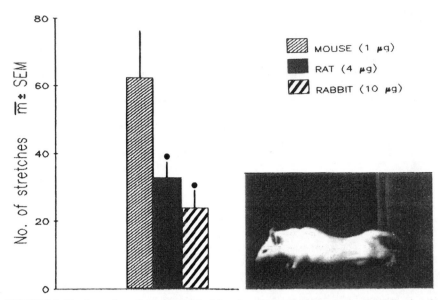

FIGURE 4. Total number of stretches scored in mice, rats and rabbits during 2 h following the i.c.v. injection of the maximal active dose of $ACTH_{1-24}$. At least 10 animals per dose were observed; ● = < 0.05 vs. mice (overall ANOVA followed by multiple comparison test).

While some spontaneous stretching is usually observed in all these animals, particularly in rats, acts of stretching increase enormously as a result of ACTH treatment. Although sometimes very frequent (almost one stretching per minute for 2 hours, in mice) they do not apparently trouble the animal. Yet, they look somewhat pleasureable, and the animal often stretches loosely lying on one side or on its back.

Penile Erections

Spontaneous penile erections can be observed in dogs, but are extremely rare in mice, rats, and rabbits. Occasionally a penile erection may occur during genital grooming, especially in rats and mice. But penile erections induced by melanopeptides are quite peculiar. They start abruptly: the animal looks as if taken by surprise, stands up, and makes some pelvic thrusting with the penis in full erection. Rats and mice then bend down and suck the penis, and continue pelvic thrusting until the emission of seminal plug. An intense genital grooming ensues, just as after copulation. Penile erection is longer-lasting in rabbits: there are bouts of pelvic thrusting interspersed with pauses, during the course of which the animal may yawn. In the rabbit penile erection is often immediately preceded by sudden face grooming. Rabbits are the most sensitive to this genital component of the melanopeptide-induced behavioral syndrome (FIG. 7), the maximal effective dose being 10 µg per animal; with this dose, up to 10–12 penile erections terminating with ejaculation can be scored in 2 hours, the mean in our present experiments being 7.71 ± 1.27. Rats are less sensitive: the maximal effective dose (4 µg) induced a mean of 3.0 ± 1.50 penile erections in 2 hours. Mice are almost insensi-

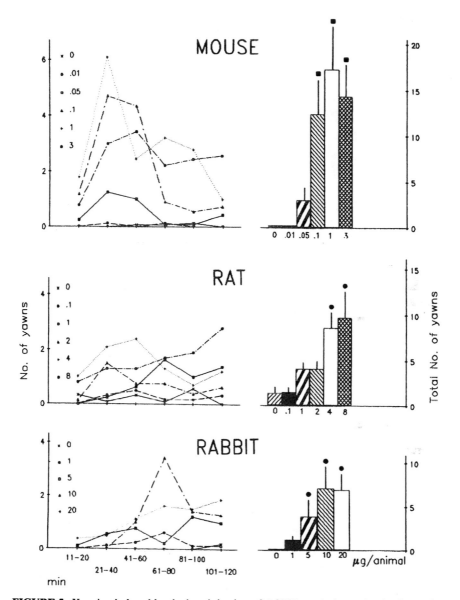

FIGURE 5. Yawning induced by the i.c.v. injection of ACTH$_{1-24}$. At least 10 animals per dose were observed. *Left*, Time course of the yawning behavior after ACTH or saline injection. *Right*, Cumulative number of yawns scored with the different doses used. Values reported are means ± SEM; ● = < 0.05, and ■ = < 0.01 vs. saline (overall ANOVA followed by Dunnett's test).

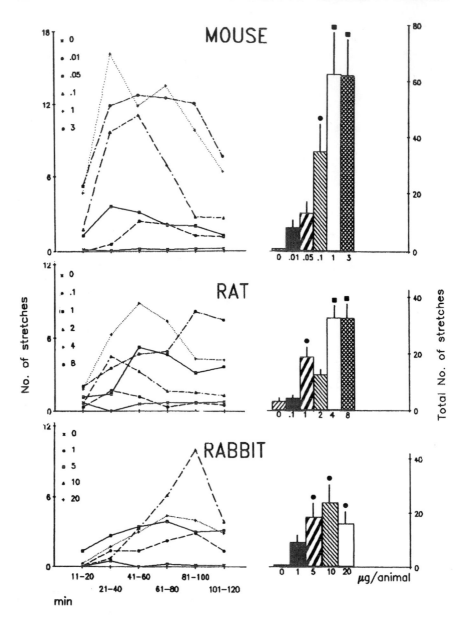

FIGURE 6. Stretching induced by the i.c.v. injection of ACTH$_{1-24}$. At least 10 animals per dose were observed. *Left*, Time course of the stretching behavior after ACTH or saline injection. *Right*, cumulative number of stretches scored with the different doses used. Values reported are means ± SEM; ● = < 0.01, and ■ = < 0.001 vs. saline (overall ANOVA followed by Dunnett's test).

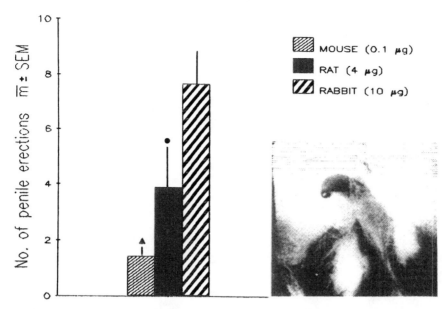

FIGURE 7. Total number of penile erections scored in mice, rats, and rabbits during 2 h following the i.c.v. injection of the maximal active dose of ACTH$_{1-24}$. At least 10 animals per dose were observed; ▲ = < 0.01, and ● = < 0.05 vs. rabbits (overall ANOVA followed by multiple comparison test).

tive: the maximal effective dose (0.1 μg) induced 1.45 ± 0.34 penile erections in 2 hours (Fig. 8).

Food Intake

Melanopeptides inhibit food intake, whether in the case of starving rats or in that of rats with food continuously available, and the effect lasts several hours.[16,17]

In the present experiments, the inhibitory activity of ACTH$_{1-24}$ on feeding was confirmed in all three species studied, the maximum anorectic effect during the 4 h of observation being obtained with a dose of 0.05 μg per animal in mice, 4 μg in rats, and 10 μg in rabbits (Fig. 9).

The duration of this anorectic effect far exceeds that of the other behavioral effects of ACTH. It thus seems unlikely that the possible inhibitory activity on feeding of the other behaviors (particularly grooming and stretching) plays an important role in the overall reduction of food intake. The fact that the rabbit is most markedly affected by the anorectic activity of ACTH, while being the least susceptible to the grooming-stimulant effect of the peptide, serves to support this observation.

Whole Body Shakes

These are often observed in the course of a grooming bout, and are usually consid-

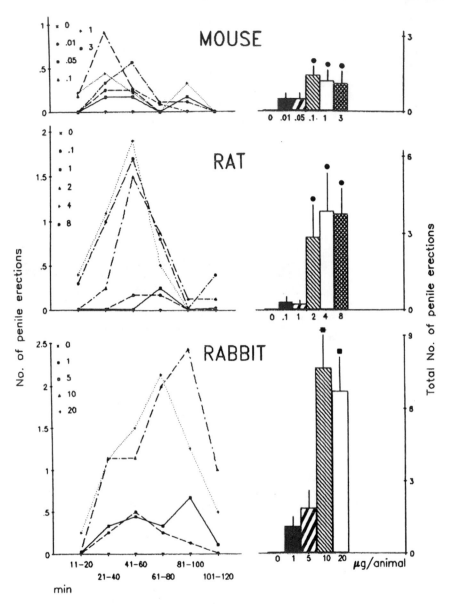

FIGURE 8. Penile erection induced by the i.c.v. injection of ACTH$_{1-24}$. At least 10 animals per dose were observed. *Left*, Time course of penile erections after ACTH or saline injection. *Right*, Cumulative number of penile erections scored with the different doses used. Values reported are means ± SEM; ● = < 0.05, and ■ = < 0.01 vs. saline (overall ANOVA followed by Dunnett's test).

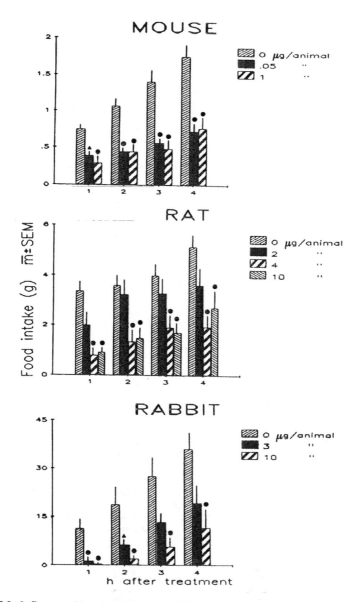

FIGURE 9. Influence of i.c.v. injection of $ACTH_{1-24}$ on food intake in mice, rats, and rabbits after 24-h fasting. At least 10 animals per dose were observed; ▲ = < 0.05, and ● = < 0.01 vs. saline (overall ANOVA followed by Dunnett's test).

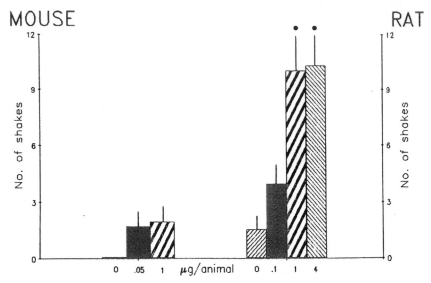

FIGURE 10. Whole-body shakes scored during 1 h following the i.c.v. injection of ACTH$_{1-24}$. At least 10 animals per dose were observed. Values are means ± SEM; ● = < 0.01 vs. saline (overall ANOVA followed by Dunnett's test).

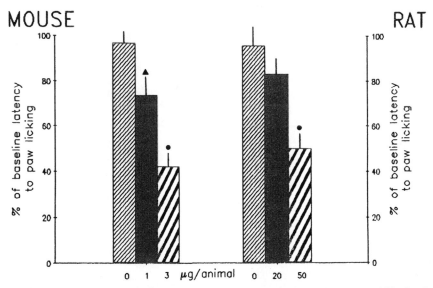

FIGURE 11. Influence of i.c.v. injection of ACTH$_{1-24}$ on pain threshold, measured 10 min after treatment. Hot plate test (45±0.2°C). At least 10 animals per dose were observed. Values are means ± SEM; ▲ = < 0.05, and ● = < 0.01 vs. saline (overall ANOVA followed by Dunnett's test).

ered as a component of grooming behavior. However, after ACTH-treatment, they also occur in isolation, so we kept them separate in this study.

They are completely absent in rabbits. In rats, we scored more than 10 whole-body shakes (10.60 ± 1.96) in 1 hour after the maximal active dose of $ACTH_{1-24}$ (1 µg/animal). In mice, the score was 2.0 ± 0.82 after the maximal active dose of 1 µg per animal (FIG. 10).

Pain Sensitivity

Adrenalectomy-induced hyperalgesia is correlated to the increased production of ACTH and prevented by hypophysectomy[43]; accordingly, hyperalgesia can be induced in rats by the i.c.v. injection of ACTH.[34,35] In the present experiments, an i.c.v. dose of 20 µg per rat of $ACTH_{1-24}$ lowered the threshold of thermal pain by 16.75%; increasing the dose to 50 µg per rat caused a 49.69% pain threshold reduction.

In mice, a dose of 1 µg per animal lowered the pain threshold by 26.31%, while a dose of 3 µg caused a reduction of 57.69% (FIG. 11). ACTH-induced hyperalgesia was highest 10 min after treatment, the pain threshold gradually returning to normal about 80 min after ACTH injection.

Of course pain sensitivity cannot be considered a behavioral effect in the strict sense of the word. However, a lowered detection threshold for painful stimuli (as well as for auditory, gustatory, and olfactory stimuli, as has been described in humans with primary adrenocortical insufficiency,[44-46] and in the adrenalectomized rat[47]) may contribute in some way to certain more strictly behavioral effects of melanopeptides.

CONCLUSIONS

The present experiments confirm with quantitative data the earlier impression[3,9] that the ACTH-induced behavioral syndrome differs qualitatively from one species to another.

In each of the three species studied, different symptoms prevail: excessive grooming in rats, penile erections and anorexia in rabbits, stretching and yawning (SYS) in mice. SYS and food intake reduction must be considered the most general behavioral signs induced by ACTH; they are clearly evident in every species examined, while other signs are almost absent in certain species (e.g., penile erections in the mouse, grooming in the rabbit).

Since it is logical to presume that the mechanism of action of the behavioral effect of ACTH is the same in every species, it is possible that the above differences depend on the species-linked prevalence of certain behavioral patterns instead of others, even though the nervous centers, neurotransmitters, and neuromodulators involved are conceivably the same in all species concerned.

Finally, the behavioral components of the ACTH-induced syndrome have different latencies, durations, and patterns.

Some appear almost immediately after injection and are of relatively short duration (excessive grooming), others have an indeterminate latency, greater duration, and a characteristic cyclic recurrence, repeatedly occurring in waves even dozens of times after one single injection of the peptide, for hours and sometimes for days[3] (yawns, stretches, penile erections), with asymptomatic intervals.

ACKNOWLEDGMENT

This work is dedicated to Professor William Ferrari on the occasion of his last year of teaching pharmacology.

REFERENCES

1. FERRARI, W. 1986. *In* Central Actions of ACTH and Related Peptides. D. de Wied & W. Ferrari, Eds.: 1–5. Liviana Press, Padova, Italy and Springer Verlag, Berlin, FRG.
2. FERRARI, W., E. FLORIS & F. PAULESU. 1955. Boll. Soc. It. Biol. Sper. **31**: 862–864.
3. FERRARI, W., G. L. GESSA & L. VARGIU. 1963. Ann. N. Y. Acad. Sci. **104**: 330–345.
4. GISPEN, W. H., V. M. WIEGANT, H. M. GREVEN & D. DE WIED. 1975. Life Sci. **17**: 645–652.
5. GISPEN, W. H. & R. L. ISAACSON. 1981. Pharmacol. Ther. **12**: 209–246.
6. FERRARI, W. 1958. Nature **181**: 925–926.
7. BERTOLINI, A. & W. VERGONI. 1968. Boll. Soc. It. Biol. Sper. **44**: 954–956.
8. BERTOLINI, A., W. VERGONI, G. L. GESSA & W. FERRARI. 1969. Nature **221**: 667–669.
9. BERTOLINI, A., G. L. GESSA & W. FERRARI. 1975. *In* Sexual Behavior: Pharmacology and Biochemistry. M. Sandler & G. L. Gessa, Eds.: 247–257. Raven. New York.
10. BERTOLINI, A. & G. L. GESSA. 1981. J. Endocrinol. Invest. **4**: 241–251.
11. HAUN, C. K. & G. C. HALTMEYER. 1975. Neuroendocrinology **19**: 201–213.
12. BALDWIN, M. D., C. K. HAUN & C. H. SAWYER. 1974. Brain Res. **80**: 291–301.
13. SAWYER, C. H., M. D. BALDWIN & C. K. HAUN. 1975. *In* Sexual Behavior: Pharmacology and Biochemistry. M. Sandler & G. L. Gessa, Eds.: 259–268. Raven. New York.
14. THODY, A. J., C. A. WILSON & D. EVERARD. 1979. Physiol. Behav. **22**: 447–450.
15. WILSON, C. A., A. J. THODY & D. EVERARD. 1979. Horm. Behav. **13**: 293–300.
16. VERGONI, A. V., R. POGGIOLI & A. BERTOLINI. 1986. Neuropeptides **7**: 153–158.
17. BERTOLINI, A., R. POGGIOLI, A. V. VERGONI, M. CASTELLI & S. GUARINI. 1986. *In* Central Actions of ACTH and Related Peptides. D. de Wied & W. Ferrari, Eds.: 207–222. Liviana Press. Padova, Italy and Springer Verlag, Berlin, FRG.
18. BOHUS, B. & D. DE WIED. 1980. *In* General, Comparative and Clinical Endocrinology of the Adrenal Cortex. I. C. Jones & I. W. Henderson, Eds.: 265–347. Academic. London.
19. DE WIED, D. & J. JOLLES. 1982. Physiol. Rev. **62**: 976–1059.
20. MIRSKY, I. A., R. MILLER & M. STEIN. 1953. Psychosom. Med. **15**: 574–588.
21. MURPHY, J. V. & R. MILLER. 1955. J. Comp. Physiol. Psychol. **48**: 47–49.
22. BOHUS, B. & E. ENDROCZI. 1965. Acta Physiol. Acad. Sci. Hung. **26**: 183–189.
23. LEY, K. F. & J. A. CORSON. 1971. Experientia **27**: 958–959.
24. BOHUS, B. & D. DE WIED. 1981. *In* Endogenous Peptides and Learning and Memory Processes. J. L. Martinez, Jr., R. A. Jensen, R. N. Messing, H. Rigter & J. L. McGaugh, Eds.: 59–77. Academic. New York.
25. MILLER, R. & N. OGAWA. 1962. J. Comp. Physiol. Psychol. **55**: 211–213.
26. DE WIED, D. 1969. *In* Frontiers in Neuroendocrinology. W. F. Ganong & L. Martini, Eds.: 97–140. Oxford Univ. Press. Oxford.
27. GOLD, P. E. & R. VAN BUSKIRK. 1976. Behav. Biol. **16**: 387–400.
28. GREVEN, H.M. & D. DE WIED. 1973. *In* Progress in Brain Research. Drug Effects on Neuroendocrine Regulation. E. Zimmermann, W. H. Gispen, B. H. Marks & D. de Wied, Eds.: 429–442. Elsevier. Amsterdam.
29. GUTH, S., S. LEVINE & J. P. SEWARD. 1971. Physiol. Behav. **7**: 195–200.
30. GARRUD, P., J. A. GRAY & C. COEN. 1977. Physiol. Behav. **18**: 813–818.
31. ISAACSON, R. L., A. J. DUNN, H. D. REES & B. WALDOCK. 1976. Physiol. Psychol. **4**: 159–162.
32. BOHUS, B., H. H. L. HENDRICKX, A. A. VAN KOLFSCHOTEN & T. G. KREDIET. 1975. *In* Sexual Behavior: Pharmacology and Biochemistry. M. Sandler & G. L. Gessa, Eds.: 269–275. Raven. New York.
33. MEYERSON, B. J. & B. BOHUS. 1976. Pharmacol. Biochem. Behav. **5**: 539–545.
34. BERTOLINI, A., R. POGGIOLI & W. FERRARI. 1979. Experientia **35**: 1216–1217.

35. BERTOLINI A, R. POGGIOLI & W. FERRARI. 1980. Adv. Biochem. Psychopharmacol. **22:** 109–117.
36. AMIR, S. 1981. Neuropharmacology **20:** 959–962.
37. JACQUET, Y. F. 1978. Science **201:** 1032–1034.
38. JACQUET, Y. F. & G. M. ABRAMS. 1982. Science **218:** 175–177.
39. BERTOLINI, A., R. POGGIOLI & W. FRATTA. 1981. Life Sci. **29:** 249–252.
40. VERGONI, A. V., O. GAFFORI & D. DE WIED. 1985. *In* Abstracts of the Fourth Capo Boi Conference on Neuroscience. p. 177.
41. SAWYER, C. H., J. W. EVERETT & J. D. GREEN. 1954. J. Comp. Neurol. **101:** 801–824.
42. PAXINOS, G. & C. WATSON. 1982. The Rat Brain in Stereotaxis Coordinates. Academic. New York.
43. HEYBACH, J. P. & J. VERNIKOS-DANELLIS. 1978. J. Physiol. **283:** 331–340.
44. HENKIN, R. L., R. E. MCGLONE, R. DALY & F. C. BARTTER. 1967. J. Clin. Invest. **46:** 429–435.
45. KOSOWICZ, J. & A. PRUSZEWICZ. 1967. J Clin. Endocrinol. Metab. **27:** 214–218.
46. HENKIN, R. I. & F. C. BARTTER. 1966. J. Clin. Invest. **45:** 1631–1639.
47. SAKELLARIS, P. C. 1972. Physiol. Behav. **9:** 495–501.

Structure-Activity Studies on the Neuroactive and Neurotropic Effects of Neuropeptides Related to ACTH

D. DE WIED AND G. WOLTERINK

Medical Faculty
Rudolf Magnus Institute for Pharmacology
University of Utrecht
3521 CG Utrecht, The Netherlands

INTRODUCTION

A causal relationship between the function of the adrenal cortex and human behavior has been suggested by frequently observed psychological aberrations in patients with adrenocortical dysfunction associated with psychiatric syndromes. The relationship between pituitary-adrenal activity and brain function was more directly demonstrated in clinical investigations in which the administration of adrenocorticotropic hormone (ACTH) or adrenal steroids resulted in deviant behavior.[1] These early findings led to the first animal experiments on the behavioral effect of ACTH by Mirsky, Miller, and Stein[2] in monkeys in which they found that the administration of ACTH reduced the effectiveness of an anxiety-producing stimulus. This effect, however, was interpreted as an influence of ACTH mediated by steroids produced by the adrenal cortex. Murphy and Miller[3] subsequently found that daily administration of ACTH during acquisition of shuttle box avoidance behavior in rats delayed extinction of the avoidance response. In contrast to ACTH, the synthetic glucocorticosteroid prednisolone appeared to facilitate extinction of shuttle box avoidance behavior (Miller and de Wied, 1958, unpublished observations). Thus, the influence of ACTH did not seem to be mediated by the adrenal cortex but to be a central nervous system effect of the hormone. Similar conclusions were drawn from a series of experiments in which ACTH was injected intracranially in dogs. This caused a behavioral stereotypy consisting of episodes of stretching and yawning,[4] discovered unexpectedly. To determine whether the ACTH-induced fall in circulating eosinophils was mediated in part by an effect in the CNS, ACTH was injected directly into the cisterna magna of dogs. The authors not only found a more pronounced eosinopenia but they also discovered this impressive and peculiar syndrome.[5] They decided to leave the original problem unsolved and to concentrate their efforts on this unusual effect of ACTH. These studies initiated numerous experiments on the central nervous system effects of ACTH and related peptides which appeared to reside in the NH_2-terminal part of the molecule. These effects are dissociated from the corticotropic influence and range from learning processes to nerve regeneration.

AVOIDANCE BEHAVIOR

One of the first discovered CNS effects of ACTH and related peptides was that on learning processes.[6] ACTH also facilitated the deficient acquisition of shuttle box avoidance behavior of hypophysectomized rats.[7,8] As mentioned before, ACTH given during acquisition delayed subsequent extinction of shuttle box avoidance behavior.[3] The effect of ACTH was more marked when given during the extinction period. In addition, α-MSH had the same effect on avoidance behavior as ACTH.[6] Fragments of ACTH virtually devoid of corticotropic effects like $ACTH_{1-10}$ or $ACTH_{4-10}$ were as effective as the parent molecule. Structure-activity studies were done for the most part on extinction of pole jumping avoidance behavior in intact rats. In this test $ACTH_{1-24}$, $ACTH_{1-10}$, $ACTH_{4-10}$, and $ACTH_{4-7}$ were equally active. Removal of the amino acid residues 4 or 7 markedly reduced the behavioral effect.[9] Peptides such as $ACTH_{7-10}$ Ac-$ACTH_{11-13}$-NH_2, and $ACTH_{11-24}$ had some residual activity. The residual potency of $ACTH_{7-10}$ could be increased to the same level as $ACTH_{4-10}$, extending the COOH-terminal sequence to $ACTH_{7-16}$-NH_2. Thus, other sequences (11–13) contain elements for behavioral activity that are present in a latent form. Substitution of phenylalanine by a D-enantiomer in $ACTH_{1-10}$, $ACTH_{4-10}$ or $ACTH_{4-7}$ had an effect opposite to that of the L-enantiomer peptides. These D-Phe-7 analogues facilitate extinction of shuttle box and pole-jumping avoidance behavior.[9] This could mean that D-Phe-7 analogues are competitive or functional antagonists of endogenous ACTH-MSH peptides. However, they are not antagonistic in all behaviors tested. Thus, $ACTH_{4-10}$ and [D-Phe-7] $ACTH_{4-10}$ both facilitate passive avoidance behavior, the latter peptide differing from the former only by its longer duration of action.[10] The effects of ACTH and related peptides on avoidance behavior do not seem to be affected by opioid antagonists.[10,11]

Several substitutions in the sequence $ACTH_{4-9}$ led to a highly selective and potent neuropeptide with a marked loss of endocrine effects. Thus, H-Met(O_2)-Glu-His-Phe-D-Lys-Phe-OH (Org 2766) appeared to be 1000 times more effective than $ACTH_{4-10}$ but to contain 1000 times less melanotropic activity. It also had a markedly reduced steroidogenic, fat-mobilizing, and opiate-like activity. These substitutions not only cause potentiation and prolongation of the activity but also the incorporation of other activities. Because of its increased resistance to enzymatic degradation, Org 2766 also appeared to be orally active.[12]

ACTH and related peptides not only affect active and passive avoidance behavior. They also facilitate food-rewarded behavior in a multiple T-maze[13] and maze performance.[14] Furthermore, these peptides delay extinction of food-motivated behavior,[14–16] sexually motivated approach behavior,[17] and conditioned taste aversion.[18] A number of studies indicated that ACTH-related neuropeptides reverse experimentally induced retrograde amnesia, suggesting an effect on memory retrieval processes.[19–22] These results on avoidance behavior and other findings were interpreted as effects on motivation. Bohus and de Wied[23] postulated that ACTH and related neuropeptides temporarily increase the motivational value of specific environmental cues, thus facilitating the occurrence of stimulus-specific behavioral responses. These may be caused by a selective arousal in limbic-midbrain structures. Electrophysiological evidence for this has been provided.[24,25] Other explanations have been offered[26] and effects have been interpreted as influences on attention and concentration. It seems clear that ACTH neuropeptides participate in learning processes that are essential prerequisites for survival.

STRETCHING AND YAWNING SYNDROME (SYS)

Intracranial injection of ACTH causes a typical syndrome in several species that was observed for the first time in dogs.[5] Half an hour after the intracranial injection of ACTH, the dogs were drowsy, yawned frequently, and after about 1 hour they started to stretch in a way they usually do when they awake from physiological sleep. The syndrome was called stretching and yawning syndrome (SYS). Structure-activity studies indicated that α-MSH was as active as ACTH. ACTH$_{4-10}$, however, which is common to both peptides, had negligible effects. Thus additional structural features were necessary to elicit the syndrome. ACTH and α-MSH, if dissolved in alkali and heated, retained both melanotropic and neurotropic activities while destroying the corticotropic effect. The authors therefore regarded the influence of ACTH and α-MSH as a neurotropic effect. From more recent structure-activity studies it was concluded that the optimal sequence for the syndrome is ACTH$_{4-12}$.[27]

In addition to repeated episodes of stretching and yawning, ACTH-derived peptides also induce repeated episodes of penile erection. This is seen in particular in rats when observed in groups but less in isolated animals. The structure for this effect seems to be similar to that of SYS, which suggests the existence of a relationship between these two phenomena.[28] However, Vergoni et al[29] showed that naltrexone pretreatment facilitates the episodes of penile erection but does not affect SYS in young and mature rats. The stretching and yawning syndrome can be elicited in monkeys, rabbits, cats, guinea pigs, and mice.[30]

EXCESSIVE GROOMING

In rodents the stretching and yawning syndrome is preceded by an enhanced display of grooming.[31-32] In contrast to effects of ACTH and related peptides on other behaviors, excessive grooming and SYS can be elicited only after intraventricular or intracisternal application. ACTH-induced excessive grooming can be reduced by pretreatment with the opioid antagonist naloxone.[33,34] Extensive structure-activity studies using doses equimolar to 3 μg of ACTH$_{1-24}$[32] found that ACTH$_{1-16}$-NH$_2$, α-MSH, and β-MSH were as active as ACTH$^{1-24}$; [Lys-17, Lys-18]ACTH$_{5-18}$-NH$_2$, ACTH$_{5-16}$-NH$_2$, ACTH$_{1-13}$-NH$_2$, and ACTH$_{5-14}$ were less effective while ACTH$_{11-24}$, ACTH$_{7-16}$-NH$_2$, ACTH$_{1-10}$, [Ac-Ser-1]ACTH$_{1-10}$ and ACTH$_{4-10}$ were inactive. Also γ$_2$-MSH, which has the sequence of MSH$_{6-9}$ in common with ACTH, and α-MSH and β-MSH, does not induce excessive grooming. Although ACTH$_{4-10}$ was not active even when given in high doses, as were the peptides ACTH$_{4-9}$, ACTH$_{4-8}$, ACTH$_{4-6}$, ACTH$_{5-7}$ and ACTH$_{7-10}$, the only sequence which induced grooming was ACTH$_{4-7}$.[35] It had, however, only 30% of the activity of ACTH$_{1-24}$. Apparently the 4–7 sequence is required for grooming but it represents only part of the active core, because for full expression COOH-terminal elongation is needed. However, whereas ACTH$_{7-16}$-NH$_2$ is not effective, ACTH$_{5-14}$ is. Interestingly, the administration of the ACTH$_{4-10}$ sequence, in which the amino acid residue phenylalanine at place 7 was replaced by a D-enantiomer, caused excessive grooming, in contrast to ACTH$_{4-10}$. The [D-Phe-7]ACTH$_{4-10}$ displayed 50–60% of the activity of ACTH$_{1-24}$. COOH-terminal elongation of the active site also leads to a marked potentiation of grooming activity. Thus ACTH$_{5-7}$ is inactive but [Lys-17, Lys-18]ACTH$_{5-18}$-NH$_2$ has considerable grooming activity. Also N-acetylation of ACTH$_{1-13}$-NH$_2$ potentiates the effect on excessive grooming.[32,36] This suggests that metabolic stability and physicochemical qualities are important factors in the expression of the CNS effects of these peptides, which cautions against the ultimate validity of structure-activity consideration. In a series of experiments in which the potentiating effect of ACTH$_{1-4}$ on excessive grooming induced by ACTH neuropeptides was studied, De Graan et al.[37] showed that α-MSH$_{5-13}$ was as active as the analogue [Lys-17, Lys-18]ACTH$_{5-18}$-NH$_2$. Thus the activity resides within the 5–13 sequence of ACTH-MSH.

AFTER-DISCHARGE AND BEHAVIORAL DEPRESSION

Cottrell *et al.*[38] showed that the after-discharge and behavioral depression that follows kindling of the dorsal hippocampus can be reduced by ACTH-related neuropeptides. A structure-activity study aimed at elucidating the sequence within the ACTH molecule to reduce after-discharge showed that $ACTH_{1-16}$, α-MSH, $ACTH_{7-16}$, [D-Phe-7]$ACTH_{7-16}$, γ_2-MSH, and [Met-4(0)-D-Lys-8, Phe-9-D-Lys-11]$ACTH_{4-16}$ *i.e.*, (Org HP-953A) were active. Behavioral depression was reduced by $ACTH_{1-16}$, [D-Phe-7]$ACTH_{7-16}$, γ_2-MSH and [D-Lys-11]$ACTH_{4-16}$. The structural requirements for the reduction of behavioral depression and that of after-discharge therefore is different. $ACTH_{4-10}$ and (D-Phe-7)$ACTH_{4-10}$ were ineffective in reducing after-discharge while α-MSH, $ACTH_{4-10}$, and $ACTH_{7-16}$ were unable to attenuate behavioral depression. It is difficult to determine from these studies which part of the ACTH-molecule contains the essential requirements for these effects. For after-discharge, the sequence 10–13 seems to be essential while that for behavioral depression lies in the 4–7 and 13–16 sequences.[39] The reducing effect of ACTH fragments on behavioral depression was mimicked by the opiate antagonist naltrexone and the functional opiate antagonist γ_2-MSH. The after-discharge induced by the kindling was not markedly influenced by opioid antagonists.

DORSAL IMMOBILITY

Behavioral inhibition is a phylogenetically old behavioral pattern in vertebrates.[40] The dorsal immobility response is a stereotypical immobility posture of the rat.[41] Dorsal immobility may mimic the transport response of the young of some mammalian species[42] and the immobility of a prey.[43] Tonic immobility, which is found in the lizard, can be attenuated by $ACTH_{4-10}$.[44] $ACTH_{4-10}$ also attenuates dorsal immobility elicited by grasping the rat by the dorsal skin at the nape of the neck and lifting it off its feet.[45] $ACTH_{7-16}$ reduced dorsal immobility but was not as potent as $ACTH_{4-10}$. Also, γ_2-MSH had a considerable effect but an inverted U-shaped dose response relationship was found; [D-Phe-7]$ACTH_{4-10}$ had the opposite effect. Endogenous opiates may be involved in tonic immobility.[46] However, neither naloxone nor naltrexone affected the immobility response. The opposite effects of L-Phe-7 and D-Phe-7 analogues argue against a direct interaction with opiate receptor sites and point to a naltrexone-independent activating (arousal) mechanism, as also holds for active avoidance behavior.[10]

INTERACTION WITH OPIATE BINDING SITES AND ANTINOCICEPTION

Interaction of ACTH neuropeptides with opiate receptor sites can be demonstrated on morphine-induced analgesia as measured on the hot plate. Gispen *et al.*[47] showed that sequences $ACTH_{1-16}$-NH_2, $ACTH_{5-16}$-NH_2, and $ACTH_{5-14}$, in addition to [D-Phe-7]$ACTH_{4-10}$ inhibited morphine-induced analgesia up to about 60%. Effects of $ACTH_{1-24}$ were equivocal. Thus also for this effect, more than one activity site seems to be present in the ACTH molecule. Injection of $ACTH_{1-24}$ into the periaqueductal gray matter of rats results in an opiate abstinence syndrome characterized by fearful hyperreactivity and explosive motor behavior.[48] Injection of α-MSH or ACTH$_{4-10}$ induce attenuated forms of this behavior, while γ$_2$-MSH on the other hand is active at much lower amounts.[49] This, and other observations, led Van Ree *et al.* to hypothesize that γ$_2$-MSH acts as a functional opiate antagonist. These behavioral effects appear to be insensitive to naloxone treatment. Unilateral microinjection of ACTH$_{1-24}$, α-MSH, or ACTH$_{4-10}$ in the locus ceruleus of the rat induces postural assymmetry and locomotor impairment.[50] This might be caused by an interaction with opiate receptor sites which are densly present in the locus ceruleus.

Opiate receptors in the rat brain have affinity for ACTH neuropeptides.[51] Structure-activity studies on the binding of [H^3]dihydromorphine and [H^3]naltrexone revealed that ACTH$_{4-10}$ and the D-Phe-7 analogue were equally active. Good activity was also found in ACTH$_{7-16}$-NH$_2$. Acetylation of the amino acid residue at position 1 led to a marked reduction in affinity.[52] Thus, there are two active sites in the ACTH molecule, one within the 4-10, the other within the 7-16 portion.

Exposure to stress elicits antinociception. This stress-induced analgesia is associated with the release of pituitary hormones like ACTH and endorphins, the adrenocortical steroids, and adrenomedullary hormones. It has been proposed that stress-induced analgesia depends on pituitary and adrenal opioid peptides.[53] A variant of stress-induced analgesia is "heat" stress-induced sedation. Rats placed on a hot plate (57°C) for 30 seconds and tested for locomotor activity show a significant reduction in locomotion[54] together with a time-dependent increase in rearing and sniffing. Hypophysectomy prevents "heat" stress-induced sedation. ACTH$_{4-10}$ does not affect "heat" stress-induced sedation. But this peptide significantly decreased locomotor activity of rats subjected to a "nonfunctional" hot plate of 21°C. Structure-activity studies showed that [D-Phe-7]ACTH$_{4-10}$, ACTH$_{7-10}$, and ACTH$_{4-9}$-NH$_2$ had the same effect as ACTH$_{4-10}$, whereas ACTH$_{4-7}$, ACTH$_{4-9}$, and ACTH$_{8-10}$ were inactive.[55] The ACTH$_{4-9}$ analogue Org 2766 and its COOH-terminal tripeptide Phe-D-Lys-Phe were also not effective. Thus the activity resides in the 7-9 sequence of ACTH, Phe-Arg-Trp-NH$_2$. Opiate antagonists do not affect "heat" stress-induced sedation but mimic hypolocomotion after exposure to the nonfunctional hot plate.[56] This effect of ACTH$_{7-10}$ therefore resembles that of an opioid antagonist. It is worth noting in this respect that γ$_2$-MSH, which has been regarded as an endogenous opioid antagonist,[49] contains the ACTH$_{6-9}$ sequence His-Phe-Arg-Trp. It might therefore be of interest to investigate whether this tetrapeptide possesses opioid antagonistic activities and thus represents the active core in the ACTH molecule for this effect.

Peripheral δ-opiate receptors are measured on the mouse vas deferens preparation. A number of ACTH neuropeptides depress the electrically induced contraction of the mouse vas deferens.[57] ACTH$_{1-24}$ is active in this assay as are ACTH$_{1-10}$ and ACTH$_{4-10}$. The activity resides in the ACTH$_{7-10}$ part of the molecule and is blocked by opiate antagonists. The ability of ACTH neuropeptides to inhibit the binding of morphine to morphine antibodies *in vitro* also is located in this sequence since it was as active as ACTH$_{1-24}$.[57] This suggests that ACTH$_{7-10}$, which is recognized by morphine antibodies, may have a structure related to morphine or opioid peptides. It is of interest to mention that Castelli *et al.*[58] showed that ACTH$_{1-24}$, ACTH$_{1-17}$, and ACTH$_{1-13}$ competitively antagonize the contractile effect of morphine on the isolated rat colon.

SOCIAL BEHAVIOR AND MOTOR ACTIVITY

ACTH$_{1-24}$ and ACTH$_{4-10}$ reduce social interaction.[59,60] This effect is seen mainly when rats are tested under intense illumination and unfamiliar conditions. The ACTH$_{4-9}$

analogue Org 2766 has opposite effects and facilitates social interaction. In another situation where social behavior was studied in dyadic encounters between short-term (7 days) isolated and nonisolated rats, social interaction of rats is markedly increased. This effect of short-term isolation is reduced by Org 2766 but not by $ACTH_{1-24}$, $ACTH_{4-10}$, or [D-Phe-7]$ACTH_{4-10}$.[56] Pretreatment with naltrexone antagonizes this effect of Org 2766. Chronically administered $ACTH_{1-24}$ reduces isolation-induced aggressive behavior in mice.[61] This effect of ACTH is extra-adrenal and may be mediated by the sequence $ACTH_{4-10}$. The environmental conditions that affect social interaction also induce changes in motor activity of rats.[56,62] Thus, group-housed rats tested under low-light conditions show a high level of motor activity as compared to rats isolated for 7 days tested under intense illumination. $ACTH_{4-10}$, and to a lesser extent $ACTH_{7-10}$ and [D-Phe-7]$ACTH_{4-10}$, enhance the effects of the environment on motor activity. Org 2766 on the other hand attenuates the environmentally induced changes in motor activity. Structure-activity studies revealed that this effect of Org 2766 is located in the COOH-terminal tripeptide Phe-D-Lys-Phe and that it is antagonized by naltrexone. Recent studies suggest that the central site of action for these effects of Org 2766 and naltrexone is the central amygdala.[63]

NERVE CRUSH REGENERATION

ACTH and related peptides accelerate recovery of sensorimotor functions after peripheral nerve damage.[64–66] These peptides have neurotropic effects and facilitate growth rate and number of outgrowing neurons in regenerating sciatic nerves. Structure-activity studies indicate that $ACTH_{1-24}$, $ACTH_{1-16}$-NH_2, α-MSH, $ACTH_{4-9}$, and $ACTH_{4-10}$ are active in this respect whereas the sequences $ACTH_{4-7}$ and $ACTH_{11-24}$ are without effect. The sequence $ACTH_{6-10}$ contains melanotropic activity for which Phe-7 and Arg-8 are key positions.[67] Thus the effects on nerve regeneration may be related to the melanotropic property rather than the behavioral or corticotropic activity of the ACTH molecule. Evidence for the generation of melanotropic compounds in the crushed nerve has been found.[68]

GENERATION OF ACTH NEUROPEPTIDES IN THE BRAIN

In the above mentioned studies, the multiple effects of ACTH neuropeptides on stretching and yawning, penile erection, excessive grooming, avoidance behavior, opiate receptor interaction, counteraction of morphine-induced analgesia, after-discharge and behavioral depression, dorsal immobility, social behavior, motor activity, stress-induced analgesia, and nerve regeneration have been briefly reviewed. Structure-activity studies suggest that neuroactive and neurotropic effects of ACTH are located within the NH_2-terminal part of the molecule, in particular of $ACTH_{1-16}$. The 1–4 sequence may not be necessary but seems under certain conditions capable of potentiating the effect of ACTH fragments on excessive grooming.[37] Although full effects on avoidance behavior can be elicited by the tetrapeptide $ACTH_{4-7}$, more than one activity site is present generally for most of the neuroactive effects of ACTH. These are confined to the $ACTH_{4-16}$ sequence. ACTH is derived from the precursor pro-opiomelanocortin (POMC). The POMC gene is expressed not only in the pituitary but also in the brain, in particular in the nucleus arcuatus. To know the processing of ACTH from the precursor may therefore reveal the structure of putative ACTH neuropeptides with neuroactive and neurotropic effects. One approach is the identification of fragments of ACTH after incubation with synaptic membrane fractions of the rat brain (for review see ref. 69). Incubation of $ACTH_{1-39}$ at different pH values indicated the generation of a variety of fragments which could well be responsible for the neuroactive and neutropic effects of ACTH neuropeptides as discussed above. At pH 7.4, fragments isolated were $ACTH_{1-38}$ as the major component and a number of minor metabolites: $ACTH_{1-34}$, $ACTH_{1-33}$, $ACTH_{1-28}$, $ACTH_{2-21}$, $ACTH_{7-21}$, and $ACTH_{7-10}$. At pH 6.2, only large COOH-terminal fragments, as $ACTH_{1-38}$, $ACTH_{1-37}$, and $ACTH_{1-36}$, were formed as a result of carboxypeptidase activity. Fragments isolated were $ACTH_{1-16}$ and $ACTH_{17-39}$ but also $ACTH_{22-39}$ and $ACTH_{3-15}$. Thus the main pathway at neutral and acidic pH concerns a carboxypeptidase-like mechanism. The formation of $ACTH_{1-16}$ may have biological significance since it contains all neuroactive and neurotropic effects found so far.[70] It is devoid of corticotropic effects.[71]

Interestingly, $ACTH_{1-16}-NH_2$, which is protected against carboxy peptidase activity, can be converted to amidated $ACTH_{3-16}$, $ACTH_{4-16}$, $ACTH_{5-16}$, or $ACTH_{7-16}$ at pH 7.4 and pH 8.5. At pH 7.4 the main metabolites are $ACTH_{4-16}-NH_2$, $ACTH_{4-10}-NH_2$, and $ACTH_{7-16}-NH_2$. Two endogenous forms of melanotropin in the brain are α-MSH and des-acetyl-α-MSH. They differ in behavioral activities,[72,73] and α-MSH appeared to be much more stable to proteolysis than des-acetyl-α-MSH. Both peptides are converted differently. The predominant type of proteolysis of des-acetyl-peptide is mediated by amino peptidase. Peptides like $ACTH/α\text{-}MSH_{3-9}$, $ACTH/α\text{-}MSH_{4-9}$, and $ACTH/α\text{-}MSH_{5-9}$ can be formed. Both the acetylated and the nonacetylated peptides[69] generate $α\text{-}MSH_{10-13}$.

DISCUSSION

The biological activity of a peptide depends on processes of absorption, distribution, metabolism and excretion, and binding to a putative receptor site followed by signal transduction.[74] Thus, the chemical structure, spatial conformation, metabolic stability, and transport properties are important factors.[75,76] An increase in metabolic stability,

TABLE 1

	Important Sequences
Avoidance behavior	$ACTH_{4-7}$, $ACTH_{7-16}$
Stretching and yawning syndrome	$ACTH_{4-12}$
Excessive grooming	$ACTH_{4-7}$, $ACTH_{5-13}{}^{a}$
Reduction of after-discharge	$ACTH_{10-13}$
Reduction of behavioral depression	$ACTH_{4-7}$, $ACTH_{13-16}{}^{a}$
Inhibition of dorsal immobility response	$ACTH_{4-10}$
Induction of postural asymmetry	$ACTH_{4-10}$
Counteraction of morphine-induced analgesia	$ACTH_{5-14}{}^{a}$
"Heat" stress-induced sedation	$ACTH_{7-9}-NH_2{}^{a}$
Opiate binding site interaction	$ACTH_{4-10}$, $ACTH_{7-16}-NH_2{}^{a}$
Other opiate effects *in vitro*	$ACTH_{7-10}{}^{a}$
Aggressive behavior	$ACTH_{4-10}$
Social behavior	$ACTH_{7-10}{}^{a}$
Motor activity	$ACTH_{7-10}{}^{a}$
Nerve crush regeneration	$ACTH_{6-10}$

^{a} Probably mediated by opiate systems.

TABLE 2

Sequence	Av. Beh.	SYS	Ex. Gr.	A/Ds	Beh. Depr.	Do. Im.	Op. Rec.	Co. Mo. An.	H. Str. Sed.	Soc. Beh.	Mo. Act.	Cr. Reg.
$ACTH_{1-39}$	++	++	++				+	+/0		+		++
$ACTH_{1-24}$	++	++	++		++		+	+				+
$ACTH_{1-16}$				+	++		+	+		+		
$ACTH_{1-16}-NH_2$		++	++				+					+
$ACTH_{1-13}-NH_2$		+	+									
α-MSH,												
[Ac-Ser-1] $ACTH_{1-13}$	++	++	++	++	0		0					++
$ACTH_{1-10}$	++		0				+					
[Ac-Ser-1]$ACTH_{1-10}$		++	0				+					
$ACTH_{4-13}$	++	0	0	0	0	++	+	+/0	++	++	++	+
$ACTH_{4-10}$	++		+	0	++	-	+	+	++	++	++	+
[D-Phe-7]$ACTH_{4-10}$			0	0			0		0		0	0
$ACTH_{4-9}$	++	+		0			0		+		0	0
$ACTH_{4-9}-NH_2$									+			
$ACTH_{4-7}$							0		0			
[Lys-17, Lys-18]												
$ACTH_{5-18}-NH_2$		+		++			++	++				
$ACTH_{5-16}-NH_2$		+		++			++					
$ACTH_{5-14}$		+										
$ACTH_{5-13}-NH_2$												
$ACTH_{7-16}$	++	0		++	0	+	+					
$ACTH_{7-16}-NH_2$		0		++	++							
[D-Phe-7]$ACTH_{7-16}$	+	0		++	++		0		+		+	
$ACTH_{7-10}$												
[Ac-Lys-11]												
$ACTH_{11-13}-NH_2$	+	0					0	0	0	-		0
$ACTH_{11-24}$	++			++	++	+						+
Org 2766	++						0	0	0	-	-	0
γ-MSH	-											+
Phe-D-Lys-Phe											-	0

Av. Beh., avoidance behavior.
SYS, stretching and yawning syndrome.
Ex. Gr., excessive grooming.
A/Ds, reduction of after-discharge.
Beh. Depr., reduction of behavioral depression.

Op. Rec., opiate receptor interaction.
Co. Mo. An., counteraction of morphine-induced analgesia.
H. Str. Sed., "heat" stress-induced sedation.
Soc. Beh., environmentally induced changes in social behavior.
Mo. Act., environmentally induced changes in motor activity.

which protects the peptide chain against proteases, can be achieved by the incorporation of D-amino acid residues. The result, however, may be a decrease or an increase in potency. It can also affect transport properties[77] or lead to a different type of activity. For example, an opposite effect is found with ACTH neuropeptides.[78] An increase in potency may be associated with a decrease in intrinsic activity which is compensated by metabolic stability[79] or an increase in intrinsic activity. Increased resistance to enzymatic degradation combined with an increased affinity for binding sites has been reported for GnRH analogues.[80,81] Some of these influences can be bypassed by using an *in vitro* assay, or by administration of the material directly into the target tissue (in the brain structure involved). Modification of the terminal amino group by acetylation, alkylation, hydroxylation desamination, amidation, or reduction also affect physicochemical, conformational, and other factors involved in peptide-cell interaction.

The various loci within the ACTH molecule responsible for the neuroactive and neurotropic effects do not indicate, as far as could be indicated with reasonable certainty, marked specificity for the respective activities (TABLES 1 and 2). $ACTH_{4-7}$ and $ACTH_{7-16}$ have specificity for avoidance behavior, $ACTH_{5-13}$ for excessive grooming, $ACTH_{4-12}$ for SYS, $ACTH_{4-10}$ and $ACTH_{7-16}-NH_2$ for interaction with opiate binding sites, $ACTH_{6-10}$ for nerve regeneration, $ACTH_{5-14}$ for morphine-induced analgesia, $ACTH_{1-16}-NH_2$ and γ_2-MSH for after-discharge and behavioral depression following kindling of the dorsal hippocampus, $ACTH_{4-10}$ for dorsal immobility, $ACTH_{4-10}$ for postural asymmetry following unilateral injection into the locus ceruleus, and $ACTH_{4-10}$ and presumably $ACTH_{7-10}$ for various opiate antagonistic effects (mouse vas deferens, colon contraction, binding to morphine antibodies, cold-plate induced hypolocomotion). This may also hold for the effect of $ACTH_{4-9}$ analogue Org 2766 on social behavior which is naltrexone sensitive and resides in the COOH-terminal tripeptide. ACTH neuropeptides which are generated in the brain depend on: the structure of the precursor $ACTH_{1-39}$, $ACTH_{1-16}-NH_2$, α-MSH or desacetylated α-MSH; the enzymes, aminopeptidases, endopeptidases, carboxypeptidase present in the tissue; the circumstances (pH) in the nerve cell or adjacent structures; and the brain structures involved in the neuroactive and neurotropic effects of these neuropeptides. Interestingly, in contrast to effects induced by $ACTH_{4-7}$, those induced by $ACTH_{7-16}$ are sensitive to opiate antagonists (TABLE 1). It may be, therefore, that the influence of the latter is mediated by opiate systems in the brain.

Further studies on the localization of peptidases, sites, and the regulation of *in vivo* metabolism are needed. In addition, more extensive structure-activity studies are necessary to compare the various neuroactive and neurotropic effects of ACTH neuropeptides. These may allow more insight into the generation of a particular neuropeptide and the induction of a given neuroactive or neurotropic effect.

REFERENCES

1. CLEGHORN, R. A. 1952. *In* Ciba Foundation Colloquia on Endocrinology, Vol. 3. G. E. W. Wolstenholm, Ed.: 187. Churchill, London.
2. MIRSKY, I. A., R. MILLER & M. STEIN. 1953. Psychosom. Med. **15**: 574–588.
3. MURPHY, J. V. & R. E. MILLER. 1955. J. Comp. Physiol. Psychol. **48**: 47–49.
4. FERRARI, W., E. FLORIS & F. PAULESU. 1955. Boll. Soc. Ital. Biol. Sper. **31**: 859–862.
5. FERRARI, W., E. FLORIS & F. PAULESU. 1955. Boll. Soc. Ital. Biol. Sper. **31**: 862–864.
6. DE WIED, D. 1969. *In* Frontiers in Neuroendocrinology. W. F. Ganong & L. Martini, Eds.: 97–140. Oxford Univ. Press. London, New York.
7. APPLEZWEIG, M. H. & G. MOELLER. 1959. Acta Psychol. **15**: 602–603.
8. DE WIED, D. 1964. Am. J. Physiol. **207**: 255–259.

9. GREVEN, H. M. & D. DE WIED. 1973. Progr. Brain Res. **39**: 429–442.
10. FEKETE, M. & D. DE WIED. 1982. Pharmacol. Biochem. Behav. **16**: 387–392.
11. GISPEN, W. H., M. E. A. REITH, P. SCHOTMAN, V. M. WIEGANT & H. M. ZWIERS. 1977. *In* Neuropeptide Influences of the Brain and Behavior. L. H. Miller, C. A. Sandman & A. J. Kastin, Eds.: 61–80. Raven. New York.
12. GREVEN, H. M. & D. DE WIED. 1977. Front. Horm. Res. **4**: 140–152.
13. ISAACSON, R. L., A. J. DUNN, H. D. REES & B. WALDVIK. 1976. Physiol. Psychol. **4**: 159–162.
14. FLOOD, J. F., M. E. JARVIK, E. L. BENNETT & A. E. ORME. 1976. Pharmacol. Biochem. Behav. **5** (Suppl. 1): 41–51.
15. GARRUD, P., J. A. GRAY & D. DE WIED. 1974. Physiol. Behav. **12**: 109–119.
16. SANDMAN, C. A., A. J. KASTIN & A. V. SCHALLY. 1969. Experientia **25**: 1001–1002.
17. BOHUS, B., H. H. L. HENDRICKX, A. A. VAN KOLFSCHOTEN & T. G. KREDIET. 1975. *In* Sexual Behavior: Pharmacology and Biochemistry. M. Sandler & G. L. Gessa, Eds.: 269–275. Raven. New York.
18. RIGTER, H. & A. POPPING. 1976. Psychopharmacologia (Berlin) **46**: 255–261.
19. RIGTER, H., H. VAN RIEZEN & D. DE WIED. 1974. Physiol. Behav. **13**: 381–388.
20. RIGTER, H., R. ELBERTSE & H. VAN RIEZEN. 1975. Progr. Brain Res. **42**: 163–171.
21. FLEXNER, J. B. & L. B. FLEXNER. 1971. Proc. Natl. Acad. Sci. **68** 2519–2521.
22. FLOOD, J. F., E. L. BENNETT & A. E. ORME. 1975. Physiol. Behav. **14**: 177–184.
23. BOHUS, B. & D. DE WIED. 1980. *In* General, Comparative and Clinical Endocrinology of the Adrenal Cortex. I. C. Jones & I. W. Henderson, Eds.: 265–347. Academic. London.
24. URBAN, I. & D. DE WIED. 1976. Exp. Brain Res. **24**: 325–344.
25. WOLTHUIS, O. L. & D. DE WIED. 1976. Pharmacol. Biochem. Behav. **4**: 273–278.
26. KASTIN, A. J., C. A. SANDMAN, L. O. STRATTON, A. V. SCHALLY & L. H. MILLER. 1975. Progr. Brain Res. **42**: 143–150.
27. SERRA G., M. COLLU, G. L. GESSA & W. FERRARI. 1986. *In* Central Actions of ACTH and Related Peptides. Fidia Research Series, Symposia in Neuroscience IV. D. de Wied & W. Ferrari, Eds.: 163–178. Liviana Press. Padova.
28. BERTOLINI, A., G. L. GESSA & W. FERRARI. 1975. *In* Sexual Behavior: Pharmacology and Biochemistry, M. Sandler & G. L. Gessa, Eds.: 247–257. Raven. New York.
29. VERGONI, A. V., O. GAFFORI & D. DE WIED. 1985. *In* Abstract. Presented at the Fourth Capo Boi Conference on Neuroscience, Villasimus, Italy, June 2–7. p. 177.
30. FERRARI, W., G. L. GESSA & L. VARGIU. 1963. Ann. N.Y. Acad. Sci. **104**: 330–345.
31. IZUMI, K., J. DONALDSON & A. BARBEAU. 1973. Life Sci. **12**: 203–210.
32. GISPEN, W. H., V. M. WIEGANT, H. M. GREVEN & D. DE WIED. 1975. Life Sci. **17**: 645–652.
33. JOLLES, J., V. M. WIEGANT & W. H. GISPEN. 1978. Neurosci. Lett. **9**: 261–266.
34. GISPEN, W. H. & R. L. ISAACSON. 1986. *In* Neuropeptides and Behavior, Vol. 1. D. de Wied, W. H. Gispen & Tj.B. van Wimersma Greidanus, Eds.: 273–312. Pergamon. Oxford.
35. WIEGANT, V. M. & W. H. GISPEN. 1977. Behav. Biol. **19**: 554–558.
36. O. DONOHUE, T. L., G. E. HANDELMANN, G. E. CHACONAS, R. L. MILLER & D. M. JACOBOWITZ. 1981. Peptides **2**: 333–344.
37. DE GRAAN, P. N. E., A. EBERLE, B. M. SPRUIJT, H. GERARD & W. H. GISPEN. 1985. Peptides **7**: 1–4.
38. COTTRELL, G. A., C. NYAKAS, B. BOHUS & D. DE WIED. 1983. *In* Integrative Neurohumoral Mechanisms. Developments in Neuroscience, Vol. 16. E. Endröczi, D. de Wied & L. Angelucci, Eds.: 91–97. Elsevier Biomedical Press. Amsterdam.
39. BOHUS, B., G. A. COTTRELL, C. NYAKAS & M. E. MEYER. 1986. *In* Central Actions of ACTH and Related Peptides. Fidia Research Studies, Symposia in Neuroscience IV. D. de Wied & W. Ferrari, Eds.: 189–198.
40. GALLUP, G. G., JR. 1974. Psychol. Rec. **21**: 41–61.
41. WEBSTER, D. G., T. H. LANTHORN, D. A. DEWSBURY & M. E. MEYER. 1981. Behav. Neural Biol. **31**: 32–41.
42. BREWSTER, J. & M. LEON. 1980. J. Comp. Physiol. Psychol. **94**: 80–88.
43. SARGEANT, A. B. & L. E. EBERHARDT.1975. Am. Midl. Nat. **94**: 108–119.
44. STRATTON, L. O. & A. J. KASTIN. 1976. Physiol. Behav. **16**: 771–774.
45. MEYER, M. E. & B. BOHUS. 1983. Behav. Neural Biol. **38**: 194–204.
46. CARLI, G. 1977. Psychol. Rec. **27**: 123–143.

47. GISPEN, W. H., J. BUITELAAR, V. M. WIEGANT, L. TERENIUS & D. DE WIED. 1976. Eur. J. Pharmacol. **39**: 393-397.
48. JACQUET, Y. F. 1978. Science **201**: 1032-1034.
49. VAN REE, J. M., B. BOHUS, K. M. CSONTOS, W. H. GISPEN, H. M. GREVEN, F. P. NIJKAMP, F. A. OPMEER, A. A. DE ROTTE, TJ. B. VAN WIMERSMA GREIDANUS, A. WITTER & D. DE WIED. 1981. Life Sci. **28**: 2875-2888.
50. JACQUET, Y. F. & G. M. ABRAMS. 1982. Science **218**: 175-177.
51. TERENIUS, L. 1975. J. Pharm. Pharmacol. **27**: 450-452.
52. TERENIUS, L., W. H. GISPEN & D. DE WIED. 1975. Eur. J. Pharmacol. **33**: 395-399.
53. LEWIS, J. W., M. G. TORDOFF, J. E. SHERMANN & J. D. LIEBESKIND. 1982. Science **217**: 557-559.
54. GALINA, Z. H., C. J. SUTHERLAND & Z. AMIT. 1983. Pharmacol. Biochem. Behav. **19**: 251-256.
55. WOLTERINK, G. & J. M. VAN REE. 1985. Proc. 26th Dutch Fed. Meeting. Amsterdam. Abstract 393.
56. VAN REE, J. M., G. WOLTERINK, Z. H. GALINA & R. J. M. NIESINK. 1986. *In* Central Actions of ACTH and Related Peptides. Fidia Research Series, Symposia in Neuroscience IV. D. de Wied and W. Ferrari, Eds.: 139-146. Liviana Press. Padova.
57. PLOMP, G. J. J. & J. M. VAN REE. 1978. Br. J. Pharmacol. **64**: 223-227.
58. CASTELLI, M., G. L. GESSA & A. BERTOLINI. 1985. Eur. J. Pharmacol. **108**: 213-214.
59. FILE, S. E. 1979. Brain Res. **171**: 157-160.
60. NIESINK, R. J. M. & J. M. VAN REE. 1984. Life Sci. **34**: 961-970.
61. BRAIN, P. F. 1972. Neuroendocrinology **10**: 371-376.
62. WOLTERINK, G. & J. M. VAN REE. 1987. Brain Res. In press.
63. WOLTERINK, G. & J. M. VAN REE. 1986. Psychopharmacology **89**: S38.
64. STRAND, F. L. & KUNG, T. T. 1980. Peptides **1**: 135-138.
65. BIJLSMA, W. A., F. G. I. JENNEKENS, P. SCHOTMAN & W. H. GISPEN. 1981. Eur. J. Pharmacol. **76**: 73-79.
66. BIJLSMA, W. A., P. SCHOTMAN, F. G. I. JENNEKENS, W. H. GISPEN & D. DE WIED. 1983. Eur. J. Pharmacol. **92**: 231-236.
67. EBERLE, A. & R. SCHWYZER. 1976. Clin. Endocrinol. **5**(Suppl.): 41S-48S.
68. EDWARDS, P. M. & W. H. GISPEN. 1985. *In* Senile Dementia of the Alzheimer Type. J. Traber & W. H. Gispen, Eds.: 231-240. Springer-Verlag. Berlin-Heidelberg.
69. BURBACH, J. P., J. L. M. LEBOUILLE & XIN-CHANG WANG. 1986. *In* Central Actions of ACTH and Related Peptides. Fidia Research Series, Symposia in Neuroscience IV. D. de Wied & W. Ferrari, Eds.: 53-67. Liviana Press. Padova.
70. DE WIED, D. & J. JOLLES. 1982. Physiol. Rev. **62**: 976-1059.
71. HOFMANN, K., T. Y. KIU, H. YAJIMA, N. YANAIHARA, C. YANAIHARA & J. L. HULMES. J. Am. Chem. Soc. **84**: 1054-1056.
72. LOH, P. Y., R. L. ESKAY & M. BROWNSTEIN. 1980. Biochem. Biophys. Res. Commun. **94**: 916-923.
73. O'DONOHUE, T. L., G. E. HANDELMANN, R. L. MILLER & D. M. JACOBOWITZ. 1982. Science **215**: 1125-1127.
74. VAN NISPEN, J. W. & H. M. GREVEN. 1986. *In* Neuropeptides and Behavior, Vol. 1. CNS effects of ACTH, MSH and Opioid Peptides. D. de Wied, W. H. Gispen & Tj. B. van Wimersma Greidanus, Eds.: 349-383. Pergamon. Oxford.
75. RUDINGER, J. 1971. *In* Drug Design, Vol. 11. E. J. Ariëns, Ed.: 319-349. Academic. New York.
76. BENNETT, H. P. J. & C. McMARTIN. 1979. Pharmacol. Rev. **30**: 247-292.
77. BAJUSZ, S. 1979. Pharmazie **34**: 352-352.
78. BOHUS, B. & D. DE WIED. 1966. Science **153**: 318-320.
79. McMARTIN, C., G. E. E. PURDON, L. SCHENKEL, P. A. DESAULLES, R. MAIER, M. BRUGGEN, W. RITTEL & P. SIEBER. 1977. J. Endocrinol. **73**: 79-89.
80. VALE, W., C. RIVIER, M. BROWN, J. LEPPALUOTO, N. LING, M. MONAHAM & J. RIVIER. 1976. Clin. Endocrinol. **5**(Suppl):261-273.
81. RIVIER, J., M. BROWN, C. RIVIER, N. LING & W. VALE. 1976. *In* Peptides 1976. A. Loffet, Ed.: 427-451. Editions de l'Université de Bruxelles. Brussel.

Transmembrane Signal Transduction and ACTH-induced Excessive Grooming in the Rat

WILLEM HENDRIK GISPEN

Division of Molecular Neurobiology
Rudolf Magnus Institute for Pharmacology
and
Institute of Molecular Biology and Medical Biotechnology
3584 CH Utrecht, the Netherlands

INTRODUCTION

In this paper I shall review some of the neurochemical events that possibly underlie ACTH-induced excessive grooming in the rat. In previous reviews I have documented this behavioral reponse to ACTH and congeners in great detail.[1,2] The response is readily seen following intracranial administration of ACTH and can be recorded using a time sampling method yielding quantitative information on occurrence and duration of grooming bouts.[3] As studied by many authors, the grooming bout consists of a rather fixed pattern of motor acts which I refer to collectively as grooming behavior. In view of some of the evidence also presented at this meeting, it is increasingly clear that different behavioral elements seen as part of the grooming bout may originate from different neural substrates and possibly may result from different ACTH brain interactions. As in most of the studies in which both behavioral and neurochemical aspects are monitored, the behavioral response to ACTH is merely measured in terms of frequency and duration of grooming as such and not in terms of the display of various behavioral elements constituting grooming behavior. I cannot describe any neurochemical event in relation to the display of a particular grooming element. Nonetheless, as will be discussed below, by looking at grooming as a whole much progress has been made in our understanding of how ACTH acts in the brain.

In the search for the neurochemical mechanism of action of ACTH a major problem has been the lack of information on stereo-specific, saturable binding sites for this peptide.[4] Although the brain has networks of neurons which express pro-opiomelanocortin (POMC), the precursor for peptides of the ACTH family, and ACTH peptides have been found in brain and liquor,[5] the absence of information on a specific ACTH receptor is surprising and puzzling. Previously it was reported that ACTH-like peptides might act as partially agonist-antagonist at central opiate receptors.[6,7] However, the high concentrations of peptides necessary to displace the labeled opiate from its receptor have called into question the physiological importance of these findings. Although opiate antagonists have been shown to block ACTH-induced excessive grooming,[8,9] I still maintain that the opiate-sensitive component in grooming behavior is a site which is distal to the primary site at which ACTH acts to initiate the grooming bout.[10]

Despite the drawback that presently no information is available on the first interaction of the peptide with neural cells, much information has been gained concerning

ACTH-sensitive processes in neuronal cell membranes normally serving a role in receptor-mediated transmembrane signal processing.[11] Both in the adrenal cortex cell[12] and in brain tissue,[11] ACTH may activate or modulate at least two different signal transduction systems, *i.e.*, the receptor-activated production of cAMP and the receptor-activated hydrolysis of phosphatidylinositol 4,5-bisphosphate (PIP_2) followed by the production of diacyglycerol (DG) and D-myoinositol 1,4,5-trisphosphate (IP_3; see below). As the structure activity of ACTH in excessive grooming is very similar to that observed in the modulation of the receptor-mediated poly-PI response (see below), the significance of this response for ACTH-induced excessive grooming will be assessed in detail.

RECEPTOR-MEDIATED POLYPHOSPHOINOSITIDE HYDROLYSIS

Activation of various hormone and neurotransmitter receptors results in phosphodiesteratic degradation by a phospholipase C (PLC) of PIP_2 to IP_3 and DG. IP_3 can be either first phosphorylated or be directly degraded to myoinositol. DG is phosphorylated to phosphatidate and then, through a liponucleotide intermediate, converted to phosphatidylinositol (PI). In two phosphorylation steps, PI is converted via phosphatidylinositol 4-phosphate (PIP) to PIP_2 (FIG. 1).[13,14] The relatively low content of PIP_2 and the transient nature of subsequent increases in IP_3, D-myoinositol 1,4-bisphosphate, and D-myoinositol 1-monophosphate, and decreases in PIP_2, PIP, and PI suggest that PIP_2 is replenished after its breakdown by PDE. Such replenishment appears to provide a mechanism to control the inositide response.[15]

Both initial PDE products have a second messenger function in the cell (FIG. 1). IP_3 or a metabolite thereof can mobilize Ca^{2+} from putative endoplasmic reticulum stores.[16,17] DG can activate protein kinase C (PKC). The activity of PKC is dependent on the presence of an acidic phospholipid such as phosphatidylserine (PS) and is calcium sensitive. DG enhances the calcium sensitivity to PKC to the concentration normally present in an ionized form in the cell.[18]

The DG derivative dioctanoylglycerol (DOG) or the phorbol diester phosphoryl 12,13-dibutyrate (PDB) mimic DG in this respect, but the latter is about 1000-fold more potent than DG and has hence been a useful tool to investigate PKC.[18] In various cell types protein substrates of PKC have been identified, although the function of most of them cannot as yet be defined. In brain, two phosphoproteins in synaptic plasma membranes (SPM) have been characterized as a substrate of PKC, *i.e.*, an 87 kDa protein[19] and the neuron-specific phosphoprotein B-50.[20] It is believed that these substrate proteins may mediate the function of PKC in the neuronal response to enhanced hydrolysis of PIP_2.

THE ROLE OF THE PKC SUBSTRATE PROTEIN B-50 IN THE POLYPHOSPHOINOSITIDE RESPONSE

The phosphoprotein B-50 (M_r 48 kDa, pI 4.5) is neuron-specific and in adult rat brain is predominantly found in presynaptic terminals, presumably associated with the inner side of the plasma membrane and vesicle membranes.[21-23] In view of the similarities in apparent M_r, pI, substrate specificity, metal requirements, peptide maps, sensitivity to modulators, phospholipids, and protease treatment,[20,24] it was concluded that B-50

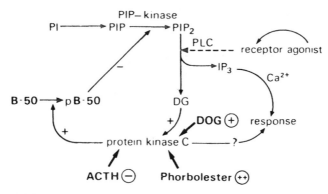

FIGURE 1. Model of the regulatory role of B-50 in receptor-mediated polyphosphoinositide hydrolysis in brain.

protein kinase is very similar if not identical to PKC. This is further substantiated by recent findings indicating that DOG and phorbol esters stimulate the phosphorylation of B-50 in synaptic plasma membranes.[25,26] These data link B-50 via its B-50 kinase/PKC to the role that enhanced production of DG plays in synaptic membrane function (see FIG. 1).

Under a variety of conditions, it was found that there is a reciprocal relationship between the degree of phosphorylation of B-50 and the activity of PIP kinase (for review see ref. 25). Partially purified PIP kinase from rat brain cytosol was tested in the presence of added purified B-50 preparations, which differed in their degree of phosphorylation. In such a reconstituted system, conditions were found under which phosphorylated B-50 proteins reduced PIP kinase activity while dephosphorylated B-50 did not.[27] It was suggested, therefore, that the B-50 protein may be an endogenous modulator of PIP kinase activity in rat SPM (FIG. 1).

As discussed above, activation of certain receptors is associated with enhanced hydrolysis of PIP$_2$. The degraded polyphosphoinositide (PPI) is replenished via PI and PIP, ultimately involving PIP kinase.[28] As enhanced phosphorylation of B-50 is accompanied by a decrease in the activity of PIP kinase, it has been proposed that the sequence DG production, PKC activation, B-50 phosphorylation, PIP kinase inhibition may represent a negative feedback control mechanism in the receptor-mediated hydrolysis of inositol phospholipid (FIG. 1).[29] In fact, several authors have recently shown that direct activation of PKC by means of phorbol esters reduces the production of inositol phosphates in response to muscarinic receptor activation.[30-32]

The proposed feedback role of B-50 in the inositol lipid response was tested in hippocampal slices using the behaviorally active ACTH$_{1-16}$-NH$_2$ to inhibit and PDB to stimulate PKC. We confirmed the results of Labarca et al.[30] and showed that preincubation of hippocampal slices with PDB diminished subsequent receptor-mediated hydrolysis of PIP$_2$ by incubation with carbamylcholine. More importantly, it was shown that preincubation with ACTH$_{1-16}$-NH$_2$, known to inhibit PKC, counteracted the phorbol effect on receptor-mediated PIP$_2$ hydrolysis. Thus, under appropriate conditions ACTH-like peptides may modulate synaptic transmembrane signal transduction through a reduction of the negative feedback role of the PKC–B-50–PIP kinase loop.[33]

ARE CHANGES IN B-50 PHOSPHORYLATION RELATED TO THE EXPRESSION OF GROOMING?

Some 10 years ago, we reported some preliminary findings that pointed to changes in the post hoc *in vitro* phosphorylation of several SPM proteins following prior *in vivo* administration of ACTH$_{1-24}$.[34] In FIG. 2, the effect on *in vivo* grooming and *in vitro* B-50 phosphorylation is presented. As expected the intraventricular administration of ACTH$_{1-24}$ induced excessive grooming in a dose-dependent manner. The doses used were 3, 30, 300, and 3000 ng per 3 μl i.c.v. and the behavior was analyzed by a time sampling procedure between 10 and 30 min following the i.c.v. injection. Immediately after termination of the behavioral recording the rats were killed by decapitation, their subcortical brain tissue was dissected out, and light synaptic plasma membranes were isolated. The endogenous phosphorylation of SPM proteins was studied by addition of [γ-^{32}P]ATP and the incorporation of phosphates into individual proteins was quantified by autoradiographic scanning of SDS-PAGE gels. It was found that several low molecular weight proteins showed an enhanced phosphorylation following ACTH administration *in vivo*. Although in the original report the quantification of an 18 kDa protein is presented, the response of the 48 kDa protein (B-50) is identical.[34] We discussed the difficulty of studying post hoc phosphorylation following an experiment like ACTH-induced grooming. Indeed, a variety of factors may have contributed to the observed change in phosphorylation. Within the context of this paper, the question is whether the observed change in phosphorylation is related to the grooming activity per se or to the interaction of ACTH with certain populations of neurons involved in the expression of the grooming response. Our conclusion at the time was that since *in vitro* addition of ACTH$_{1-24}$ leads to changes in the phosphorylation of the same low molecular weight bands, the latter possibility seemed the more likely.

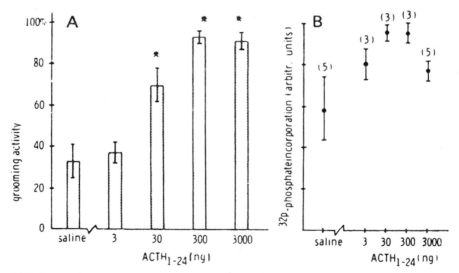

FIGURE 2. (A) ACTH-induced excessive grooming in the rat. Mean ± SEM, $n = 4$, * $p < 0.05$ (t-test, two-tailed). (B) [^{32}P] incorporation into B-50 in SPM after *in vivo* administration of ACTH$_{1-24}$ (30 min; numbers in parentheses represent number of experiments).

Our neurochemical tools in the analysis of B-50 phosphorylation have improved tremendously yet we never went back and repeated this experiment. Thus, although presently more accurate data could be collected addressing this issue more properly, I am convinced at least as far as B-50 is concerned that the conclusion that *in vivo* administration of ACTH may lead to a change in the endogenous phosphorylation as measured in a post hoc assay is valid, for we and others have employed this approach extensively in a variety of experimental designs.

Our interest in the function of the B-50 phosphoprotein originates from our observation that ACTH and congeners *in vitro* inhibit the phosphorylation of this protein by reducing the activity of PKC.[35,36] In a series of experiments the peptide structure requirements for inducing excessive grooming *in vivo* and for inhibiting PKC in SPM *in vitro* were studied. As can be seen in FIG. 3, there is a striking similarity in structural requirements. For instance, the full sequence $ACTH_{1-24}$ is the most active whereas the constituting fragments $ACTH_{1-10}$ and $ACTH_{11-24}$ are virtually inactive in both assays. Even the equimolar combination of $ACTH_{1-10}$ and $ACTH_{11-24}$ is without effect on grooming and PKC. To my knowledge this is one of the best examples of a correlation between a behavioral and a neurochemical activity of a given peptide. I have interpreted these data to underscore the importance of the stereo-confirmation on the effect of the ACTH molecule on the brain. Furthermore, these data show that the presumed ACTH-sensitive sites in the brain *in situ* respond to the information encoded in the ACTH peptide in the same way as the membrane-bound enzyme PKC does when

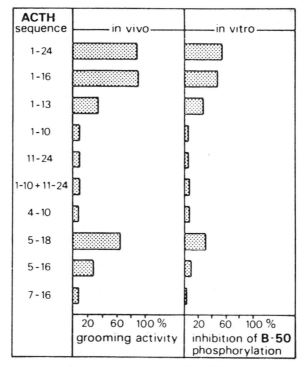

FIGURE 3. Comparison of structural requirements of ACTH for inducing excessive grooming and inhibition of B-50 phosphorylation.

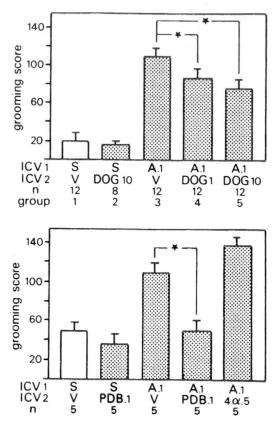

FIGURE 4. The effect of 1,2-dioctanoyl glycerol (*DOG*, top panel) and 4β-phorbol 12,13-dibutyrate (*PDB*, bottom panel) on ACTH-induced excessive grooming: *ICV1, ICV2*, first and second intracerebroventricular injection; *S*, saline; *V*, 0.5% ethanol in saline; *A.1*, 0.1 μg $ACTH_{1-24}$ in saline; *DOG 1, DOG 10*, 1 or 10 μg DOG in 0.5% ethanol in saline; *PDB.1*, 0.1 μg PDB in 0.5% ethanol in saline; *4α.5*, 0.5 μg 4α-phorbol in 0.5% ethanol in saline; *n* = number of rats; *bars*, mean ± SEM; *asterisks*, significantly different from group 3 ($p < 0.05$).

measured *in vitro*, again suggesting that the latter may be involved in the mechanism by which ACTH induces grooming.

If inhibition of PKC is part of the mechanism by which ACTH induces grooming, it was reasoned that concomitant stimulation of this enzyme by phorbol diesters should reduce this behavioral response to ACTH. Rats were first given $ACTH_{1-24}$ (i.c.v. 0.1 μg/3 μl) followed by either 0.5% ethanol in saline (vehicle) or DOG, PDB, PMA, or 4α-PDB, respectively. In order to compare the relative potency of the three phorbol esters to DOG, equimolar amounts to 1 μg DOG were used. As can be seen from Fig. 4, the biochemically nonactive 4α-PDB did not reduce the grooming in response to ACTH. The potency order by which the other compounds inhibit ACTH-induced excessive grooming is PDB > PMA > DOG. This potency order is similar to that found for the activation of PKC by this compound in SPM.

CONCLUDING REMARKS

Although the evidence on the involvement of PKC inhibition in ACTH-induced excessive grooming in increasing, great caution is still warranted in interpreting the significance of our findings.

First of all, most of our work concerned broken cell preparations rather than intact cells or circuits. Only recently have we been able to demonstrate effects of ACTH in intact systems that support the role in altering the negative feedback in the PPI response as was suggested from the data obtained in *in vitro* studies.[33,37] However, even if the peptide was indeed operative via PKC inhibition, how is this accomplished across the membrane as PKC is localized at the inner side of the cell envelope? Previously, Gysin and Schwyzer[38] have discussed the possibility that the ACTH-like peptides might influence membrane function by interactions other than classical receptor activation. The NH_2 terminus of ACTH known to contain all biological activity of the molecule has been shown to form an amphipathic helix[39] that could be inserted into the membrane bilayer upon proper interaction of the positively charged region of the peptide with negatively charged domains in the cell membranes.[38] Such an insertion might bring the biologically important NH_2 terminus in the close vicinity of membrane-bound PKC.

Secondly, the doses at which the peptide inhibits PKC in SPM are considerably larger (IC_{50} 10^{-7} M) than those necessary to activate the signal transduction pathways in the peripheral target cell.[12] This and other issues have led us to propose that the suggested mechanism of action of ACTH on SPM protein phosphorylation would only be of functional significance in brain regions rich in projections of neurons that produce melanocortins, for in such peptidergic synaptic clefts the peptide concentration would be sufficiently high to allow for the train of events depicted in FIG. 1.

Given such a regional specificity of the neurochemical effect of ACTH, what region do we know of that contains PKC and B-50, peptidergic terminals, and is part of the neural substrate underlying behavioral activity of ACTH?

If grooming behavior is considered as a behavioral response to ACTH then the periaqueductal gray is of singular importance. Following reports on opiate effects of implantation of extremely high doses of ACTH into the periaqueductal gray[40] and on the role of periaqueductal gray in bombesin-induced grooming in rats,[41] Spruijt *et al.*[42] concluded that the PAG is essential for the display of ACTH-induced grooming and suggested that this structure was the primary target for the peptide to act upon. Other transmitter systems such as those containing DA and GABA would modulate the output of the PAG, *i.e.*, ACTH-induced excessive grooming.[42] Furthermore, Oestreicher *et al.*[43] demonstrated that the PAG contains relatively high concentrations of B-50. Finally, it is well-established that this brain region receives peptidergic projections containing peptides of the melanocortin family.[44] We are therefore currently studying the significance of ACTH modulation of synaptic transmembrane signal transduction systems in the PAG in relation to induction of excessive grooming. Thus, in addition to the behavioral study of the significance and structure of peptide-induced grooming per se, my laboratory has also used this behavioral response to ACTH as a reliable and relatively simple model system to unravel some of the molecular aspects underlying the behavioral activity of ACTH.

REFERENCES

1. GISPEN, W. H. & R. L. ISAACSON. 1981. ACTH-induced excessive grooming in the rat. Pharmacol. Ther. **12:** 209–246.

2. GISPEN, W. H. & R. L. ISAACSON. 1986. Excessive grooming in response to ACTH. *In* Neuropeptides and Behavior, Vol. 1. D. de Wied, W. H. Gispen & Tj. B. van Wimersma Greidanus, Eds.: 273–312. Pergamon. Oxford.

3. GISPEN, W. H., V. M. WIEGANT, H. M. GREVEN & D. DE WIED. 1975. The induction of excessive grooming in the rat by intraventricular application of peptides derived from ACTH: Structure-activity studies. Life Sci. **17:** 645–652.

4. WITTER, A., W. H. GISPEN & D. DE WIED. 1981. Mechanisms of action of behaviorally active ACTH-like peptides. *In* Endogenous Peptides and Learning and Memory Processes. J. L. Martinez, R. A. Jensen, R. B. Messing, H. Rigter & J. L. McGaugh, Eds.: 37–57. Academic. New York.

5. DE KLOET, E. R., M. PALKOVITS & E. MEZEY. 1986. Opioid peptides: Localization, source and avenues of transport. *In* Neuropeptides and Behavior, Vol. 1. D. de Wied, W. H. Gispen & Tj. B. van Wimersma Greidanus, Eds. Pergamon. New York.

6. TERENIUS, L., W. H. GISPEN & D. DE WIED. 1975. ACTH-like peptide and opiate receptors in the rat brain. Structure-activity studies. Eur. J. Pharmacol. **33:** 395–399.

7. TERENIUS, L. 1976. Somatostatin and ACTH are peptides with partial antagonist-like selectivity for opiate receptors. Eur. J. Pharmacol. **38:** 211–213.

8. GISPEN, W. H. & V. M. WIEGANT. 1976. Opiate antagonists suppress ACTH$_{1-24}$-induced excessive grooming in the rat. Neurosci. Lett. **2:** 159–164.

9. DUNN, A. J., S. R. CHILDERS, N. R. KRAMACY & J. W. VILLIGER. 1981. ACTH-induced grooming involves high-affinity opiate receptors. Behav. Neural. Biol. **31:** 105–109.

10. ALOYO, V. J., B. M. SPRUIJT, H. ZWIERS & W. H. GISPEN. 1983. Peptide-induced excessive grooming in the rat: The role of opiate receptors. Peptides **4:** 833–836.

11. WIEGANT, V. M., H. ZWIERS, A. B. OESTREICHER & W. H. GISPEN. 1986. ACTH and brain cAMP and phosphoproteins. *In* Neuropeptides and Behavior, Vol. 1. D. de Wied, W. H. Gispen & Tj. B. van Wimersma Greidanus, Eds.: 189–210. Pergamon. Oxford.

12. FARESE, R. V., N. ROSIE, J. BABISCHKIN, M. G. FARESE, R. FOSTER & J. S. DAVIS. 1986. Dual activation of the inositol-trisphosphate calcium and cyclic nucleotide intracellular signalling systems by adrenocorticotropin in rat adrenal cells. Biochem. Biophys. Res. Commun. **3:** 742–748.

13. FISCHER, S. K., L. L. A. VAN ROOIJEN & B. W. AGRANOFF. 1984. Renewed interest in the polyphosphoinositides. Trends Biochem. Sci. **9:** 53–56.

14. NISHIZUKA, Y. 1984. Turnover of inositol phospholipids and signal transduction. Science **225:** 1365–1370.

15. GISPEN, W. H., P. N. E. DE GRAAN, L. H. SCHRAMA & L. L. A. VAN ROOIJEN. 1987. Control mechanism in receptor-mediated polyphosphoinositide hydrolysis. Biochem. Pharmacol. In press.

16. EICHBERG, J. & L. V. BERTI-MATTERA. 1986. The role of inositolphosphate in intracellular calcium mobilization. Progr. Brain Res. **69:** 15–28.

17. MICHELL, B. 1986. Cellular signalling. A second messenger function for inositol tetracisphosphate. Nature **324:** 613.

18. NISHIZUKA, Y. 1984. The role of protein kinase C in cell surface signal transduction and tumor promotion. Nature **308:** 693–697.

19. ALBERT, K. A., S. I. WALRAS, J. K-T. WANG & P. GREENGARD. 1986. Wide-spread occurrence of '87 Kda', a major specific substrate for protein kinase C. Proc. Natl. Acad. Sci. **83:** 2822–2826.

20. ALOYO, V. J., H. ZWIERS & W. H. GISPEN. 1982. B-50 protein kinase and kinase C in rat brain. Progr. Brain Res. **56:** 303–315.

21. ZWIERS, H., P. SCHOTMAN & W. H. GISPEN. 1980. Purification and some characteristics of an ACTH-sensitive protein kinase and its substrate protein in rat brain membranes. J. Neurochem. **34:** 1689–1699.

22. KRISTJANSSON, G. I., H. ZWIERS, A. B. OESTREICHER & W. H. GISPEN. 1982. Evidence that the synaptic phosphoprotein B-50 is localized exclusively in nerve tissue. J. Neurochem. **39:** 371–378.

23. GISPEN, W. H., J. L. M. LEUNISSEN, A. B. OESTREICHER, A. J. VERKLEIJ & H. ZWIERS. 1985. Presynaptic localization of B-50 phosphoprotein: The ACTH-insensitive protein kinase substrate involved in rat brain polyphosphoinositide metabolism. Brain Res. **328:** 381–385.

24. ALOYO, V. J., H. ZWIERS & W. H. GISPEN. 1983. Phosphorylation of B-50 protein by calcium-activated, phospholipid-dependent protein kinase and B-50 protein kinase. J. Neurochem. 41: 649–653.

25. DE GRAAN, P. N. E., A. B. OESTREICHER, L. H. SCHRAMA & W. H. GISPEN. 1986. Phosphoprotein B-50: Localization and function. Progr. Brain Res. 69: 37–50.

26. EICHBERG, J., P. N. E. DE GRAAN, L. H. SCHRAMA & W. H. GISPEN. 1986. Dioctanoylglycerol and phorbol diesters enhance phosphorylation of phosphoprotein B-50 in native synaptic plasma membranes. Biochem. Biophys. Res. Commun. 136: 1007–1012.

27. VAN DONGEN, C. J., H. ZWIERS, P. N. E. DE GRAAN & W. H. GISPEN. 1985. Modulation of the activity of purified phosphatidylinositol 4-phosphate kinase by phosphorylated and dephosphorylated B-50 protein. Biochem. Biophys. Res. Commun. 8: 1219–1227.

28. DOWNES, C. P. & M. WÜSTERMAN. 1983. Breakdown of polyphosphoinositide and phospholipid inositol accounts for muscarinic agonist-stimulated inositol phospholipid metabolism in rat parotid glands. Biochem. J. 216: 633–640.

29. GISPEN, W. H. 1986. Phosphoprotein B-50 and phosphoinositides in brain synaptic plasma membranes: A possible feedback relationship. Trans. Biochem. Soc. U.K. 14: 163–165.

30. LABARCA, R., A. JANOWSKY, J. PATEL & S. M. PAUL. 1984. Phorbol esters inhibit agonist-induced [^{3}H]-inositol-1-phosphate accumulation in rat hippocampal slices. Biochem. Biophys. Res. Commun. 123: 703–709.

31. VINCENTINI, L. M., F. DI VIRGILIO, A. AMBROSINI, T. POZZAN & J. MELDOLESI. 1985. Tumor promoter phorbol 12-myristate, 13-acetate inhibits phosphoinositide hydrolysis and cytosolic calcium rise induced by the activation of muscarinic receptors in PC12 cells. Biochem. Biophys. Res. Commun. 127: 310–317.

32. ORELLANA, S. A., P. A. SOLSKI & J. H. BROWN. 1985. Phorbol ester inhibits phosphoinositide hydrolysis and calcium mobilization in cultured astrocytoma cells. J. Biol. Chem. 260: 5236–5239.

33. SCHRAMA, L. H., P. N. E. DE GRAAN, J. EICHBERG & W. H. GISPEN. 1986. Feedback control of the inositol phospholipid response in rat brain is sensitive to ACTH. Eur. J. Pharmacol. 121: 403–404.

34. ZWIERS, H., V. M. WIEGANT, P. SCHOTMAN & W. H. GISPEN. 1977. Intraventricular administered ACTH and changes in rat brain protein phosphorylation. A preliminary report. In Mechanism, Regulation and Special Functions of Protein Synthesis in the Brain. S. Roberts, A. Lajtha & W. H. Gispen, Eds.: 267–272. Elsevier/North-Holland Biomedical. Amsterdam.

35. ZWIERS, H., D. VELDHUIS, P. SCHOTMAN & W. H. GISPEN. 1976. ACTH, cyclic nucleotides and brain protein phosphorylation in vitro. Neurochem. Res. 1: 669–677.

36. ZWIERS, H., V. M. WIEGANT, P. SCHOTMAN & W. H. GISPEN. 1978. ACTH-induced inhibition of endogenous rat brain protein phosphorylation in vitro: structure activity. Neurochem. Res. 3: 455–463.

37. DE GRAAN, P. N. E., L. V. DEKKER, M. DE WIT, L. H. SCHRAMA & W. H. GISPEN. 1987. Modulation of B-50 phosphorylation and polyphosphoinositide metabolism in synaptic plasma membranes by protein kinase C, phorbol diesters and ACTH. J. Rec. Res. In press.

38. GYSIN, B. & R. SCHWYZER. 1984. Hydrophobic and electrostatic interactions between adrenocorticotropin-(1-24) tetracosapeptide and lipid vesicles. Amphiphilic primary structures. Biochemistry 23: 1811–1818.

39. VERHALLEN, P. J. E., R. A. DEMEL, H. ZWIERS & W. H. GISPEN. 1984. Adrenocorticotropic hormone (ACTH)-lipid interactions. Implications for involvement of amphipathic helix formation. Biochim. Biophys. Acta 775: 246–254.

40. JACQUET, J. F. 1978. Opiate effects after adrenocorticotropin or β-endorphin injection in the periaqueductal gray matter of rats. Science 205: 425.

41. GMEREK, D. E. & A. COWAN. 1982. Classification of opioids on the basis of their ability to antagonize bombesin-induced grooming in rats. Life Sci. 31: 2229–2232.

42. SPRUIJT, B. M., A. R. COOLS & W. H. GISPEN. 1986. The periaqueductal gray: A prerequisite for ACTH-induced excessive grooming. Behav. Brain Res. 20: 19–25.

43. OESTREICHER, A. B., L. V. DEKKER & W. H. GISPEN. 1986. A radioimmunoassay for the phosphoprotein B-50: Distribution in rat brain. J. Neurochem. 46: 1366–1369.

44. WATSON, S. J., C. W. RICHARD III & J. D. BARCHAS. 1978. Adrenocorticotropin in rat brain: Immunochemical localization in cells and axons. Science 275: 226–228.

Studies on the Neurochemical Mechanisms and Significance of ACTH-induced Grooming[a]

ADRIAN J. DUNN

Department of Neuroscience
University of Florida College of Medicine
Gainesville, Florida 32610

In this paper I shall address three topics: (1) the sites of ACTH action that elicit grooming, and the neurochemical systems that are involved in the grooming response; (2) the identification of neurochemical events that correlate with the occurrence of ACTH-induced grooming; and (3) the physiological significance of ACTH-induced grooming.

GENERAL METHODS

The rats used were male Sprague-Dawleys, bred at the University of Florida Health Center Animal Research Department Facility. The mice used in most experiments were male CD-1 mice from Charles River, but male C57B1/6J mice from Jackson Laboratories were used for the studies on protein synthesis. For the experiments with intracerebroventricular (i.c.v.) injections, mice were implanted with plastic cannulae (Clay-Adams PE-50) in each lateral ventricle as described by Guild and Dunn,[1] based on the methods of Brakkee *et al.*[2] for rats and of Delanoy *et al.*[3] for mice. Rats were implanted with 23 gauge stainless steel cannulae in each lateral ventricle as previously described.[4,5] After cannulation rats and mice were housed one per cage on a 12h-12h light-dark cycle (lights on at 7 a.m.). Peptides were dissolved in isotonic saline containing 0.001 M HCl. Solutions used were fresh or were stored frozen in small aliquots for injection. They were injected by hand into both cannulae, in a volume of either 2 or 3 µl for each cannula.[1,5] Immediately following injection, the rats or mice were placed in clean Plexiglas cages. The behaviors were recorded by an observer unaware of the nature of the injectate, using a 15-second time sampling procedure as previously described.[3,6] The grooming scores presented are the actual or the percentage scores recorded for the interval 15–60 min after injection of peptide. Experiments with different doses of peptides or drugs were conducted in a randomized repeated measures design, such that each animal received each dose of peptide or drug, and there was only one scoring session each day. We did not observe any significant habituation or sensitization to repeated daily doses of i.c.v. $ACTH_{1-24}$.

Unless otherwise specified, the ACTH used in most experiments was $ACTH_{1-24}$, a gift from Organon International, BV. $ACTH_{4-10}$ and $[D-Phe-7]ACTH_{4-10}$ were ob-

[a] The research reported here was supported by a grant from the National Institute of Mental Health (no. MH25486).

tained from the same source. All other peptides and drugs were obtained from Sigma, with the following exceptions: human β-endorphin, N-acetyl-β-endorphin, α-MSH, desacetyl-α-MSH, and CRF from Peninsula, β-endorphin, dynorphin$_{1-17}$, α-MSH, desacetyl-α-MSH, and Tyr-D-Ala-Gly-MePhe-NH(CH$_2$)$_2$OH from Cambridge Research Biochemicals (CRB). Dihydroergotamine methanesulfonate (DHEMS) was from Sandoz.

PEPTIDES ACTIVE IN ELICITING GROOMING BEHAVIOR

Because our experiments depend in part on the use of different peptides with differing abilities to elicit grooming behavior, I shall first summarize our data on the ability of various ACTH analogues and opioid peptides to induce grooming in rats and mice.

As found earlier by Gispen et al.[6] in rats, ACTH$_{1-24}$ was the most potent of the peptides in mice; ACTH$_{4-10}$ was essentially inactive, although [D-Phe-7]ACTH$_{4-10}$ was active[7] (FIG. 1). In our studies in rats, α-MSH([N-Ac-Ser-1]ACTH$_{1-13}$ amide) was as potent as ACTH$_{1-24}$, but in agreement with O'Donohue et al., desacetyl-α-MSH (i.e., ACTH$_{1-13}$) was inactive[8] (FIG. 2). With regard to the opioid peptides, β-endorphin was active at low doses, but became inhibitory at higher doses (FIG. 2). These results are consistent with those of Gispen's group,[9,10] except that we do not find β-endorphin to be any more potent than ACTH$_{1-24}$. This was true for samples of β-endorphin from Sigma, Peninsula, and CRB. N-Acetyl-β-endorphin was completely inactive (data not shown). We found no grooming reponse to the enkephalins, Met-enkephalin (see also refs. 9 and 10), Leu-enkephalin, and D-Ala,D-Leu-enkephalin (DADLE), and only a

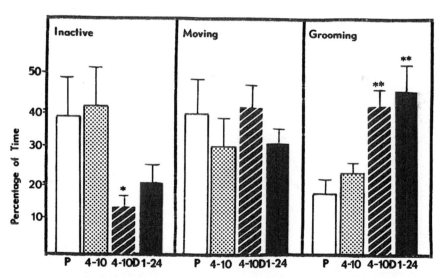

FIGURE 1. Effect of ACTH analogues on grooming behavior in C57Bl/6J mice; 2.5 pmol ACTH$_{4-10}$ (4–10), [D-Phe-7]ACTH$_{4-10}$ (4–10D), ACTH$_{1-24}$ (1–24), or placebo (P, an amino acid mixture corresponding to the constituents of ACTH$_{4-10}$) were injected intracerebroventricularly. Inactivity (quiet plus sleeping) scores were significantly depressed by [D-Phe-7]ACTH$_{4-10}$, and grooming was significantly increased by [D-Phe-7]ACTH$_{4-10}$ and ACTH$_{1-24}$ (Dunnett's test). Asterisks in this and subsequent figures: *$p < 0.05$, **$p < 0.01$, ***$p < 0.001$. (Data from Rees et al.[7])

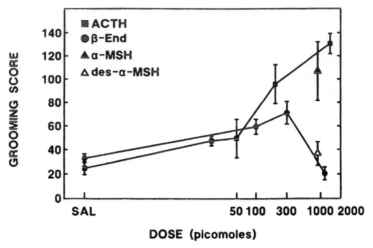

FIGURE 2. The effect of i.c.v. ACTH, α-MSH, and β-endorphin (Sigma) on grooming behavior in rats. The same rats ($n = 8$) were injected i.c.v. with various doses of ACTH$_{1-24}$ (*squares*), β-endorphin (*circles*), α-MSH (*solid triangle*) or desacetyl-α-MSH (*open triangle*), and scored behaviorally for 60 minutes. Grooming scores are for the period 15–60 min postinjection.

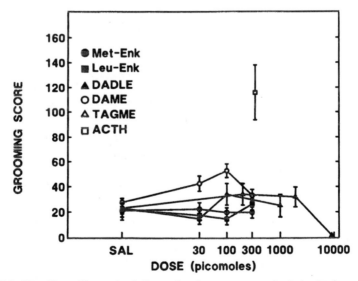

FIGURE 3. The effect of i.c.v. enkephalins and analogues on grooming behavior in rats: [Met-5] enkephalin (Met-Enk, *solid circles*), [Leu-5]enkephalin (Leu-Enk, *solid squares*), [D-Ala-2,D-Leu-5]enkephalin (DADLE, *solid triangles*), [D-Ala-2,Met-5]enkephalinamide (DAME, *open circles*), and Tyr-D-Ala-Gly-MePhe-NH(CH$_2$)$_2$OH (TAGME, *open triangles*), were injected i.c.v. at various doses as in FIGURE 2. For comparison, the response to 1 μg of ACTH$_{1-24}$ (*open square*) in the same animals is shown.

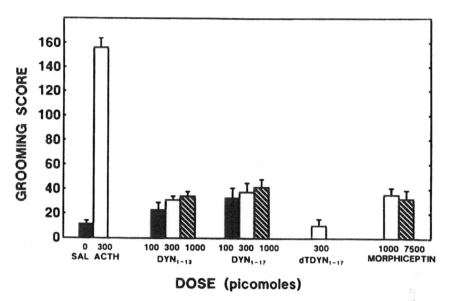

FIGURE 4. The effect of i.c.v. dynorphin and morphiceptin on grooming behavior in rats. Dynorphin$_{1-13}$ (*DYN$_{1-13}$*), dynorphin$_{1-17}$ (*DYN$_{1-17}$*), [des-Tyr-1]dynorphin$_{1-17}$ (*dTDYN$_{1-17}$*), or morphiceptin (β-casomorphin) were injected i.c.v. as in FIGURE 2. For comparison the effect of ACTH$_{1-24}$ in the same animals is shown.

very slight response to D-Ala-Met-enkephalinamide (DAME) (FIG. 3). The activity of β-endorphin but not the enkephalins suggested the involvement of a μ-receptor, therefore we tested several other drugs known to be μ-agonists. Morphiceptin (β-casomorphin) had very little activity (FIG. 4). Tyr-D-Ala-Gly-MePhe-NH(CH$_2$)$_2$OH (TAGME), the potent μ-agonist,[11] had a little activity at low doses, but like β-endorphin was inhibitory at high ones (FIG. 3). In contrast to the results of Gispen's group,[13,14] we observed very little response to dynorphin$_{1-13}$, dynorphin$_{1-17}$, or des-Tyr-dynorphin$_{1-17}$ (FIG. 4).

In addition to the endorphins, many other peptides have been reported to elicit grooming following i.c.v. injection. These include vasopressin,[13,14] bombesin,[15-17] eledoisin,[15] substance P[18], prolactin,[19] and CRF.[20,21] In mice, we have observed grooming behavior in response to vasopressin,[3] substance P, and prolactin, but not to bombesin. We observe a very small response to CRF in mice, although we do observe a substantial grooming response to CRF in rats.[22] In our experience, none of these peptides elicits grooming to the degree that ACTH can. One microgram of ACTH$_{1-24}$ injected into the lateral ventricle of a rat can cause the animal to groom for more than 90% of the time over the next hour.[4,23] In the case of vasopressin and CRF the grooming may be due to secondary release of ACTH. In other cases the grooming elicited is qualitatively different from ACTH-induced grooming; the grooming elicited by vasopressin[3,14] and bombesin[16,17] is more of a compulsive scratching. Thus the grooming behavior induced by ACTH may be special in that it is indistinguishable from natural grooming.[23]

CEREBRAL SYSTEMS INVOLVED IN
ACTH-INDUCED GROOMING

Lesion studies have not been particularly fruitful in determining the brain structures and sites involved in ACTH-induced grooming. Lesions of the septal complex, medial and lateral preoptic areas, anterior hypothalamus, mammillary bodies, amygdala, posterior thalamus, and dorsal or ventral hippocampus did not alter ACTH-induced grooming.[24] The only lesions that successfully inhibited ACTH-induced grooming were the total removal (by aspiration) of the hippocampus,[24] or the destruction of the periaqueductal gray.[25] (Gispen and Isaacson cite an unpublished observation that substantia nigra lesions inhibited ACTH-induced grooming.[23]) Superior colliculus lesions did not alter ACTH-induced grooming, but GABA antagonists in this area did suppress it.[26] Localization of the primary site of action of ACTH in eliciting grooming has been attempted using direct local injections. Early experiments by Gessa et al.[27] suggested periventricular hypothalamic sites to be effective for ACTH to elicit the stretching and yawning syndrome (SYS). Later experiments have suggested the substantia nigra but not the striatum as a potential locus of action.[28] There are conflicting data on the nucleus accumbens; Wiegant et al.[28] did not find grooming from injections of $ACTH_{1-24}$ into the accumbens, whereas Ryan and Isaacson[29] did.

We have studied the localization of the grooming-promoting action of ACTH using a different procedure; intracerebroventricular injection of cold cream to block access of injected ACTH to specific brain sites. The results[30,31] indicated clearly that the ventricular site of action of ACTH is in the third ventricle, and not in the lateral or fourth ventricles. Experiments involving coating specific parts of the ependymal surface of the third ventricle with cold cream indicated further that it is the anteroventral quadrant of the third ventricle in the region of the organum vasculosum of the lamina terminalis (OVLT) that is critical for the action of ACTH. This of course does not prove that the OVLT is the site of action of ACTH; the peptide could be taken up in the OVLT and transported to another brain site. However, the very short latency of grooming behavior following i.c.v. injection of $ACTH_{1-24}$ (less than 5 minutes) suggests that the crucial site cannot be too far away. It may be relevant that studies of angiotensin-induced drinking behavior[32] and of bradykinin-induced increases in blood pressure[33] also localize to the OVLT, which may, therefore, have a rather special role in the uptake and response to peptides in the ventricular system.

THE PHARMACOLOGY OF ACTH-INDUCED GROOMING

A second approach to mechanism is a pharmacological one. In an early study, Gispen and Wiegant[34] observed that grooming in rats could be prevented by pretreatment with the opiate antagonists, naloxone and naltrexone, and that low doses of morphine could elicit grooming. Subsequently, it was shown that i.c.v. β-endorphin, but not α-endorphin, γ-endorphin, or Met-enkephalin, elicit grooming.[9] The ability of ACTH analogues to elicit grooming closely parallels their ability to counteract morphine-induced antinociception in the tail-flick test, and to compete for dihydromorphine binding to presumed opiate receptors.[10]

We confirmed the ability of naloxone to antagonize ACTH-induced grooming in mice, and furthermore showed that the irreversible opiate antagonist, naloxazone, inhibited grooming as well as morphine-induced antinociception[35] (FIG. 5). This inhibition was present at 16 h post injection when the acute effects of naloxazone had worn off. However, at 96 h after naloxazone, a time when opiate-binding sites return (pre-

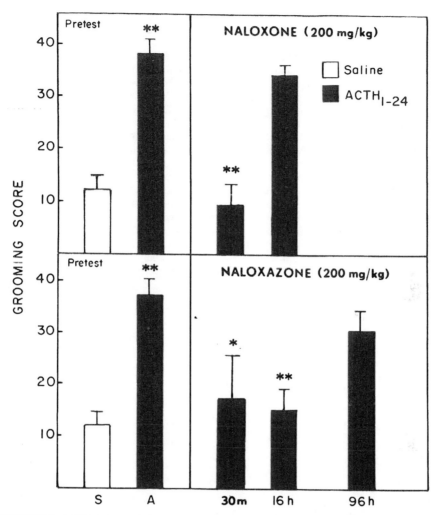

FIGURE 5. ACTH-induced grooming after naloxone and naloxazone treatment of CD-1 mice. Grooming behavior was studied for a 30-min period commencing 15 min following i.c.v. injection of saline or 1 μg ACTH$_{1-24}$, scoring every 30 seconds. Mice were pretreated i.p. with naloxone or naloxazone (200 mg/kg), and ACTH$_{1-24}$ was injected at various times after this. The same mice were also tested for morphine-induced antinociception in the tail-immersion test (for details, see ref. 35). In the pretest (*left panels*), ACTH$_{1-24}$ significantly increased grooming behavior compared to saline controls (paired *t*-test). Thirty minutes after naloxone or naloxazone (*right panels*) ACTH-induced grooming was significantly decreased. At 16 h after drug injection, the naloxone-treated group resembled control animals, while ACTH-induced grooming was still significantly lower than pretest scores in the naloxazone-treated group. At this time, morphine-induced antinociception had recovered in the naloxone-, but not the naloxazone-treated group. By 96 h after naloxazone, both morphine-induced antinociception and ACTH-induced grooming had recovered. (Data from Dunn *et al.*[35])

sumably from new synthesis), ACTH-induced grooming recovered simultaneously with the recovery of morphine-induced antinociception, as determined in the tail-immersion test.[35]

This parallellism between the recovery of opiate-binding, ACTH-induced grooming, and morphine-induced antinociception suggests that μ-receptors are involved in ACTH-induced grooming. ACTH does have a low affinity for opiate receptors.[36] Nevertheless, the idea that ACTH acts directly on an opiate receptor to elicit grooming is not widely accepted. The behavioral pattern elicited by i.c.v. β-endorphin is distinct from that induced by ACTH. After i.c.v. ACTH, the increased grooming is reflected in an increased length of grooming episodes (bouts), with only a small increase in the number of bouts.[4,23] According to Gispen and Isaacson,[23] β-endorphin does not increase the bout length, but does increase the number of bouts. (Our data are not entirely in agreement with this, but we obtain a lower maximal grooming response to β-endorphin.) Moreover, β-endorphin, unlike ACTH, does not elicit SYS. These discrepancies could be explained by action of β-endorphin on receptors in addition to those on which ACTH is effective. Because of the lack of activity of the enkephalins[9,10] (see above) and especially DADLE, a δ-receptor would appear to be excluded. Dynorphin, a κ-agonist, is active according to the work of Gispen's group but not our own (FIG. 4). Activity of β-endorphin would suggest a μ- or ε-receptor. However, our studies (FIGS. 3 and 4) indicate little activity of known μ-agonists, such as morphiceptin and TAGME, even though our naloxazone data[35] suggest the involvement of a μ-receptor in grooming. Therefore, we, like most other authors,[13,23] consider the opiate receptor involvement not to be primary, but rather "downstream" from the primary ACTH receptor.

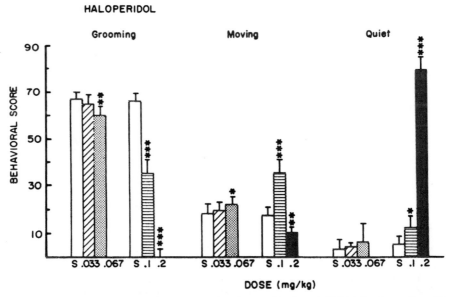

FIGURE 6. Effects of haloperidol on behaviors induced by i.c.v. ACTH. Mice (n = 11 or 12) were treated with the indicated dose of haloperidol intraperitoneally immediately after 1 μg ACTH$_{1-24}$ administered intracerebroventricularly. Behavior was scored for 15–60 minutes postinjection. *Bars* are the medians plus interquartile ranges; *S,* saline vehicle (From Guild and Dunn.[1] Reprinted with permission from *Pharmacology, Biochemistry and Behavior.*)

Other pharmacological agents whose effects on grooming have been clearly established are the dopaminergic antagonists. Haloperidol is effective in both rats[28,37,38] and mice[1]; ergometrine,[38] fluphenazine,[28] and clozapine in rats,[37] and pimozide, butaclamol, and metoclopramide in mice.[1] Because dopamine (DA) receptor antagonists are well known to decrease motor activity, it is important that Guild and Dunn found that at certain doses of haloperidol, ACTH-induced grooming was decreased while locomotor activity in the same animals increased[1] (FIG. 6). Interestingly, low doses of metoclopramide, which may act preferentially on presynaptic receptors, can potentiate ACTH-induced grooming, whereas higher doses, with a presumed postsynaptic action, inhibit it.[1] In support of a dopaminergic mechanism of action, apomorphine stimulated grooming at low doses of ACTH in mice.[1] Cools *et al.* found that haloperidol was more effective in inhibiting ACTH-induced grooming when injected into the striatum than into nucleus accumbens, while the reverse was true for ergometrine.[38] A dopaminergic mechanism for ACTH-induced grooming is also supported by the activity of ACTH injected directly into the substantia nigra but not the striatum.[28] However, in our experience, lesions with 6-hydroxydopamine (6-OHDA) of the prefrontal cortex dopaminergic projections, or those of the nucleus accumbens, both failed to alter ACTH-induced grooming.[4] Another study found an inhibition of ACTH-induced grooming with 6-OHDA lesions of nucleus accumbens.[39] The reasons for the discrepancy are not clear, but the earlier study used repeated injections of 6-OHDA and obtained lesser depletions of nucleus accumbens DA, and thus there may have been nonspecific damage to nondopaminergic systems.

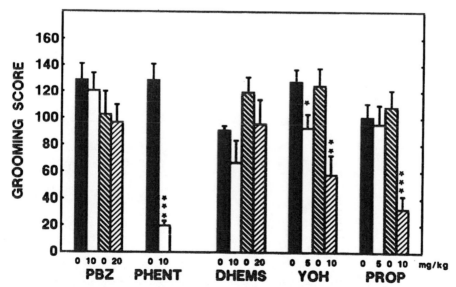

FIGURE 7. The effect of adrenergic antagonists on ACTH-induced grooming. Rats ($n = 8$) were pretreated (15 min) with phenoxybenzamine (*PBZ*, 10 or 20 mg/kg), phentolamine (*PHENT*, 10 mg/kg), dihydroergotamine methanesulfonate (*DHEMS*, 10 or 20 mg/kg), yohimbine (*YOH*, 5 or 10 mg/kg), or propranolol (*PROP*, 5 or 10 mg/kg), before receiving 1 µg ACTH$_{1-24}$ intracerebroventricularly. Each dose of each drug was compared with saline or vehicle in a 2-day design. ACTH-induced grooming was significantly reduced by phentolamine, yohimbine, and the 10 mg/kg dose of propranolol (paired *t*-test).

We have found a limited effect of noradrenergic antagonists on ACTH-induced grooming. Whereas the relatively less specific α-antagonists, phenoxybenzamine and dihydroergotamine, had very little effect, phentolamine, a more specific α-antagonist, and yohimbine, an α_2-antagonist, did inhibit ACTH-induced grooming (FIG. 7). Propranolol (β-antagonist) was effective only at a relatively high dose. These data may explain the observations of Lassen[40] suggesting an adrenergic involvement in amphetamine-induced grooming.

Muscarinic cholinergic antagonists are, however, rather effective in inhibiting ACTH-induced grooming. Both atropine and scopolamine inhibit ACTH-induced grooming (FIG. 8). This action is almost certainly central because the methyl derivatives of these drugs which cross into the brain very poorly do not inhibit ACTH-induced grooming when administered whereas they are effective when administered intracerebroventricularly.[5] Hemicholinium, which depletes CNS acetylcholine when given i.c.v. also prevents ACTH-induced grooming.[5]

An important question is the relationship between ACTH-induced grooming and SYS. Higher doses of $ACTH_{1-24}$ appear to be necessary to elicit SYS as compared to grooming.[6,7,23] Colbern et al.[24] dissociated SYS from grooming on the basis of lesion studies (amygdaloid and hippocampal lesions were reported to increase SYS, but not to affect grooming). Nevertheless, ACTH-induced grooming pharmacologically resembles MSH-induced SYS, which is inhibited by both muscarinic and dopaminergic antagonists.[41]

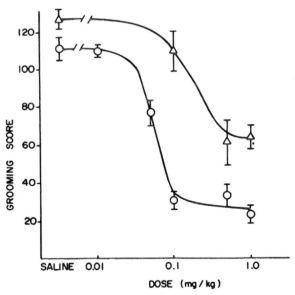

FIGURE 8. The effect of atropine and scopolamine on ACTH-induced grooming in rats. Rats ($n = 8$) were pretreated with various doses of atropine (*triangles*) or scopolamine (*circles*) immediately before 1 μg $ACTH_{1-24}$ i.c.v. in a randomized repeated measures design. (From Dunn and Vigle. Reprinted with permission from *Neuropharmacology*.)

GLUCOSE UTILIZATION STUDIES

A further method for elucidating the brain structures involved in the grooming response is the deoxyglucose method of Sokoloff[42] for determining glucose utilization (GU) and hence cerebral metabolic activity. Using a microdissection procedure followed by scintillation counting, we have studied presumed glucose utilization following i.c.v. administration of ACTH. $ACTH_{1-24}$, which elicits grooming, significantly decreased GU in the prefrontal and pyriform cortices, and increased it in the thalamus[43,44] (Fig. 9). Statistically significant correlations between the changes in GU and the grooming scores were observed in the olfactory bulbs, pyriform cortex (negative), and thalamus and cerebellum (positive).[44] No such changes were observed with $ACTH_{4-10}$[44] (Fig. 9), or with peripherally injected ACTH,[45] neither of which elicit grooming. Naloxone, which antagonized the grooming behavior, appeared to attenuate the changes in GU.[44] The changes in GU may thus reflect changes in cerebral activity involved in the expression of grooming, or of sensory responses to the behavior, rather than the direct action of ACTH. These experiments do not enable such distinctions to be made, therefore the activity changes in pyriform cortex, which is strongly linked to the olfactory system, may be associated with olfactory stimuli perceived during grooming.

CEREBRAL PROTEIN SYNTHESIS

Several years ago we observed small but consistent and statistically significant increases of the incorporation of [³H]lysine into protein (i.e., brain protein synthesis) after con-

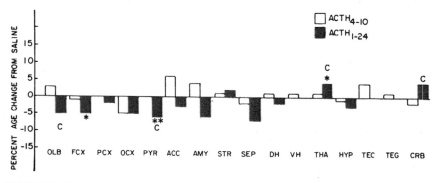

FIGURE 9. Regional glucose utilization in CD-1 mice after $ACTH_{4-10}$ or $ACTH_{1-24}$. $ACTH_{4-10}$ or $ACTH_{1-24}$ (1 µg i.c.v.) were injected 10 min before [³H]2-deoxyglucose (10 µCi s.c.). Mice were scored behaviorally during the glucose uptake period of 40 min. Radioactivity per milligram of tissue was determine by scintillation counting following freehand microdissection of the brain regions, and normalized by dividing by the specific activity of the entire brain. *OLB*, olfactory bulbs; *FCX*, prefrontal cortex; *PCX*, parietal cortex; *OCX*, occipital cortex; *PYR*, pyriform cortex; *ACC*, nucleus accumbens; *AMY*, amygdala; *STR*, striatum; *SEP*, septum; *DH*, dorsal hippocampus; *VH*, ventral hippocampus; *THA*, thalamus; *HYP*, hypothalamus; *TEC*, tectum; *TEG*, tegmentum; *CRB*, cerebellum. $ACTH_{1-24}$ significantly decreased glucose utilization in prefrontal and pyriform cortices, and increased it in the thalamus. There was a statistically significant correlation (*C*) between the grooming behavior induced by $ACTH_{1-24}$ and the glucose utilization in the olfactory bulb, pyriform cortex, thalamus, and cerebellum. No statistically significant changes were observed following $ACTH_{4-10}$. (Data from Dunn and Hurd.[44])

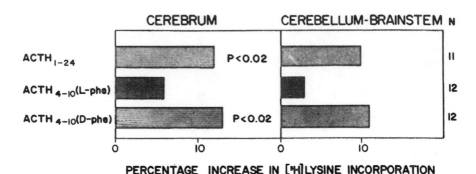

PERCENTAGE INCREASE IN [³H]LYSINE INCORPORATION

FIGURE 10. Brain protein synthesis following i.c.v. injection of ACTH analogues. $ACTH_{4-10}$, [D-Phe-7]$ACTH_{4-10}$, or $ACTH_{1-24}$ (2.5 pmol each) were injected into C57Bl/6J mice i.c.v. 75 minutes before 10 μCi [³H]lysine (s.c.). Brain protein synthesis was estimated by the incorporation of [³H]lysine into protein over a 10-min period corrected for the total amino acid uptake. $ACTH_{1-24}$ and [D-Phe-7]$ACTH_{4-10}$ significantly increased the corrected incorporation of [³H]lysine into protein in the cerebrum (forebrain), but not in the cerebellum and brain stem. The behavioral data from these animals are shown in FIGURE 1. (Data from Rees *et al.*[7])

ditioned avoidance training in mice[46] and rats.[47] Subsequent work indicated that this neurochemical response was due at least in part to the release of ACTH during stress. The effect of footshock in mice was prevented by prior hypophysectomy,[48] but not adrenalectomy,[49] and peripheral injections of $ACTH_{1-24}$ could mimic the effects of footshock.[50] Intracerebroventricular administration of $ACTH_{1-24}$ or [D-Phe-7]$ACTH_{4-10}$, which elicit grooming, but not $ACTH_{4-10}$, which does not, also increased brain protein synthesis[7] (FIG. 10). Thus the neurochemical response paralleled the elicitation of grooming behavior, and there was a statistically significant correlation between the increase of protein synthesis and grooming behavior in the ACTH-injected animals.

DOPAMINERGIC SYSTEMS

A further suggestion of a relationship between ACTH and DA systems became apparent from studies in brain slices. Intracerebroventricular injection of $ACTH_{1-24}$ activated the synthesis of DA in slices prepared from the prefrontal cortex, but not from any other brain region.[51] The effects of ACTH *in vivo* could not be mimicked by $ACTH_{1-24}$ or $ACTH_{4-10}$ *in vitro*. An effect similar to that of i.c.v. $ACTH_{1-24}$ was also produced by prior footshock treatment in intact or adrenalectomized mice.[52] Thus it seems likely that either ACTH released during stress initiates the prefrontal cortex DA response, and/or that the i.c.v. injection of $ACTH_{1-24}$ is stressful. The relationship between this prefrontal cortex DA response and grooming behavior is unclear, because under strong stressful conditions such as footshock grooming behavior is not enhanced.[23] Moreover, as indicated above, depletion of prefrontal cortex DA with 6-OHDA did not alter ACTH-induced grooming.[4]

NOVELTY-INDUCED GROOMING: IS ACTH INVOLVED?

The last point I wish to discuss is the potential physiological significance of the grooming response to ACTH. Grooming has been described as a displacement behavior, which is a term applied to behavioral patterns that appear to be out of context with the behaviors that closely precede or follow them.[53,54] The behavior may appear in conflict situations where it seems inappropriate.[53,55]

An experimental tool to study this phenomenon became available with the observation of Colbern et al.[56] that the mere exposure to a novel environment significantly increased the grooming scores of rats. This was in fact a rediscovery of effects reported earlier.[57,58] The question was whether secretion of ACTH was responsible for this increase in grooming elicited in a novel environment. First we found that, like ACTH-induced grooming, novelty-induced grooming was antagonized by haloperidol and by naloxone at low doses (0.2 mg/kg SC for each) that did not alter general activity.[59] This pharmacological similarity was encouraging but not conclusive. Therefore, we examined the grooming response to a novel environment in hypophysectomized rats. Hypophysectomy does not impair ACTH-induced grooming,[6] but in our laboratory it markedly decreased the increased grooming in a novel cage[60] (FIG. 11). This implicates pituitary ACTH. However, others have failed to reproduce our findings,[61] and the reasons for the discrepancy are unresolved.[23,62] A more definitive experiment was to inject the rats i.c.v. with antiserum specific to ACTH before placing them in the

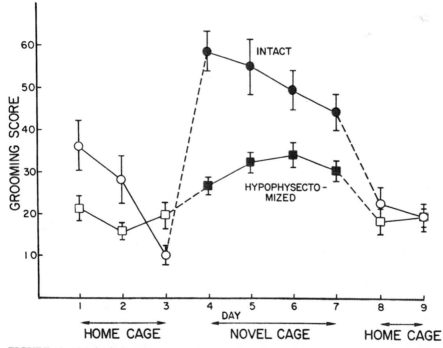

FIGURE 11. Novelty-induced grooming in hypophysectomized rats. Hypophysectomized (*n* = 16) or intact (*n* = 8) rats were scored for grooming in their home cage on days 1–3 and on days 8 and 9, and in a novel cage on days 4–7. Grooming was significantly increased in the novel cage on days 4–7 compared with day 3 in intact but not in hypophysectomized rats (although when the mean grooming score on days 4–7 was compared to that on day 3, there was a statistically significant difference). (From Dunn et al.[60] Reprinted with permission from *Science*.)

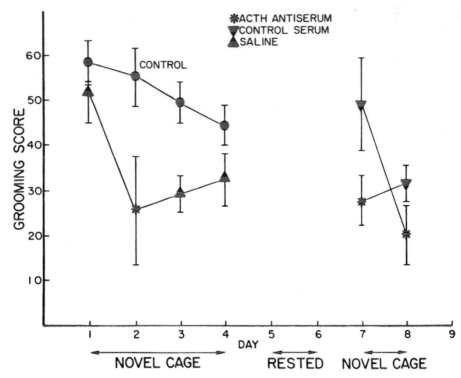

FIGURE 12. The effect of i.c.v. antiserum to ACTH on novelty-induced grooming behavior in rats. Rats were tested in novel cages on days 1–4 and on days 7 and 8. On days 1, 3, and 4, they received a saline injection i.c.v., and on day 2 an injection of antiserum to ACTH (*). On day 7, half of the rats received ACTH antiserum and half received control (preimmune) serum, with the complementary injection on day 8. ACTH antiserum significantly reduced grooming on day 2 compared to day 1, and there was a statistically significant difference between ACTH antiserum and control serum on day 7. (From Dunn *et al.*[60] Reprinted with permission from *Science*.)

novel environment. The antiserum significantly decreased the grooming response in the novel cage[60] (FIG. 12). Such an effect was not observed with preimmune serum or saline injections. These results suggest strongly that the incresed grooming in a novel environment is due to secretion within the brain of ACTH.

The source of this ACTH is more difficult to determine. In addition to the pituitary, small populations of neurons in the brain synthesize pro-opiomelanocortin (POMC), the precursor to α-MSH, ACTH, and β-endorphin,[63] and thus these neurons could be the source of the peptides involved in novelty-induced grooming. A selective destruction of the POMC neurons in brain occurs following neonatal treatment with monosodium glutamate (MSG).[64,65] Therefore we used this technique to determine whether destruction of the cerebral POMC neurons would alter novelty-induced grooming. In our MSG-treated rats, hypothalamic β-endorphin was depleted by 84%, and ACTH by 97%. Nevertheless, the grooming response in a novel cage was not altered.[66] This might be accounted for by a corrective supersensitivity of the postsynaptic receptors to MSH or ACTH. However, we did not observe any increase in the sensi-

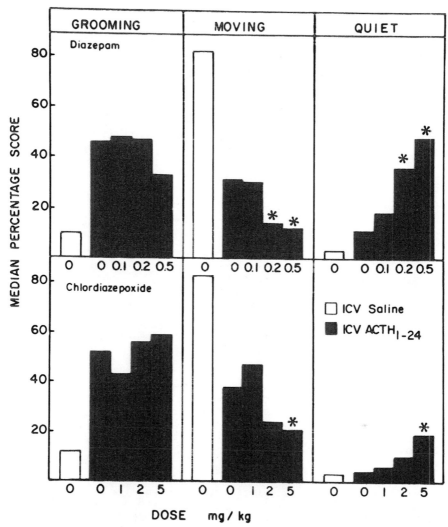

FIGURE 13. The effect of benzodiazepine treatment on ACTH-induced grooming. Mice were injected with $ACTH_{1-24}$ (1 µg i.c.v.), 15 min before diazepam (DZP) or chlordiazepoxide (CDP). Motor activity in the period 15–45 min after i.c.v. ACTH was significantly reduced by DZP at 0.2 or 0.5 mg/kg and by CDP at 5 mg/kg, and there were corresponding significant increases in inactivity scores. ACTH-induced grooming scores were not depressed by any of the doses of DZP or CDP. (Data from Dunn et al.[67])

tivity of the grooming response to various doses of ACTH.[66] Thus the results of these experiments do not support the hypothesis that cerebral POMC neurons are the source of the ACTH involved in novelty-induced grooming.

Benzodiazepines are known to decrease the responsiveness of the hypothalamic-pituitary-adrenal system in stress,[67-69] therefore we tested whether these "anxiolytics"

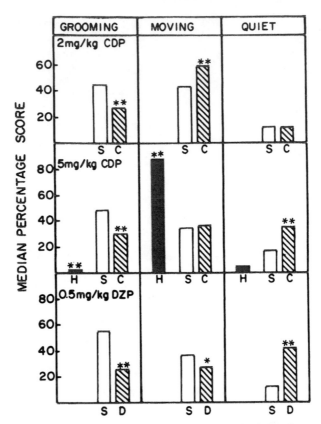

FIGURE 14. The effect of benzodiazepine pretreatment on novelty-induced grooming. Novelty-induced grooming was significantly decreased by 2 or 5 mg/kg CDP (*C*), and by 0.5 mg/kg DZP (*D*) compared with saline (*S*), 10 minutes before placement in the novel cage; 2 mg/kg CDP significantly increased motor activity, whereas 0.5 mg/kg DZP significantly decreased it. Quiet scores were significantly elevated by 5 mg/kg CDP and by 0.5 mg/kg DZP. Home cage scores (*H*) are presented for saline-treated animals. (Data from Dunn *et al.*[67])

would decrease novelty-induced grooming. We determined doses of chloriazepoxide and diazepam that did not alter ACTH- or β-endorphin-induced grooming (FIG. 13) Both 2 and 5 mg/kg chlordiazepoxide decreased novelty-induced grooming without decreasing locomotor activity, in fact 2 mg/kg increased it[70] (FIG. 14). Diazepam (0.5 mg/kg) also reduced novelty-induced grooming at doses that did not alter ACTH-induced grooming (FIGS. 13 and 14). These findings are also consistent with a pituitary source for the ACTH involved in novelty-induced grooming. Therefore, I tentatively conclude that the most likely explanation for novelty-induced grooming is the release of pituitary ACTH into the brain.

THE SIGNIFICANCE OF ACTH-INDUCED GROOMING

I speculate that the significance of the grooming response in conflict situations is an attempt to conceal indecision (which would signal weakness) by emitting a normal (although out of context) behavior that has neutral significance. Grooming, which signals neither threat nor fear to a potential adversary, is the perfect kind of behavior to conceal the intent of the subject (*i.e.,* whether to fight or flee) until a decision is made. I further speculate that ACTH is released from the pituitary into the brain during the stress of the conflict situation and initiates the grooming. This inappropriate behavior can be overridden if a more compelling need is evident,[71] or when the animal decides what action to take. This hypothesis is consistent with our data on the effects of hypophysectomy, ACTH antiserum, and benzodiazepines on novelty-induced grooming, and with the pharmacological similarities between ACTH- and novelty-induced grooming. It is to be noted, however, that the grooming response in stress is rather specific, and does not occur during or after all forms of stress.[23,72] Intense prolonged stress can decrease grooming.[73] Thus the grooming response may be confined to mild stress, or perhaps to conflict situations. This may be because in more stressful situations more compelling behaviors are appropriate.

ACKNOWLEDGMENTS

I gratefully acknowledge the participation in the work reported here of Dr. Howard D. Rees, Dr. P. Michael Iuvone, Dr. Richard L. Delanoy, Dr. Ron Tintner, Dr. Robert L. Isaacson, Dr. Edward J. Green, Dr. Suzanne Steelman, Dr. Neal R. Kramarcy, Russell Hurd, Marsha Ware, Audrey Guild, and Greg Vigle.

REFERENCES

1. Guild, A. L. & A. J. Dunn. 1982. Dopamine involvement in ACTH-induced grooming behavior. Pharmacol. Biochem. Behav. **17:** 31–36.
2. Brakkee, J. H., V. M. Wiegant & W. H. Gispen. 1979. A simple technique for rapid implantation of a permanent cannula into the rat brain ventricular system. Lab. Anim. Sci. **29:** 78–81.
3. Delanoy, R. L., A. J. Dunn & R. Tintner. 1978. Behavioral responses to intracerebroventricularly administered neurohypophyseal peptides in mice. Horm. Behav. **11:** 348–362.
4. Dunn, A. J., J. E. Alpert & S. D. Iversen. 1984. Dopamine denervation of frontal cortex or nucleus accumbens does not affect ACTH-induced grooming behavior. Behav. Brain Res. **12:** 307–315.
5. Dunn, A. J. & G. Vigle. 1985. Grooming behavior induced by ACTH involves cerebral cholinergic neurons and muscarinic receptors. Neuropharmacology **24:** 329–331.
6. Gispen, W. H., V. M. Wiegant, H. M. Greven & D. de Wied. 1975. The induction of excessive grooming in the rat by intraventricular application of peptides derived from ACTH: Structure-activity studies. Life Sci. **17:** 645–652.
7. Rees, H. D., A. J. Dunn & P. M. Iuvone. 1976. Behavioral and biochemical responses of mice to the intraventricular administration of ACTH analogs and lysine vasopressin. Life Sci. **18:** 1333–1339.
8. O'Donohue, T. L., G. E. Handelmann, T. Chaconas, R. L. Miller & D. M. Jacobowitz. 1981. Evidence that N-acetylation regulates the behavioral activity of α-MSH in the rat and human central nervous system. Peptides **2:** 333–344.
9. Gispen, W. H., V. M. Wiegant, A. F. Bradbury, E. C. Hulme, D. G. Smyth, C. R. Snell & D. de Wied. 1976. Induction of excessive grooming in the rat by fragments of lipotropin. Nature **264:** 794–795.
10. Wiegant, V. M., W. H. Gispen, L. Terenius & D. de Wied. 1977. ACTH-like peptides and morphine: Interaction at the level of the CNS. Psychoneuroendocrinology. **2:** 63–70.
11. Kosterlitz, H. W. & S. J. Paterson. 1981. Tyr-D-Ala-Gly-MePhe-NH(CH$_2$)$_2$OH is a selective ligand for the μ-opiate binding site. Br. J. Pharmacol. **73:** 299P.

12. ZWIERS, H., V. J. ALOYO & W. H. GISPEN. 1981. Behavioral and neurochemical effects of the new opioid peptide dynorphin-(1-13): Comparison with other neuropeptides. Life Sci. **28:** 2545–2551.
13. ALOYO, V. J., B. SPRUIJT, H. ZWIERS & W. H. GISPEN. 1983. Peptide-induced excessive grooming in the rat: The role of opiate receptors. Peptides **4:** 833–836.
14. MEISENBERG, G. & W. H. SIMMONS. 1982. Behavioral effects of intracerebroventricularly administered neurohypophyseal hormone analogs in mice. Pharmacol. Biochem. Behav. **16:** 819–825.
15. KATZ, R. 1980. Grooming elicited by intracerebroventricular bombesin and eledoisin in the mouse. Neuropharmacology **19:** 143–146.
16. GMEREK, D. E. & A. COWAN. 1983. Studies on bombesin-induced grooming in rats. Peptides **4:** 907–913.
17. VAN WIMERSMA GREIDANUS, T. B., D. K. DONKER, F. F. M. VAN ZINNICQ BERGMANN, R. BEKENKAMP, C. MAIGRET & B. SPRUIJT. 1985. Comparison between excessive grooming induced by bombesin or by ACTH: The differential elements of grooming and development of tolerance. Peptides **6:** 369–372.
18. KATZ, R. J. 1979. Central injection of substance P elicits grooming behavior and motor inhibition in mice. Neurosci. Lett. **12:** 133–136.
19. DRAGO, F., P. L. CANONICO, R. BITETTI & U. SCAPAGNINI. 1980. Systemic and intraventricular prolactin induces excessive grooming. Eur. J. Pharmacol. **65:** 457–458.
20. MORLEY, J. E. & A. S. LEVINE. 1982. Corticotrophin releasing factor, grooming and ingestive behavior. Life Sci. **31:** 1459–1464.
21. BRITTON, D. R., G. F. KOOB, J. RIVIER & W. VALE. 1982. Intraventricular corticotropin-releasing factor enhances behavioral effects of novelty. Life Sci. **31:** 363–367.
22. DUNN, A. J., C. W. BERRIDGE, Y. I. LAI, T. L. YACHABACH & S. E. FILE. 1987. Excessive grooming behavior in rats and mice induced by corticotropin-releasing factor. This volume.
23. GISPEN, W. H. & R. L. ISAACSON. 1981. ACTH-induced excessive grooming in the rat. Pharmacol. Ther. **12:** 209–246.
24. COLBERN, D., R. L. ISAACSON, B. BOHUS & W.H. GISPEN. 1977. Limbic-midbrain lesions and ACTH-induced excessive grooming. Life Sci. **21:** 393–401.
25. SPRUIJT, B. M., A. R. COOLS & W. H. GISPEN. 1986. The periaqueductal gray: A prerequisite for ACTH-induced excessive grooming. Behav. Brain Res. **20:** 19–25.
26. SPRUIJT, B. M., B. ELLENBROEK, A. R. COOLS & W. H. GISPEN. 1986. The colliculus superior modulates ACTH-induced excessive grooming. Life Sci. **39:** 461–470.
27. GESSA, G. L., M. PISANO, L. VARGIU, F. CRABAI & W. FERRARI. 1967. Stretching and yawning movements after intracerebral injection of ACTH. Rev. Can. Biol. **26:** 229–236.
28. WIEGANT, V. M., A. R. COOLS & W. H. GISPEN. 1977. ACTH-induced excessive grooming involves brain dopamine. Eur. J. Pharmacol. **41:** 343–345.
29. RYAN, J. P. & R. L. ISAACSON. 1983. Intra-accumbens injections of ACTH induce excessive grooming in rats. Physiol. Psychol. **11:** 54–58.
30. DUNN, A. J. & R. W. HURD. 1986. ACTH acts via an anterior ventral third ventricular site to elicit grooming behavior. Peptides **7:** 651–657.
31. DUNN, A. J. & R. W. HURD. 1987. Location of the ventricular site of ACTH action to elicit grooming behavior. This volume.
32. HOFFMAN, W. E. & M. I. PHILLIPS. 1976. Regional study of cerebral ventricle sensitive sites to angiotensin II. Brain Res. **110:** 313–330.
33. LEWIS, R. E. & M. I. PHILLIPS. 1984. Localization of the central pressor action of bradykinin to the cerebral third ventricle. Am. J. Physiol. **247:** R63–R68.
34. GISPEN, W. H. & V. M. WIEGANT. 1976. Opiate antagonists suppress ACTH$_{1-24}$-induced excessive grooming in the rat. Neurosci. Lett. **2:** 159–164.
35. DUNN, A. J., S. R. CHILDERS, N. R. KRAMARCY & J. W. VILLIGER. 1981. ACTH-induced grooming involves high-affinity opiate receptors. Behav. Neural Biol. **31:** 105–109.
36. TERENIUS, L., W. H. GISPEN & D. DE WIED. 1975. ACTH-like peptides and opiate receptors in the rat brain: Structure-activity studies. Eur. J. Pharmacol. **33:** 395–399.
37. TRABER, J., H. R. KLEIN & W. H. GISPEN. 1982. Actions of antidepressant and neuroleptic drugs on ACTH- and novelty-induced behavior in the rat. Eur. J. Pharmacol. **80:** 407–414.

38. Cools, A. R., V. M. Wiegant & W. H. Gispen. 1978. Distinct dopaminergic systems in ACTH-induced grooming. Eur. J. Pharmacol. **50:** 265–268.
39. Springer, J. E., R. L. Isaacson, J. P. Ryan & J. H. Hannigan. 1983. Dopamine depletion in nucleus accumbens reduces ACTH$_{1-24}$-induced excessive grooming. Life Sci. **33:** 207–211.
40. Lassen, J. B. 1977. Evidence for a noradrenergic mechanism in the grooming produced by (+)-amphetamine and 4,α-dimethyl-m-tyramine (H77/77) in rats. Psychopharmacology **54:** 153–157.
41. Yamada, K. & T. Furukawa. 1981. The yawning elicited by α-melanocyte-stimulating hormone involves serotonergic-dopaminergic-cholinergic neuron link in rats. Naunyn-Schmiedebergs Arch. Pharmakol. **316:** 155–160.
42. Sokoloff, L., M. Reivich, C. Kennedy, M. H. Des Rosiers, C. S. Patlak, K. D. Pettigrew, O. L. Sakurada & M. Shinohara. 1977. The [^{14}C]deoxyglucose method for the measurement of local cerebral glucose utilization: Theory, procedure, and normal values in the conscious and anesthetized albino rat. J. Neurochem. **28:** 897–916.
43. Dunn, A. J., S. Steelman & R. Delanoy. 1980. Intraventricular ACTH and vasopressin cause regionally specific changes in cerebral deoxyglucose uptake. J. Neurosci. Res. **5:** 485–495.
44. Dunn, A. J. & R. W. Hurd. 1982. Regional glucose metabolism in mouse brain following ACTH peptides and naloxone. Pharmacol. Biochem. Behav. **17:** 37–41.
45. Delanoy, R. L. & A. J. Dunn. 1978. Mouse brain deoxyglucose uptake after footshock, ACTH analogs, α-MSH, corticosterone or lysine vasopressin. Pharmacol. Biochem. Behav. **9:** 21–26.
46. Rees, H. D., L. L. Brogan, D. J. Entingh, A. J. Dunn, P. G. Shinkman, T. Damstra-Entingh, J. E. Wilson & E. Glassman. 1974. Effect of sensory stimulation on the uptake and incorporation of radioactive lysine into protein of mouse brain and liver. Brain Res. **68:** 143–156.
47. Rees, H. D., D. J. Entingh & A. J. Dunn. 1977. Stimulation increases incorporation of [^{3}H]lysine into rat brain and liver proteins. Brain Res. Bull. **2:** 243–245.
48. Dunn, A. J. & P. Schotman. 1981. Effects of ACTH and related peptides on cerebral RNA and protein synthesis. Pharmacol. Therap. **12:** 353–372.
49. Rees, H. D. & A. J. Dunn. 1977. The role of the pituitary-adrenal system in the footshock-induced increase of [^{3}H]lysine incorporation into mouse brain and liver proteins. Brain Res. **120:** 317–325.
50. Dunn, A. J., H. D. Rees & P. M. Iuvone. 1978. ACTH and the stress-induced changes of lysine incorporation into brain and liver proteins. Pharmacol. Biochem. Behav. **8:** 455–465.
51. Delanoy, R. L., N. R. Kramarcy & A. J. Dunn. 1982. ACTH$_{1-24}$ and lysine vasopressin selectively activate dopamine synthesis in frontal cortex. Brain Res. **231:** 117–129.
52. Kramarcy, N. R., R. L. Delanoy & A. J. Dunn. 1984. Footshock treatment activates catecholamine synthesis in slices of mouse brain regions. Brain Res. **290:** 311–319.
53. Delius, J. D. 1967. Displacement activities and arousal. Nature **214:** 1259–1260.
54. Hinde, R. A. 1970. Animal Behaviour: A Synthesis of Ethology and Comparative Psychology, 2d edit. New York. McGraw-Hill.
55. Fentress, J. C. 1968. Interrupted ongoing behaviour in two species of vole (*Microtus agrestis and Clethrionomys britannicus*). II. Extended analysis of motivational variables underlying fleeing and grooming behaviour. Anim. Behav. **16:** 154–167.
56. Colbern, D. L., R. L. Isaacson, E. J. Green & W. H. Gispen. 1978. Repeated intraventricular injections of ACTH 1–24: The effects of home or novel environments on excessive grooming. Behav. Biol. **23:** 381–387.
57. Bindra, D. & N. Spinner. 1958. Response to different degrees of novelty: The incidence of various activities. J. Exp. Anal. Behav. **1:** 341–350.
58. Doyle, G. & E. P. Yule. 1959. Grooming activities and freezing behaviour in relation to emotionality in albino rats. Anim. Behav. **7:** 18–22.
59. Green, E. J., R. L. Isaacson, A. J. Dunn & T. H. Lanthorn. 1979. Naloxone and haloperidol reduce grooming occurring as an aftereffect of novelty. Behav. Neural Biol. **27:** 546–551.

60. DUNN, A. J., E. J. GREEN & R. L. ISAACSON. 1979. Intracerebral adrenocorticotropic hormone mediates novelty-induced grooming in the rat. Science **203**: 281–283.
61. JOLLES, J., J. ROMPA-BARENDREGT & W. H. GISPEN. 1979. Novelty and grooming behavior in the rat. Behav. Neural Biol. **25**: 563–572.
62. GISPEN, W. H., J. H. BRAKKEE & R. L. ISAACSON. 1980. Hypophysectomy and novelty-induced grooming in the rat. Behav. Neural Biol. **29**: 481–486.
63. AKIL, H. & S. J. WATSON. 1983. β-Endorphin and biosynthetically related peptides in the central nervous system. In Handbook of Psychopharmacology, vol. 16: Neuropeptides. L. L. Iversen, S. D. Iversen & S. H. Snyder, Eds: 209–253. Plenum. New York.
64. KRIEGER, D. T., A. S. LIOTTA, G. NICHOLSEN & J. S. KIZER. 1979. Brain ACTH and endorphin reduced in rats with monosodium glutamate-induced arcuate nuclear lesions. Nature **278**: 562–563.
65. ESKAY, R. L., M. J. BROWNSTEIN & R. T. LONG. 1979. α-Melanocyte-stimulating hormone: Reduction in adult rat brain after monosodium glutamate treatment in neonates. Science **205**: 827–829.
66. DUNN, A. J., E. W. WEBSTER & C. B. NEMEROFF. 1985. Neonatal treatment with monosodium glutamate does not alter grooming behavior induced by novelty or adrenocorticotropic hormone. Behav. Neural Biol. **44**: 80–89.
67. DUNN, A. J., A. L. GUILD, N. R. KRAMARCY & M. D. WARE. 1981. Benzodiazepines decrease grooming in response to novelty but not ACTH or β-endorphin. Pharmacol. Biochem. Behav. **15**: 605–608.
68. JOLLES, J., J. ROMPA-BARENDREGT & W. H. GISPEN. 1979. ACTH-induced excessive grooming in the rat: The influence of environmental and motivational factors. Horm. Behav. **12**: 60–72.
69. LAHTI, R. A. & C. BARSUHN. 1974. The effect of minor tranquilizers on stress-induced increases in rat plasma corticosteroids. Psychopharmacology **35**: 215–220.
70. LE FUR, G., F. GUILLOUX, N. MITRANI, J. MIZOULE & A. UZAN. 1979. Relationship between plasma corticosteroids and benzodiazepines in stress. J. Pharmacol. Exp. Ther. **211**: 305–308.
71. FILE, S. E. 1982. The rat corticosterone response: Habituation and modification by chlordiazepoxide. Physiol. Behav. **29**: 91–95.
72. COHEN, J. A. & E. O. PRICE. 1979. Grooming in the Norway rat: Displacement activity or "boundary-shift"? Behav. Neural Biol. **26**: 177–188.
73. STONE, E. A. 1978. Possible grooming deficit in stressed rats. Res. Commun. Psychol. Psychiat. Behav. **3**: 109–115.

Influence of ACTH$_{1-24}$ on Grooming and Sociosexual Behavior in the Rat

BENGT J. MEYERSON,[a] URBAN HÖGLUND,[a]
AND BERRY SPRUIJT[b]

[a] *Department of Medical Pharmacology*
University of Uppsala
S-751 24 Uppsala, Sweden

[b] *Division of Molecular Neurobiology*
Institute of Molecular Biology
University of Utrecht
3584 CH Utrecht, The Netherlands

Any goal-directed behavior is likely to be preceded by general exploratory activity in which the animal learns about the environmental situation. This internal mapping is an important part of the animal's behavior. In our working hypothesis we assume that this exploratory activity leads to a process of habituation that precedes goal-directed behavior such as sociosexual contacts and self-oriented behavior such as grooming. Data obtained in various test situations suggest that ACTH and ACTH fragments may influence the aforementioned functions.[1-3] In addition, ACTH peptides have been found to have effects on exploratory activity, sociosexual behavior and grooming.[4-12]

The following presentation is concerned with the influence of ACTH$_{1-24}$ on various elements of exploratory and sociosexual behavior. The possibility has been investigated that ACTH influences the attention paid to various environmental stimuli and hence the subsequent process of habituation, with consequences for goal-directed performance. Exploratory activity, different forms of sociosexual behavior, and grooming were measured in a series of tests, the results of which together provided a multivariate behavioral profile from which the acceptability of our hypothesis could be evaluated.

TEST PROCEDURE

Three dimensions of attention were considered, namely attention to the physical environment, social attention to conspecific subjects, and self-oriented attention, the latter measured in terms of grooming behavior. (FIG. 1)

Grooming and Exploratory Behavior

The test employed for this part of the study measures the various forms of behavior a laboratory rat displays in a novel environment. With respect to the latency of onset, the exploratory activity consists mainly of two sets of behavior, namely behaviors (ambulating, sniffing, rearing) with a short latency before onset (1–100 s) and those (investigating, resting, intense sniffing) with a longer latency (>100 s). During

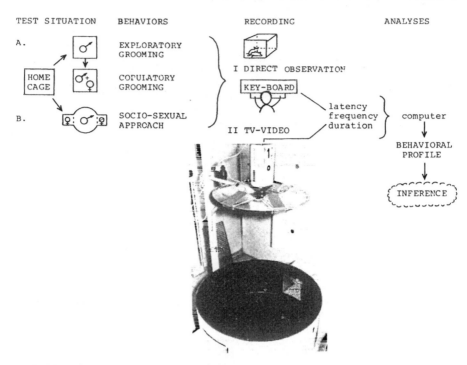

FIGURE 1. Overview of the experimental procedures. $ACTH_{1-24}$ (1 μg in 10 μl artificial CSF) or 10 μl CSF was infused into one of the lateral cerebral ventricles. The following time schedules were used: **(A)** ACTH or CSF infusion at 0 min; home cage, 0–15 min; observation cage, exploratory 15–25 min, copulatory 25–50 min. **(B)** ACTH or CSF infusion at 0 min; sociosexual approach, 0–60 min. Observation cage for direct observation measured 40 × 60 × 40 cm, had a Plexiglas front panel and wood shavings covering the floor. TV-recorded observation arena measured 90 cm in diameter with two openings located in the walls opposite each other. Two incentive cages with front panels of wire bars were inserted into these openings. The recording was carried out by an infrared-sensitive videocamera-microcomputer device.[16]

repeated test procedures the short-latency behaviors decrease in duration and frequency and the long-latency ones increase. We consider this to reflect a process of habituation, in which stimuli of unimportant consequences elicit less general exploratory activity (*e.g.*, ambulation, sniffing, rearing) and more time is devoted to goal-directed performance. Grooming, under nondrug conditions, has an average latency of less than 100 s. For a detailed description of the different forms of behavior and the recording techniques, see refs. 4 and 14. Briefly, they may be defined as follows:

1. Sniffing: rapid movements of whiskers while the animal explores.
2. Intensive sniffing: sniffing directed at a particular object.
3. Rearing: standing on the hindlegs.
4. Grooming: licking, gently biting or scratching different areas of the fur or genitals.
5. Investigation: intensive sniffing directed at a particular object such as a wood shaving, a fecal bolus, etc., which is picked up by the animal.
6. Resting: lying or standing still without any particular behavior.
7. Scanning: standing still, moving the head from side to side.

Copulatory Behavior

This test followed directly upon that for exploratory activity. An estrous female (ovariectomized and treated with estradiol benzoate 25 µg/kg and 48 h later with 1 mg of progesterone per animal) was transferred to the observation cage. The test sequence meant that in the copulatory test situation the male had become well adapted to the cage when the test began.

The copulatory repertoire in the male rat consists of:

1. Mounting: clasping the flanks of the female and performing pelvic thrusts.
2. Intromission: the mount ending with a vigorous backward lunge.
3. Ejaculation: a prolonged mount with intense clasping of the female followed by slow dismounting and subsequent genital licking.

Other forms of behavior scored in this test were grooming, pursuit (following or chasing the female), and nonspecific social contact. The time between ejaculation and a subsequent mount with intromission is the postejaculatory interval.

Sociosexual Approach Behavior

An automatic laboratory testing procedure has been established at our laboratory to measure sociability under defined experimental conditions.[15,16] The method permits measurements of various elements of social approach behavior in the laboratory rat. The elements investigated are the general willingness to seek contact with another animal, the specific goal direction of the social motivation, the behavioral pattern in terms of the willingness to stay for a certain length of time in the vicinity of an incentive animal, and the number of visits to the goal area. The latter measure has to be judged in relation to the general locomotor activity displayed by the experimental animal. To allow the animal to become adapted before the experiment it was placed in the apparatus for 30 minutes on three consecutive days, with ovariectomized females as incentives. The incentive situation was then "sexual"; the experimental animal (male) had a choice between an estrous female (for treatment, see above) and an anestrous female (ovariectomized, untreated). The aim of the procedure was to record the "appetitive" component of the social behavior, and therefore the contact between the experimental animal and the incentive animal was restricted (by wire bars, see FIG. 1). Immediately after an intracerebroventricular (i.c.v.) injection of ACTH or CSF, the male was placed in the observation arena. (For further details of the protocol and methods, see FIGS. 1 and 4). After 30 minutes the incentive animals were removed and the test was continued for another 30 minutes.

STATISTICS

Experimental and control groups were compared by means of the Mann-Whitney U test. The combined data from the different experiments per group yielded two data matrices comprising 12 animals and 12 behavioral elements. Calculation of Spearman rank correlation coefficients between all pairs of behavior categories yielded a correlation matrix. This matrix was processed by an interactive method.[17] The highest correlation coefficient was collected from the matrix and the two categories that were linked by this correlation coefficient were clustered. From the condensed matrix the next largest correlation coefficient was obtained, and so on. This procedure was also used to find out which forms of behavior were most negatively correlated. The positive and negative correlations are depicted in FIGURE 6.

RESULTS

Exploratory and Grooming Behavior

Administration of $ACTH_{1-24}$ resulted in a significant decrease in the duration of short-latency exploratory behavior such as sniffing and rearing ($z = 3.09$ and 3.01, respectively; $p < 0.01$). The duration of intensive sniffing increased ($z = 3.44$, $p < 0.001$), whereas that of investigational behavior remained unchanged. Grooming increased significantly ($z = 2.40$, $p < 0.05$, FIG. 2). An almost significant ($p < 0.06$) increase in the duration of grooming was also noted when the animal was still in the home cage during the first 15 minutes after ACTH treatment.

Copulatory and Grooming Behavior

All animals showed copulatory activity and reached ejaculation. The frequency of mounting was significantly decreased by the ACTH treatment. This was not due to reduced latency before ejaculation. The ejaculation latency was not significantly changed, but if anything it was rather increased. A tendency toward an increased postejaculatory interval was observed ($z = 1.81$), but other copulatory variables were not significantly altered. The duration of grooming was clearly increased, however ($z = 3$, $p < 0.01$, FIG. 3).

Sociosexual Approach Behavior

The results concerning various elements of sociosexual approach behavior are given in FIGURES 4 and 5 for time periods of 15 minutes. The behavior in the last time period,

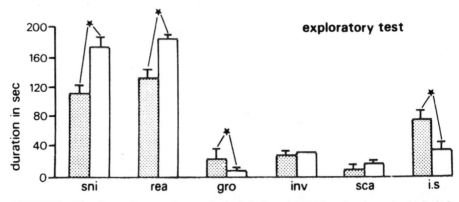

FIGURE 2. The effects of intracerebroventricular infusion of $ACTH_{1-24}$, 1 µg per animal (*shaded bars*), and of artificial CSF (*open bars*) on different forms of behavior as displayed in an exploratory test situation 15–25 min after infusion. Data from Spruijt.[12]; $n = 12$ per group; *sni*, sniffing; *rea*, rearing; *gro*, grooming; *inv*, investigating; *sca*, scanning; *i.s.*, intensive sniffing.

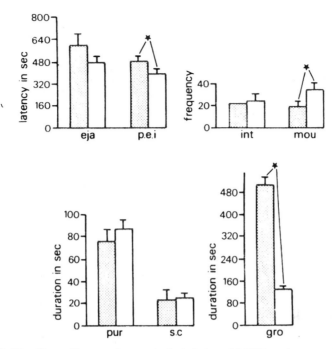

FIGURE 3. The effects of intracerebroventricular infusion of ACTH$_{1-24}$, 1 μg per animal (*shaded bars*), and of artificial CSF (*open bars*) on copulatory behavior 25–50 min after infusion (*n* = 12 per group): *eja*, ejaculation; *p.e.i.*, postejaculatory interval; *int*, intromission; *mou*, mounting; *pur*, pursuit; *s.c.*, social contact; *gro*, grooming. (Data from ref. 12.)

45–60 min after treatment, is the period after the incentive females had been removed, which was done at 30 minutes. On comparison of the first two 15-min periods with regard to the behavior of the CSF-treated males it was found that quantitatively the social performance was the same, but the preference for the estrous female increased progressively during the 30 minutes of the incentive test. This meant that less time was spent close to the anestrous female and more time in the area close to the estrous female. The visits to the anestrous female decreased significantly, as did the locomotor activity. After removal of the incentive females, the most distinct change in performance was the decrease in preference pattern and in locomotion. The male now more randomly distributed its time among the various areas.

The behavior of the ACTH-treated males very closely followed the pattern of the controls in the incentive test situation. However, during the extinction phase of the test period differences were noted. The change in the distribution of time between the areas close to the goal cages which was so clearly evident in the controls only occurred to a minor extent and was not statistically significant after ACTH treatment. Another effect of the latter treatment was an increase in the number of entries into the incentive area formerly adjacent to the estrous female goal cage. This contrasted with the decrease in corresponding entries by the control animals.

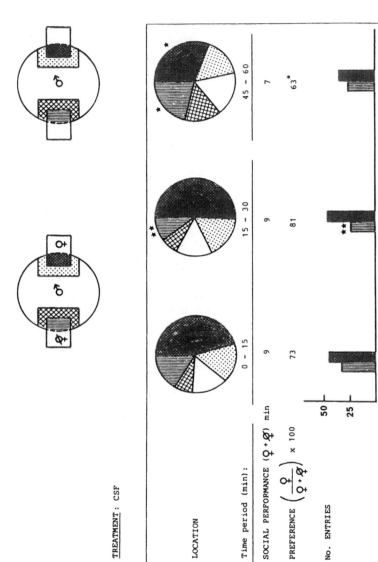

FIGURE 4. The effects of intracerebroventricular infusion of artificial CSF, 10 µl per animal, on sociosexual approach behavior. For a further description of the procedure, see TEST PROCEDURE and the legend to FIGURE 1. Differences between data from consecutive time periods (0–15/15–30 and 15–30/45–60 min): *$p < 0.05$, **$p < 0.01$. Wilcoxon matched-pairs signed ranks test ($n = 18$).

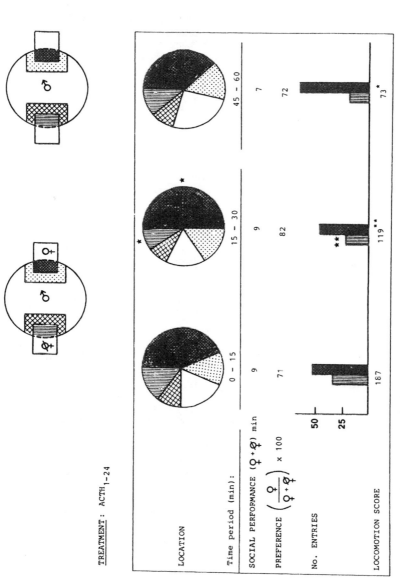

FIGURE 5. The effects of intracerebroventricular infusion of ACTH$_{1-24}$, 1 µg per animal, on sociosexual approach behavior. For a further description of the procedure, see TEST PROCEDURE and the legend to FIGURE 1. Differences between data from consecutive time periods (0–15/15–30 and 15–30/45–60 min): *$p < 0.05$, **$p < 0.01$. Wilcoxon matched-pairs signed ranks test ($n = 18$).

Cluster Analysis

This analysis shows whether behaviors are closely linked or competitive. For the controls (FIG. 6) no significant correlation was found between exploratory and copulatory behavior. There was a strong negative correlation between rearing and sniffing and a less strong negative correlation between grooming and investigation. Intromission and grooming were positively correlated. The ACTH-treated males exhibited a different picture. Grooming and investigation were now positively correlated ($r = 0.63$), as were the ejaculation latency and the postejaculatory interval; moreover, the two clusters — recorded in different tests — were also correlated ($r = 0.59$) with each other. Mounting and intromission were most strongly correlated ($r = 0.81$); the positively correlated cluster ejaculation and postejaculatory interval were negatively correlated with rearing and sniffing ($r = -0.62$); grooming during sexual performance was also negatively correlated with rearing and sniffing. The intensively sniffing animal tended to show a high frequency of intromission and mounting ($r = 0.59$).

DISCUSSION

The different forms of behavior following administration of ACTH were recorded under three different conditions, namely in a test arena without a social encounter (exploratory, novel environment), in a direct sexual situation (consumatory, familiar environment), and in a test situation which permitted indirect contact with a sexually active versus a sexually inactive partner of the opposite sex (appetitive, familiar environment). In a novel environment the animal is initially aroused by certain stimuli, but it ceases to respond to some of these stimuli if they are of no or secondary importance to the animal. Operationally we use the concept of habituation for this phenomenon. After such an initial period characterized by high general locomotor activity combined with exploratory behavior, types of behavior directed more specifically toward certain goal objects become more frequent and of longer duration.[13,14] Grooming was measured in the exploratory and copulatory tests and increased significantly following the ACTH treatment, indicating that the ACTH dose chosen was effective in eliciting a response under the present conditions. In the test situation with a novel environment, treatment with ACTH resulted in a decrease in forms of behavior with a short latency of onset. Thus, the time the animal devotes to exploratory activity such as sniffing and rearing became shorter. In contrast, some behaviors which normally have a relatively longer latency of onset changed in accordance with the hypothesis that ACTH facilitates a process of habituation, for example in the explorative situation intensive sniffing and grooming were enhanced.

Another interesting effect was the correlation observed in the cluster analyses following ACTH treatment between grooming behavior and investigative activity. The latter activity reflects a habituation to the environment. Self-directed behavior such as grooming also presupposes a state of habituation. As we have demonstrated previously, the latency of the investigative activity decreases and the frequency and duration of this activity increase under test conditions of repeated exposure to the same environment.[13] Although in the present experiment investigative activity was not altered by administration of ACTH, it showed a strong positive correlation to grooming behavior.

The copulatory behavior in the male rat consists of repeated bouts of pursuing and mounting of the female. After such a bout of activity, the male often ceases to respond to the repeated stimulus from the female. In this connection grooming, espe-

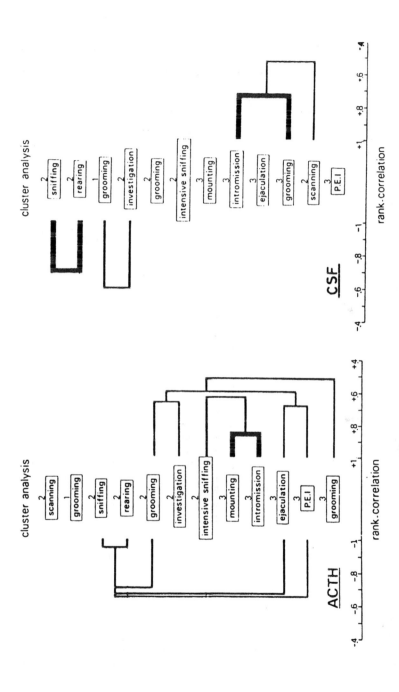

FIGURE 6. Behaviors recorded in the three different tests are linked by bridges on the right-hand side for positive correlations and on the left-hand side for negative correlations. The numbers 1, 2, and 3 refer, respectively, to the grooming test in the home cage, the exploratory test in the novel box, and the copulatory behavior test. The thickness of the line represents the degree of correlation, which can be read from the correlation coefficients at the bottom of the figure. If $r > 0.63$, then $p < 0.005$. (Data from ref. 12.)

cially genital grooming, occurs. For a certain period of time the stimulus offered by the female seems to be of no consequence to the male. Interestingly, ACTH facilitated the self-oriented activity grooming (not only genital) during such periods, which may imply a relationship between ACTH facilitation of grooming and habituation to the repeated sexual stimulation. The frequency of initiation of mounting of the estrous female decreased following ACTH treatment, and other variables indicating the threshold for sexual excitement — such as the latency before ejaculation, the postejaculatory interval, the time devoted to pursuing the female, and the preference for and number of entries into the estrous female area in the approach test — were unchanged or altered in a way which indicated an elevated rather than a decreased threshold.

The most marked effect of ACTH in the sociosexual approach test was the slow extinction of the sexual preference pattern when the incentive females had been removed. The influence of ACTH on extinction behavior in rats in various test situations is well established.[1] The present results showed that ACTH treatment was effective in producing a change not only in grooming, but also in the retention of an appetitive response. This suggests that if ACTH influences the elements of sexual behavior tested in the present study we should have been able to detect this. So far we have not found any excitatory effects of ACTH treatment in the copulatory or more appetitive test situation. In view of the reported erective responses induced by i.c.v. administration of ACTH,[5,6] an increase in sexual performance could be expected. The present results are consistent with the observed effects of a shorter fragment $ACTH_{4-10}$ on various elements of sexual performance.[9,10] $ACTH_{4-10}$ increased the latency before intromission and before ejaculation, and also the postejaculatory interval. The urge to seek contact with an estrous female in a runway test was preserved when the female was removed from the goal box. In females trained to cross an electrified grid to reach a male the response was increased to the stimulus when $ACTH_{4-10}$ was given during a preexperimental period of adaptation to the test procedure. These findings support the conclusion that ACTH, or at least certain parts of the molecule, modify sexual performance indirectly by mechanisms related to habituation to environmental stimuli.

It is concluded that $ACTH_{1-24}$ decreases behavioral parameters which are positively associated with an increased arousal and increases forms of behavior usually observed late in the test situation or during periods of refractoriness to a certain stimulus. This raises the possibility that ACTH accelerates habituation to external stimuli, with a consequent increase in the occurrence of behaviors seen after habituation, such as grooming and the goal-directed behaviors observed in the exploratory test. When the animal had become well familiar with the test situation (sociosexual approach test, copulatory test), ACTH did not facilitate goal-directed behaviors but clearly delayed the extinction of the response after removal of the incentive animal. The effects of ACTH in different learning paradigms have been explained in terms of enhanced attention to external stimuli. Our findings are very much in line with this view. By an enhanced capacity for attention the animal learns about its environment, a process which is an integrated element of the process of habituation. This is of basic importance for the establishment of adequate behavior, including sociosexual performance.

SUMMARY

The effects of intraventricular administration of ACTH 1-24 on grooming, exploratory and socio-sexual behaviors in male rats were studied. The various behavioral elements were analyzed in terms of latency, frequency, and duration, which provided a multivariate behavioral profile. Behaviors that occur early during the test procedure had decreased, whereas behaviors displayed after the animal had spent some time in the test situation increased. Sociability, sexual approaches, and sexual preference in a sociosexual test situation were not altered by this treatment. The extinction of the response after removal of the incentive animals was delayed, however. It is concluded that $ACTH_{1-24}$ did not specifically influence goal-directed behaviors. The obtained change in the behavioral profile, including the grooming pattern, rather suggest that $ACTH_{1-24}$ affects mechanisms associated with attention and habituation.

REFERENCES

1. DE WIED, D. & J. JOLLES. 1982. Neuropeptides derived from proopiocortin: Behavioral, physiological and neurochemical effects. Physiol. Rev. **62**: 967–1059.
2. GISPEN, W. H., J. M. VAN REE & D. DE WIED. 1977. Lipotropin and the central nervous system. Int. Rev. Neurobiol. **20**: 209–250.
3. SANDMAN, C. A., B. E. BECKWITH & A. J. KASTIN. 1980. Are learning and attention related to the sequence of amino acids in ACTH/MSH peptides? Peptides **1**: 277–280.
4. GISPEN, W. H., V. M. WIEGANT, H. GREVEN & D. DE WIED. 1975. The induction of excessive grooming in the rat by intraventricular application of peptides derived from ACTH: Structure activity studies. Life Sci. **17**: 645–652.
5. BERTOLINI, A. & M. BARALDI. 1975. Anabolic steroids: Permissive agents of ACTH-induced penile erections in rats. Life Sci. **17**: 306–310.
6. BERTOLINI, A., G. L. GESSA & W. FERRARI. 1975. Penile erection and ejaculation: A central effect of ACTH-like peptides in mammals. *In* Sexual Behavior: Pharmacology and Biochemistry. M. Sandler & G. L. Gessa, Eds.: 247–257. Raven. New York, NY.
7. CATANZARO, D., S. D. GRAY & B. B. GORZALKA. 1979. Effects of acute central and peripheral $ACTH_{1-24}$ administration on lordosis behavior. Physiol. Behav. **26**: 207–213.
8. HAUN, C. K. & G. C. HALTMEYER. 1975. Effects of an intraventricular injection of ACTH on plasma testosterone, progestrone and LH levels and on sexual behavior in male and female rats. Neuroendocrinology **19**: 201–213.
9. BOHUS, B., H. H. L. HENDRICKX, A. A. VAN KOLFSCHOTEN & T. G. KREDIET. 1975. Effect of $ACTH_{4-10}$ on copulatory and sexually motivated approach behavior in the male rat. *In* Sexual Behavior: Pharmacology and Biochemistry. M. Sandler & G. L. Gessa, Eds.: 269–276. Raven. New York. NY.
10. MEYERSON, B. J. & B. BOHUS. 1976. Effects of $ACTH_{4-10}$ on copulatory behavior and on the response in a test for sociosexual motivation in the female rat. Pharmacol. Biochem. Behav. **5**: 539–545.
11. WILSON, C. A., A. J. THODY & D. EVERARD. 1979. Effects of ACTH analogues on lordosis behavior in the female rat. Horm. Behav. **13**: 293–300.
12. SPRUIJT, B. M., U. HÖGLUND, W. H. GISPEN & B. J. MEYERSON. 1985. Effects of $ACTH_{1-24}$ on male rat behavior in an exploratory, copulatory and socio-sexual approach test. Psychoneuroendocrinology **10**: 431–438.
13. HOGLUND, A. U. & B. J. MEYERSON. 1982. Effects of lysine-vasopressin in an exploratory behaviour test situation. Physiol. Behav. **29**: 189–193.
14. MEYERSON, B. J. & A. U. HÖGLUND. 1981. Exploratory behavior and socio-sexual behavior in the male laboratory rat: A methodological approach for the investigation of drug action. Acta Pharmacol. Toxicol. **48**: 168–180.
15. HETTA, J. & B. J. MEYERSON. 1978. Sexual motivation in the male rat: A methodological study of sex-specific orientation and the effects of gonadal hormones. Acta Physiol. Scand. Suppl. **453**: 1–68.
16. HÖGLUND, A. U., J. -E. HÄGGLUND & B. J. MEYERSON. 1982. A video-interface for behavioral recordings with applications. Physiol. Behav. **30**: 489–492.
17. VAN HOOFF, J. A. R. A. M. 1982. Handbook of Methods in Nonverbal Behavior Research. Cambridge University Press. Cambridge.

ACTH-induced Grooming Behaviors and Body Temperature

Temporal Effects of Neurotensin, Naloxone, and Haloperidol[a]

DEBORAH L. COLBERN[b] AND DENNIS A. TWOMBLY[c]

[b]Department of Physiology and Biophysics
University of Illinois at Chicago
College of Medicine
Chicago, Illinois 60680

[c]Department of Pharmacology
Northwestern University School of Medicine
Chicago, Illinois 60611

INTRODUCTION

Grooming is an essential component of an animal's behavioral repertoire. It obviously serves to maintain the integrity of the fur and keep it free from dirt and parasites. However, animals also groom in unfamiliar environments,[1,2] in conflict situations, or in anticipation of punishment.[3] Some consider grooming an "irrelevant" behavior in these circumstances. Others believe that grooming may act to calm or "dearouse" a stressed animal and thus may be very important to its psychological and physical well-being.[4]

In 1975, Gispen et al.[5] reported that rats exhibit a dramatic increase in grooming behaviors when injected intracerebroventricularly (i.c.v.) with various fragments of the pituitary hormone ACTH (adrenocorticotropic hormone). The behaviors induced by this peptide are qualitatively similar to those observed when rats are placed into a novel environment. Since these initial discoveries, many investigators have sought to uncover the neural substrates underlying "ACTH-induced grooming," as well as the grooming induced by novelty and other stressful situations. There is now considerable information regarding the neurochemical and neuroanatomical systems involved in these behaviors. As yet, however, the adaptive significance of grooming under these circumstances is not known.

In addition to its role in hygiene and perhaps dearousal, grooming behavior in rodents serves thermoregulatory functions (see papers by W. Roberts and D. Thiessen in this volume). When exposed to warm environments, rodents spread saliva on the fur and on vascularized areas of their body such as the paws, ears, and tail.[6-10] As the moisture evaporates, the body temperature of the animal decreases. In cold environments, certain rodents spread a pigmented lipid material that insulates and darkens the fur, thereby increasing body temperature through the absorption of radiant heat.[11] Handling and/or exposure to novel stimuli have been shown to increase the body tem-

[a]These studies were sponsored in part by the Deutsch Foundation (Los Angeles, CA), which provided the resources to obtain the biotelemetry system, and by the Rudolf Magnus Institute for Pharmacology and Institute for Molecular Biology, University of Utrecht, The Netherlands.

perature of rats, a response referred to as "emotional hyperthermia" by some investigators.[12,13] Under these conditions, ACTH is released from the pituitary and animals exhibit increased grooming behavior. Because of the association between thermoregulation and grooming, we were interested in whether the well-known grooming response to exogenously administered ACTH is related to changes in body temperature.

In the present study, we used a biotelemetry system to obtain concurrent measurements of grooming behavior and body temperature in undisturbed, freely moving rats. In addition to ACTH, we tested the effects of naloxone,[14] haloperidol,[15] and neurotensin,[16] agents known to inhibit ACTH-induced grooming. Until recently, we typically used the term "grooming" to refer to composite behavioral data, which included head grooming, body grooming, anogenital grooming, scratching, stretching, and yawning. As more peptides have been studied, it has become clear that "grooming," used in this manner, is not sufficiently specific.[17,18] Moreover, certain elements of the behavioral response to ACTH, such as head grooming and body grooming, are more likely than others to serve thermoregulatory functions. It was particularly important, therefore, to study changes in body temperature in relation to the specific types of behaviors performed.

EXPERIMENTAL PROCEDURES

Animals

Male Wister rats (original stock from TNO, Zeist, The Netherlands) were born and raised in Prof. W. H. Gispen's laboratory at the Institute for Molecular Biology (University of Utrecht, The Netherlands). The animals were group-housed until the day of surgery, after which they were housed individually on sawdust bedding in small plastic cages. Food and water were available ad libitum. Lights in the colony room were on from 0800 to 2000 h; all experiments were conducted in the afternoon between 1300 and 1700 h. Rats weighed 215–235 g at the time of the experiment.

Transmitter Implantation

In these experiments, core body temperature was monitored with a thermosensitive AM transmitter device (VM-FH Mini-mitter, Sunriver Oregon; 1 cm diameter, 1.5 cm long, 2.3 g, sensitive to changes within $0.02°C$) surgically implanted into the peritoneal cavity of each rat. Prior to implantation, each transmitter was coated with an inert, waterproof coating (Paraffin/Elvax; DuPont, Wilmington, DE), and calibrated in a controlled water bath ($\pm0.02°C$). Rats were anesthetized with 1 cc/kg Hypnorm (10 mg/ml fluanison and 0.2 mg/ml fentanyl; Duphar, Amsterdam), and the fur was shaved from the abdomen. A 1.5-cm midline incision was made through the skin and the linea alba between the abdominal muscles, and the transmitter was inserted into the peritoneal cavity. The muscle was sutured with silk thread before closing the skin with wound clips.

Intracerebroventricular Cannulation

Immediately following implantation of the transmitter, a polyethylene cannula was placed into the right lateral ventricle of the brain according to procedures described by Brakkee et al.[19] A small hole was drilled in the skull, 1 mm lateral to the sagittal suture, abutting the posterior edge of the coronal suture. The cannula (o.d. = 0.8 mm, i.d. = 0.4 mm) was inserted 4.5 mm into the brain, which positioned the tip in the lateral ventricle. Dental cement and skull screws were used to anchor the cannula permanently to the skull. A stereotaxic apparatus was not required for this procedure. Animals were housed individually after surgery, and allowed to recover for 8-9 days prior to behavioral observation. At the end of the experiment, methylene blue (2 µl) was injected through each cannula to verify its correct placement in the lateral ventricle. Cannulation was considered successful if the dye was found throughout the cerebroventricular system 5 min after injection.

Body Temperature Measurements

Body temperatures of rats were measured before and during behavioral observations by monitoring signals emitted from the implanted transmitters. These signals, whose frequencies were proportional to body temperature, were received with an AM radio (Realistic 12-117) and decoded with a frequency counter (Fluke 8060A). After the experiment, the recorded frequencies were converted to body temperatures according to the calibration curve calculated for each transmitter.

Rats were transported in their homecages to the experimental room just prior to the behavioral observation. In the quiet colony room, the transmitter frequency of each rat was determined by briefly placing the cage over the AM radio. Each measurement was completed within 5 s of removing the cage from its home position. Cages were placed on a cart and transported to the experimental room (transport time: 2-3 min). Upon arrival, transmitter frequencies were measured again. The experiment began 10 to 20 min after transport. Each rat was removed from its home cage, injected i.c.v. (followed by a s.c. injection in some cases), and placed in a clean glass observation cage (24 × 12.5 × 14 cm) unfamiliar to the animal. Immediately thereafter, the observation cage was positioned above the radio to record transmitter frequency. Each cage was then placed on a shelf inside a walk-in, sound-attenuated, experimental chamber (23-24.5°C). Cages were visually isolated from each other with foil-lined partitions. Wire loop antennas were oriented in three planes around each cage. The antennas were connected to individual coaxial cables, which led outside the experimental chamber to the AM radio and frequency counter. Beginning 10 min after injection and continuing for 70 min, the frequency emitted from each rat was recorded once within every 2-min period.

Behavioral Measurements

In the present experiment, behaviors were operationally defined according to the following criteria:

- Head grooming. Scored when the rat rubbed his mouth, nose, eyes, face, ears, or head with his forepaws. Vibration of the forelimbs (referred to as "high-frequency forelimb flailing" by J. Fentress, this volume) was also included in this category, since it consistently occurs just prior to head-grooming behaviors. Our category of head grooming included what others have termed "face washing," "head washing," and "forepaw licking."
- Body grooming. Recorded when the animal was observed to lick, bite, or use its paws to manipulate the fur or skin on the shoulders, back, abdomen, forelimbs, hindlimbs, or tail.

- Anogenital grooming. Scored if the animal was grooming the penis, anus, scrotum, or the areas immediately adjacent to these structures. Erection and/or ejaculation were sometimes observed, but they were not differentiated within this category.
- Stretch-yawn syndrome. Scored when the rat stretched, yawned, or exhibited both behaviors simultaneously.
- Scratching. Recorded each time the animal scratched its body, head, or ears with a hindlimb. Immediately following a bout of scratching, rats almost always lick the hindpaw that is used to scratch. For this reason, licking of the hindpaws was also counted as scratching.
- Moving. Scored when the animal was walking, rearing, sniffing, or moving in some fashion, but not grooming as defined for the previous behaviors.
- Resting. An animal was considered to be resting if it was sitting or lying still with its eyes open or closed.

The behavior and body temperatures of ten rats were monitored in each experimental session. While one experimenter recorded radio-frequency transmissions, another recorded the behavior of the animals using a computer-assisted, time-sampling technique. Rats were observed through a one-way window for 70 min beginning 10 min after injection. Every 15 seconds the behavior of each animal was coded and entered into the computer as head grooming, body grooming, anogenital grooming, scratching, stretch-yawn syndrome, moving, or resting. Thus, during the 70-min observation period, 280 activity scores and 35 temperature readings were recorded for each rat.

Statistical Analysis

For each behavior and for body temperature, treatment differences were tested with a two-factor mixed design analysis of variance with repeated measures on one factor. Duncan's multiple-range test was used to further analyze significant differences among the main effect means. Unless noted otherwise, discussion of results will pertain to differences reaching significance at or below $p = .05$.

Drug treatments were randomly assigned and counterbalanced across all experimental sessions; hence data from different groups can be compared directly. Results are described according to preplanned comparisons between different treatment groups.

EFFECT OF TRANSPORT, HANDLING, AND INJECTION ON BODY TEMPERATURE

Body temperature was measured in the colony room before transport, and again in the experimental room immediately after transport. As shown in TABLE 1, there were

TABLE 1. Effect of Transport, Injection, and Novelty on Core Body Temperature in the Rat

Handling Condition	Body Temperature (°C, Mean ± SEM, $n = 54$)
Colony (homecage)	37.17 ± 0.04
Posttransport (homecage)	37.21 ± 0.04
Postinjection (novel cage)	37.60 ± 0.04^a

[a] Significantly increased compared to colony, posttransport, or mean baseline temperatures, $p < .0001$.

no differences in body temperature due to transport of the animals. Therefore, the mean of the pre- and posttransport values was used as the "baseline temperature" for each rat. All temperature measurements were expressed as change from this baseline. Although transport did not affect temperature, the procedures of handling, injecting, and placing rats into novel observation cages resulted in an immediate increase in body temperature of 0.41 ± .04°C (mean ± SEM). For this initial measurement, there were no significant differences due to the type of injection given.

NEUROTENSIN EFFECTS ON ACTH-INDUCED BEHAVIORS AND BODY TEMPERATURE

Drug Treatment

Neurotensin is a brain-gut tridecapeptide known to affect dopamine systems in the brain.[20] When it is administered centrally, neurotensin is effective in blocking ACTH-induced grooming behaviors.[17,18] To test the effects of neurotensin on the ACTH-induced changes in behavior and temperature, rats received one of the following i.c.v. injections: 3 µg ACTH$_{1-24}$ (Organon, Oss, The Netherlands) in 4 µl of sterile saline; an equal volume of saline alone; 2 µg neurotensin (R4322; Bachem, Torrance, California); or 3 µg ACTH$_{1-24}$ with 2 µg neurotensin.

Behavioral Analysis

Rats injected i.c.v. with 3 µg ACTH$_{1-24}$ exhibited significantly more grooming behaviors than rats injected i.c.v. with saline. This is the phenomenon often referred to as "ACTH-induced excessive grooming."[5] Neurotensin (2 µg, i.c.v.) blocked the increase in grooming induced by ACTH, as has been reported previously.[17,18] The effects of these peptides on specific behavioral elements and their time courses are presented in FIGURE 1. For each 10-min period of the 70-min observation, the figure shows the mean percentage of time each of the seven behavioral elements was observed.

Body Grooming. ACTH increased body grooming throughout the entire observation period compared to saline-treated rats. Neurotensin blocked this increase during the first 10 to 40 min of the observation and also from 50 to 60 min, but not from 40 to 50 or 60 to 80 min after i.c.v. injection. There were no differences between groups treated with saline or neurotensin alone.

Head Grooming. Compared to saline, ACTH increased head grooming for only the first 20 min of the observation. Neurotensin blocked the ACTH-induced increase in head grooming during this early period. From 40 to 50 and 70 to 80 min, in contrast, these neurotensin-treated rats exhibited significantly more head grooming than rats receiving ACTH alone. By itself, neurotensin had no effect on head grooming compared to rats injected with saline.

Anogenital Grooming. There were no differences in anogenital grooming among any of the groups tested.

Scratching. ACTH-treated animals scratched more than saline animals from 20 to 40 min and 50 to 60 min after injection. Neurotensin prevented this ACTH-induced increase in scratching. There were no differences in scratching behavior between rats treated wtih saline and those treated with neurotensin.

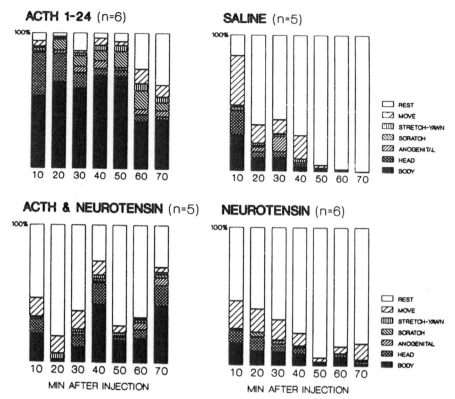

FIGURE 1. Specific behavioral responses to i.c.v. injections of ACTH$_{1-24}$ (3 µg) and neurotensin (2 µg). The time courses of various behaviors after i.c.v. injection of peptide(s) or saline (4 µl) are illustrated in the four panels. Behavior of each animal was observed at 15-s intervals during the 70-min postinjection period. Data were compiled for each 10-min period after injection (abscissa), based on a total of 40 samples per animal. The mean number of observations for each type of behavior, expressed as a percentage of the total number of observations, is represented by a specific bar pattern (see key). "Grooming" behaviors (body grooming, head grooming, anogenital grooming, scratching, stretching and yawning) are located at the bottom of the bar. The category of behavior defined as moving (but not grooming) is shown above the grooming categories. Resting behavior is shown by the open bar at the top. Significant effects of drugs on different behaviors are discussed in the text.

Stretching and Yawning. ACTH induced significant increases in stretching and yawning 50–60 min after injection; neurotensin also blocked this increase.

Moving. During the first 10 min of the observation period, saline-treated animals engaged in more nongrooming movement than did the other treatment groups. There were no differences among groups for any of the other time periods.

Resting. Except for the first 10-min period, saline-treated rats spent more of their time resting than did ACTH-treated rats, who were usually grooming. Neurotensin increased the probability of resting in saline-treated rats during the first 10-min period.

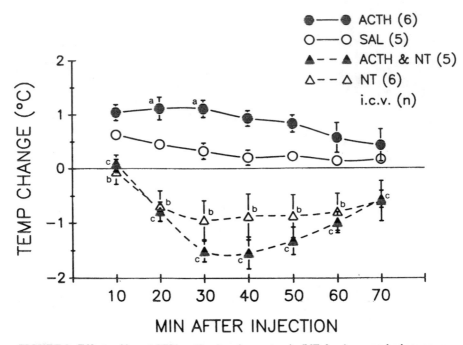

FIGURE 2. Effects of i.c.v. $ACTH_{1-24}$ (3 µg) and neurotensin (NT, 2 µg) on core body temperature. Temperatures were recorded at 2-min intervals after i.c.v. injection of peptide(s) or saline (SAL, 4 µl). Scores for each 10-min interval after injection are expressed as mean change from baseline (±SEM). Significant group differences for each interval are indicated as follows: *a*, ACTH vs. SAL; *b*, NT vs. SAL; *c*, ACTH and NT vs. ACTH (Duncan's Multiple Range test, $p \leqslant .05$).

In ACTH-treated animals, neurotensin increased resting 10–40 and 50–60 min after injection. After 60 min, when ACTH animals were beginning to rest more, the effect of neurotensin on resting was not significant.

Body Temperature

Body temperatures were compiled for each 10-min period of the observation session. In FIGURE 2, temperatures are plotted as mean change from baseline (± SEM). TABLE 2 reports significance levels for mean temperatures versus baseline for each 10-min period. As mentioned previously (TABLE 1), injection and placement of rats into novel cages resulted in immediate increases in temperature, regardless of the substance injected. In both ACTH- and saline-treated rats, body temperature was significantly increased from baseline 10 min after i.c.v. injection and placement in the novel observation chamber. Temperatures of saline-treated rats gradually declined, and were not significantly different from baseline by 30 min after injection. The temperatures of ACTH-treated rats remained elevated for up to 60 min after injection. The increased temperatures of ACTH-treated rats were significantly higher than those of saline rats from 20–40 min after injection.

Neurotensin effectively reduced body temperatures. Within 10 min of injection, the posthandling hyperthermia was no longer evident. Temperatures subsequently

TABLE 2. Effects of Intracerebroventricular Injections of $ACTH_{1-24}$ and/or Neurotensin on Core Body Temperature

Treatment	Min after Injection						
(i.c.v.)	10–19	20–29	30–39	40–49	50–59	60–69	70–79
ACTH (3 μg)	+ + +	+ +	+ + +	+ + +	+ +		
ACTH and Neurotensin		– –	– – –	– –	– –	– –	–
Saline (4 μl)	+ + +	+ +					
Neurotensin (2 μg)		–					

Significant increases (+) or decreases (–) from baseline for 10-min periods throughout the 70-min observation session are indicated as follows: + or –, $p < .05$; + + or – – $p < .01$; + + + or – – – $p < .001$. Data are presented in FIGURE 2.

declined from baseline, but the change was only significant for the 30–40 min interval. Compared to saline-treated rats, however, temperatures of the neurotensin groups were significantly reduced for all but the last 10 min of the observation session. Combined treatment with ACTH and neurotensin produced an even greater decrease in body temperature. Beginning 20 min after injection, and continuing throughout the experimental session, temperatures were signficantly decreased compared to baseline. At their lowest point, 40–50 min after injection, the mean (± SEM) temperature of these rats was decreased by 1.54 ± 0.29°C. The temperatures were significantly lower than those of ACTH and saline groups for the entire observation period.

Correlations of Body Temperature with Specific Behaviors

TABLE 3 lists Spearman rank-order correlation coefficients between body temperature and specific behaviors for each 10-min interval. Body temperature in saline-treated rats was highly correlated with head grooming, body grooming, stretching and yawning, and nongrooming movements; temperature was negatively correlated with resting. Neurotensin disrupted the relationships between body temperature and these behaviors. After neurotensin treatment, the only behavior that remained significantly correlated with temperature was body grooming, although the strength of the relationship was

TABLE 3. Spearman Rank-Order Correlation Coefficients between Body Temperature (FIG. 2) and Specific Behaviors (FIG. 1)

	ACTH	SAL	ACTH and NT	NT
n	6	5	5	6
Paired obs.	42	35	35	41
Specific behaviors				
Head	.18	[.62^f]	– .19	.26
Body	.04	[.70^f]	– .12	[.32^a]
Anogenital	– .04	.19	.04	.11
Scratch	– .14	.30	[– .33^a]	– .02
Stretch-yawn	– .15	[.43^c]	– .31	.28
Move	[– .43^c]	[.70^f]	.01	– .05
Rest	.25	[– .71^f]	.16	– .12

Rats were injected i.c.v. with 4 μl saline (SAL), 3 μg $ACTH_{1-24}$, and/or 2 μg neurotensin (NT). Brackets denote a significant correlation with temperature at 10-min intervals throughout the 70-min observation period: $^a p \leqslant .05$; $^b p \leqslant .01$; $^c p \leqslant .005$; $^d p \leqslant .001$; $^e p \leqslant .0005$; $^f p \leqslant .0001$.

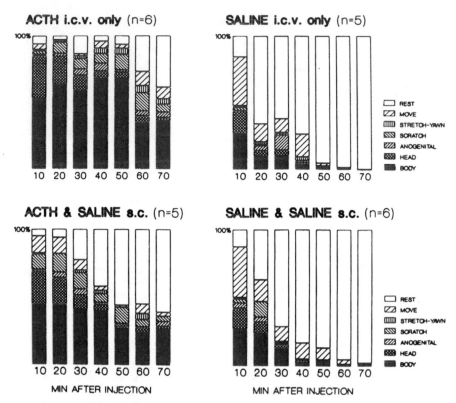

FIGURE 3. Effects of subcutaneous saline injections on specific behavioral responses to i.c.v. ACTH$_{1-24}$ (3 µg) and i.c.v. saline (4 µl). The time courses of various behaviors after i.c.v. injections alone are illustrated in the top panels; the bottom panels show data from groups that also received s.c. saline injections (2 cc/kg). Behavior of each animal was observed at 15-s intervals during the 70-min postinjection period. Data were compiled for each 10-min period after injection (abscissa), based on a total of 40 samples per animal. The mean number of observations for each type of behavior, expressed as a percentage of the total number of observations, is represented by a specific bar pattern (see key).

diminished. In ACTH-treated rats, grooming behaviors showed no correlation with body temperature. Nongrooming movements, however, were negatively correlated. When neurotensin was combined with ACTH, the negative correlation with movement disappeared, while a mild negative correlation between scratching and temperature emerged. The poor correlation between temperature and ACTH-induced grooming may have resulted from sustained effects of ACTH on both of these variables. A longer observation period or lower dose of ACTH may have allowed a relationship to be detected. This is suggested by data from ACTH-treated rats, described below, in which submaximal grooming scores showed quite good correlations between temperature and grooming behaviors.

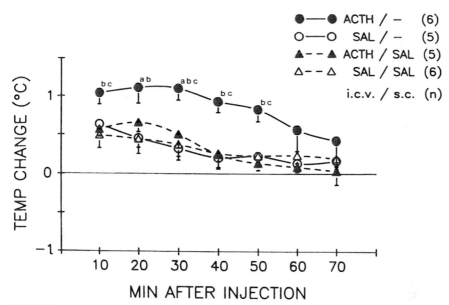

FIGURE 4. Effects of subcutaneous saline injections (*SAL*, 2 cc/kg) on temperature responses to i.c.v. $ACTH_{1-24}$ (3 μg) and i.c.v. saline (4 μl). Temperatures were recorded at 2-min intervals after i.c.v. injection alone (*ACTH/–, SAL/–*) or after i.c.v. and s.c. injections together (*ACTH/SAL, SAL/SAL*). Scores for each 10-min interval after injection are expressed as mean change from baseline (±SEM). Significant group differences for each interval are indicated as follows: *a*, ACTH/– vs. SAL/–; *b*, ACTH/– vs. SAL/SAL; *c*, ACTH/– vs. ACTH/SAL (Duncan's Multiple Range Test, $p \leqslant .05$).

EFFECT OF SUBCUTANEOUS SALINE INJECTION

Naloxone and haloperidol are much more effective in counteracting ACTH-induced grooming when they are administered systemically rather than intracerebroventricularly.[17] Therefore, to examine the effects of these agents on specific behaviors and body temperatures, naloxone-HCl (1 mg/kg) and haloperidol (0.05 mg/kg) were dissolved in sterile physiological saline and injected s.c. (2 cc/kg) immediatley after an i.c.v. injection of 3 μg $ACTH_{1-24}$ or 4 μl saline. Saline (2 cc/kg) was also administered s.c. after i.c.v. injections of ACTH or saline, to control for possible effects of the additional peripheral injection.

Behavioral Analysis

Surprisingly, the subcutaneous injection of saline significantly altered the behavioral and temperature responses of ACTH-treated rats. FIGURE 3 shows the effects of s.c. saline on the behavior of groups receiving i.c.v. saline and i.c.v. ACTH. The panels on the left show data for ACTH-treated animals, with (bottom) or without (top) an additional s.c. saline injection; their respective saline control groups are shown on the

TABLE 4. The Effects of Subcutaneous Injection of Saline (2 cc/kg) on Core Body Temperature of Rats Injected Intracerebroventricularly with $ACTH_{1-24}$ (3 µg) or Saline (4 µl).

	Min after Injection						
Injections	10–19	20–29	30–39	40–49	50–59	60–69	70–79
i.c.v. only							
ACTH	+ + +	+ +	+ + +	+ + +	+ +		
Saline	+ +	+					
i.c.v./s.c.							
ACTH/saline	+ +	+ +	+				
Saline/saline	+						

Significant increases (+) from baseline for 10-min periods throughout the 70-min observation session are indicated as follows: $+$, $p < 0.05$. $+ +$, $p < 0.01$. $+ + +$, $p < 0.001$. Data are presented in FIGURE 4.

right. As described above, ACTH-treated rats (no s.c. injections) exhibited enhanced body grooming throughout the entire 70-min observation session. In contrast, rats receiving i.c.v. ACTH and s.c. saline only showed significant body grooming during the period between 20 and 50 min after injection. Comparison of the two ACTH groups indicated that s.c. injection decreased the amount of ACTH-induced body grooming from 40 to 60 min after injection.

Subcutaneous saline attenuated other responses to ACTH as well. Without s.c. injections, head grooming was significantly increased from 10 to 30 min after injection. With the s.c. injections, head grooming was increased only from 20 to 30 minutes. Stretching and yawning were not significantly elevated at all in animals with s.c. saline injections. In summary, the general effect of s.c. injection on ACTH-treated animals was to reduce grooming behaviors later in the session while increasing the amount of resting. Other than the increased scratching observed from 20 to 30 min, s.c. saline injections did not significantly affect the behavior of rats receiving i.c.v. saline.

Body Temperature

The body temperatures of rats receiving s.c. saline and ACTH were also altered (FIG. 4). Whereas temperatures of ACTH-treated rats (no s.c. injection) were elevated for 60 min after injection, the temperatures of rats that also received s.c. saline were initially lower, and declined faster. Their temperatures were increased from baseline from 10 to 40 min after injection, rather than from 10 to 60 min after injection as was the case for rats given ACTH only (TABLE 4). In this sense, their pattern of temperature change resembled that of saline groups more than that of the i.c.v. ACTH group. In fact, there were no significant differences in temperature between the s.c. injected ACTH group and either saline group.

EFFECTS OF NALOXONE

Naloxone, a specific opioid receptor antagonist, blocks novelty-induced [21] and ACTH-induced grooming.[14] This agent was tested for its effects on the time course of specific behaviors and body temperature. Rats received i.c.v. injection of 3 µg $ACTH_{1-24}$ or

4 µl sterile saline, as well as a s.c. injection of naloxone-HCl (1 mg/kg in sterile saline) or an equal volume of saline (2 cc/kg). Specific behavioral responses to these treatments are shown in FIGURE 5. To allow comparisons among the various groups, data from FIGURE 3 are reproduced in the top two panels.

Behavioral Analysis

Body Grooming. ACTH and saline groups both exhibited more grooming in the early part of the observation session than in the later part. The grooming in ACTH-treated animals exceeded that of saline animals during the 30–60 min period after injection. Naloxone suppressed body grooming in both groups, after a delay of about 20 min. Thus, naloxone reduced body grooming in the ACTH group from 20 to 50 min after injection. Body grooming in the saline group was reduced between 20 and 30 minutes.

Head Grooming. Rats treated with ACTH exhibited more head grooming than saline-treated rats 20–30 after injection. Naloxone decreased head grooming in ACTH-treated rats between 20 and 40 min after injection, and in saline-treated rats from 20 to 30 min after injection.

Anogenital Grooming. There were no differences in anogenital grooming among any of the groups tested.

Stretching and Yawning. There were no differences in stretching and yawning among any of the groups tested.

Scratching. ACTH induced more scratching than did saline from 30 to 50 min after injection. Naloxone reduced scratching in ACTH-treated rats 20–50 min after injection; in saline-treated rats, scratching was reduced between 20 and 30 minutes. Thus, naloxone was very effective in decreasing the scratching behavior in both i.c.v. treatment groups.

Moving. Saline-treated rats exhibited more nongrooming movements in the first 10-min observation period than did ACTH-treated rats. Naloxone had no effect on this movement of saline animals.

Resting. In ACTH and saline groups with s.c. saline injection, animals displayed significant grooming early in the session. This declined as resting became more frequent (see previous section on s.c. saline injections). In the ACTH-treated rats, naloxone produced significantly more resting during the 20–40 min interval after injection. Naloxone also increased resting from 10 to 30 min after injection in i.c.v. saline-injected animals.

Body Temperature

The effects of naloxone on body temperature are illustrated in FIGURE 6 (top panel). The significance levels for changes from baseline temperature are presented in TABLE 5.

As described previously, ACTH and saline-treated groups showed elevated temperatures when injected and placed in the novel observation cages. There were no differences in body temperature between these two groups. These animals were all injected s.c. with saline to control for the s.c. injection of naloxone. Naloxone reversed the initial hyperthermic response to handling and injection in rats injected with i.c.v. saline. This effect was evident within 20 min after injection. The temperatures of these animals eventually dropped below baseline, but the decreases were not significant. Naloxone

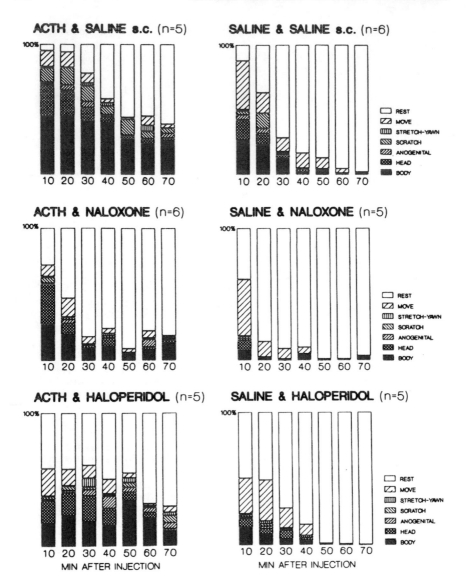

FIGURE 5. Effects of naloxone and haloperidol on specific behavioral responses to i.c.v. ACTH$_{1-24}$ (3 µg) and i.c.v. saline (4 µl). The time courses of various behaviors are illustrated for groups receiving i.c.v. injections together with s.c. saline (2 cc/kg, top panels), s.c. naloxone (1 mg/kg, middle panels), or s.c. haloperidol (0.05 mg/kg, bottom panels). Behavior of each animal was observed at 15-s intervals during the 70-min postinjection period. Data were compiled for each 10-min period after injection (abscissa), based on a total of 40 samples per animal. The mean number of observations for each type of behavior, expressed as a percentage of the total number of observations, is represented by a specific bar pattern (see key).

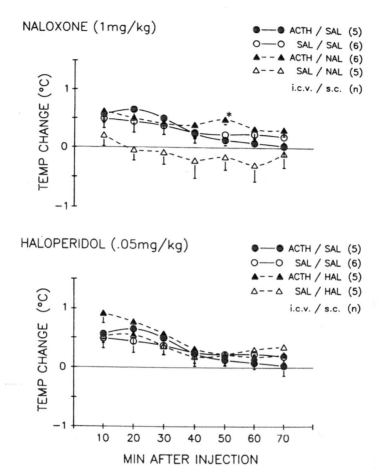

FIGURE 6. Effects of naloxone and haloperidol on temperature responses to i.c.v. ACTH$_{1-24}$ and i.c.v. saline. Temperatures were recorded at 2-min intervals after i.c.v. injection of ACTH (3 µg) or saline (SAL, 4 µl), together with s.c. naloxone (NAL, 1 mg/kg), s.c. haloperidol (HAL, 0.05 mg/kg), or s.c. saline (SAL, 2 cc/kg). Scores for each 10-min interval after injection are expressed as mean change from baseline (±SEM). Significant group differences are discussed in the text.

did not block the hyperthermia that occurred in ACTH-injected rats. In fact, temperatures during the last 30 min of the observation appeared to remain slightly higher in rats treated with ACTH and naloxone than in rats treated with ACTH alone.

Correlations of Body Temperature with Specific Behaviors

In animals receiving i.c.v. ACTH and s.c. saline together, body temperature was highly correlated with head grooming and body grooming; body temperature was nega-

TABLE 5. Effects of Subcutaneous Injections of Saline (2 cc/kg), Haloperidol (0.05 mg/kg), or Naloxone (1 mg/kg) on Core Body Temperature

Treatment	Min after Injection						
	10–19	20–29	30–39	40–49	50–59	60–69	70–79
ACTH$_{1-24}$ (i.c.v.) and							
Saline	+ +	+ +	+				
Haloperidol	+ +	+	+ +				
Naloxone	+				+ +		+
Saline (i.c.v.) and							
Saline	+						
Haloperidol	+ +	+				+	+ +
Naloxone							

Rats were injected i.c.v. with ACTH$_{1-24}$ (3 µg) or saline (4 µl). Significant increases (+) from baseline for 10-min periods throughout the 70-min observation session are indicated as follows: +, $p < .05$; + +, $p < .01$; + + +, $p < .001$. Data are presented in FIGURE 6.

tively correlated with resting behavior (TABLE 6). This contrasts with the behavior of animals treated only with i.c.v. ACTH, who failed to show correlations between temperature and head and body grooming (TABLE 3). In these animals, grooming activity and temperatures were markedly elevated for most of the session. But because scores for head grooming, body grooming, and temperature did not undergo much change, the correlations were minimal. In rats treated with ACTH and s.c. saline, however, grooming scores declined steadily during the observation session, allowing a correlation to be detected.

When naloxone was injected into ACTH-treated rats, the very strong correlation of temperature with head grooming became insignificant. The correlation with body grooming was reduced slightly by naloxone, while the correlation with resting remained unchanged. Two other behaviors, scratching and nongrooming movement, were not significantly correlated with temperature in animals receiving ACTH alone. With the additional injection of naloxone, these behaviors became more strongly correlated with body temperature.

TABLE 6. Spearman Rank-Order Correlation Coefficients between Body Temperature (FIG. 6) and Specific Behaviors (FIG. 5)

	i.c.v./s.c.					
	ACTH/SAL	SAL/SAL	ACTH/NAL	SAL/NAL	ACTH/HAL	SAL/HAL
n	5	6	6	5	5	5
Paired obs.	34	41	41	33	35	35
Specific behaviors						
Head	[.74^f]	.29	.26	.09	.28	.26
Body	[.49^c]	.11	[.32^a]	− .07	.16	.18
Anogenital	− .04	.09	.19	.03	.07	.09
Scratch	.27	− .04	[.41^b]	− .11	− .29	− .31
Stretch-yawn	.09	.15	.00	− .26	− .19	[.46^c]
Move	.27	.24	[.47^c]	.15	.27	[.40^a]
Rest	[− .51^c]	− .14	[− .46^c]	− .06	− .11	[− .35^a]

Rats were injected intracerebroventricularly (i.c.v.) with 3 µg ACTH$_{1-24}$ or 4 µl saline (SAL), and subcutaneously (s.c.) with 2 cc/kg saline, 1 mg/kg naloxone (NAL), or 0.05 mg/kg haloperidol (HAL). Brackets denote a significant correlation with temperature at 10-min intervals throughout the 70-min observation period: ^{a} $p \leqslant .05$; ^{b} $p \leqslant .01$; ^{c} $p \leqslant .005$; ^{d} $p \leqslant .001$; ^{e} $p \leqslant .0005$; ^{f} $p \leqslant .0001$.

Naloxone alone did not produce any significant correlation between temperature and any of the behaviors observed. Except for the first 10-min period, these animals rested almost exclusively during the entire session, and displayed few other behaviors. Their temperatures were noticeably low, but showed little variation.

EFFECTS OF HALOPERIDOL

The effects of haloperidol on specific behaviors are shown in the bottom two panels of FIGURE 5. At the low dose of 0.05 mg/kg, haloperidol was only mildly effective in reducing the grooming induced by i.c.v. injection of ACTH.

Behavioral Analysis

Body Grooming. During the first 10-min interval, as well as 30–40 min after injection, haloperidol significantly decreased body grooming in ACTH-treated rats (compared to ACTH alone).

Head Grooming. Haloperidol did not affect ACTH-induced head grooming.

Anogenital Grooming. Anogenital grooming was increased 40–50 min after injection in ACTH-treated animals that received haloperidol.

Scratching. Haloperidol suppressed the scratching induced by ACTH 30–50 min after injection.

Stretching and Yawning. Compared to saline, haloperidol increased stretching and yawning 20–30 min after injection. It also increased stretching and yawning in ACTH-treated rats from 30 to 40 min after injection.

Moving. There was no significant effect of haloperidol on nongrooming movements.

Resting. Haloperidol increased resting in saline- and ACTH-treated rats during the first 10-min period.

Body Temperature

The effects of haloperidol on body temperature are shown in the bottom panel of FIGURE 6 and in TABLE 5. Haloperidol had no effect on the hyperthermia observed in saline- or ACTH-treated rats. Although initial temperatures of the haloperidol group were higher than those of the other groups, the differences were not significant.

Correlations of Body Temperature with Specific Behaviors

In the group treated with both i.c.v. ACTH and s.c. saline, there was a negative correlation of temperature with resting, and strong positive correlations with head grooming and body grooming (TABLE 6). When haloperidol was administered to ACTH-treated animals, the correlations with head and body grooming were no longer present. In fact, temperature was not significantly correlated with any of the behaviors in this group.

The group receiving haloperidol alone showed positive correlations of temperature with nongrooming movements, and with stretching and yawning. Temperature was negatively correlated with resting. These correlations were not evident in the saline-treated controls.

DISCUSSION

Grooming behaviors occur under a variety of conditions, and they undoubtedly serve many different physiological and ethological functions. Maintenance of the external body surface is just one of these functions. Grooming activity often appears in response to novelty, conflict, or other stressful situations. The adaptive significance of grooming under these conditions remains obscure. Grooming can also be induced pharmacologically, by injection of ACTH or certain other neuroactive peptides. In these cases as well, it is not clear which functional systems are responsible for the grooming response.

In rodents, grooming has been reported to play a role in thermoregulation. Moreover, stimuli that induce grooming, such as novelty or handling, also elevate body temperature. These same stimuli initiate the release of ACTH. The objective of our studies was to assess more directly the relationship between grooming activity and body temperature under conditions that either elicit or suppress the grooming response. The results of these experiments indicate that, in general, body temperature does indeed react to treatments that affect grooming.

Handling, injection procedures (i.c.v. and/or s.c.), and the placement of rats into a novel observation chamber resulted in an immediate increase in body temperature. This increase occurred in all treatment groups. On the other hand, temperature measurements before and after transport showed that when rats remained in their home cages, transport did not affect body temperature. Similarly, Jolles et al.[2] found transport to be a relatively minor factor in the induction of grooming, compared to the much greater effects of handling and placement into novel chambers.

Injection of $ACTH_{1-24}$ (3 μg, i.c.v.) resulted in a dramatic increase in grooming behavior and a significant elevation in body temperature. These changes were observed within 10–20 min after injection. Some treatments that reduced ACTH-induced grooming, such as i.c.v. injection of neurotensin (2 μg) or the subcutaneous injection of saline (2 cc/kg), reduced the hyperthermia observed in ACTH-treated animals. Naloxone (1 mg/kg, s.c.) also inhibited ACTH-induced grooming, but it did not affect ACTH-induced hyperthermia. However, naloxone prevented the hyperthermic response to saline injection and placement in the novel environment. Haloperidol, at a relatively low dose (0.05 mg/kg, s.c.), mildly reduced the grooming produced by ACTH. This dose did not alter the temperatures of ACTH- or saline-treated animals.

Subcutaneous injections of saline reduced the grooming of rats injected with i.c.v. ACTH and attenuated their rises in body temperature. The mechanisms underlying this suppression are not known, but the phenomenon of "acute tolerance" might be involved. Administration of ACTH,[22] endorphin,[23] or dynorphin[24] can suppress the grooming response to a subsequent injection of ACTH for up to 12 hours. In our experiment, the pain of the s.c. saline injection may have resulted in endogenous release of these substances, thereby reducing the behavioral and hyperthermic responses to the injection of ACTH.

Of the few relevant studies in the literature, there is no consensus as to the central effects of ACTH on body temperature. In rabbits, ACTH has most often been reported to induce hypothermia.[25,26] In rats, Lin et al.[27] found that i.c.v. injection of 4 μg ACTH increased rectal temperatures as well as grooming behavior. These authors attributed the higher rectal temperatures to increased metabolism and cutaneous vasoconstriction of the tail and feet. However, they reported only the peak changes in foot, tail, and rectal temperatures for the 90-min period after injection, and did not report temporal changes in the variables they measured. Thornhill and Wilfong[28] found little change in rectal temperatures after i.c.v. ACTH, but tail temperatures rose noticeably beginning 30 min after injection. Injection of 1 μg ACTH into the preoptic-anterior hypothalamic region (POAH), an area of brain implicated in heat-dissipation mechanisms,[29,30] produced even greater effects on tail temperature. From the time course of these responses, it appeared that core temperatures were being regulated through increases in tail temperature. Interestingly, grooming was also observed after injection or push-pull perfusion of ACTH into the POAH region.

β-Endorphin has also been linked to thermoregulatory systems in the rat. When injected i.c.v. or into the POAH, low doses of β-endorphin increase body temperature.[28] Furthermore, plasma β-endorphin levels rise dramatically after handling and exposure to a novel environment.[31] Since naloxone has been shown to block the "emotional hyperthermia" of rats in a novel environment, endogenous opioids have been implicated in this response.[31,32] In the present study, naloxone prevented the hyperthermia that developed after saline injection and placement in the novel cage, while it did not affect the hyperthermia observed in ACTH-treated rats. These results suggest that ACTH may contribute to novelty-induced hyperthermia through a nonopioid mechanism.

Neurotensin and naloxone both prevented the hyperthermic response to saline injection and placement in the novel cage. However, the temperatures of ACTH-treated rats were differentially affected by these two agents. Naloxone did not affect the hyperthermia in this group, while neurotensin converted the ACTH-induced hyperthermia into a hypothermic response. The hypothermia produced by the combination of neurotensin and ACTH exceeded that produced by neurotensin alone. These results do not readily explain the effects of naloxone and neurotensin on grooming behavior; however, they may shed light on the different effects of these drugs on paper-shredding behavior after water immersion. Paper shredding is operationally related to nest building, a behavior performed by rats in response to cold. After water immersion, paper shredding and grooming are increased. Both behaviors are blocked by naloxone.[33] Neurotensin also blocks grooming, but it significantly increases paper shredding.[34] The stress-related release of ACTH that occurs after water immersion may potentiate the hypothermic effects of neurotensin, thereby producing enhanced paper-shredding behavior.

In addition to ACTH, many other peptides increase grooming behaviors when injected into the cerebroventricular system of the rat. These include α-MSH, β-endorphin, dynorphin, cholecystokinin, TRH, vasopressin, oxytocin, prolactin, bombesin, substance P, and somatostatin. Of the peptides listed, only α-MSH,[35] dynorphin[24] and perhaps cholecystokinin (see paper by Kulkosky in this volume) produce behavior that is qualitatively similar to that observed after novelty- or ACTH-induced grooming. The most frequent behaviors exhibited by these animals are head and body grooming, although increases in scratching, anogenital grooming, and stretching and yawning are also observed. In contrast, scratching is the most prominent behavior displayed by rats after central injection of vasopressin, bombesin, or substance P (see papers by van Wimersma Greidanus, Wilcox, and Meisenberg, this volume). Injections of prolactin and oxytocin enhance anogenital grooming (see papers by Drago and Pedersen in this volume).

All of the above peptides have been reported to affect body temperature.[36,37] However, the relationships between specific grooming behaviors and body temperature have yet to be determined. Head and body grooming are more likely than other behaviors to be involved in the reduction of body temperature, since the spreading of saliva can result in evaporative cooling.[6] In the present study, the particular elements of the grooming repertoire that were most highly correlated with body temperature were head grooming and body grooming. The strength of these correlations was diminished in animals treated with naloxone, neurotensin, and haloperidol. Thus, it seems reasonable to suspect that head and body grooming are participating in thermoregulation, presumably via saliva spreading. Another possible function of grooming is to shift attention away from novel or threatening stimuli. The resulting decrease in autonomic arousal could also assist in lowering body temperature.

In conclusion, certain elements of grooming, either in a novel environment or after i.c.v. ACTH, appear to occur as part of a thermoregulatory response. Our data suggest that the systems controlling grooming and temperature are in some way associated, although the relationship is not at all simple. This is probably because grooming serves other functions as well, and because body temperature can be regulated through several different physiological and behavioral mechanisms. As yet, we do not have sufficient data to determine whether changes in body temperature are a cause or consequence of altered grooming, or are merely concomitant with these behaviors. Despite such questions, we suggest that thermoregulation be at least one of the systems considered when assigning a functional role for grooming in various contexts.

ACKNOWLEDGMENTS

We wish to thank Mr. Jan Brakkee for his expertise and excellent assistance during the course of these experiments.

REFERENCES

1. COLBERN, D. L., R. L. ISAACSON, E. J. GREEN & W. H. GISPEN. 1978. Repeated intraventricular injection of $ACTH_{1-24}$: The effects of home or novel environments on excessive grooming. Behav. Biol 23: 381–387.
2. JOLLES, J., J. ROMPA-BARENDREGT & W. H. GISPEN. 1979. Novelty and grooming behavior in the rat. Behav. Neural Biol. 25: 563–572.
3. HANNIGAN, J.H., JR. & R. L. ISAACSON. 1982. Conditioned excessive grooming in the rat after footshock: Effect of naloxone and situational cues. Behav. Neural Biol. 33: 280–292.
4. DELIUS, J. D. 1970. Irrelevant behaviour, information processing and arousal homeostasis. Psychol. Forsch. 33: 165–188.
5. GISPEN, W. H., V. M. WIEGANT, H. J. GREVEN & D. DE WIED. 1975. The induction of excessive grooming in the rat by intraventricular application of peptides derived from ACTH: Structure-activity studies. Life Sci. 17: 645–652.
6. HAINSWORTH, F. R. 1969. Saliva spreading, activity, and body temperature regulation in the rat. Am. J. Physiol. 212: 1288–1292.
7. THIESSEN, D. D., M. GRAHAM, J. PERKINS & S. MARCKS. 1977. Temperature regulation and social grooming in the Mongolian gerbil (Meriones unguiculatus). Behav. Biol. 19: 270–288.
8. ROBERTS, W. W. & R. D. MOONEY. 1974. Brain areas controlling thermoregulatory grooming, prone extension, locomotion, and tail vasodilation in rats. J. Comp. Physiol. Psych. 86: 470–480.
9. ROBERTS, W. W. & A. B. FROL. 1979. Interaction of central and superficial peripheral thermosensors in control of thermoregulatory behaviors of the rat. Physiol. Behav. 23: 503–512.
10. TANAKA, H., K. KANOSUE, T. NAKAYAMA & Z. SHEN. 1986. Grooming, body extension,

and vasomotor responses induced by hypothalamic warming at different ambient temperatures in rats. Physiol. Behav. **38**: 145–151.

11. THIESSEN, D., M. PENDERGRASS & A. E. HARRIMAN. 1982. The thermoenergetics of coat colour maintenance by the Mongolian gerbil. (*Meriones unguiculatus*). J. Therm. Biol. **7**: 51–56.

12. BRIESE, E. & M. G. DE QUIJADA. 1970. Colonic temperature of rats during handling. Acta Physiol. Latinoam. **20**: 97–102.

13. SINGER, R., C. T. HARKER, A. J. VANDER & M. J. KLUGER. 1986. Hyperthermia induced by open-field stress is blocked by salicylate. Physiol. Behav. **36**: 1179–1182.

14. GISPEN, W. H. & V. M. WIEGANT. 1976. Opiate antagonists suppress $ACTH_{1-24}$-induced excessive grooming in the rat. Neurosci. Lett. **2**: 159–164.

15. WIEGANT, V. M., A. R. COOLS & W. H. GISPEN. 1977. ACTH-induced grooming involves brain dopamine. Eur. J. Pharmacol. **41**: 343–345.

16. VAN WIMERSMA GREIDANUS, TJ. B. & G. J. E. RINKEL. 1983. Neurotensin suppresses ACTH-induced grooming. Eur. J. Pharmacol. **88**: 117–120.

17. VAN WIMERSMA GREIDANUS, TJ. B., C. MAIGRET, J. A. TEN HAAF, B. M. SPRUIJT & D. L. COLBERN. 1986. The influence of neurotensin, naloxone, and haloperidol on elements of excessive grooming behavior induced by ACTH. Behav. Neural Biol. **46**: 137–144.

18. VAN WIMERSMA GREIDANUS, TJ. B., D. K. DONKER, R. WALHOF, J. C. A. VAN GRAFHORST, N. DE VRIES, S. J. VAN SCHAIK, C. MAIGRET, B. M. SPRUIJT & D. L. COLBERN. 1985. The effects of neurotensin, naloxone and haloperidol on elements of excessive grooming behavior induced by bombesin. Peptides **6**: 1179–1183.

19. BRAKKEE, J. H., V. M. WIEGANT & W. H. GISPEN. 1979. A simple technique for rapid implantation of a permanent cannula into the rat brain ventricular system. Lab. Animal Sci. **29**: 78–81.

20. NEMEROFF, C. B., D. LUTTINGER, D. E. HERNANDEZ, R. B. MAILMAN, G. A. MASON, S. D. DAVIS, E. WIDERLÖV, G. D. FRYE, C. A. KILTS, K. BEAUMONT, G. R. BREESE & A. J. PRANGE, JR. 1983. Interactions of neurotensin with brain dopamine systems: Biochemical and behavioral studies. J. Pharmacol. Exp. Therap. **225**: 337–345.

21. GREEN, E. J., R. L. ISAACSON, A. J. DUNN & T. H. LANTHORN. 1979. Naloxone and haloperidol reduce grooming occurring as an aftereffect of novelty. Behav. Neural Biol. **27**: 546–551.

22. JOLLES, J., V. M. WIEGANT & W. H. GISPEN. 1978. Reduced behavioral effectiveness of $ACTH_{1-24}$ after a second administration: Interaction with opiates. Neurosci Lett. **9**: 261–266.

23. WIEGANT, V. M., J. JOLLES & W. H. GISPEN. 1978. β-Endorphin grooming in the rat: Single dose tolerance. *In* Characteristics and Functions of Opioids. Developments in Neuroscience, vol. 4. J. M. van Ree & L. Terenius, Eds.: 447–450. Elsevier/North Holland Biomedical Press. Amsterdam.

24. ALOYO, V. J., B. SPRUIJT, H. ZWIERS & W. H. GISPEN. 1983. Peptide-induced excessive grooming behavior: The role of opiate receptors. Peptides **4**: 833–836.

25. LIPTON, J. M. & J. R. GLYN. 1980. Central administration of peptides alters thermoregulation in the rabbit. Peptides **1**: 15–18.

26. LIPTON, J. M., J. R. GLYN & J. A. ZIMMER. 1981. ACTH and α-melanotropin in central temperature control. Fed. Proc. **40**: 2760–2764.

27. LIN M. T., L. T. HO & W. N. UANG. 1983. Effects of anterior pituitary hormones and their releasing hormones on physiological and behavioral functions in rats. J. Steroid Biochem. **19**: 433–438.

28. THORNHILL, J. A. & A. WILFONG. 1982. Lateral cerebral ventricle and preoptic-anterior hypothalamic area infusion and perfusion of β-endorphin and ACTH to unrestrained rats: Core and surface temperature responses. Can. J. Physiol. Pharmacol. **60**: 1267–1274.

29. CRAWSHAW, L. I. & H. J. CARLISLE. 1974. Thermoregulatory effects of electrical brain stimulation. J. Comp. Physiol. Psychol. **87**: 440–448.

30. ROBERTS, W. W., E. H. BERGQUIST & T. C. L. ROBINSON. 1969. Thermoregulatory grooming and sleep-like relaxation induced by local warming of preoptic area and anterior hypothalamus in opossum. J. Comp. Physiol. Psychol. **67**: 182–188.

31. BLÄSIG, J., V. HOLLT, U. BÄUERLE & A. HERZ. 1978. Involvement of endorphins in emo-

tional hyperthermia of rats. Life Sci **23**: 2525–2532.
32. STEWART, J. & R. EIKELBOOM. 1979. Stress masks the hypothermic effect of naloxone in rats. Life Sci **25**: 1165–1172.
33. COLBERN, D. L., R. L. ISAACSON, J. H. HANNIGAN & W. H. GISPEN. 1981. Water immersion, excessive grooming, and paper shredding in the rat. Behav. Neural. Biol. **32**: 428–437.
34. VAN WIMERSMA GREIDANUS, TJ. B., F. F. M. VAN ZINNICQ BERGMAN & D. L. COLBERN. 1985. Neurotensin diminishes grooming and stimulates paper shredding behavior induced by water immersion of rats. Eur. J. Pharmacol. **108**: 201–203.
35. SPRUIJT, B. M., P. N. E. DE GRAAN, A. N. EBERLE & W. H. GISPEN. 1985. Comparison of structural requirements of α-MSH and ACTH for inducing excessive grooming and pigment dispersion. Peptides **6**: 1185–1189.
36. BLATTEIS, C. M. 1981. Hypothalamic substances in the control of body temperature: General characteristics. Fed. Proc. **40**: 2735–2740.
37. CLARK, W. G. & J. M. LIPTON. 1983. Brain and pituitary peptides in thermoregulation. Pharmac. Ther. **22**: 249–297.

Bombesin and Ceruletide-induced Grooming and Inhibition of Ingestion in the Rat[a]

PAUL J. KULKOSKY

Department of Psychology
University of Southern Colorado
Pueblo, Colorado 81001

Grooming is elicited by a surprising number of structurally diverse, yet specific peptides. The phenomenon of peptide-induced excessive grooming was first described as a property of peptides derived from pro-opiomelanocortin (POMC) that were administered intracisternally in mammals.[1] Early work focused on the grooming activity of POMC-derived peptides,[2-4] but subsequent research has revealed that a wide variety of bioactive peptides can potently elicit grooming and scratching responses when applied to the central nervous system. Recent reviews of the growing set of peptides that induce grooming cite at least eight non-POMC-related peptides.[5,6]

We encountered peptide-induced grooming when researching the hypothesis that certain peptides found in gut and brain function as ingestion-produced "satiety signals" in the short-term control of feeding behavior.[7-9] Excessive grooming was measured in the course of an initial search for neural sites of action of neuropeptides in the elicitation of postprandial satiety. These studies utilized peptides that were isolated originally from amphibian skin, but have structurally related counterparts in mammalian gastrointestinal and brain tissues. These peptides are referred to as "brain-gut peptides," and are subjects of extensive research as candidate hormones, neurotransmitters, and neuromodulators.[10] Amino acid sequences of two extensively studied brain-gut peptides, bombesin (BBS) and ceruletide (CER), are displayed in TABLE 1, along with sequences of the endogenous mammalian relative of CER, sulfated cholecystokinin octapeptide (CCK-8), and the prototypical stimulus of grooming, adrenocorticotropic hormone fragment 1–24 (ACTH$_{1-24}$). Comparison reveals no common sequence among these three peptide types that elicit grooming. One of the intriguing mysteries of chemically induced grooming is that markedly diverse molecules can elicit grooming, and yet display a high degree of structural specificity. A single amino acid substitution or desulfation can abolish excessive grooming elicited by BBS or CCK-8, respectively. This finding initially suggests multiple or convergent neuroendocrine mechanisms of peptide-induced grooming. Discovery of non-POMC-related peptide-induced grooming broadens the generality of the neurochemical control of body surface maintenance behaviors, and challenges simple models of the neural mechanisms and biological meanings of grooming. Brain-gut peptide control of grooming also clearly demonstrates that putative feeding satiety signals may function in the regulation of several classes of adaptive behavior in addition to ingestion at different sites in the nervous system.

[a] This research was supported by a University of Southern Colorado Faculty Research Grant and a grant from the National Institutes of Health (no. RR-08197-06).

BOMBESIN-INDUCED GROOMING

Bombesin is a tetradecapeptide isolated from anuran skin that exerts a wide range of neuropharmacological effects when injected parenterally in mammals.[11-13] Bombesin is the prototype of a group of structurally related peptides that have been identified and localized in mammalian gut and brain.[14,15] Central nervous system administration of bombesin elicits physiological effects that include alteration of endocrine secretions[16] and gut function,[17] hypothermia,[18] and hyperglycemia.[19] Intracerebroventricular (i.c.v.) injection of BBS also potently elicits a variety of behavioral effects in the rat, including suppression of feeding and drinking,[7,20,21] inhibition of resting and sleep,[21,22] analgesia,[23] and stimulation of locomotion,[23-25] rearing,[25] and grooming behaviors.[18,21,26,27]

Peripherally administered bombesin suppresses feeding in the rat in a specific manner characteristic of postprandial satiety signals.[7,28,29] Of particular interest is the finding that intraperitoneal (i.p.) BBS elicits the complete behavioral sequence of satiety in the rat,[28] which consists of the cessation of feeding, followed by exploratory and maintenance behaviors, including a short bout of grooming, and finally resting or sleep.[30,31] This pattern of behaviors is observed after a normal meal or a meal shortened by administration of nutrients or gut hormones, but not after feeding terminated by bitter tastes or psychoactive drugs.[31] Therefore, one criterion for verifying a genuine role of a neuropeptide in the satiation process is that administration of the peptide elicits a particular behavioral display that includes only a low frequency of grooming.[9]

In an initial assessment of possible sites of action of systemic BBS in the induction of satiety, we compared feeding suppressions and behavioral displays after a range of doses of intracerebroventricular bombesin.[21] Rats were deprived of food for 18 h and given access to liquid diet for 30 min. Behavior of each rat was observed at a tone-cued 0.6-s interval, once each minute. Behaviors were classified into categories of feeding, grooming, resting, and 15 other categories.[21,31] Each of the latter categories typically occurred at a low frequency during a meal, and they were combined into one designated "others" for purposes of graphic display. Grooming was recorded when the rat was observed during the 0.6-s interval in the following activities: biting or licking coat, paws, genitals or tail, scratching or stroking head, face or vibrissae with one or both front paws, or scratching head or body with hind limb. Subcategories of grooming recorded were: grooming head and/or front paws, grooming body, grooming genitals, and grooming tail. Results from this time-sampling observational characterization of meal-related behaviors are depicted in FIGURE 1, for rats receiving i.c.v. bombesin or vehicle (artificial cerebrospinal fluid, CSF). The i.c.v. route of administration of bombesin produced decreases in feeding and resting and a large increase in grooming, in contrast to i.p. BBS, which decreased feeding, increased resting, and did not alter grooming. Excessive grooming after i.c.v. BBS had been noticed previously in the rat[18,20] and mouse.[32] Grooming increases and resting decreases were produced at 0.01 μg of BBS, accompanied by declines in sniffing and standing behaviors. Feeding was suppressed at doses of 0.1 μg or higher, and this grooming-associated inhibition of feeding clearly did not resemble the normal termination of a meal or the shortening of a meal by i.p. BBS. Thus, it was concluded that although bombesin-like immunoreactivity is found in cerebrospinal fluid,[33] the satiation effect of systemic BBS could not be mediated solely by an increase in cerebroventricular bombesin. Subsequent work has indicated a peripheral gut afferent site of action of systemic bombesin in the elicitation of satiety.[29,34]

Bombesin-induced excessive grooming was also observed during an 18-h water-deprivation-induced drinking session, again at 0.01 μg, but in this motivational context, drinking and drinking-associated feeding were suppressed at the same threshold,

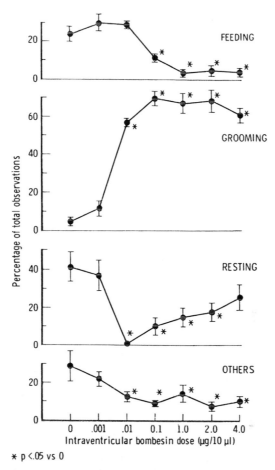

FIGURE 1. Mean (± standard error, SE) percentage of total observations of feeding, grooming, resting, and all other categories combined, for food-deprived rats ($n = 5$) as a function of i.c.v. bombesin dose. (From Kulkosky et al.[21] Reprinted with permission from *Physiology and Behavior*.)

as shown in FIGURE 2. A feeding satiety signal, such as systemic BBS, does not suppress drinking at the same doses,[9] so this is further evidence against an exclusively periventricular site of action of BBS in food satiation. Motivated opportunity for ingestive acts did not affect the amount of grooming elicited by i.c.v. BBS (0.1 µg), as seen in FIGURE 3, which presents data from rats previously fed and watered ad libitum receiving i.c.v. BBS or CSF in the absence of food or water. Therefore, grooming elicited by i.c.v. bombesin appeared robust, prepotent, and independent of motivational context. Large changes in ingestion, resting, and other behaviors are neither necessary stimuli to, nor necessary concomitants of, BBS-induced grooming. An earlier study of the motivational contexts of ACTH-induced grooming[35] supports a general conclusion that i.c.v. peptide-induced grooming is difficult to influence by changes in food or water deprivation conditions.

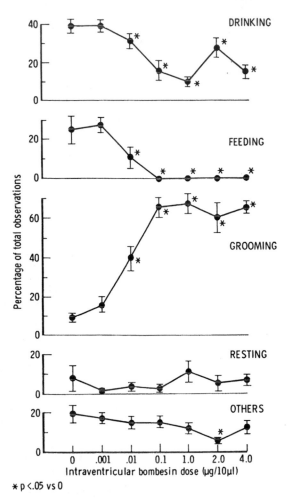

FIGURE 2. Mean (± SE) percentage of total observations of drinking, feeding, grooming, resting, and all other categories combined, for water-deprived rats ($n = 6$) as a function of i.c.v. bombesin dose. (From Kulkosky *et al.*[21] Reprinted with permission from *Physiology and Behavior*.)

The effects of prior experience with bombesin and the structural specificity of this peptide effect were examined in a subsequent experiment.[36] Repeated i.c.v. bombesin injections (1.0 µg) did not result in significant changes in feeding suppression or excessive grooming across a series of nine injections spaced 48 h apart, as shown in FIGURES 4 and 5. A similar result for grooming was reported with repeated injections of ACTH$_{1-24}$ outside the context of a meal.[37] Combined injections of i.p. and i.c.v. BBS also did not change the behavioral display, but injection of [D-Trp-8] bombesin (1.0 µg), a non-bioactive tetradecapeptide analogue of BBS,[19,38] did not change feeding or grooming. Thus, a single amino acid substitution abolished the behavioral effects of bombesin. Structurally related bioactive bombesin-like peptides, litorin (LIT), an amphibian-

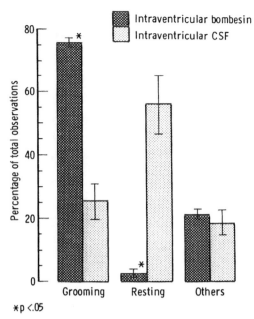

FIGURE 3. Mean (± SE) percentage of total observations of grooming, resting, and all other categories combined, for previously ad lib fed and watered rats (*n* = 6) tested in the absence of food and water. (From Kulkosky *et al.*[21] Reprinted with permission from *Physiology and Behavior.*)

derived nonapeptide,[39] and gastrin-releasing peptide (GRP), a mammalian 27 amino acid peptide representative of the bombesin family,[40] also induced excessive grooming in these rats. Intracerebroventricularly administered LIT (1.0 µg) elicited excessive grooming comprising 41.5 + 8.9% (mean + standard error) of the 30-min observation period and slightly reduced food intake (−19.3 + 7.2%). Peripheral injection of litorin also suppresses feeding, but it does not elicit excessive grooming.[39] Intracerebroventricular GRP (1.0 µg) also elicited excessive grooming (54.8 + 3.8%) and suppressed food intake (−53.5 + 14.8%). Peripheral injection of GRP also suppresses feeding without eliciting grooming.[41] A study of GRP structure-activity relations demonstrated that a wide variety of BBS-like peptides can centrally elicit excessive grooming in the rat, although the action of GRP appeared less potent and less persistent than BBS.[42]

Tetradecapeptide bombesin and biologically active bombesin-like peptides induce excessive grooming in the rat independently of motivational context and prior experience. BBS-like peptide-induced grooming can be observed with or without concomitant inhibition of food or water intake or resting. These properties of BBS-induced grooming are incompatible with a simple displacement motivation model of peptide-induced excessive grooming, as was previously discussed in relation to ACTH-induced grooming.[35] Deprivation-induced consumption of food and water does not greatly affect BBS- or ACTH-induced grooming, so an explanation of peptide-induced excessive grooming in terms of energy displaced from competing behaviors is not tenable.

An alternative explanation of BBS-like peptide-induced grooming may be derived

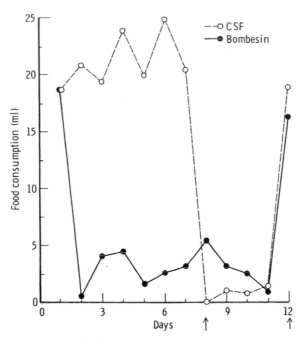

FIGURE 4. Mean liquid diet consumption of food-deprived rats after 1.0 μg i.c.v. CSF (n = 3) or bombesin (n = 4). On days 8–11 both groups received i.c.v. BBS. On day 11, BBS (4.0 μg) was also injected (i.p.). On day 12, D-Trp-8 BBS (1.0 μg) was injected (i.c.v.). (From Kulkosky et al.[36] Reprinted with permission from *Brain Research*.)

from a topographic analysis of excessive grooming and scratching in terms of the anatomical region of the body that is groomed, regardless of the specific motor act performed in grooming. A recent immunocytochemical and radioimmunoassay study demonstrated BBS-containing neurons in spinal cord laminae of the dorsal horn of cat, rat, and mouse.[43] In that study, bombesin injected into mouse spinal cord induced grooming responses. These findings led to the hypothesis that BBS-like peptides are neurotransmitters of primary sensory afferents.[43] When the BBS-induced grooming data of FIGURE 3 were analyzed topographically, the proportions of body surfaces groomed did not differ significantly after BBS or CSF, as shown in FIGURE 6. The observed body surface proportions groomed are remarkably similar to the corresponding proportions of body surfaces represented on the rat primary somatosensory cortex animunculus.[44] However, areas of body surface groomed after i.c.v. BBS or CSF are not similar to the areas of body represented in rat primary motor cortex.[44] The close relationship between body surface groomed and corresponding area of somatosensory cortex representation, along with neuronal localization data,[43] favor a novel explanation of BBS-induced grooming in terms of alterations in skin sensation. In turn, these considerations are not compatible with a displacement model of grooming or a direct motor stimulation of grooming by BBS. The new model of BBS-induced grooming is in accord with a "boundary-shift" model of induced grooming, in which responsivity to the ordinary stimuli to grooming is enhanced.[45,46] If verified, an altered-

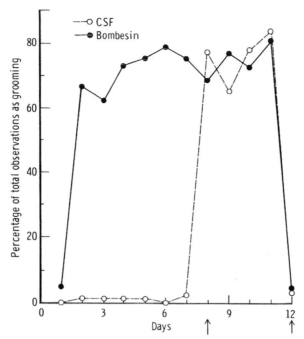

FIGURE 5. Mean percentage of total observations of behavior in the category of grooming for rats and treatments described in FIGURE 4. (From Kulkosky et al.[36] Reprinted with permission from *Brain Research*.)

sensation hypothesis of BBS-induced grooming would further validate its use as an animal model of the clinical disorder pruritis, which is characterized by excessive itching accompanied by scratching.[47]

In summary, bombesin-induced grooming in the rat is a robust phenomenon, appearing independently of, or concomitantly with, alterations in ingestive behaviors and prior experience with bombesin. The grooming effect of the neuropeptide bombesin is dependent on structure and administration route, and may be explained as a result of alteration of neurotransmission in somatosensory neurons. Bombesin also elicits route- and structure-specific changes in feeding, drinking, resting, and other behaviors. The well-documented, behaviorally specific satiety effect of peripherally administered bombesin indicates that neural bombesin may have multiple, separable, and site-specific functions in the regulation of behavior.

CERULETIDE-INDUCED GROOMING

Ceruletide (caerulein diethylammonium hydrate) is an example of bioactive peptide that elicits grooming, but bears little amino acid sequence similarity to either ACTH$_{1-24}$ or BBS (TABLE 1). Ceruletide is a decapeptide isolated from anuran skin that shares amino acid sequencing and biological effects with the mammalian gut hormone and

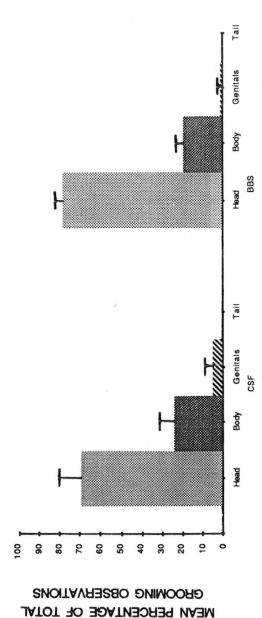

FIGURE 6. Mean (+ SE) percentage of total observations of grooming directed to the head, body, genitals, or tail, for rats described in FIGURE 4 after i.c.v. CSF or BBS (1 μg/10 μl).

TABLE 1. Amino Acid Sequences of Peptides That Induce Grooming

Bombesin	pGlu-Gln-Arg-Leu-Gly-Asn-Gln-Trp-Ala-Val-Gly-His-Leu-Met-NH$_2$
Ceruletide	pGlu-Gln-Asp-Tyr(SO$_3$)-Thr-Gly-Trp-Met-Asp-Phe-NH$_2$
CCK-8	Asp-Tyr(SO$_3$)-Met-Gly-Trp-Met-Asp-Phe-NH$_2$
ACTH$_{1-24}$	Ser-Tyr-Ser-Met-Glu-His-Phe-Arg-Trp-Gly-Lys-Pro-Val-Gly-Lys-Lys-Arg-Arg-Pro-Val-Lys-Val-Tyr-Pro-OH

neuropeptide cholecystokinin (CCK).[48,49] The neuropharmacological effects of CER and bioactive CCK-like peptides include elicitation of food satiety, sedation, catalepsy, ptosis, hypothermia, antistereotypic, anticonvulsant and tremorolytic activity, inhibition of drinking, rearing, and locomotion, analgesia, and alteration of conditioned

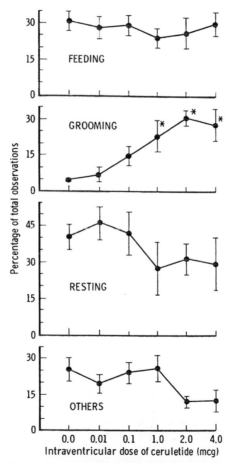

FIGURE 7. Mean (± SE) percentage of total observations of feeding, grooming, resting, and all other categories combined, for food-deprived rats ($n = 5$) as a function of i.c.v. ceruletide dose (in micrograms). (From Smith et al.[51] Reprinted with permission from *Peptides*.)

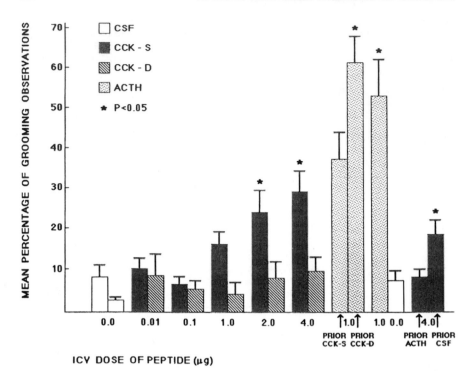

FIGURE 8. Mean (+ SE) percentage of total observations made in the category of grooming behavior for rats injected i.c.v. with CSF, sulfated (n = 6) or desulfated (n = 6) CCK-8, or ACTH$_{1-24}$, as a function of dose of peptide. Rats received either sulfated or desulfated CCK-8 (4.0 µg) 3 h prior to the first trial with ACTH. The second trial with ACTH was conducted 21 h later, and followed 3 h later by CCK-8-S (4.0 µg). (From Kulkosky *et al.*[79] Reprinted with permission from *Experimental Neurology*.)

avoidance and self-stimulation responses.[48,49] We were interested in the finding that peripheral administration of ceruletide inhibits food intake in a manner that satisfies many of the criteria for a satiety signal, as expected from its similarity to cholecystokinin.[50,51] However, intracerebroventricular injection of ceruletide did not inhibit feeding, although it did elicit excessive grooming in 18-h food-deprived rats given 30 min access to liquid diet, as shown in FIGURE 7.[51] This finding was consistent with the hypothesis of a peripheral, vagus-nerve-dependent site of action of CCK-like peptides in the elicitation of postprandial satiety.[8,51] The data also reveal a new neuropharmacological effect of i.c.v. CER, the induction of excessive grooming in the rat. Excessive grooming was the only behavioral effect of i.c.v. ceruletide observed at the doses tested, which further demonstrates that peptide-induced grooming may be dissociated from changes in ingestive behavior, resting, or other behaviors. Route of administration specificity and independence from ingestion changes argue that neuropeptides function site specifically and perform multiple, separable roles in control of behavior.

If grooming elicited by i.c.v. ceruletide depends on a structurally specific, biological action of cholecystokinin-like peptides, then it may be expected that other bioactive CCK-like peptides have grooming properties. We have recently demonstrated that sulfated cholecystokinin octapeptide (CCK-8-S, TABLE 1) induced grooming in the rat.[79] Sulfated CCK-8 is the predominant form of cholecystokinin found in mammalian gut and brain tissues, and desulfation of CCK-8 reduces or abolishes membrane receptor binding and most or all of CCK's physiological and behavioral actions.[52-55] We found that i.c.v. CCK-8-S, but not desulfated CCK-8, elicits excessive grooming in the rat and exhibits a short-term cross tolerance to ACTH (ACTH$_{1-24}$, TABLE 1). FIGURE 8 displays observations of grooming behavior during a 30-min session of ad lib fed and watered, home-caged rats after i.c.v. CCK-8-S or desulfated CCK-8. Grooming is elicited only by the sulfated, biologically active octapeptide, at doses in the range of ceruletide-induced grooming. This excessive grooming syndrome occurs at a lower frequency than the more obviously excessive grooming induced by ACTH or BBS, and is consistent with a previous observation of grooming behavior after injection of CCK-8 into caudate, substantia nigra, or nucleus accumbens of rat.[56] Prior (-3 h) administration of i.c.v. CCK-8-S (4.0 µg) inhibited the excessive grooming produced by ACTH (1.0 µg), and conversely, prior injection of ACTH abolished the ability of CCK-8-S to elicit grooming, as is displayed in FIGURE 8. Therefore, a bidirectional, acute cross-tolerance exists between CCK-8 and the prototypical peptide stimulus of grooming, ACTH. Short-term cross-tolerance among POMC-related peptides in the induction of excessive grooming has been reported previously,[5,57,58] but such cross-tolerance has not been observed before with two peptides of unrelated structures. The time courses and response topographies of CCK and ACTH-induced grooming were also found to be similar.

Cross-tolerance of peptides in control of grooming may suggest a neuroendocrine mechanism of CCK-like peptides in elicitation of grooming. CCK-8 and corticotropin-releasing factor (CRF) are found co-localized in neurons of the hypothalamic paraventricular nucleus,[59] and CCK-8 or CER administration stimulates release of ACTH and corticosterone.[60-63] CRF also elicits grooming in the rat.[64] Thus, it is possible that CCK-like peptides stimulate grooming by causing ACTH or CRF release. This putative mechanism of CCK-induced grooming suggests an example of convergent neuroendocrine mechanism of dissimilar peptides in control of grooming, and indicates a novel neuropharmacological effect of brain cholecystokinin, the elicitation of grooming.

MULTIPLE MECHANISMS OF PEPTIDE-INDUCED GROOMING

Peptides having very different structures elicit grooming behavior when applied to the central nervous system of rats. This finding indicates either multiple, independent neural mechanisms of grooming elicitation, or converging neuroendocrine processes that ultimately act at a common neural locus. Cross-tolerance effects with structurally dissimilar peptides are consistent with a convergence upon a final common receptor. An explanation of CCK-8-induced grooming is possible in terms of the stimulation of ACTH-release. Further evidence for a convergent action of CCK-like and ACTH-like peptides would be provided by evidence of common blockage of their grooming effects by neuropharmacological agents such as naloxone.

However, it also appears there are multiple and largely independent mechanisms for peptide-induced grooming. Bombesin has been reported not to exhibit short-term cross-tolerance with ACTH or neurohypophyseal hormones in elicitation of grooming.[65,66] Also, BBS-induced grooming is not blocked by naloxone, unlike ACTH, although dopaminergic antagonists, lesion of mesolimbicocortical dopamine neurons, benzomorphan κ-receptor agonists, anxiolytics, and spantide, a bombesin receptor

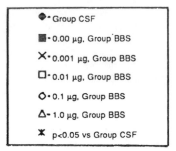

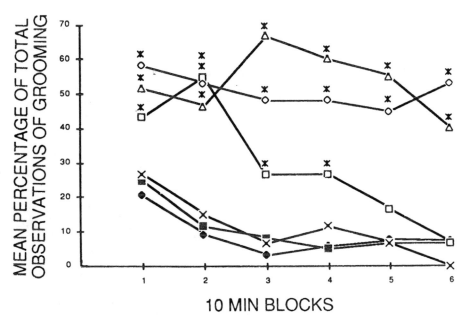

FIGURE 9. Mean percentage of total observations made in the category of grooming behavior of golden hamsters after i.c.v. injections of bombesin ($n = 6$) or CSF ($n = 5$) across consecutive 10-min intervals.

blocker, can antagonize BBS-induced grooming.[25,27,32,57,58,67-73] Further, the response topographies of excessive grooming syndromes of BBS and ACTH are quite different. Bombesin is reported to produce an excessive hindleg scratching syndrome in the rat,[27,65] whereas ACTH elicits a proportionally normal grooming and scratching display.[4,5] Bombesin also does not affect the release of ACTH or corticosterone in the rat,[74] so it appears less likely that all peptide-induced grooming can be explained as converging on a single, common receptor in the rat brain. The presently described peptide-induced grooming syndromes differ greatly in their response topographies, neuropharmacological sensitivities, and in exhibition of cross-tolerance with ACTH.

We (Kulkosky, Foderaro, Glazner, Niichel, and Schnur) have recently found that

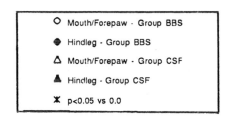

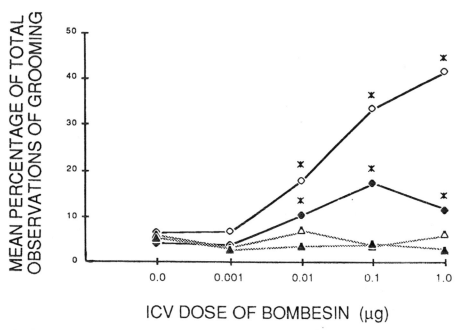

FIGURE 10. Mean percentage of total observations made in the categories of forepaw and mouth grooming or hindleg scratching for golden hamsters receiving doses of i.c.v. bombesin or repeated i.c.v. injections of CSF.

bombesin elicits excessive grooming in ad lib fed and watered golden hamsters, as was previously noted in food-deprived hamsters.[75] The threshold-eliciting dose of i.c.v. BBS was the same in hamsters and rats (0.01 µg), as can be seen in Figure 9, which presents grooming observations made across a 60-min session to illustrate the time course of the effect. However, unlike the rat, the hamster shows increases largely in forelimb and mouth grooming behaviors, and not hindleg scratching, as shown in Figure 10, which shows separately the increases in forepaw and mouth grooming and hindleg scratching as a function of dose of bombesin. This species-specificity of the topography of BBS-induced grooming in rodents further indicates that neuroendocrine mechanisms of peptide-induced grooming may also differ among species.[76]

Further research will determine comparative topographies, neuropharmacological sensitivities, and neuroanatomical loci of the various peptide-induced grooming syndromes. Many peptides may activate common receptors in common loci to produce qualitatively similar excessive grooming displays, as may be the case with CCK-8 and ACTH. Nevertheless, it is also clear that there may be multiple independent receptors and pathways involved in qualitatively distinct peptide-induced grooming syndromes, as the bombesin syndrome indicates. Finally, we might expect species specificities in both response predispositions[77] and neuroendocrine mechanisms in the peptide elicitation of grooming.

SUMMARY

Peptides of diverse structure stimulate grooming in rodents and other mammals. Peptide-induced grooming may be observed in several motivational contexts, with or without strong alternative response tendencies. Bombesin-like peptides elicit grooming route dependently in the rat and hamster, independently of, or concomitantly with, changes in ingestive behaviors or resting. The pattern of body surfaces groomed after i.c.v. BBS is in proportion to the representation of body surfaces in somatosensory but not motor cortex of rat. A bombesin-like peptide may be a neurotransmitter in somatosensory afferent processing, and grooming after i.c.v. BBS may reflect a response to alteration of cutaneous sensation. Bombesin is a putative satiety signal in the control of feeding and ethanol intake,[78] but the satiation effects of systemic BBS can be dissociated from the grooming effect of central BBS. Thus, bombesin may perform independent and site-specific functions in the control of behavior. Grooming produced by BBS is not affected by naloxone, involves a different proportion of motor acts than is observed in normal or ACTH-induced grooming, and no cross-tolerance has been reported between ACTH and BBS in the rat. These properties of bombesin-induced grooming indicate multiple, separable mechanisms of peptide-induced grooming and scratching.

Cholecystokinin-like peptide-induced grooming is observed after central injection in the rat and is unaccompanied by changes in feeding or resting. The well-documented satiety action of systemic CCK-like peptides is not accompanied by excessive grooming, so multiple, site-specific behavioral roles are also indicated for CCK-like peptides in control of behavior. CCK-8 exhibits short-term cross-tolerance with ACTH in elicitation of grooming, and central CCK-8 is co-localized with CRF and stimulates ACTH and corticosterone release in the rat. Thus, CCK-8 may induce grooming by increasing CRF or ACTH activity. These properties of CCK-like peptide-induced grooming indicate convergent neuroendocrine mechanisms that may explain some, but not all, peptide-induced grooming syndromes. Further characterization of the qualitative topographic, neuropharmacological, and neuroanatomical differences and species specificities of peptide-induced excessive grooming should provide a basis for understanding how brains coordinate grooming. Knowledge of the processes of neuropeptide control of grooming may provide potential peptide-based controls of grooming-related clinical disorders such as pruritis and allergic reactions. Neuropeptides are potent tools for examining how the brain manages the many complex motor acts involved in the maintenance of the body surface.

ACKNOWLEDGMENTS

The author thanks G. P. Smith, J. Gibbs, R. L. Isaacson, P. Schnur, C. L. Molello, and M. A. Foderaro for their contributions to this work.

REFERENCES

1. FERRARI, W., G. L. GESSA & L. VARGIU. 1963. Behavioral effects induced by intracister-nally injected ACTH and MSH. Ann. N.Y. Acad. Sci. **104:** 330–345.

2. GISPEN, W. H., V. M. WIEGANT, H. M. GREVEN & D. DE WIED. 1975. The induction of excessive grooming in the rat by intraventricular application of peptides derived from ACTH: Structure-activity studies. Life Sci. **17:** 645–652.

3. GISPEN, W. H., V. M. WIEGANT, A. F. BRADBURY, E. C. HULME, D. G. SMYTH, C. R. SNELL & D. DE WIED. 1976. Induction of excessive grooming in the rat by fragments of lipotropin. Nature **264:** 794–795.

4. GISPEN, W. H. & R. L. ISAACSON. 1981. ACTH-induced excessive grooming in the rat. Pharmacol. Ther. **12:** 209–246.

5. GISPEN, W. H. & R. L. ISAACSON. 1986. Excessive grooming in response to ACTH. Encycl. Pharmacol. Exp. Ther. **117:** 273–312.

6. ISAACSON, R. L. & W. H. GISPEN. Neuropeptides and the issue of stereotypy in behavior. *In* Neurobiology of Behavioral Stereotypy. S. J. Cooper & C. T. Dourish, Eds. Oxford Univ. Press. Oxford, England. In press.

7. GIBBS, J., P. J. KULKOSKY & G. P. SMITH. 1981. Effects of peripheral and central bombesin on feeding behavior of rats. Peptides **2** (Suppl. 2): 179–183.

8. SMITH, G. P., J. GIBBS & P. J. KULKOSKY. 1982. Relationships between brain-gut peptides and neurons in the control of food intake. *In* The Neural Basis of Feeding and Reward. B. G. Hoebel & D. Novin, Eds.: 149–165. Haer Institute. Brunswick, ME.

9. KULKOSKY, P. J. 1985. Conditioned food aversions and satiety signals. Ann. N.Y. Acad. Sci. **443:** 330–347.

10. WALSH, J. H., Ed. 1981. The Brain-Gut Axis: A New Frontier. Peptides **2** (Suppl. 2).

11. ERSPAMER, V. 1980. Peptides of the amphibian skin active on the gut. II. Bombesin-like peptides: Isolation, structure, and basic functions. *In* Gastrointestinal Hormones. G. B. Jerzy Glass, Ed.: 343–361. Raven. New York.

12. NEMEROFF, C. B., D. LUTTINGER & A. J. PRANGE, JR. 1983. Neurotensin and bombesin. *In* Handbook of Psychopharmacology, Vol. 16: Neuropeptides. L. L. Iversen, S. D. Iversen & S. H. Snyder, Eds.: 363–466. Plenum. New York.

13. WALSH, J. H. 1983. Bombesin-like peptides. *In* Brain Peptides. D. T. Krieger, M. J. Brownstein & J. B. Martin, Eds.: 941–960. John Wiley & Sons. New York.

14. MCDONALD, T. J. 1981. Non-amphibian bombesin-like peptides. *In* Gut Hormones. 2d edit. S. R. Bloom & J. M. Polak, Eds.: 407–412. Churchill Livingstone. New York.

15. SPINDEL, E. 1986. Mammalian bombesin-like peptides. Trends Neurosci. **9:** 130–133.

16. RIVIER, C., J. RIVIER & W. VALE. 1978. The effect of bombesin and related peptides on prolactin and growth hormone secretion in the rat. Endocrinology **102:** 519–522.

17. TACHE, Y., W. VALE, J. RIVIER & M. BROWN. 1981. Brain regulation of gastric acid secretion in rats by neurogastrointestinal peptides. Peptides **2** (Suppl. 2): 51–55.

18. BROWN, M., J. RIVIER & W. VALE. 1977. Bombesin: Potent effects on thermoregulation in the rat. Science **196:** 998–1000.

19. BROWN, M., Y. TACHE & D. FISHER. 1979. Central nervous system action of bombesin: Mechanism to induce hyperglycemia. Endocrinology **105:** 660–665.

20. DE CARO, G., M. MASSI & L. G. MICOSSI. 1980. Effect of bombesin on drinking induced by angiotensin II, carbachol, and water deprivation in the rat. Pharmacol. Res. Commun. **12:** 657–666.

21. KULKOSKY, P. J., J. GIBBS & G. P. SMITH. 1982. Behavioral effects of bombesin administration in rats. Physiol. Behav. **28:** 505–512.

22. RASLER, F. E. 1984. Behavioral and electrophysiological manifestations of bombesin: Excessive grooming and elimination of sleep. Brain Res. **321:** 187–191.

23. PERT, A., T. W. MOODY, C. B. PERT, L. A. DEWALD & J. RIVIER. 1980. Bombesin: Receptor distribution in brain and effects on nociception and locomotor activity. Brain Res. **193:** 209–220.

24. HAWKINS, M. F. & D. D. AVERY. 1983. Effects of centrally-administered bombesin and adrenalectomy on behavioral thermoregulation and locomotor activity. Neuropharmacology **22:** 1249–1255.

25. MERALI, Z., S. JOHNSTON & S. ZALCMAN. 1983. Bombesin-induced behavioural changes:
     Antagonism by neuroleptics. Peptides 4: 693–697.
26. CANTALAMESSA, F., G. DE CARO, M. MASSI & L. G. MICOSSI. 1982. A study on behavioural
     alterations induced by intracerebroventricular administration of bombesin to rats. Phar-
     macol. Res. Commun. 14: 163–173.
27. GMEREK, D. E. & A. COWAN. 1983. Studies on bombesin-induced grooming in rats. Pep-
     tides 4: 907–913.
28. GIBBS, J., D. J. FAUSER, E. A. ROWE, B. J. ROLLS, E. T. ROLLS & S. P. MADDISON. 1979.
     Bombesin suppresses feeding in rats. Nature 282: 208–210.
29. GIBBS, J. 1985. Effect of bombesin on feeding behavior. Life Sci. 37: 147–153.
30. BOLLES, R. C. 1960. Grooming behavior in the rat. J. Comp. Physiol. Psychol. 53: 306–310.
31. ANTIN, J., J. GIBBS, J. HOLT, R. C. YOUNG & G. P. SMITH. 1975. Cholecystokinin elicits
     the complete behavioral sequence of satiety in rats. J. Comp. Physiol. Psychol. 89: 784–790.
32. KATZ, R. 1980. Grooming elicited by intracerebroventricular bombesin and eledoisin in the
     mouse. Neuropharmacology 19: 143–146.
33. YAMADA, T., M. S. TAKAMI & R. H. GERNER. 1981. Bombesin-like immunoreactivity in
     human cerebrospinal fluid. Brain Res. 223: 214–217.
34. STUCKEY, J. A., J. GIBBS & G. P. SMITH. 1985. Neural disconnection of gut from brain
     blocks bombesin-induced satiety. Peptides 6: 1249–1252.
35. JOLLES, J., J. ROMPA-BARENDREGT & W. H. GISPEN. 1979. ACTH-induced excessive grooming
     in the rat: The influence of environmental and motivational factors. Horm. Behav. 12:
     60–72.
36. KULKOSKY, P. J., J. GIBBS & G. P. SMITH. 1982. Feeding suppression and grooming repeat-
     edly elicited by intraventricular bombesin. Brain Res. 242: 194–196.
37. COLBERN, D. L., R. L. ISAACSON, E. J. GREEN & W. H. GISPEN. 1978. Repeated intraven-
     tricular injections of ACTH₁₋₂₄: the effects of home or novel environments on excessive
     grooming. Behav. Biol. 23: 381–387.
38. MOODY, T. W., C. PERT, J. RIVIER & M. BROWN. 1978. Bombesin: Specific binding to
     rat brain membranes. Proc. Natl. Acad. Sci. U.S.A. 75: 5372–5376.
39. KULKOSKY, P. J. & J. GIBBS. 1982. Litorin suppresses food intake in rats. Life Sci. 31: 685–692.
40. TACHE, Y., W. MARKI, J. RIVIER, W. VALE & M. BROWN. 1981. Central nervous system
     inhibition of gastric secretion in the rat by gastrin-releasing peptide, a mammalian
     bombesin. Gastroenterology 81: 298–302.
41. STEIN, L. J. & S. C. WOODS. 1982. Gastrin releasing peptide reduces meal size in rats. Pep-
     tides 3: 833–835.
42. GIRARD, F., C. AUBE, S. ST-PIERRE & F. B. JOLICOEUR. 1983. Structure-activity studies
     on neurobehavioral effects of bombesin (BB) and gastrin releasing peptide (GRP). Neu-
     ropeptides 3: 443–452.
43. O'DONOHUE, T. L., V. J. MASSARI, C. J. PAZOLES, B. M. CHRONWALL, C. W. SHULTS,
     R. QUIRION, T. N. CHASE & T. W. MOODY. 1984. A role for bombesin in sensory pro-
     cessing in the spinal cord. J. Neurosci. 4: 2956–2962.
44. WOOLSEY, C. N. 1958. Organization of somatic sensory and motor areas of the cerebral
     cortex. In Biological and Biochemical Bases of Behavior. H. F. Harlow & C. N. Woolsey,
     Eds.: 63–81. Univ. Wisconsin Press. Madison, WI.
45. FENTRESS, J. C. 1977. The tonic hypothesis and the patterning of behavior. Ann. N.Y. Acad.
     Sci. 290: 370–395.
46. COHEN, J. A. & E. O. PRICE. 1979. Grooming in the Norway rat: Displacement activity
     or "boundary-shift"? Behav. Neural Biol. 26: 177–188.
47. GMEREK, D. E. & A. COWAN. 1983. An animal model for preclinical screening of systemic
     antipruritic agents. J. Pharmacol. Meth. 10: 107–112.
48. ZETLER, G. 1985. Neuropharmacological profile of cholecystokinin-like peptides. Ann. N.Y.
     Acad. Sci. 448: 110–120.
49. ZETLER, G. 1985. Caerulein and its analogues: Neuropharmacological properties. Peptides
     6 (Suppl. 3): 33–46.
50. GIBBS, J., R. C. YOUNG & G. P. SMITH. 1973. Cholecystokinin decreases food intake in
     rats. J. Comp. Physiol. Psychol. 84: 488–495.

51. SMITH, G. P., C. JEROME, P. KULKOSKY & K. J. SIMANSKY. 1984. Ceruletide acts in the abdomen, not in the brain, to produce satiety. Peptides 5: 1149–1157.
52. BEINFELD, M. C. 1983. Cholecystokinin in the central nervous system: A minireview. Neuropeptides 3: 411–427.
53. DOCKRAY, G. J. 1983. Cholecystokinin. In Brain Peptides. D. T. Kreiger, M. J. Brownstein & J. B. Martin, Eds.: 851–869. John Wiley & Sons. New York.
54. EMSON, P. C. & P. D. MARLEY. 1983. Cholecystokinin and vasoactive intestinal polypeptide. In Handbook of Psychopharmacology, vol. 16: Neuropeptides. L. L. Iversen, S. D. Iversen & S. H. Snyder, Eds.: 255–306. Plenum. New York.
55. VARGAS, F., O. FREROT, M. D. T. TUONG, K. ZUZEL, C. ROSE & J.-C. SCHWARTZ. 1985. Sulfation and desulfation of cerebral cholecystokinin. Ann. N.Y. Acad. Sci. 448: 110–120.
56. DIAMOND, B. I., G. PASINETTI, A. HITRI & R. L. BORISON. 1984. Extrapyramidal dopamine regulation by cholecystokinin and its role in Parkinson's disease. Adv. Neurol. 40: 483–488.
57. JOLLES, J., V. M. WIEGANT & W. H. GISPEN. 1978. Reduced behavioral effectiveness of ACTH$_{1-24}$ after a second administration: Interaction with opiates. Neurosci. Lett. 9: 261–266.
58. ALOYO, V. J., B. SPRUIJT, H. ZWIERS & W. H. GISPEN. 1983. Peptide-induced excessive grooming in the rat: The role of opiate receptors. Peptides 4: 833–836.
59. MEZEY, E., T. D. REISINE, L. SKIRBOLL, M. BEINFELD, & J. Z. KISS. 1985. Cholecystokinin in the medial parvocellular subdivision of the paraventricular nucleus: Co-existence with corticotropin-releasing hormone. Ann. N.Y. Acad. Sci. 448: 152–156.
60. ITOH, S., R. HIROTA, G. KATSUURA & K. ODAGUCHI. 1979. Adrenocortical stimulation by a cholecystokinin preparation in the rat. Life Sci. 25: 1725–1730.
61. PORTER, J. R. & L. D. SANDER. 1981. The effect of cholecystokinin on pituitary-adrenal hormone secretion. Regul. Pept. 2: 245–252.
62. REISINE, T. & R. JENSEN. 1986. Cholecystokinin-8 stimulates adrenocorticotropin release from anterior pituitary cells. J. Pharmacol. Exp. Ther. 236: 621–626.
63. BASSO, N., A. MATERIA, V. D'INTINOSANTE, A. GINALDI, V. PONA, P. REILLY, S. RUGGERI & M. FIORAVANTI. 1981. Effect of ceruletide on pituitary-hypothalamic peptides and on emotion in man. Peptides 2 (Suppl. 2): 71–75.
64. MORLEY, J. E. & A. S. LEVINE. 1982. Corticotrophin releasing factor, grooming and ingestive behavior. Life Sci. 31: 1459–1464.
65. VAN WIMERSMA GREIDANUS, TJ. B., D. K. DONKER, F. F. M. VAN ZINNICQ BERGMANN, R. BEKENKAMP, C. MAIGRET & B. SPRUIJT. 1985. Comparison between excessive grooming induced by bombesin or by ACTH: The differential elements of grooming and development of tolerance. Peptides 6: 369–372.
66. MEISENBERG, G. 1982. Short-term behavioural effects of neurohypophyseal hormones: Pharmacological characteristics. Neuropharmacology 21: 309–316.
67. DUNN, A. J., S. R. CHILDERS, N. R. KRAMARCY & J. W. VILLIGER. 1981. ACTH-induced grooming involves high-affinity opiate receptors. Behav. Neural Biol. 31: 105–109.
68. GISPEN, W. H. & V. M. WIEGANT. 1976. Opiate antagonists suppress ACTH$_{1-24}$-induced excessive grooming in the rat. Neurosci. Lett. 2: 159–164.
69. GMEREK, D. E. & A. COWAN. 1982. Classification of opioids on the basis of their ability to antagonize bombesin-induced grooming in rats. Life Sci. 31: 2229–2232.
70. GMEREK, D. E. & A. COWAN. 1984. In vivo evidence for benzomorphan-selective receptors in rats. J. Pharm. Exp. Ther. 230: 110–115.
71. MERALI, Z., S. JOHNSTON & J. SISTEK. 1985. Role of dopaminergic system(s) in mediation of the behavioural effects of bombesin. Pharmacol. Biochem. Behav. 23: 243–248.
72. YACHNIS, A. T., J. N. CRAWLEY, R. T. JENSEN, M. M. McGRANE & T. W. MOODY. 1984. The antagonism of bombesin in the CNS by substance P analogues. Life Sci. 35: 1963–1969.
73. CRAWLEY, J. N. & T. W. MOODY. 1983. Anxiolytics block excessive grooming behavior induced by ACTH$_{1-24}$ and bombesin. Brain Res. Bull. 10: 399–401.
74. BROWN, M. R., J. RIVIER & W. W. VALE. 1977. Bombesin affects the central nervous system to produce hyperglycemia in rats. Life Sci. 21: 1729–1734.
75. MICELI, M. O. & MALSBURY, C. W. 1985. Effects of putative satiety peptides on feeding and drinking behavior in golden hamsters (Mesocricetus auratus). Behav. Neurosci. 99: 1192–1207.

76. COWAN, A., P. KHUNAWAT, X. ZU ZHU & D. E. GMEREK. 1985. Effects of bombesin on behavior. Life Sci. **37**: 135–145.
77. BOLLES, R. C. 1984. Species-typical response predispositions. *In* The Biology of Learning. P. Marler & H. S. Terrace, Eds.: 435–446. Springer-Verlag. New York.
78. KULKOSKY, P. J., M. R. SANCHEZ, N. CHIU & G. W. GLAZNER. 1987. Characteristics of bombesin-induced inhibition of ethanol intake. Neuropharmacology **26**: 1211–1216.
79. KULKOSKY, P. J., C. L. MOLELLO & R. L. ISAACSON. 1987. CCK-8 elicits grooming: Cross tolerance to $ACTH_{1-24}$. Exp. Neurol. **97**: 697–703.

Comparison of Bombesin-, ACTH-, and β-Endorphin-induced Grooming

Antagonism by Haloperidol, Naloxone, and Neurotensin

Tj. B. VAN WIMERSMA GREIDANUS, F. VAN DE BRUG,
L. M. DE BRUIJCKERE, P. H. M. A. PABST, R. W. RUESINK,
R. L. E. HULSHOF, B. N. M. VAN BERCKEL, S. M. ARISSEN,
E. J. P. DE KONING, AND D. K. DONKER

Rudolf Magnus Institute for Pharmacology
Medical Faculty
University of Utrecht
3521 GD Utrecht, The Netherlands

Bombesin, a tetradecapeptide originally isolated from amphibian skin[1] and presently known as a potent peptide in the central nervous system,[2] induces various effects on the central nervous system and on behavior (for review see ref. 3) such as reduction of ambulation in an open field,[4] increased locomotor activity,[5,6] suppression of food intake,[7-11] and decrease in drinking behavior.[8] Moreover, it is one of the numerous peptides inducing excessive grooming behavior. However, the various grooming-inducing peptides have been reported to induce structurally different grooming bouts.[12,13] ACTH, MSH, and dynorphin produce grooming bouts comparable to normally occurring grooming behavior,[14-18] whereas excessive grooming induced by the opioid peptide β-endorphin, by substance P, and by bombesin is characterized by the predominant display of scratching.[12-16,19,20,21] In addition, development of tolerance was observed for the grooming-inducing effects of ACTH and β-endorphin,[13,16,22] but not for those of bombesin,[13] and it has been concluded that the grooming displayed by animals treated with $ACTH_{1-24}$ or with bombesin is of a completely different nature.[13]

It has been claimed that dopaminergic as well as opioid systems are involved in peptide-induced grooming, and dopamine as well as opiate-receptor antagonists have been shown to suppress peptide-induced grooming. In fact, the grooming response depends on the integrity of neural pathways with dopaminergic and/or opiate receptors (for review see ref. 15). In addition to the suppressive effects of dopamine and opiate antagonists on excessive grooming, neurotensin has also been reported to reduce excessive grooming induced by a new environment, by water immersion, or by administration of $ACTH_{1-24}$ or bombesin.[18,21,23,24]

In the present study we compared the composition of excessive grooming behavior as induced by ACTH, bombesin, and β-endorphin and investigated the effects of haloperidol, naloxone, and neurotensin on excessive grooming so induced. In particular, we studied the effects of these drugs on separately distinguished elements of excessive grooming behavior as induced by the peptides.

MATERIAL AND METHODS

Male rats from an inbred Wistar strain (TNO, Zeist, The Netherlands) weighing approximately 150 g were used. Cannulation of the right lateral ventricle of the brain was performed 5 days prior to the behavioral experiments. Animals were kept in single cages under a standard light-dark condition (light on 6:00 A.M., off 8:00 P.M.) with food and water *ad libitum*.

Rats were allocated to different treatment groups and received intracerebroventricular (i.c.v.) injections of 2 µl. Each animal was used only once. $ACTH_{1-24}$ (Organon International b.v., Oss, The Netherlands) was i.c.v. injected in a dose of 1 µg. Bombesin (B 610; UCB, The Hague, The Netherlands) was i.c.v. injected in doses grading from 10 ng, 30 ng, to 100 ng; β-endorphin and des-tyrosine β-endorphin (Organon International b.v., Oss, The Netherlands) were injected in grading doses of 3 ng, 10 ng, 30 ng, 100 ng, 300 ng, 1 µg, to 3 µg. Pretreatment with naloxone was performed at 20 min prior to the i.c.v. injection of the peptide. The doses of naloxone were 1 mg/kg for subcutaneous (s.c.) and 1 µg and 10 µg for i.c.v. injections. Pretreatment with haloperidol was also performed at 20 min prior to the i.c.v. injection of the peptide. The doses of this drug applied were 0.02 mg/kg, 0.05 mg/kg, or 0.1 mg/kg for s.c. injection and 1 µg or 10 µg for i.c.v. administration. Pretreatment with neurotensin (R 4322, Bachem, Torrance, CA) was performed immediately prior to i.c.v. administration of ACTH, bombesin, or β-endorphin, in a dose of 1 µg i.c.v. Artificial CSF was used as placebo for i.c.v. injections, whereas saline was used as placebo for s.c. administration.

For local application of bombesin or ACTH into restricted brain areas, rats were bilaterally equipped by use of a stereotaxic instrument with a stainless steel cannula in the nucleus accumbens, in the striatum or in the periaqueductal gray. The animals were allowed to recover from the operation for a time period of 5-6 days.

For the experiments with rats bearing lesions in certain brain structures, the animals were operated 6 days prior to behavioral experimentation. Using a stereotaxic instrument, electrolytical lesions were made in the nucleus accumbens, the substantia nigra, or the periaqueductal gray (PAG). During the same surgical procedure, a polyethylene cannula was also placed in the lateral ventricle for i.c.v. administration of bombesin or ACTH.

The grooming test was performed according to Gispen *et al.*[25] Rats were placed into perspex observation boxes (26 × 20 × 13 cm) immediately after administration of ACTH, bombesin, β-endorphin, or placebo. Scoring of grooming frequencies was started 10 min later, using a 15-s sampling procedure for a period of 50 min, resulting in a maximally possible grooming score (T) of 200. The following elements of grooming were scored: head washing (H), bodily grooming (B), paw licking (P), sexual (anogenital) grooming (A), and scratching (S). Results are expressed as mean (± SEM) grooming score per treatment group per 50 min. Statistical evaluation of the data was performed by analysis of variance (ANOVA) in combination with a supplemental t-test.

RESULTS

The mean grooming frequency displayed by placebo-treated rats amounted to approximately 35. Bombesin (3–300 ng), $ACTH_{1-24}$ (100 ng–2 µg), and β-endorphin (30–300 ng) induced a dose-dependent increase in grooming behavior. Administration of the highest doses of bombesin and of $ACTH_{1-24}$ resulted in grooming scores approaching

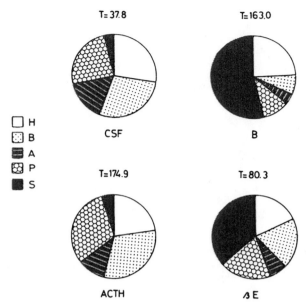

FIGURE 1. Relative distribution of grooming elements following i.c.v. administration of bombesin (*B*, 100 ng), ACTH (2 μg), and β-endorphin (*βE*, 100 ng): T, total grooming frequency; *H*, head washing; *B*, bodily grooming; *A*, anogenital (sexual) grooming; *P*, paw licking; *S*, scratching.

the maximally possible score of 200 by inducing total grooming scores of 178.8 (± 5.5) and 174.9 (± 5.4), respectively. Intracerebroventricular injection of β-endorphin resulted in increased frequencies of grooming, amounting to scores varying from 60 to 110 when doses of 30–100 ng were applied. Administration of higher doses of this opioid peptide did not further increase total grooming scores, mainly due to the immobility displayed by the animals during the first 25 min following injection of 1 μg or 3 μg β-endorphin. Administration of des-tyrosine β-endorphin did not result in increased grooming behavior. The grooming frequencies scored following i.c.v. injection of the various doses of this peptide never exceeded 43, which is not significantly different from the grooming displayed by placebo-treated rats. Lower doses of bombesin induced a more general type of compulsive grooming, whereas higher amounts of bombesin induced a shift towards the scratching element at the cost of bodily grooming, sexual (anogenital) grooming, and paw licking. Scratching was also predominantly displayed by animals treated with β-endorphin. In contrast, ACTH$_{1-24}$ induced a dose-dependent increase of all elements of grooming. Consequently bombesin and β-endorphin cause a shift in the relative distribution of grooming elements, whereas the composition of the grooming behavior following ACTH administration is similar to that displayed by placebo-treated rats (see FIG. 1).

Subcutaneous pretreatment with naloxone resulted in a suppression of excessive grooming induced by ACTH, bombesin, and β-endorphin. Intracerebroventricular administration of a high dose (10 μg) of naloxone resulted generally in a similar suppression of peptide-induced grooming. The naloxone-induced suppression concerned generally all elements of ACTH-induced excessive grooming. In contrast, the suppressive

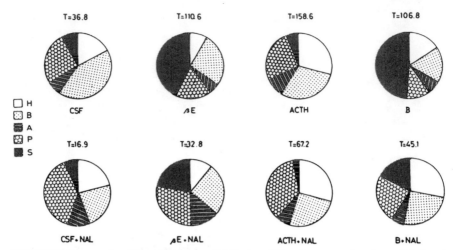

FIGURE 2. Suppressive effects of naloxone (*NAL*, 1 mg/kg, s.c.) on grooming elements induced by i.c.v. injection of β-endorphin (*βE*, 100 ng), ACTH (1 µg), or bombesin (*B*, 10 ng). For other abbreviations see legend to FIGURE 1.

effect of naloxone on bombesin-induced excessive grooming was mainly due to a suppression of the scratching and bodily grooming elements. This differential suppression of various elements of bombesin-induced compulsive grooming behavior by naloxone was also observed in β-endorphin-induced excessive grooming. Also in this case, scratching and bodily grooming appeared to be the main elements involved in the suppressive effect of naloxone. In addition to its suppressive effects on peptide-induced grooming, naloxone also suppressed the grooming displayed by placebo-treated animals. As was observed for the suppression by naloxone of ACTH-induced grooming, generally all elements were involved in the suppression by naloxone of the novelty-induced grooming displayed by the placebo-treated rats.

Because of the differential suppression by naloxone of various elements of excessive grooming induced by bombesin and by β-endorphin, the pretreatment with this opiate antagonist resulted in a strong shift in the relative distribution of grooming elements. In fact, naloxone pretreatment caused a reduction in scratching that was accompanied in the bombesin-treated rats by a relative (but not absolute) increase in the paw licking and head washing elements and in the β-endorphin-treated rats by a relative increase in paw licking and anogenital grooming (see FIG 2).

Pretreatment s.c. and i.c.v. with the higher doses of haloperidol resulted in a marked reduction of grooming-behavior in placebo-treated rats, as well as in rats treated with bombesin, ACTH, or β-endorphin. All elements were more or less equally involved in this reduction of total grooming scores following haloperidol pretreatment, and consequently the relative distribution of the elements of grooming behavior was not significantly different in the various groups of rats pretreated with haloperidol or with its placebo (see FIG. 3).

Neurotensin suppressed basal grooming by reducing most elements more or less equally. It also suppressed ACTH-induced grooming, and this suppression concerned three out of the five grooming elements. However, the suppressive effect of neurotensin on bombesin- and β-endorphin-induced excessive grooming is mainly due to a marked

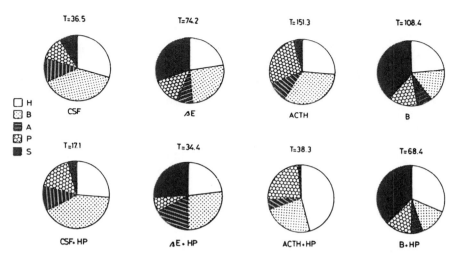

FIGURE 3. Suppressive effects of haloperidol (*HP*, 0.05 mg/kg, s.c.) on grooming elements induced by i.c.v. injection of β-endorphin (*βE*, 100 ng), ACTH (1 μg), or bombesin (*B*, 10 ng). For other abbreviations see legend to FIGURE 1.

reduction of the scratching element. Consequently neurotensin administration results in a change in the pattern of grooming behavior displayed by rats treated with bombesin or β-endorphin. In bombesin-treated rats the neurotensin-induced reduction of scratching is accompanied by a relative increase in the head washing, anogenital grooming, and paw licking elements, whereas in β-endorphin-treated animals the reduced scratching is accompanied by a relative increase in bodily grooming (see FIG. 4).

Preliminary studies in which animals with lesions in the periaqueductal gray (PAG), the nucleus accumbens, or the substantia nigra were used revealed that lesions in the nucleus accumbens did not affect bombesin-induced excessive grooming. However, partial damage of the PAG, and lesions in the substantia nigra, reduced bombesin-induced excessive grooming, mainly by a reduction of the scratching element. Lesions in the substantia nigra also resulted in a (weak) reduction of ACTH-induced excessive grooming, mainly by a significant reduction of the bodily grooming, whereas lesions in the nucleus accumbens did not affect ACTH-induced excessive grooming.

Local application of bombesin (2 × 10 ng) in the nucleus accumbens induced a significant increase of scratching, whereas a lower dose (2 × 1 ng) was ineffective. A similar effect was observed following local application of bombesin in the substantia nigra. Local application of bombesin (2 × 10 ng) close to or in the PAG induced excessive grooming and scratching, and as far as the relative distribution of elements is concerned, this grooming was similar to that displayed by rats i.c.v. injected with 30 ng bombesin. No effect was observed following local application of bombesin in the striatum or in the ventral tegmental area.

DISCUSSION

ACTH, bombesin, and β-endorphin-induce excessive grooming of different natures. Whereas ACTH-induced excessive grooming involves most elements more or less equally,

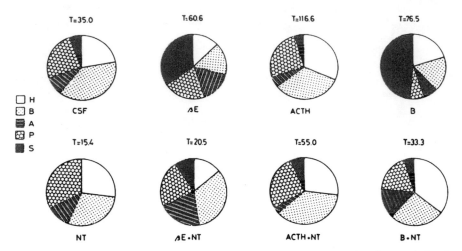

FIGURE 4. Suppressive effects of neurotensin (*NT*, 1 µg, i.c.v.) on grooming elements induced by i.c.v. injection of β-endorphin (*βE*, 100 ng), ACTH (1 µg), or bombesin (*B*, 10 ng). For other abbreviations see legend to FIGURE 1.

bombesin-induced excessive grooming and, albeit to a lesser extent, β-endorphin-induced grooming are characterized by a predominant display of scratching. This marked difference in the relative distribution of grooming elements following administration of these grooming-inducing peptides strongly suggests that the underlying substrate involved in ACTH-induced excessive grooming differs from that of bombesin- and β-endorphin-induced grooming. The presence or absence of development of tolerance for the grooming-inducing effects of ACTH or bombesin[13] also points to such a difference.

The differential suppression of elements of peptide-induced excessive grooming by the opiate antagonist naloxone, by the dopamine antagonist haloperidol, and by neurotensin also illustrates the difference in substrates underlying the excessive grooming induced by the peptides used. All three drugs (naloxone, haloperidol, and neurotensin) suppress basal (novelty-induced) grooming as well as excessive grooming induced by ACTH, bombesin, or β-endorphin, confirming results of previous studies by others and by us.[15,23,26-28] However, there is a marked difference in the composition of peptide-induced grooming behavior following pretreatment with these drugs. Whereas haloperidol suppresses the various elements more or less equally, with no resulting shift in the pattern of grooming, pretreatment with naloxone or with neurotensin results in a marked change in the distribution of the grooming elements of bombesin- and of β-endorphin-induced excessive grooming. This is mainly due to the preferential suppression by naloxone and neurotensin of the most pronounced element of the grooming induced by these peptides, *i.e.*, scratching. The results suggest that the scratching element is the component of grooming behavior mainly displayed by activation of opiate-receptor systems, whereas the role of dopaminergic systems can be regarded as more generally modulating. In addition, the observation that des-tyrosine β-endorphin does not induce excessive grooming supports the view that opiate receptors are involved, at least in the β-endorphin-induced excessive grooming. Since the suppressive influences of naloxone and neurotensin on peptide-induced excessive grooming appear to be very similar, it is tempting to assume that in addition to its capability of modulating dopamine transmission in the brain,[29] neurotensin interferes with opiate-receptor systems as well.

It has been put forward previously[30] that bombesin may mediate its behavioral effects through dopaminergic systems, and the present data on the reduction of bombesin-induced excessive grooming by haloperidol support this view. In addition, it has been suggested[31] that bombesin acts through receptors in certain terminal regions (nucleus accumbens) of the mesolimbic dopamine system when affecting locomotor behavior. Also, the suppressive influence of neurotensin on hyperactivity is thought to occur by an action on efferent outputs of mesolimbic stimulation.[32] In contrast, the present findings that lesions in the substantia nigra, but not in the nucleus accumbens, affected bombesin-induced grooming suggest that bombesin affects the nigrostriatal dopamine system rather than the mesolimbic system when inducing excessive grooming and scratching. However, it may also be that the GABAergic striato-nigral pathway, which can be activated by dopamine, is involved in bombesin-induced excessive grooming. In fact the involvement of the striato-nigro-collicular pathway in the modulation of excessive grooming has been demonstrated,[33] but whether or not this pathway is indeed involved in excessive grooming as induced by bombesin is not clear as yet, and further experimentation is needed. Interestingly, the PAG is a site of action of ACTH with respect to excessive grooming behavior.[33] Since the PAG contains opiate receptors, receives afferents from the colliculus superior, and because its output is among other things modulated by dopaminergic structures,[33] it is tempting to assume that bombesin-induced excessive grooming (suppressed by dopamine antagonists) and scratching (suppressed by opiate antagonists) is regulated by this structure. The observations that lesions in the PAG reduce bombesin-induced excessive grooming and that local application of bombesin in the PAG elicits excessive grooming support this assumption.

ACKNOWLEDGMENT

The technical assistance of Carla Maigret is gratefully acknowledged.

REFERENCES

1. ANASTASI, A., V. ERSPAMER & M. BUCCI. 1971. Isolation and structure of bombesin and alytesin, two analogous active peptides from the skin of the European amphibians Bombina and Alytes. Experientia **27:** 166–167.
2. MOODY, T. W., J. N. CRAWLEY & R. T. JENSEN. 1982. Pharmacology and neurochemistry of bombesin-like peptides. Peptides **3:** 559–563.
3. ERSPAMER, V. & P. MELCHIORRI. 1983. Actions of amphibian skin peptides on the central nervous system and the anterior pituitary. *In* Neuroendocrine Perspectives, Vol. 2. E. E. Müller & R. M. MacLeod, Eds.: 37–106. Elsevier. Amsterdam.
4. VAN WIMERSMA GREIDANUS, TJ.B., J. A. SCHIJFF, J. L. NOTEBOOM, M. C. SPIT, L. BRUINS, M. VAN ZUMMEREN & G. J. E. RINKEL. 1984. Neurotensin and bombesin, a relationship between their effects on body temperature and locomotor activity? Pharmacol. Biochem. Behav. **21:** 197–202.
5. HAWKINS, M. F. & D. D. AVERY. 1983. Effects of centrally administered bombesin and adrenalectomy on behavioral thermoregulation and locomotor activity. Neuropharmacology **22:** 1249–1255.
6. MERALI, Z., S. JOHNSTON & S. ZALCMAN. 1983. Bombesin-induced behavioural changes: Antagonism by neuroleptics. Peptides **4:** 693–697.
7. GIBBS, J., D. J. FAUSER, E. A. ROWE, B. J. ROLLS, E. T. ROLLS & S. P. MADDISON. 1979. Bombesin suppresses feeding in rats. Nature **282:** 208–210.

8. KULKOSKY, P. J., J. GIBBS & G. P. SMITH. 1982. Behavioral effects of bombesin adminis-
    tration in rats. Physiol. Behav. 28: 505-512.
9. KULKOSKY, P. J., J. GIBBS & G. P. SMITH. 1982. Feeding suppression and grooming repeatedly
    elicited by intraventricular bombesin. Brain Res. 242: 194-196.
10. MARTIN, C. F. & J. GIBBS. 1980. Bombesin elicits satiety in sham feeding rats. Peptides
    1: 131-134.
11. MORLEY, J. E. & A. S. LEVINE. 1981. Bombesin inhibits stress-induced eating. Pharmacol.
    Biochem. Behav. 14: 149-151.
12. COWAN, A., P. KHUNAWAT, X. ZU ZHU & D. E. GMEREK. 1985. Current concepts. IV.
    Effects of bombesin on behavior. Life Sci. 37: 135-145.
13. VAN WIMERSMA GREIDANUS, TJ. B., D. K. DONKER, F. F. M. VAN ZINNICQ BERGMANN,
    R. BEKENKAMP, C. MAIGRET & B. SPRUIJT. 1985. Comparison between excessive grooming
    induced by bombesin or by ACTH: The differential elements of grooming and develop-
    ment of tolerance. Peptides 6: 369-372.
14. ALOYO, V. J., B. SPRUIJT, H. ZWIERS & W. H. GISPEN. 1983. Peptide-induced excessive
    grooming in the rat: The role of opiate receptors. Peptides 4: 1-4.
15. GISPEN, W. H. & R. L. ISAACSON. 1981. ACTH-induced excessive grooming in the rat.
    Pharmacol. Ther. 12: 209-246.
16. GMEREK, D. E. & A. COWAN. 1983. ACTH-(1-24) and RX 336-M induce excessive grooming
    in rats through different mechanisms. Eur. J. Pharmacol. 88: 339-346.
17. SPRUIJT, B. M. & W. H. GISPEN. 1983. ACTH and grooming behavior in the rat. In Hor-
    mones and Behaviour in Higher Vertebrates. J. Balthazert, E. Pröve & R. Gilles, Eds.:
    118-136. Springer-Verlag. Berlin.
18. VAN WIMERSMA GREIDANUS, TJ. B., C. MAIGRET, J. A. TEN HAAF, B. M. SPRUIJT & D. L.
    COLBERN. 1986. The influence of neurotensin, naloxone and haloperidol on elements
    of excessive grooming behavior induced by ACTH. Behav. Neural Biol. 46: 137-144.
19. KATZ, R. J. 1980. Substance P elicited grooming in the mouse: Behavioral and pharmaco-
    logical characteristics. Int. J. Neurosci. 10: 187-189.
20. KATZ, R. 1980. Grooming elicited by intracerebroventricular bombesin and eledoisin in the
    mouse. Neuropharmacology 19: 143-146.
21. VAN WIMERSMA GREIDANUS, TJ. B., D. K. DONKER, R. WALHOF, J. C. A. VAN GRAFHORST,
    N. DE VRIES, S. J. VAN SCHAIK, C. MAIGRET, B. M. SPRUIJT & D. L. COLBERN. 1985.
    The effects of neurotensin, naloxone and haloperidol on elements of excessive grooming
    behavior induced by bombesin. Peptides 6: 1179-1183.
22. JOLLES, J., V. M. WIEGANT & W. H. GISPEN. 1978. Reduced behavioral effectiveness of
    ACTH$_{1-24}$ after a second administration: Interaction with opiates. Neurosci. Lett. 9:
    261-266.
23. VAN WIMERSMA GREIDANUS, TJ. B. & G. J. E. RINKEL. 1983. Neurotensin suppresses ACTH-
    induced grooming. Eur. J. Pharmacol. 88: 117-120.
24. VAN WIMERSMA GREIDANUS, TJ. B., F. F. M. VAN ZINNICQ BERGMANN & D. L. COLBERN.
    1985. Neurotensin diminishes grooming and stimulates paper shredding behavior induced
    by water immersion of rats. Eur. J. Pharmacol. 108: 201-203.
25. GISPEN, W. H., V. M. WIEGANT, H. M. GREVEN & D. DE WIED. 1975. The induction of
    excessive grooming in the rat by intraventricular application of peptides derived from
    ACTH. Structure-activity studies. Life Sci. 17: 645-652.
26. COLBERN, D. L., R. L. ISAACSON & J. H. HANNIGAN, JR. 1981. Water immersion, excessive
    grooming, and paper shredding in the rat. Behav. Neural Biol. 32: 428-437.
27. GREEN, E. J., R. L. ISAACSON, A. J. DUNN & T. H. LANTHORN. 1979. Naloxone and
    haloperidol reduce grooming occurring as an after effect of novelty. Behav. Neural Biol.
    27: 546-551.
28. TRABER, J., H. R. KLEIN & W. H. GISPEN. 1982. Actions of antidepressant and neuroleptic
    drugs on ACTH- and novelty-induced behavior in the rat. Eur. J. Pharmacol. 80: 407-414.
29. NEMEROFF, C. B., P. W. KALIVAS & A. J. PRANGE, JR. 1984. Interaction of neurotensin
    and dopamine in limbic structures. In Catecholamines: Neuropharmacology and Cen-
    tral Nervous System—Theoretical Aspects. E. Usdin, A. Carlsson, A. Dahlström & J.
    Engel, Eds.: 199-206. Alan R. Liss. New York.

30. MERALI, Z., S. JOHNSTON & J. SISTEK. 1985. Role of dopaminergic system(s) in mediation of the behavioural effects of bombesin. Pharmacol. Biochem. Behav. **23:** 243–248.
31. SCHULZ, D. W., P. W. KALIVAS, C. B. NEMEROFF & A. J. PRANGE, JR. 1984. Bombesin-induced locomotor hyperactivity: Evaluation of the involvement of the mesolimbic dopamine system. Brain Res. **304:** 377–382.
32. JOLICOEUR, F. B., R. RIVEST, S. ST.-PIERRE, M. A. GAGNÉ & M. DUMAIS. 1985. The effects of neurotensin and [D-Tyr^{11}]-NT on the hyperactivity induced by intra-accumbens administration of a potent dopamine receptor agonist. Neuropeptides **6:** 143–156.
33. SPRUIJT, B. M. & W. H. GISPEN. 1986. ACTH and grooming. *In* Central Actions of ACTH and Related Peptides. D. de Wied & W. Ferrari, Eds.: 179–187. Fidia Research Series. Liviana Press. Padova.

Pharmacological Studies of Grooming and Scratching Behavior Elicited by Spinal Substance P and Excitatory Amino Acids[a]

GEORGE L. WILCOX

Department of Pharmacology
University of Minnesota Medical School
Minneapolis, Minnesota 55455

INTRODUCTION

Rodents respond to noxious cutaneous chemical stimulation with behaviors not unlike the grooming and scratching behaviors described elsewhere in this volume. The behaviors elicited are of short duration (1 to 5 minutes for short-lived irritants) and are directed at the site of cutaneous stimulation.[1-4] We have used such behaviors, which can be termed counter-irritative or nocisponsive, in studies of the spinal circuitry subserving the transmission and modulation of nociceptive information. Our overall goal has been to identify chemicals possibly involved as neurotransmitters in the spinal relay or modulation of nociceptive information.

Two criteria are important for establishment of an agent as a neurotransmitter: (1) the agent should be contained in neural structures appropriate for release (*e.g.,* dorsal root ganglion cells and their axon terminals in the spinal cord dorsal horn); and (2) application of that agent in an appropriate site should mimic the physiologic action of the agent when released in that site. We have applied this logic to investigations of putative neurotransmitters thought to excite or inhibit neurons carrying nociceptive information in the spinal cord. Our approach has been to observe the behavior in mice given intrathecal injections of putative spinal nociceptive or antinociceptive neurotransmitters or their analogues (see FIG. 1).

Neurokinins,[5,6] excitatory amino acids[7,8] (EAAs), and somatostatin[9,10] have been proposed as transmitters in the central projection of nociceptive information. We have tested the hypothesis that spinal administration of these agents would elicit behavior like that following the presentation of some noxious stimulus. In general, we have not investigated the behavioral activity of an intrathecally injected agent until immunohistochemical, electrophysiological, or other evidence has indicated its presence in the spinal cord dorsal horn. We have found that most agents contained in small primary afferent neurons, including the neurokinins,[1] EAAs[11,12] and somatostatin,[13,14] have indeed elicited counter-irritative behaviors which include caudally directed or oro-genital

[a] This work was supported by USPHS grants DA-01933 (to GLW), DA-4274 (to GLW), and DA-4090 (to Alice Larson, Alvin Beitz, and GLW), and by a grant from the Proctor and Gamble Company (to GLW).

grooming, or hindlimb scratching directed at the abdomen or thorax. These results are consistent with our hypothesis and have been confirmed by other investigators for the neurokinins,[15] EAAs,[16] and somatostatin.[17]

Monoamines,[15-17] opioids,[18,19] and neurotensin[20] have been proposed as inhibitory transmitters in the spinal relay of nociceptive information. We have tested the hypothesis that spinal administration of these agents would inhibit the reaction to some types of noxious stimulation; these included peripheral stimuli such as chemical or thermal irritation of the skin and central stimuli such as the spinal administration of a behavior-eliciting agent. We have reported that most agents contained predominantly in descending systems (e.g., norepinephrine[2] and serotonin[21,22]) or in dorsal horn interneurons (e.g., neurotensin[23] and enkephalins[2]) inhibit escape reactions to thermal stimulation and/or counter-irritative behaviors elicited by chemical stimuli. These results are similarly consistent with our hypothesis.

The present paper describes investigations conducted over the past 7 years in this laboratory. Our primary interest is in nociception rather than in the nature of the specific behaviors. In a sense, the behaviors observed in these studies are epiphenomena with respect to the spinal nociceptive system under study. Nonetheless, the behaviors displayed in response to supposedly noxious stimuli resemble those reported by others in this volume who have not presented noxious stimuli to the experimental animals. It therefore seems appropriate to review the results we have obtained in our studies so that they can be compared and contrasted with those observed in very different experimental conditions.

Three differences bear emphasis at this point: first, all our behavioral observations were carried out within 5 min of stimulus presentation while most studies included in this volume involve observations which began approximately 10–15 min after pharmacological or physiological manipulation of the animal; second, few other studies in this volume involve the intrathecal administration of agents to test for behavioral activity; third, most other studies in this volume involve rats whereas the present studies used mice. These procedural differences make comparison of these results with those of others in this volume difficult. Perhaps the most important point upon which to base comparison is the similarity of the behavioral repertoire of mice and rats in situations of stress, adversity, and novelty. It may not be an accident that the behaviors elicited by supraspinal administration of a hormone correlated with stress[24,25] (i.e., ACTH) include behaviors elicited by spinal administration of a neurohormone correlated with sensation and/or pain.

METHODS

In all the studies reported here, mice were given drug injections into the spinal subarachnoid space by percutaneous lumbar puncture. The mice were injected with compounds that elicit caudally directed licking and scratching behavior or modulate its expression. The subjects were 17–27 g male Swiss-Webster-derived mice (Biolab, St. Paul, MN; Laboratory Animal Supply, Indianapolis, IN; or Harlan Sprague-Dawley, Madison, WI). Mice were maintained in cages containing no more than 10 subjects and had free access to food and water for at least 24 h before experimentation (University of Minnesota Research Animal Resources facilities). Mice were used only once and were killed by cervical dislocation.

Intrathecal (i.t.) injections were carried out using 1 cm 30 ga disposable needles with plastic luer hubs (B-D Yale) mated to 50 µl Hamilton syringes. Each needle was used for no more than 10 injections. Each mouse was held unanesthetized by the hips

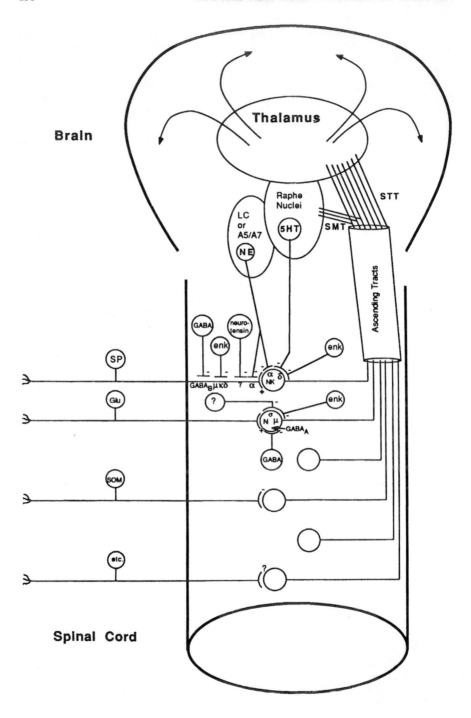

using the thumb and index finger while its head was kept under a clean towel in the cup of the injector's hand. The needle was introduced at an angle of about 30° above horizontal through the tautly held skin of the back at a point midway between the injector's thumb and index finger. With experience, one to five palpations with the needle are required to allow easy penetration of the intervertebral space. After penetration of this space, puncture of the dura is almost inevitable and in any case is reliably indicated by a flick of the mouse's tail. The injection volume is always 5 µl. When more than one agent is administered, the same total injection volume is mixed in the syringe. Less than 5% of mice vocalize upon penetration to the subarachnoid space and injection of isotonic saline or 0.01 N acetic acid in normal saline (pH 3, vehicle for neurokinins) elicits no overt behavioral signs of reaction, escape, or discomfort; these observations indicate that the procedure is humane and that the observed behaviors are in fact elicited by active agents, not by the injection procedure.[26]

After injection, mice were observed in 2-l glass beakers for 1 min and the number of abdominally directed licks and scratches together with the number of episodes of hind limb scratching were counted by a single observer. Experiments involving serotonin and somatostatin involved observation periods of 1.5 to 5 min, and episodes of biting or licking were counted separately from episodes of hind limb scratching. Some observations were carried out blind, but results of comparable experiments never differed between nonblinded and blinded designs. Raw behavioral data were analyzed by analysis of variance with regression (ANOVAR) across multiple doses of the agent eliciting the behavior. Alternatively, when a single dose of agent eliciting the behavior was challenged by various doses of inhibitory agents, derived data (percent inhibition relative to a control group run the same day) were analyzed by ANOVAR. In all cases significant dose-related effects are reported only for the significance level $p < 0.01$.

RESULTS

The results are presented in four sections below, each dealing with a class of agents that elicit specific grooming behaviors in mice. The classes are distinguished both by the chemistry and pharmacology of the agents and by the nature of the behavior elicited. All results presented here were observed in mice, but similar activity has been observed

---

FIGURE 1. A speculative schematic of neurotransmitter systems thought to be involved in transmission of information concerning noxious stimuli through the spinal cord dorsal horn to supraspinal structures. The circles to the left of the spinal cord represent neurons in dorsal root ganglia containing immunoreactivity for the named putative neurotransmitter (*SP*, substance P; *Glu*, glutamate or other excitatory amino acid; *SOM*, somatostatin; see text for references). *Circles* within the spinal cord represent neurons of the dorsal horn containing immunoreactivity for the named putative neurotransmitter (*GABA*, γ-amino butyric acid, *Enk*, Leu- Met- or extended enkephalin, and neurotensin) or for receptors hypothetically involved in the observed modulation of behavior reported here (*NK*, neurokinin; κ, κ opioid; µ, µ opioid; δ, δ opioid; α, $α_2$ adrenergic; σ, PCP/σ "opioid"; and *N*, *N*-methyl-D-aspartate or NMDA). Receptors possibly located on presynaptic terminals of primary afferent neurons (again hypothetical) appear below the line representing the axon of the topmost dorsal root ganglion neuron. Positive and negative signs near synaptic connections indicate excitatory or inhibitory action, respectively. Midbrain systems that send axons containing norepinephrine (*NE*) or serotonin (*5HT*) are also diagrammed. The ascending tracts include the spinothalamic tract (*STT*) and the spinomesencephalic tract (*SMT*).

for a few of the compounds injected i.t. in rats.[20,28] Within each class, agents from other classes that inhibit the behavior are also discussed.

Neurokinins: Substance P, Neurokinin A, and Neurokinin B Analogues and Antagonists

This laboratory[1-3,27] and others[4,28,29] have found that all members of the neurokinin family of peptides, active in various peripheral neuromuscular or neurosecretion assays, elicit a characteristic behavior. This behavior has been called "caudally directed biting and scratching behavior," "abdominal licking and scratching behavior," and "reciprocal hind limb scratching." The behavior is both qualitatively and quantitatively similar to behavior elicited by the intradermal injection of acetic acid (0.1 N) or hypertonic saline (6%), and these peripherally evoked behaviors have been extensively studied.[2,3] For this reason, some investigators, including the author, have suggested that the i.t. injection of neurokinins cause the perception of a noxious stimulus,[1-4] although this suggestion has been disputed.[30] It should be pointed out that no dose of neurokinin tested produces vocalization like that produced by some excitatory amino acids,[11] suggesting that EAAs are more relevant to nociception than neurokinins. The doses of neurokinins required to elicit 25 total biting, licking, and scratching behaviors in 1 minute immediately following injection are quite low. In the case of substance P (SP), the effective dose is about 5 ng or about 5 pmol; this dose represents a concentration of about 1 μM in the injected solution, indicating high pharmacological potency and close correlation with expected physiological concentrations during release from presynaptic terminals. An equal dose of neurokinin A or B has lower efficacy,[27] but effective doses are again below 10 ng. A derivative of substance P having an NH_2 terminal similar to enkephalin[31] has lower potency than native substance P, and in addition its action is potentiated by the opioid antagonist naloxone.[27]

In general, agents with analgesic efficacy after i.t. injection[17,19] inhibit behavior elicited by substance P.[2] Adrenergic agonists, including norepinephrine and clonidine, inhibit the behavior elicited by substance P; this action is best blocked by α_2 adrenergic antagonists, suggesting the involvement of these receptors in the spinal inhibition of neurons excited by substance P. Likewise, opioid agonists active at μ (D-Ala^2-N-Met-Phe^4-Gly^5-ol enkephalin, DAMGO), δ (D-penicillamine^{2,5}-enkephalin, DPDPE), and κ(U-50488H) receptors inhibit this behavior; the inhibitory activity of these opioids is antagonized by the opioid antagonist naloxone. However, PCP/σ agonists—such as PCP, ketamine, and (+)-N-allyl-normetazocine, (+)NAM—are inactive against SP-elicited behavior. Noradrenergic heterocyclic antidepressants (HCAs, which predominantly block NE reuptake by nerve terminals) such as protriptyline and desipramine were found to be uniformly antinociceptive and to act synergistically with morphine (actions similar to NE itself), while serotonergic HCAs, such as citalopram and fluoxetine, were not.[32,33] This profile of activity is similar to the profile of spinally administered opioid and adrenergic agonists in thermal nociceptive tests,[17,19] suggesting that this substance-P-elicited behavior is relevant to the study of nociceptive mechanisms.

We also use a test involving behavior directed at the site of intradermal injection of hypertonic saline (HS test). We have suggested that the SP and HS behavioral tests can be used to distinguish pre- from postsynaptic inhibitory mechanisms at the first central sensory synapse.[3] Neurotensin[23] and the GABA$_B$ agonist, baclofen,[34] seem to act presynaptically since pretreatment with these agents inhibits behavior in the HS test but not in the SP test. Such an action would be expected if these agents blocked the release of a transmitter from the primary afferent fiber but did not interact postsynaptically with a receptor that counters the action of the i.t. administered SP. On the other hand, most other agents tested that are active in one test will be active in the other.

Somatostatin

Somatostatin elicits a different behavioral repertoire in mice than any of the other classes of agent described in this paper. The incidence of biting or licking is much lower, and the hind limb scratching is not reciprocal. That is, mice will attempt to scratch a dorsal area of their thorax with long episodes of unilateral scratching. Scratching behavior is elevated for up to 5 min after injection, in contrast to neurokinins, whose action is terminated within 1 min. Whereas neurokinin-elicited behavior resembles the response to intradermal, perigenital injection of irritants, the diffuse dorsally directed nature of somatostatin-elicited scratching resembles the activity of animals with dorsal dermatitis.

When somatostatin is coadministered with substance P, interference or antagonism between the two agents is observed. That is, the different behavioral profiles do not add; rather they subtract from each other. This effect was examined in a blind experiment where 64 groups of five mice each were administered combinations of eight substance P doses (0, 0.3, 1, 3, 10, 30, 100, 300 ng) and eight somatostatin doses (0, 3, 10, 30, 100, 300, 1000, 3000 ng). Across the dose ranges tested, there were significant substance-P-related, somatostatin-related, and interaction-related effects; post hoc tests showed a reliable antagonism whenever somatostatin doses were approximately ten times substance P doses.[14]

Serotonin, Tryptamine, and Antagonists

Serotonin, when injected i.t., also elicits behaviors, with hindlimb scratching directed at the abdomen predominating.[21] Tryptamine has similar activity and interacts synergistically with serotonin.[22] In this regard, serotonin is similar to somatostatin. Another similarity with somatostatin is that serotonin in low doses antagonized SP-elicited behaviors while higher doses elicited behaviors even in the absence of SP. The behavior-inhibiting effects of serotonin were blocked by cyproheptadine; in addition, cyproheptadine, cinanserin, and methysergide blocked the scratch-inducing action of serotonin but not that of SP. The difference in the nature of the behaviors and in the pharmacological agents that block the behavior together suggest that serotonin and SP act through different mechanisms.[21] Although we were unable to block the scratch-inducing action of serotonin with an early SP antagonist (D-Pro²-D-Trp⁷-D-Phe⁹-SP), other investigators have blocked this serotonin action with a newer SP antagonist[35] (D-Pro²-D-Trp⁷,⁹-SP). These latter results may indicate that serotonin acts through release of SP to produce behaviors.

Excitatory Amino Acids and Antagonists

Excitatory amino acids (EAAs) administered intrathecally in the mouse elicit a caudally directed biting and scratching behavior similar to that elicited by neurokinins except that there is a more notable incidence of tail biting.[11,12] This action differs from that of the neurokinins in that at high doses the animals vocalize, perhaps indicating the activation of ascending nociceptive spinal systems. All studies of EAA-induced grooming were conducted at doses below those necessary to produce vocalization. We found that NMDA, but not kainate or quisqualate, decreased the response latency in two thermal nociceptive tests, suggesting the induction of hyperalgesia at doses producing scratching. Adrenergic agonists active at α receptors (phentolamine-reversible NE) and opioid agonists active at μ (DAMGO) or PCP/σ (PCP, (+)-N-allyl-normetazocine, (+)NAM) receptors blocked scratching at high doses; μ and PCP/σ agonists blocked the thermal hyperalgesia at lower doses; pure δ (DPDPE) or κ(U-50488H) opioid agonists were inactive in both tests.

DISCUSSION

In general, the results of these behavioral studies on the spinal action of putative spinal sensory neurotransmitters have supported the simple hypotheses presented here. That is, when an agent is found by immunohistochemical analysis to reside in neuronal terminals in the spinal cord dorsal horn, intrathecal injection of that substance or of an analogue changes some aspect of spinal transmission. The scientific method imposes a sampling bias on these studies, because those putative neurotransmitters not localized in or near the spinal cord have not been investigated. Nonetheless, the results presented here indicate that excitatory agents generally elicit counterirritative behaviors while inhibitory agents (*e.g.*, opioids[36] and adrenergic agonists[16]) inhibit the expression of these behaviors. The excitatory agents that conform to this hypothesis include all the neurokinins and the EAAs; the inhibitory agents that conform to this hypothesis include the opioid and adrenergic agonists. A most dramatic difference has been found between excitatory effects of EAAs and neurokinins: whereas neurokinin-elicited behavior is inhibited by δ or μ or α_2 agonists, EAA-elicited behavior is inhibited only by μ or PCP/σ agonists. The ability or inability of an inhibitory agent to reduce the effects of an excitatory agent may suggest the locus of action of the agent or the power of inhibition mediated by that agent. These studies have thus served as a preliminary screen for future electrophysiological studies of neural transmission in the spinal cord.

Not conforming to the hypothesis are serotonin and somatostatin. Serotonin, a monoamine which often manifests more than one effect, and somatostatin, a peptide which inhibits release of other hormones, both elicited scratching behavior. Of these agents, only somatostatin is found in primary afferent sensory fibers. Thus, serotonin is the only agent tested which elicits behavior and which is not contained in primary afferent sensory fibers. The significance of this finding is unclear at this time. The clear message from this second group of findings is that the simple hypothetical framework used to design these studies has served its purpose. More encompassing hypotheses accompanied by detailed electrophysiological studies of spinal cord neural activity are now needed.

SUMMARY

Compounds that produce depolarization of nociceptive neurons in the dorsal horn of the spinal cord also elicit a rather specific kind of caudally directed biting, licking, and/or scratching behavior when they are injected intrathecally in mice. We sought to use this elicited grooming behavior as a test for compounds that might inhibit the neurons excited by the excitatory agents. All three neurokinins — substance P, neurokinin A (substance K), neurokinin B (neuromedin K) — and excitatory amino acids active at *N*-methyl-ᴅ-aspartate (NMDA) or quisqualate receptors produce similar behaviors, which last for 1 minute after i.t. injection. Our data indicate that μ opioid agonists or α adrenergic agonists block both neurokinin-elicited behavior and EAA-elicited behavior; δ opioid agonists block only neurokinin-elicited behavior; and PCP/σ "opioid" agonists block only EAA-elicited behavior. Somatostatin and serotonin produce qualitatively different behaviors by themselves and, when administered with neurokinins, partially block neurokinin-elicited behavior.

ACKNOWLEDGMENTS

The author wishes to acknowledge the assistance of several colleagues whose work is included in this review. Dr. Janice L. K. Hylden was instrumental in the early development of the intrathecal injection technique in mice and in the early work with substance P, serotonin, opioids, and norepinephrine. Subsequently, Drs. Sutaeg Hwang, Lin Aanonsen, and Sizheng Lei refined studies of naturally elicited, neurokinin-elicited, and excitatory amino-acid-elicited grooming behavior.

REFERENCES

1. HYLDEN J. L. K. & G. L. WILCOX. 1981. Intrathecal substance P elicits a caudally directed biting and scratching behavior in mice. Brain Res. **217:** 212–215.

2. HYLDEN J. L. K. & G. L. WILCOX. 1983. Pharmacological characterization of substance-P-induced nociception in mice: Modulation by opioid and noradrenergic agonists at the spinal level. J. Pharmacol. Exp. Ther. **226:** 398–404.

3. HWANG, A. S. & G. L. WILCOX. 1986. Intradermal hypertonic saline-induced behavior as a nociceptive test. Life Sci. **38:** 2389–2396.

4. PIERCEY, M. F., L. A. SCHROEDER, K. FOLKERS, J. C. HU & J. HORIG. 1981. Sensory and motor functions of spinal cord substance P. Science **214:** 1361–1363.

5. NICOLL, R. A., C. SCHENKER & S. E. LEAMAN. 1980. Substance P as a transmitter candidate. Annu. Rev. Neurosci. **3:** 227–268.

6. PERNOW, B. 1984. Substance P. Pharmacol. Rev. **35:** 85–141.

7. JOHNSON, J. L. 1978. The excitant amino acids glutamic and aspartic acid as transmitter candidates in the vertebrate central nervous system. Progr. Neurobiol. **10:** 155–202.

8. CAHUSAC, P. M. B., R. H. EVANS, R. G. HILL, R. E. RODRIQUEZ & D. A. S. SMITH. 1984. The behavioural effects of an *N*-methylaspartate receptor antagonist following application to the lumbar spinal cord of conscious rats. Neuropharmacology **23:** 719–724.

9. HOKFELT, T., R. ELDE, O. JOHANSSON, R. LUFT, G. NILSSON & A. ARIMURA. 1976. Immunohistochemical evidence for separate populations of somatostatin-containing and substance P-containing primary afferent neurons in the rat. Neuroscience **1:** 131–136.

10. RANDIC, M. & V. MILETIC. 1978. Depressant actions of methionine-enkephalin and somatostatin in cat dorsal horn neurons activated by noxious stimuli. Brain Res. **152:** 196–202.

11. AANONSEN, L. M. & G. L. WILCOX. 1986. Phencyclidine selectively blocks a spinal action of *N*-methyl-ᴅ-aspartate in mice. Neurosci. Lett. **67:** 191–197.

12. AANONSEN, L. M. & G. L. WILCOX. 1987. Nociceptive action of excitatory amino acids in the mouse: Effects of spinally administered opioids, phencyclidine and σ agonists. J. Pharmacol. Exp. Ther. In press.

13. SEYBOLD, V. S., J. L. K. HYLDEN & G. L. WILCOX. 1982. Intrathecal substance P and somatostatin in rats: Behaviors indicative of sensation. Peptides **3:** 49–54.

14. WILCOX, G. L. 1983. Antagonism between two behaviorally active peptides in rodent spinal cord. Proc. Intl. Union Physiol. Sci. **15:** 108.

15. DALSTROM, A. & K. FUXE. 1965. Experimentally induced changes in the interneuronal amine levels of bulbospinal neuron systems. Acta Physiol. Scand. Suppl. **247:** 1–36.

16. HEADLEY, P. M., A. W. DUGGAN & B. T. GRIERSMITH. 1978. Selective reduction by noradrenaline and 5-hydroxytryptamine of nociceptive responses of cat dorsal horn neurones. Brain Res. **145:** 185–189.

17. REDDY, S. V. R., J. L. MADERDRUT & T. L. YAKSH. 1980. Spinal cord pharmacology of adrenergic agonist-mediated antinociception. J. Pharmacol. Exp. Ther. **213:** 525–533.

18. RUDA, M. A. 1982. Opiates and pain pathways: Demonstration of enkephalin synapses on dorsal horn projection neurons. Science **215:** 1523–1525.

19. YAKSH, T. L. & R. NOUEIHED. 1985. The physiology and pharmacology of spinal opiates. Annu. Rev. Pharmacol. Toxicol. **25:** 433–462.

20. SEYBOLD, V. S. & R. P. ELDE. 1982. Neurotensin immunoreactivity in the superficial laminae of the dorsal horn of the rat. I. Light microscopic studies of cell bodies and proximal dendrites. J. Comp. Neurol. **205:** 89–100.

21. HYLDEN, J. L. K. & G. L. WILCOX. 1983. Intrathecal serotonin in mice: Analgesia and inhibition of a spinal action of substance P. Life Sci. **33:** 789–795.

22. LARSON, A. A. & G. L. WILCOX. 1984. Synergistic behavioral effects of intrathecal serotonin and tryptamine in mice. Neuropharmacology **23:** 1415–1418.

23. HYLDEN J. L. K. & G. L. WILCOX. 1984. Antinociceptive action of intrathecal neurotensin in mice. Peptides **4:** 517–520.

24. GIPSEN, W. H. & R. L. ISAACSON. 1981. ACTH-induced excessive grooming in the rat. Pharmacol. Ther. **12:** 209–246.

25. COWAN, A., P. KHUNAWAT, X. ZU ZHU & D. E. GMEREK. 1985. Effects of bombesin on behavior. Life Sci. **37:** 135–145.

26. HYLDEN, J. L. K. & G. L. WILCOX. 1980. Intrathecal morphine in mice: A new technique. Eur. J. Pharmacol. **67:** 313–316.

27. LEI, S., A. W. LIPKOWSKI & G. L. WILCOX. 1986. Structure activity relationships for opioid and tachykinin activities of substance P analogs in the mouse spinal cord. Soc. Neurosci. Abstr. **12:** 151.

28. VAUGHT, J. L., L. J. POST, H. I. JACOBY & D. WRIGHT. 1984. Tachykinin-like central activity of neuromedin K in mice. Eur. J. Pharmacol. **103:** 355–357.

29. VAUGHT, J. L. & R. SCOTT. 1987. Species differences in the behavioral toxicity produced by intrathecal substance P antagonists: Relationship to analgesia. Life Sci **40:** 175–181.

30. BOSSUT, D., H. FRENK & D. J. MAYER. 1986. Substance P does not appear to be a primary afferent transmitter of pain in the spinal cord. Soc. Neurosci. Abstr. **12:** 228.

31. LIPKOWSKI, A. W., B. OSIPIAK & W. S. GUMULKA. 1983. An approach to the self regulatory mechanism of substance P actions. II. Biological activity of new synthetic peptide analogs related both to enkephalin and substance P. Life Sci. **33**(Suppl 1.): 141–144.

32. KEHL, L. J. & G. L. WILCOX. 1984. Antinociceptive effect of tricyclic antidepressants following intrathecal administration. Anesth. Prog. **31:** 82–84.

33. HWANG, A. S. & G. L. WILCOX. 1987. Analgesic properties of intrathecally administered heterocyclic antidepressants. Pain **28:** 343–355.

34. HWANG, A. S. & G. L. WILCOX. 1986. Involvement of GABA$_B$ receptors in the antinociceptive activity of intrathecal baclofen in mice. Fed. Proc. **45:** 1053.

35. FASMER, O. B., O. G. BERGE & K. HOLE. 1983. Similar behavioral effects of 5-HT and substance P injected intrathecally in mice. Neuropharmacology **22:** 405–487.

36. YOSHIMURA, M. & R. A. NORTH. 1983. Substantia gelatinosa neurones in vitro hyperpolarized by enkephalin. Nature **305:** 529–530.

The Role of Prolactin in Rat Grooming Behavior

FILIPPO DRAGO

Institute of Pharmacology
University of Catania Medical School
95125 Catania, Italy

INTRODUCTION

Prolactin (PRL) is a phylogenetically ancient pituitary hormone that possesses a wide range of biological effects in many animal species ranging from fish to mammals.[1] Also, PRL exerts endocrine effects that mimic those of the other pituitary hormones and concern practically all body organs including breast, kidney, gut, adrenal gland, heart, and blood vessels. The brain has also been considered a target organ for PRL on the basis of numerous studies conducted in the past decade both in animals and humans. Neurochemical effects of PRL include alterations in the concentration and turnover of neurotransmitters and receptor modifications in various brain areas.[2,3] PRL also induces behavioral changes that may be related to its neurochemical effects.[4]

The first indication that a humoral factor involved in lactation might cause behavioral effects came from the early studies of Lienhart,[5] who described the induction of parental behavior in hens after the injection of serum from incubating animals of the same species. From these first observations considerable evidence has accumulated indicating that PRL profoundly influences behavioral expression in different animal species, including humans. The range of behavioral effects of PRL includes parental behavior, migration, feeding behavior, learning and memory, drug-induced behavioral and adaptive responses, and sexual behavior.[4] Thus, in addition to exerting a multiplicity of endocrine effects in the periphery, PRL plays multiple roles in the modulation of animal behavior.

The behavioral effects of PRL were first studied by using impure preparations of the hormone.[1] Ovine or bovine PRL were mostly used for experiments in rats, and this raised the question as to whether heterologous PRL would exert behavioral effects similar to those of the homologous hormone.[4] The model of endogenous hyperprolactinemia induced by adenohypophyseal homografts under the kidney capsule represents a suitable tool to study the effect of PRL on the brain, provided that this hormone can cross the blood-brain barrier and enter the brain.[6,7] The availability of rat PRL has made possible studies in the rat with homologous hormone.

One of the major behavioral changes observed in hyperprolactinemic rats is the enhancement of grooming.[8] This was an accidental observation made in 1977 during an experiment on the "open field" behavior of hyperprolactinemic rats.[4] Since then numerous experiments have been carried out on this topic, and this behavioral effect of PRL is now well-known and well characterized. This paper reviews all the findings concerning PRL-enhanced grooming in the rat. These data have been obtained both with animals injected with rat PRL and hyperprolactinemic rats bearing adenohypophyseal homografts. In addition, data will be presented on grooming of rats with

237

hyperprolactinemia induced by mastectomy or vagotomy. Some of the experiments described in this paper were carried out at the Rudolf Magnus Institute for Pharmacology, University of Utrecht Medical School, Utrecht (The Netherlands), under the supervision of B. Bohus and D. de Wied.

EFFECTS OF PROLACTIN ON GROOMING BEHAVIOR

In our experiments, grooming behavior was studied according to the method described by Gispen *et al.*[9] Briefly, the rats were placed individually into boxes with transparent walls in a low-noise room. The behavior of the rats was sampled every 15 s, beginning immediately after the animals were placed into the boxes. The occurrence of the following behavioral elements of grooming was recorded: washing (vibrating movements of the forepaws in front of the snout and licking of the same paws leading to a series of strokes along the snout and semicircular movements over the top of the head), licking (licking of the body fur, limbs, and tail), scratching (scratching of the body by one of the limbs), and genital grooming (licking of the genital area).

Both endogenous hyperprolactinemia induced by adenohypophyseal homografts under the kidney capsule and intracerebroventricular (i.c.v.) injection of PRL enhance grooming behavior in male rats.[8] Similar behavioral changes were observed in female rats, suggesting that this effect is not influenced by sex (Drago, unpublished observation). A time-course study in hyperprolactinemic rats revealed that the grooming behavior of these animals reached the maximum level (score: 64.5 ± 5.1) about 12 days after the implantation of pituitary homografts, and then declined to normal values (score: 24.6 ± 1.8) by day 26.[10] Furthermore, a frequency analysis of grooming elements revealed that on days 10 and 12 after surgery hyperprolactinemic rats showed a slight, statistically not significant, decrease in the percent occurrence of all elements except for genital grooming. The occurrence of this element appeared significantly higher in hyperprolactinemic rats than in control animals. Forty days after surgery, the grooming behavior of hyperprolactinemic rats was not qualitatively different from that shown by control animals.[10] Although licking appears to be the predominant element, the high incidence of genital grooming suggests that PRL-enhanced grooming may be genitally oriented. The decline of grooming activity in long-term hyperprolactinemic rats did not depend on a reduction in plasma PRL levels, which remained elevated for at least 4 weeks. A possible explanation of this phenomenon is the reduced activity of central neurotransmitter systems (*e.g.*, dopaminergic) after prolonged stimulation. Interestingly, long-term hyperprolactinemic rats show, besides a reduced level of grooming behavior, an inhibition of sexual activity; other behavioral modifications observed shortly after the beginning of hyperprolactinemia (*e.g.*, facilitated acquisition of avoidance behaviors, reduced responsiveness to electrical footshock, enhanced drug-induced stereotypy) remain unchanged.[11] These findings suggest that in long-term hyperprolactinemic rats modulatory factors develop that may influence the behavioral effects of PRL.

Intracerebroventricular injection of rat PRL enhances grooming behavior of male rats in a dose-dependent manner.[12] The dose-response curve of this effect is of an inverted U shape type, 3 μg being the most potent dose in a range of 1–10 μg. The dose range for PRL-enhanced grooming after i.c.v. administration is apparently 10 times higher than that found for the excessive grooming induced by $ACTH_{1-24}$.[13] However, the molecular weight of PRL is ten times larger than that of $ACTH_{1-24}$, and thus on a molecular basis the difference is small.

The enhanced grooming of hyperprolactinemic rats depends on the actual presence of high amounts of this hormone in the brain. In fact, the i.c.v. injection of anti-PRL serum 12 days after pituitary homografts suppressed the excessive grooming of hyperprolactinemic rats.[14] Grooming behavior of these animals was again increased 2 days after injection of the antiserum.

Right-side and bilateral mastectomy induce hyperprolatinemia and also enhance grooming behavior, while left-side mastectomy is followed by a decrease in plasma PRL levels and in grooming behavior.[15] Both left- and right-side vagotomy result in a significant increase in plasma PRL levels, but only left-side vagotomized rats exhibit increased grooming activity. These results suggest that changes in plasma PRL levels induced by surgical manipulations can affect grooming activity in rats.

INVOLVEMENT OF CENTRAL NEUROTRANSMISSION IN PROLACTIN-ENHANCED GROOMING

Since central dopaminergic neurotransmission is involved in grooming behavior,[16–18] the significance of this neurotransmitter in PRL-enhanced grooming has been studied in detail. In particular, the possible involvement of the nigrostriatal and meso-accumbens dopaminergic systems was examined, since these pathways have been linked to ACTH-induced excessive grooming.[13,19] It was found that systemic or intrastriatal administration of the dopamine antagonist, haloperidol, inhibited the excessive grooming induced by either i.c.v. injection of rat PRL or endogenous hyperprolactinemia.[20] Similar effects were found after bilateral injection of haloperidol of 6-hydroxydopamine into the nucleus accumbens.[21] It is rather unlikely that the inhibiting effect of haloperidol on PRL-enhanced grooming is due to a general depressant action since the dose of the drug used did not affect locomotor activity of the rats.

Several lines of evidence show that PRL affects dopaminergic systems in the brain other than the tubero-infundibular system. Peripheral injection of PRL enhances the cellular fluorescence response of dopaminergic neurons in the substantia nigra, suggesting an increased dopaminergic neurotransmission.[22] Furthermore, PRL stimulates dopamine turnover in the medial caudatus *in vivo*[23] and the release of dopamine from the neostriatum *in vitro*, possibly by a presynaptic modulation.[24] Acute peripheral administration of rat PRL into hypophysectomized rats increases the dopamine turnover in the posterior part of the nucleus accumbens.[25] Recently, Chen *et al.*[26] have shown that rat PRL increases the 3,4-dihydroxyphenylacetic acid (DOPAC) output from the caudate nucleus as measured by high-performance liquid chromatography (HPLC) in rats bearing a push-pull cannula. No significant changes were found in homovanillic acid (HVA) and dopamine output. In this study, the behaviors of the animals were recorded and scored 1 hour before, during, and 1 hour after PRL administration. It was found that PRL locally perfused into the caudate nucleus (10 ng per µl per min for 20 min) evoked an intense grooming behavior. Similar effects were found after subcutaneous (s.c.) injection of the hormone (Laping and Ramirez, unpublished data).

The effects of PRL on dopamine and noradrenaline utilization in various brain areas have been studied in rats with short-term endogenous hyperprolactinemia.[2] Inhibition of catecholamine synthesis by α-methyl-*p*-tyrosine (α-MPT) was used to measure utilization. In the nigrostriatal dopaminergic pathway, hyperprolactinemia decreased the utilization of dopamine in the cell-body region (substantia nigra) and increased it in the terminal projection (caudate nucleus). In the ventral tegmental area (mesolimbic dopaminergic projection), hyperprolactinemia decreased the utilization of dopamine. Hyperprolactinemia increased the utilization of noradrenaline in the locus ceruleus, the cell-body region of the dorsal noradrenergic bundle, but decreased it in

TABLE 1. Effect of Naloxone on Grooming Behavior Enhanced by i.c.v. Injection of Rat PRL or Endogenous Hyperprolactinemia Induced by Pituitary Homografts

Groups	Grooming Scores	
	Saline	Naloxone
Saline (i.c.v.)	14.0 ± 1.7 (n = 12)	12.7 ± 1.6 (12)
PRL (3 μg, i.c.v.)	56.0 ± 9.2 (6)	31.6 ± 6.5^a (6)
Control	13.1 ± 1.6 (9)	12.6 ± 1.2 (9)
Hyperprolactinemic	61.8 ± 3.0 (12)	21.2 ± 1.7^a (12)

Values are mean ± SEM. Saline or naloxone (1 mg/kg) were injected s.c. 15 min prior to initiating the behavioral test.

^{a} Significantly different as compared to saline-injected group ($p < 0.05$, Student's t-test).

some terminal projections of the same pathway (*e.g.*, the cingulate gyrus). These findings suggest that PRL is capable of modifying the activity of dopamine neurotransmission in those brain areas that have been concerned with grooming behavior, and this may be releated to the behavioral effects of the hormone.

Not only the concentration and turnover of dopamine are influenced by PRL, but also the sensitivity of central dopamine receptors. In fact, peripheral administration of PRL increases the density of striatal dopamine receptors in intact and hypophysectomized male rats.[3] These results suggest that PRL might be the common mediator of the increase in striatal dopamine receptor density produced by either estrogen or haloperidol administration.

Since brain opioid systems have also been implicated in grooming behavior,[13,27] the possible involvement of opioid transmission in PRL-enhanced grooming was investigated.[28] The results show that peripheral administration of the opiate receptor antagonist naloxone attenuated excessive grooming induced either by i.c.v. injection of rat PRL or by endogenous hyperprolactinemia (TABLE 1). Thus, both central dopaminergic and opioid neurotransmission are implicated in PRL-enhanced grooming, as has already been shown for ACTH-induced excessive grooming.[13]

Prolactin can affect the activity of other neurotransmitters in the brain, such as GABA, serotonin, and acetylcholine (see ref. 29), but the significance of these neurotransmitters in PRL-enhanced grooming is unknown.

PROLACTIN-ENHANCED GROOMING: INTERACTION WITH OTHER PEPTIDES

The interaction between PRL and ACTH$_{1-24}$ in enhancing grooming behavior after i.c.v. administration was investigated in intact and endogenously hyperprolactinemic rats.[12] In intact rats, 4 hours after the i.c.v. injection of rat PRL or ACTH, a second administration of ACTH or rat PRL, respectively, did not induce the excessive grooming observed after the first injection. This result suggests that no cross-tolerance exists between PRL and ACTH$_{1-24}$, and hence that the same putative receptor mechanisms are not involved. In hyperprolactinemic rats that exhibited excessive grooming 12 days after pituitary homografts, i.c.v. injection of rat PRL failed to enhance further the grooming activity, whereas this behavior was substantially enhanced by i.c.v. injection of ACTH$_{1-24}$. Twenty-six days after surgery, when the grooming activity of hyperprolactinemic rats was of the same level of control animals, the i.c.v. injection of rat PRL was effective in inducing excessive grooming in control but not in hyperprolactinemic

TABLE 2. Oxytocin-enhanced Grooming: Interaction with an Early Subsequent i.c.v. Injection of Prolactin, $ACTH_{1-24}$, or β-Endorphin

Experimental Groups	n	Grooming Score
Saline + saline	18	12.4 ± 1.3
Oxytocin + saline	8	82.0 ± 4.2^a
Prolactin + saline	8	68.1 ± 3.3^a
$ACTH_{1-24}$ + saline	9	102.2 ± 5.6^a
β-endorphin + saline	8	62.2 ± 4.2^a
Oxytocin + prolactin	8	81.1 ± 3.9^a
Oxytocin + $ACTH_{1-24}$	8	$118.2 \pm 4.7^{a,b}$
Oxytocin + β-endorphin	7	$104.2 \pm 4.1^{a,b}$

All substances were injected i.c.v. at a constant dose of 1 µg/3 µl of normal saline. Five min later, the animals were reinjected with saline or with prolactin, $ACTH_{1-24}$, or β-endorphin. Grooming test started 10 min later and lasted 30 min. Values are mean ± SEM.

[a] Significantly different as compared to the group of animals injected with saline + saline ($p < 0.05$, Dunnett's test for multiple comparisons).

[b] Significantly different as compared to the group of animals injected with oxytocin + saline ($p < 0.05$, Dunnett's test for multiple comparisons).

animals. In contrast, at this time the i.c.v. injection of $ACTH_{1-24}$ again induced excessive grooming in both hyperprolactinemic and control rats. These findings support the hypothesis that behaviorally competent central mechanisms become hyposensitive to PRL under conditions of hyperprolactinemia, while the responsiveness to $ACTH_{1-24}$ remains preserved. These results also suggest that independent neural mechanisms are involved in PRL- and ACTH-induced grooming.

Despite some similarities, there are differences between PRL- and ACTH-induced

TABLE 3. Oxytocin-enhanced Grooming: Interaction with a Late Subsequent i.c.v. Injection of Prolactin, $ACTH_{1-24}$, or β-Endorphin

Experimental Groups	n	Grooming Score After the First Injection	After the Second Injection
Saline + saline	10	16.8 ± 1.5	16.2 ± 1.4
Oxytocin + saline	6	85.6 ± 3.6	16.0 ± 1.2^a
Prolactin + saline	6	70.2 ± 4.0	18.2 ± 1.4^a
$ACTH_{1-24}$ + saline	7	98.2 ± 4.6	17.2 ± 1.3^a
β-endorphin + saline	7	66.4 ± 2.6	18.4 ± 1.8^a
Oxytocin + oxytocin	7	82.1 ± 4.0	84.8 ± 3.6
Prolactin + prolactin	8	72.6 ± 3.6	76.5 ± 4.0
$ACTH_{1-24}$ + $ACTH_{1-24}$	7	96.6 ± 4.1	32.2 ± 2.4^a
β-endorphin + β-endorphin	7	70.1 ± 3.0	36.4 ± 2.1^a
Oxytocin + prolactin	8	82.5 ± 3.8	73.8 ± 3.9^a
Oxytocin + $ACTH_{1-24}$	8	85.6 ± 4.2	94.2 ± 5.8
Oxytocin + β-endorphin	8	82.7 ± 4.7	70.2 ± 4.0

All substances were injected i.c.v. at a constant dose of 1 µg/3 µl of normal saline. Four hours later, the animals were reinjected with saline, or with prolactin, $ACTH_{1-24}$, or β-endorphin. Grooming test was performed either 10 min after the first injection or 10 min after the second injection. Values are mean ± SEM.

[a] Significantly different as compared to the grooming score obtained after the first injection ($p < 0.05$, Student's t-test, two-tailed).

excessive grooming. Prolactin- and ACTH-induced grooming are qualitatively different.[4] High levels of plasma PRL are accompanied by excessive grooming,[10] while ACTH only promotes vigorous grooming after i.c.v. administration.[13] ACTH-induced grooming is totally suppressed by peripheral administration of naloxone,[27] while this drug only attenuates PRL-induced excessive grooming.[28]

Recent experiments have been devoted to the study of interaction betwen PRL and oxytocin in inducing excessive grooming.[30] It was found that the i.c.v. administration of rat PRL in rats injected 5 min earlier with oxytocin failed to enhance further the grooming activity of these animals, while this behavior was substantially enhanced by i.c.v. injection of $ACTH_{1-24}$ or β-endorphin (TABLE 2). Furthermore, $ACTH_{1-24}$ and β-endorphin showed a cross-tolerance and a single dose tolerance after a second injection, while neither cross-tolerance nor single dose tolerance was found for PRL- and oxytocin-induced excessive grooming (TABLE 3).

These results, taken together, suggest that PRL-enhanced grooming resembles that induced by oxytocin in many aspects. The grooming enhanced by $ACTH_{1-24}$ or β-endorphin possesses different characteristics and represents another type of behavior.

CONCLUDING REMARKS

Prolactin-induced excessive grooming resembles the behavioral modification occurring after the application of stress stimuli. Thus, it can be considered as an aspect of the behavioral arousal induced by stress and mimicked by the administration of various substances. Other pituitary hormones, such as ACTH, MSH, endorphins, and vasopressin, which are secreted under stress conditions, share with PRL various effects on adaptive behavior and the autonomic nervous system. In fact, all these hormones enhance grooming, facilitate the acquisition of avoidance responses, and induce analgesia and hypothermia.[31,32] Thus, PRL modulates the biological responses to stress together with other hormonal factors, and the enhancement of grooming in animals subjected to stress may be the result of the specific influence of all these factors on neural mechanisms responsible for this behavior.

REFERENCES

1. NICOLL, C. S. 1974. Physiological actions of prolactin. *In* Handbook of Physiology, Vol. 7. American Physiological Society. Washington, D.C. pp. 253–292.
2. KOVACS, G. L., F. DRAGO, L. ACSAI, A. TIHANYI, U. SCAPAGNINI & G. TELEGDY. 1984. Catecholamine utilization in specific rat brain nuclei after short-term hyperprolactinemia. Brain Res. **324:** 29–34.
3. HRUSKA, R. E., K. T. PITMAN, E. K. SILBERGELD & L. M. LUDMER. 1982. Prolactin increases the density of striatal dopamine receptors in normal and hypophysectomized male rats. Life Sci. **30:** 547–552.
4. DRAGO, F. 1982. Prolactin and Behavior. Ph.D. thesis, University of Utrecht Medical School.
5. LIENHART, R. 1927. Contribution à l'étude de l'incubation. C. R. Soc. Biol. **97:** 1296–1297.
6. ASSIES, J., A. P. M. SCHELLEKENS & J. L. TOUBER. 1978. Prolactin in human cerebrospinal fluid. J. Clin. Endocrinol. Metab. **46:** 576–586.
7. PARDRIDGE, W. M., H. J. FRANK, E. M. CORNFORD, L. D. BRAUN, P. D. CRANE & W. H. OLDENDORF. 1981. Neuropeptides and the blood–brain barrier. *In* Neurosecretion and Brain Peptides. J. B. Martin, S. Reichlin & K. L. Bick, Eds.: 321–328. Raven. New York.
8. DRAGO, F., P. L. CANONICO, R. BITETTI & U. SCAPAGNINI. 1980. Systemic and intraventricular prolactin induces excessive grooming. Eur. J. Pharmacol. **65:** 457–458.
9. GISPEN, W. H., V. M. WIEGANT, H. M. GREVEN & D. DE WIED. 1975. The induction of

excessive grooming in the rat by intraventricular application of peptides derived from ACTH: structure-activity studies. Life Sci. **17**: 645–652.

10. DRAGO, F. & B. BOHUS. 1981. Hyperprolactinaemia-induced excessive grooming in the rat: time course and element analysis. Behav. Neur. Biol. **33**: 117–122.

11. DRAGO, F., B. BOHUS, J. M. VAN REE, U. SCAPAGNINI & D. DE WIED. 1982. Behavioral responses of long-term hyperprolactinaemic rats. Eur. J. Pharmacol. **79**: 323–327.

12. DRAGO, F., B. BOHUS, W. H. GISPEN, U. SCAPAGNINI & D. DE WIED. 1983. Prolactin-enhanced grooming behavior: interaction with ACTH. Brain Res. **263**: 277–282.

13. GISPEN, W. H. & R. L. ISAACSON. 1981. ACTH-induced excessive grooming in the rat. Pharmacol. Ther. **12**: 209–237.

14. DRAGO, F., B. BOHUS, R. BITETTI, U. SCAPAGNINI, J. M. VAN REE & D. DE WIED. 1986. Intracerebroventricular injection of anti-prolactin serum suppresses excessive grooming of pituitary homografted rats. Behav. Neur. Biol. **46**: 99–105.

15. GERENDAI, I., F. DRAGO, G. CONTINELLA & U. SCAPAGNINI. 1984. Effects of mastectomy and vagotomy on grooming behavior of the rat: possible involvement of prolactin. Physiol. Behav. **33**: 1–4.

16. AYHAN, I. H. & A. RANDRUP. 1973. Behavioural and pharmacological studies on morphine-induced excitation of rats. Possible relation to brain catecholamines. Psychopharmacologia **29**: 317–328.

17. COOLS, A. R., V. M. WIEGANT & W. H. GISPEN. 1978. Distinct dopaminergic systems in ACTH-induced grooming. Eur. J. Pharmacol. **50**: 265–268.

18. CHESHER, G. B. & D. M. JACKSON. 1980. Post-swim grooming in mice inhibited by dopamine receptor antagonists and by cannabinoids. Pharmacol. Biochem. Behav. **13**: 479–481.

19. WIEGANT, V. M., A. R. COOLS & W. H. GISPEN. 1977. ACTH-induced excessive grooming involves brain dopamine. Eur. J. Pharmacol. **41**: 343–345.

20. DRAGO, F., B. BOHUS, P. L. CANONICO & U. SCAPAGNINI. 1981. Prolactin induces grooming in the rat: possible involvement of nigrostriatal dopaminergic system. Pharmacol. Biochem. Behav. **15**: 61–63.

21. DRAGO, F., G. L. KOVACS, G. CONTINELLA & U. SCAPAGNINI. 1984. Dopamine in the nucleus accumbens: its role in prolactin-enhanced grooming of the rat. Biogen. Amines **1**: 75–81.

22. LICHTENSTEIGER, W. & R. LIENHART. 1975. Central action of a-MSH and prolactin: simultaneous responses of hypothalamic and mesencephalic dopamine systems. *In* Cellular and Molecular Bases of Neuroendocrine Processes. E. Endroczi, Ed.: 211–221. Akademiai Kiado. Budapest.

23. FUXE, K., K. ANDERSSON, T. HOKFELT & L. F. AGNATI. 1978. Prolactin-monoamine interactions in the rat brain and their importance in regulation of LH and prolactin secretion. *In* Progress in Prolactin Physiology and Pathology. C. Robin & M. Harter, Eds.: 95–109. Elsevier/North-Holland and Biomedical Press. Amsterdam.

24. PERKINS, N. A. & T. C. WESTFALL. 1978. The effect of prolactin on dopamine release from rat striatum and medial basal hypothalamus. Neuroscience **3**: 59–63.

25. FUXE, K., P. ENEROTH, J.-A. GUSTAFSSON, A. LOFSTROM & P. SKETT. 1977. Dopamine in the nucleus accumbens: preferential increase of DA turnover by rat prolactin. Brain Res. **122**: 177–182.

26. CHEN, J. C., A. D. RAMIREZ & V. D. RAMIREZ. 1985. Prolactin effect on *in vivo* striatal dopaminergic activity as estimated with push-pull perfusion in male rat. *In* Prolactin: Basic and Clinical Correlates. R. M. MacLeod, M. O. Thorner & U. Scapagnini, Eds., Fidia Research Series, Vol. 1: 523–532. Liviana Press. Padua.

27. WIEGANT, V. M., W. H. GISPEN, L. TERENIUS & D. DE WIED. 1977. ACTH-like peptides and morphine: interaction at the level of the CNS. Psychoneuroendocrinology **2**: 63–69.

28. DRAGO, F., W. H. GISPEN & B. BOHUS. 1981. Behavioral effects of prolactin: involvement of opioid receptors. *In* Advances in Endogenous and Exogenous Opioids. H. Takagi & E. J. Simon, Eds.: 335–337. Kodansha Ltd. Tokyo.

29. DRAGO, F. & U. SCAPAGNINI. 1983. Prolactin and behavior: a neurochemical substrate. *In* Recent Advances in Male Reproduction. R. D'Agata & L. Bardin, Eds.: 299–303. Raven Press. New York.

30. DRAGO, F., C. A. PEDERSEN, J. D. CALDWELL, G. CONTINELLA, U. SCAPAGNINI & A. J.

PRANGE, JR. 1986. Oxytocin-enhanced grooming: interaction with other peptides. Peptides. In press.

31. DRAGO, F., S. AMIR, G. CONTINELLA, M. C. ALLORO & U. SCAPAGNINI. 1984. Effects of endogenous hyperprolactinaemia on adaptive responses to stress. *In* Prolactin: Basic and Clinical Correlates. R. M. MacLeod, M. O. Thorner & U. Scapagnini, Eds., Fidia Research Series, Vol. 1: 609–614. Liviana Press. Padua.

32. DRAGO, F. & S. AMIR. 1984. Effects of hyperprolactinaemia on core temperature of the rat. Brain Res. Bull. 12: 355–358.

Grooming Behavioral Effects of Oxytocin

Pharmacology, Ontogeny, and Comparisons with Other Nonapeptides[a]

CORT A. PEDERSEN,[b-d] JACK D. CALDWELL,[b,c]
FILIPPO DRAGO,[e] LINDA R. NOONAN,[c,d] GARY PETERSON,[b,c]
LORI E. HOOD,[b,c] AND ARTHUR J. PRANGE, JR.[b-d]

[b]Department of Psychiatry
[c]Biological Sciences Research Center
[d]The Neurobiology Curriculum
University of North Carolina School of Medicine
Chapel Hill, North Carolina 27514

[e]Institute of Pharmacology
University of Catania Medical School
95125 Catania, Italy

The neurohypophyseal peptides oxytocin (OXY) and arginine vasopressin (AVP) are widely distributed within the brain.[1,2] Central administration of OXY or AVP have numerous behavioral and physiological effects.[3,4] Others[5,6] have reported that central administration of neurohypophyseal peptides (nonapeptides) increases grooming behavior in mice and rats. We have confirmed these observations in rats and extended our investigations in several new directions.[7-10] The grooming effects of nonapeptides have been compared. The mechanisms of OXY-induced grooming as well as the ontogeny of sensitivity to the grooming effects of OXY have also been investigated. Our findings in these areas are presented and discussed below.

THE PHARMACOLOGY OF OXYTOCIN INDUCTION OF GROOMING BEHAVIOR

Our initial investigations focused on the effects of intracerebroventricular (i.c.v.) administration of OXY on the display of maternal behavior and sexually receptive behavior (lordosis) in female rats.[11-14] In the course of these studies we noticed that most rats appeared to groom themselves more after administration of OXY than after administration of most other substances. To quantify this effect we infused OXY i.c.v. in rats and scored grooming behavior using the method of Gispen[15] during a 30-min period beginning 25 minutes after administration. As portrayed in FIGURE 1, grooming behavior increased in a linear dose-related manner following infusion of doses of OXY between 0.1 µg and 10.0 µg.[7] No stretching or yawning was observed after OXY administration. OXY was as potent in increasing grooming behavior in intact male rats

[a]This work was supported by NIMH grant nos. MH-33172, MH-22536 (to AJP), and HD-16159. FD was supported by a NATO Grant for Medicine.

TABLE 1. Analysis of Behavioral Components of Grooming That Were Increased after i.c.v. Infusion of Oxytocin in Male and Female Rats

	Males		Females	
	NS	OXY	NS	OXY
Total grooming (total no. observations)	23.1 ± 2.1	64.2 ± 3.5[a]	21.8 ± 2.8	63.6 ± 3.0[a]
Face washing (% of total grooming)	37.1 ± 3.3	35.8 ± 3.0	38.3 ± 3.4	37.1 ± 2.9
Body grooming (% of total grooming)	45.2 ± 6.3	42.9 ± 3.9	44.7 ± 5.0	43.1 ± 3.9
Scratching (% of total grooming)	10.4 ± 2.4	8.9 ± 2.0	10.9 ± 1.9	8.9 ± 2.1
Genital grooming (% of total grooming)	3.4 ± 1.4	8.0 ± 1.2[a]	4.2 ± 1.2	7.9 ± 1.3[a]

Values are means ± SEM. Each experimental group included six rats. OXY = oxytocin (1.0 µg/5 µl rat). NS = normal saline (5.0 µl/rat). Infusions were made i.c.v. via plastic cannulae. Similar results were obtained in rats infused i.c.v. with 0.5 µg or 5.0 µg of OXY.

[a] Significantly different as compared to controls infused with NS ($p < 0.05$, Dunnett's test for multiple comparisons).

as in females, suggesting that the facilitating effect of OXY on grooming behavior is not estrogen dependent in contrast to the facilitating effects of OXY on maternal behavior and lordosis.[11,14] Grooming behavior remained significantly elevated for approximately 50 minutes in rats infused i.c.v. with 1 µg of OXY 25 minutes prior to the beginning of behavioral measurements. Central administration of OXY increased several of the components of grooming behavior described by Gispen,[15] but increased genital grooming more than other components (see TABLE 1). Subcutaneous adminis-

TABLE 2. Effects of i.p. Injection of Haloperidol or Naloxone on Grooming Behavior Induced by i.c.v. Infusion of Oxytocin (µg/5 µl) in Female Rats

i.p. Treatment	i.c.v. Treatment	n	Mean Grooming Scores (± SEM)
NS	NS	18	12.1 ± 2.6
HALO (0.1 mg/kg)	NS	6	8.6 ± 2.1
HALO (0.5 mg/kg)	NS	6	7.8 ± 2.2
NALO (1 mg/kg)	NS	6	8.0 ± 2.4
NALO (5 mg/kg)	NS	6	8.6 ± 2.1
NS	OXY	12	70.4 ± 4.1[b]
HALO (0.1 mg/kg)	OXY	6	12.1 ± 2.8
HALO (0.5 mg/kg)	OXY	6	10.2 ± 2.4
NALO (1 mg/kg)	OXY	6	46.2 ± 6.2[a]
NALO (5 mg/kg)	OXY	6	40.6 ± 4.8[a]

Haloperidol (HALO), naloxone (NALO), or normal saline (NS) was injected i.p. 60 min prior to grooming measurements. Oxytocin (OXY) or NS was infused i.c.v. 25 min prior to grooming measurements, which lasted 30 min.

[a] Significantly different as compared to group 1 ($p < 0.05$, Dunnett's test for multiple comparisons).

[b] Significantly different as compared to group 1 ($p < 0.01$, Dunnett's test for multiple comparisons).

TABLE 3. Effects of Bilateral Microinjections of 6-Hydroxydopamine (6-OHDA) or Haloperidol (HALO) into the Nucleus Accumbens on Oxytocin-induced Grooming in the Rat

Treatment Groups			Mean Grooming
Intra-accumbens	i.c.v.	*n*	Scores ($\pm$ SEM)
Vehicle	NS	12	15.3 $\pm$ 2.7
Vehicle	OXY	12	84.2 $\pm$ 7.6[a]
6-OHDA	NS	12	14.9 $\pm$ 2.5
6-OHDA	OXY	12	19.0 $\pm$ 2.8
HALO	NS	12	15.8 $\pm$ 2.6
HALO	OXY	2	17.8 $\pm$ 2.6

Injections of 6-OHDA (3 µg/1 µl) or vehicle (1 µl) into the nucleus accumbens were performed 5 days prior to behavioral testing. Injections of haloperidol (1 µg/1 µl) or vehicle (1 µl) into the nucleus accumbens were performed 30 min prior to behavioral testing. Grooming observations were started 5 min after i.c.v. infusions (5 µl) of 3 µg oxytocin (OXY) or normal saline (NS) and lasted 30 min.

[a] Significantly different as compared to group 1 ($p < 0.01$, Dunnett's test for multiple comparison).

tration of OXY increased grooming behavior but only at doses considerably higher (100 and 200 µg) than were effective centrally.[7]

Grooming behavior induced by OXY is mediated by dopaminergic and opioid neurotransmission.[7] Administration of haloperidol completely blocked OXY-induced grooming. Naloxone treatments significantly decreased but did not abolish OXY-induced grooming (see TABLE 2). OXY facilitation of uterine contractions may be mediated in part by release of prostaglandins.[16] Intraperitoneal administration of the prostaglandin synthesis inhibitor, indomethacin (5 mg/kg), completely blocked OXY-induced grooming.[10] Thus the grooming behavioral effect of OXY may also be mediated by release of prostaglandins.

The facilitating effect of OXY administration on grooming behavior involves activation of dopamine neurotransmission in the mesolimbic pathway. Bilateral infusions of 6-OHDA or haloperidol into the nucleus accumbens completely blocked OXY-induced grooming behavior[8] (see TABLE 3). Drs. Kaltwasser and Crawley have observed a dose-related increase in grooming behavior following infusion of OXY into the ventral tegmental area (VTA) of the midbrain (personal communication). OXY may regulate grooming behavior mediated by mesolimbic dopamine neurotransmission, just as other neuropeptides regulate locomotor behavior mediated by mesolimbic dopamine neurotransmission.[17]

The potencies of OXY and other neuropeptides in increasing grooming behavior have been compared.[8,9] Equimolar doses of other nonapeptides — AVP, lysine vasopressin (LVP), and arginine vasotocin (AVT) — were as potent as 1 µg of OXY for increasing grooming (see below for other comparisons between the nonapeptides). Equimolar doses of tocinoic acid (the ring structure of the OXY molecule) and Pro-Leu-Gly-NH$_2$ (the tail structure of the OXY molecule) were ineffective in increasing grooming behavior. As summarized in TABLE 4, OXY (1 µg) was approximately as potent as the same dose of other neuropeptides — ACTH$_{1-24}$, prolactin (PRL), and β-endorphin (β-END).[18-20] However, because the molecular weight of OXY is lower than the molecular weights of these other neuropeptides, OXY may be less potent on an equimolar basis.

Some of the pharmacological properties of OXY induction of grooming are similar

TABLE 4. Total Grooming after Two i.c.v. Treatments Separated by 5 Minutes

Treatment Groups	n	Mean Grooming Scores ($\pm$ SEM)
NS + NS	18	12.4 $\pm$ 1.3
OXY + NS	8	82.0 $\pm$ 4.2[a]
PRL + NS	8	68.1 $\pm$ 3.3[a]
ACTH$_{1-24}$ + NS	9	102.2 $\pm$ 5.6[a]
β-END + NS	8	62.2 $\pm$ 4.2[a]
OXY + PRL	8	81.1 $\pm$ 3.9[a]
OXY + ACTH$_{1-24}$	8	118.2 $\pm$ 4.7[a,b]
OXY + β-END	7	104.2 $\pm$ 4.1[a,b]

One µg of oxytocin (OXY), prolactin (PRL), ACTH$_{1-24}$, β-endorphin (β-END), or normal saline (NS) was infused i.c.v. in a volume of 3 µl. Five min later, rats were again infused i.c.v. with NS, PRL, ACTH$_{1-24}$, or β-END. Grooming measurements started 10 min later and lasted 30 min.

[a] Significantly different as compared to the group of rats infused with NS + NS ($p < 0.05$, Dunnett's test for multiple comparisons).

[b] Significantly different as compared to the group of rats infused with OXY + NS ($p < 0.05$, Dunnett's test for multiple comparisons).

to those of PRL induction of grooming but are different from those of ACTH$_{1-24}$ or β-END induction of grooming. When administered i.c.v. 5 minutes after i.c.v. administration of OXY, ACTH$_{1-24}$ and β-END both had significant additive effects on the amount of grooming displayed. However, when PRL was administered i.c.v. 5 minutes after OXY there was no additive effect (see TABLE 4). Thus OXY and PRL induction of grooming may involve activation of the same neural substrate, while ACTH$_{1-24}$ and β-END induction of grooming may involve activation of other neural substrates. Rats

TABLE 5. Total Grooming after Each of Two i.c.v. Treatments Separated by 4 Hours

Treatment Groups	n	Mean Grooming Scores ($\pm$ SEM) After the First Treatment	After the Second Treatment
NS + NS	10	16.8 $\pm$ 1.5	16.2 $\pm$ 1.4
OXY + NS	6	85.6 $\pm$ 3.6	16.0 $\pm$ 1.2[a]
PRL + NS	6	70.2 $\pm$ 4.0	18.2 $\pm$ 1.4[a]
ACTH$_{1-24}$ + NS	7	98.2 $\pm$ 4.6	17.2 $\pm$ 1.3[a]
β-END + NS	7	66.4 $\pm$ 2.6	18.4 $\pm$ 1.8[a]
OXY + OXY	7	82.1 $\pm$ 4.0	84.8 $\pm$ 3.6
PRL + PRL	8	72.6 $\pm$ 3.6	76.5 $\pm$ 4.0
ACTH$_{1-24}$ + ACTH$_{1-24}$	7	96.6 $\pm$ 4.1	32.2 $\pm$ 2.4[a]
β-END + β-END	7	70.1 $\pm$ 3.0	36.4 $\pm$ 2.1[a]
OXY + PRL	8	82.5 $\pm$ 3.8	73.8 $\pm$ 3.9
OXY + ACTH$_{1-24}$	8	85.6 $\pm$ 4.2	94.2 $\pm$ 5.8
OXY + β-END	8	82.7 $\pm$ 4.7	70.2 $\pm$ 4.0

One µg of oxytocin (OXY), prolactin (PRL), ACTH$_{1-24}$, β-endorphin (β-END), or normal saline (NS) was infused i.c.v. in a volume of 3 µl. Four hours later, rats were again infused with NS, PRL, ACTH$_{1-24}$, or β-END. Grooming measurements began 10 min after the first infusion and 10 min after the second infusion and lasted 30 min.

[a] Significantly different as compared to the grooming score obtained after the first infusion ($p < 0.05$, Student's t-test, two-tailed).

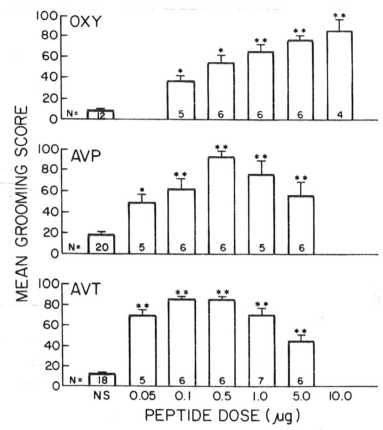

FIGURE 1. A comparison of the grooming behavior dose-response relationships of oxytocin (*OXY*), arginine vasopressin (*AVP*), and arginine vasotocin (*AVT*). Nonapeptides were infused intracerebroventricularly (i.c.v.) 25 min prior to 30-min observation periods during which grooming behaviors were scored. OXY significantly increased grooming scores across a 0.1–10.0 μg dose range. AVP and AVT significantly increased grooming scores across a 0.05–5.0 μg dose range. Grooming scores are presented as means ± SEM; * = $p < 0.05$; ** = $p < 0.01$, significantly different as compared to controls receiving normal saline (*NS*) as determined by Dunnett's test for multiple comparisons.

developed tolerance to the grooming behavioral effects of $ACTH_{1-24}$ and β-END but developed no tolerance to the grooming behavioral effects of OXY and PRL (see TABLE 5). Others reported that facilitation of grooming behavior by both $ACTH_{1-24}$ and β-END were completely blocked by opioid antagonism but only attenuated by 6-hydroxydopamine (6-OHDA) or haloperidol administration into the nucleus accumbens.[18] In contrast, OXY and PRL facilitation of grooming was attenuated by opioid antagonism but completely inhibited by 6-OHDA or haloperidol administration into the nucleus accumbens (ref. 21, and see above). Central mechanisms mediating OXY and PRL induction of grooming are, therefore, very similar and differ substantially from mechanisms mediating $ACTH_{1-24}$ or β-END induction of grooming.

COMPARISONS OF THE GROOMING BEHAVIORAL EFFECTS OF THE NONAPEPTIDES

We compared the effects of OXY with the effects of other nonapeptides on grooming behavior.[9] The amount of grooming produced by of 1 µg of OXY was equivalent to the amount of grooming produced by equimolar doses of other nonapeptides (see above). Intracerebroventricular administration of 1 µg of AVP or AVT (doses almost equimolar to 1 µg of OXY because of the similar molecular weights of the nonapeptides) 25 minutes before the beginning of a 30-minute observation period produced significant increases in the amount of grooming behavior of the same magnitude in each of the 5-minute segments of the observation period. The magnitude and consistency of the grooming behavioral effect of AVP and AVT were equivalent over this period of observation to what we had previously measured following i.c.v. administration of OXY.[7] Thus, the duration of AVP- or AVT-induced grooming may be comparable to that of OXY-induced grooming.

After establishing that OXY, AVP, and AVT each produced a consistent elevation in grooming behavior over the 30-minute period described above, we compared their dose-response relationships in this time frame (see Fig. 1). The maximum efficacy of each of these nonapeptides was similar. In contrast to OXY, AVP and AVT displayed an inverted U dose-response curve over the dose range tested.[9] AVT and AVP produced maximum increases in grooming at doses that were much lower than doses of OXY that produced comparable increases in grooming. AVT appears to be the most potent of these nonapeptides at lower doses. Our results are consistent with the finding of Meisenberg[22] that AVP significantly increased grooming in mice at doses at which OXY was ineffective. Because the maximally effective doses of AVP and AVT were considerably lower than the maximally effective dose of OXY, we compared the behavioral components of grooming in rats infused i.c.v. with 0.5 µg of AVP, 0.5 µg of AVT, or 5.0 µg of OXY. The three nonapeptides had similar effects on the amount of each of the components of grooming behavior.[9]

THE STRUCTURAL SPECIFICITY OF THE GROOMING BEHAVIORAL EFFECTS OF OXYTOCIN AND ARGININE VASOPRESSIN

Numerous analogues have been synthesized as tools with which to study the structure-function relationships of the OXY molecule.[23] The potency of these analogues as agonists in producing the effects of OXY (e.g., uterotonic, antidiuretic, and vasopressor activity) has been quantified using bioassay methods.[23,24] The potency of some analogues in antagonizing the effects of OXY have also been determined. We have begun to investigate the potency of analogues of OXY as agonists and antagonists in the induction of grooming.

Five analogues of OXY with a wide range of potencies in increasing uterine contractions and insignificant vasopressor activities were compared with each other and with OXY to determine their potencies in increasing grooming.[10] FIGURE 2 summarizes this comparison. Deamino-OXY, the most potent uterotonic agonist tested, significantly increased grooming. Deamino-OXY has very little antidiuretic activity.[24] The other agonist analogues of OXY that we tested, which are relatively weak uterotonic agonists, failed to increase grooming significantly. Our results suggest that the structural properties of the OXY molecule that contribute to its uterotonic activity also contribute to its grooming behavioral effects. However, deamino-OXY, which is 50%

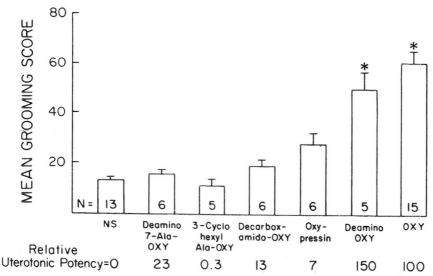

FIGURE 2. The grooming behavior potency of oxytocin (*OXY*) analogues is related to their uterotonic potency. One μg of OXY or equimolar doses of OXY analogues were infused intracerebroventricularly 25 min prior to 30-min observation periods during which grooming behaviors were scored. The uterotonic potency of each analogue is expressed as a percentage of the potency of OXY. Grooming scores are presented as means ± SEM; * = $p < 0.05$, significantly different as compared to controls receiving normal saline (*NS*), determined by Tukey's studentized range test after ANOVA. OXY was significantly more potent in inducing grooming than all other treatments except deamino-OXY. Deamino-OXY was significantly more potent than NS, deamino 7-Ala-OXY, or 3-cyclohexyl-Ala-OXY ($p < 0.05$, Tukey's standardized range test).

more potent than OXY in increasing uterine contractions,[24] was somewhat (but not significantly) less potent than OXY in increasing grooming. Perhaps receptors mediating the uterotonic and grooming effects of OXY differ to some degree in their ligand structural requirements. Alternatively, the relatively lower potency of deamino-OXY in increasing grooming behavior may reflect differences between OXY and deamino-OXY in their metabolism or distribution within the CNS after i.c.v. administration.

We have investigated the effects of a specific uterotonic antagonist, [Pen-1, Phe-2, Thr-4, Δ-3,4, Pro-7, Orn-8] OXY ([PPTPO]OXY), on the grooming behavioral effects of OXY and AVP. This antagonist was synthesized and generously provided by Dr. Victor Hruby. As summarized in FIGURE 3, i.c.v. coadministration of 1 μg of [PPTPO]OXY completely blocked the grooming behavioral effect of 1 μg of OXY but did not diminish the grooming behavioral effect of 1 μg of AVP.[10] Previous comparisons (see above) of the dose-response relationships of OXY- and AVP-induced grooming demonstrated that the grooming potency of 1 μg of OXY was closer to the grooming potency of 0.1 μg AVP. We therefore tested the effect of coadministration of 1 μg of [PPTPO]OXY on the grooming effect of this lower dose of AVP. Again, [PPTPO]OXY had no significant effect on AVP-induced grooming.[10] Our results strongly suggest that OXY, but not AVP, stimulates grooming behavior at receptors within the CNS resembling the receptors that mediate the uterotonic effects of nonapeptides (uterotonic-like receptors). Thus two nonapeptides that are structurally very similar

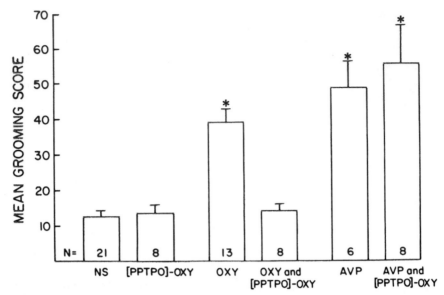

FIGURE 3. Intracerebroventricular coadministration of the uterotonic antagonist [PPTPO]OXY inhibited oxytocin (*OXY*) but not arginine vasopressin (*AVP*) induction of grooming behavior. Rats were infused with 1 μg of OXY or AVP or normal saline (*NS*) 25 minutes prior to 30-min observation periods during which grooming behaviors were scored. Rats infused with OXY, AVP, or AVP + [PPTPO]OXY had significantly elevated grooming scores compared to rats receiving other treatments (* = $p < 0.05$; Tukey's standardized range test after ANOVA). Grooming scores are presented as means ± SEM.

(the amino acid sequences of OXY and AVP differ at only two positions) appear to stimulate the same behavior (grooming) at different receptor sites.

Based on their results obtained in mice, others[25] have hypothesized that nonapeptides increase grooming behavior by acting at receptors in the CNS similar to receptors that mediate the vasopressor effects of nonapeptides (vasopressor-like receptors). Our results in rats clearly demonstrate that OXY induces grooming by stimulating uterotonic-like receptors, while AVP induces grooming by stimulating other types of receptors. Further studies involving i.c.v. coadministration of AVP and specific antagonists of vasopressor and antidiuretic activity should clarify the receptor types mediating AVP induction of grooming. AVT is structurally similar to both OXY and AVP and has considerable uterotonic, vasopressor, and antidiuretic activity.[26] AVT may, therefore, induce grooming behavior by acting simultaneously at different receptors that separately mediate OXY- and AVP-induced grooming. This may explain why AVT stimulates maximal amounts of grooming at lower doses than OXY or AVP.

THE ONTOGENY OF SENSITIVITY TO THE GROOMING BEHAVIORAL EFFECTS OF OXY

Very young rat pups spontaneously display some components of grooming behavior. These components mature and others emerge so that by approximately age 20 days

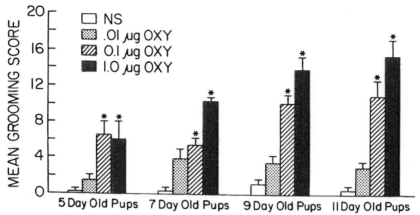

FIGURE 4. The effects of three doses of oxytocin (*OXY*) infused intracisternally on the amount of grooming behavior displayed by infant rat pups of various ages. Grooming scores are presented as the mean number of 1-min time periods ± SEM in which grooming was observed, out of a possible 21 periods; * = $p < 0.05$, significantly different as compared to normal saline (*NS*) group (Dunnett's test for multiple comparisons).

rat pups display the adult pattern of grooming.[27,28] We have begun to investigate the ontogeny of sensitivity to the grooming behavioral effects of OXY in young rats. As displayed in FIGURE 4, intracisternal administration of OXY significantly increased total grooming behavior in a dose-related manner in rat pups of 5, 7, 9, or 11 days of age. OXY was significantly more potent in increasing grooming in older (9 and 11 day old) compared to younger (5 and 7 day old) rat pups. OXY administration increased all components of grooming that were scored (face grooming, forelimb licking, hindlimb licking, scratching with the hindlimb) at all ages. Doses of OXY of 1.0 and 0.1 µg were effective in increasing grooming, whereas 0.01 µg was ineffective. Some components of grooming were stimulated precociously by OXY. For example, hindleg licking, which normally emerges around 11 days of age,[28] was stimulated by OXY in 5-day-old pups.

Other oral behaviors (licking of other pups and objects, yawning, and stereotypic mouthing movements) were also significantly increased by OXY but to a lesser magnitude than the increase in grooming (data not shown). OXY was effective in increasing

TABLE 6. Effects of Methysergide on Oxytocin-induced Grooming and Oral Behaviors in Infant Rats

Treatment Groups	Mean Scores (± SEM)	
	Grooming	Oral Behaviors
OXY + Vehicle	6.5 ± 2.5	6.5 ± 0.5
OXY + METH	0.1 ± 0.1	0.5 ± 0.3
NS + METH	0.0 ± 0.0	0.3 ± 0.3

Behaviors are expressed as the number of time periods that the behaviors occurred ± SEM out of a possible 21 periods. Oxytocin (OXY) was infused intracisternally at a dose of 0.1 µg/µl. Methysergide (METH) was injected at doses of 0.5 µg/µl intracisternally or 2.5 mg/kg/5 ml subcutaneously. Oral behaviors included mouthing, licking, and yawning. Pups were either 8 or 15 days of age.

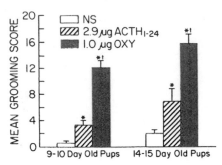

FIGURE 5. A comparison of the amount of grooming exhibited by infant rats following a 1 µl intracisternal infusion of normal saline (*NS*) or equimolar doses of oxytocin (*OXY*) or $ACTH_{1-24}$. Grooming scores indicate the mean number of 1-min time periods ± SEM in which grooming occurred, out of a possible 21 periods; * = $p < 0.05$, significantly different as compared to NS; † $p < 0.05$, significantly different as compared to $ACTH_{1-24}$ (Student's *t*-test after ANOVA).

these oral behaviors at doses of 0.01 and 0.1 µg but not at 1.0 µg. Administration of 1 µg of OXY, but not lower doses, significantly increased treading (alternate hindlimb movements), paddling (alternate forelimb movements), and rolling (side-to-side movements of the body). Thus the increase in limb and body movements produced by administration of 1 µg of OXY appeared to compete with other oral behaviors but not with the grooming behavior induced by OXY.

Others[29,31] have reported that suckling behavior and other oral behaviors in young rat pups were inhibited by administration of the serotoninergic antagonist methysergide. We found that similar doses of methysergide abolished OXY induction of grooming as well as OXY induction of other oral behaviors (see TABLE 6). Thus, serotoninergic mechanisms that are known to be involved in suckling and other oral behaviors in pups may also mediate the grooming and oral behavioral effects of OXY in rat pups.

Intracerebroventricular administration of 1 µg of OXY in adult rats produced approximately the same amount of grooming behavior as 1 µg of $ACTH_{1-24}$ (see above). On an equimolar basis, however, $ACTH_{1-24}$ appeared to be more potent than OXY in adult rats. We compared the grooming behavioral potencies of 1 µg of OXY and an equimolar dose of $ACTH_{1-24}$ (2.9 µg) in young rat pups. As is summarized in FIGURE 5, OXY and $ACTH_{1-24}$ significantly increased grooming in rat pups that were 9–10 and 14–15 days old. However, OXY was significantly more potent than $ACTH_{1-24}$ in each of these age groups. ACTH is nearly ineffective in stimulating corticosterone secretion from the adrenal gland during the first 2 or 3 weeks of life in rats.[32,33] Thus, compared to adult rats, pups are insensitive to both the behavioral and endocrine effects of ACTH. Our results suggest that nonapeptides may play a more central role than ACTH in the ontogeny of grooming behavior.

CONCLUDING REMARKS

A number of lines of evidence support the hypothesis that central OXY plays a physiologically relevant role in the activation of grooming behavior. There is general agreement that acute stress increases grooming behavior in rats.[16] Other investigators[34,35] have reported that OXY is released in response to various types of stressors. Genital grooming increases around parturition and copulation.[36,37] Blood levels of OXY also increase at these times.[19,38] As discussed above, genital grooming increased more than other components of grooming behavior following i.c.v. administration of OXY.[7] Pathways containing OXY within the CNS project to the ventral tegmental area (VTA) and to the substantia nigra,[1,2] brain regions giving rise to dopamine pathways that mediate grooming behavior.[16,21] These considerations will guide future investigations of the grooming behavioral effects of OXY.

ACKNOWLEDGMENTS

The authors wish to thank Ms. Sharon Powell for her excellent clerical assistance in preparing this manuscript.

REFERENCES

1. BUIJS, R. M., G. J. DEVRIES, F. W. VAN LEEUWEN & D. F. SWABB. 1983. Vasopressin and oxytocin: distribution and putative functions in the brain. *In* The Neurohypophysis: Function and Control. Progress in Brain Research, Vol. 60. B. A. Cross & G. Leng, Eds.: 115–122. Elsevier. Amsterdam/New York.
2. SOFRONIEW, M. V. 1983. Morphology of vasopressin and oxytocin neurones and their central and vascular projections. *In* The Neurohypophysis: Function and Control. Progress in Brain Research, Vol. 60. B. A. Cross & G. Leng, Eds.: 101–114. Elsevier. Amsterdam/New York.
3. MEISENBERG, G. & W. H. SIMMONS. 1983. Centrally mediated effects of neurohypophyseal hormones. Neurosci. Biobehav. Rev. 7: 263–280.
4. AMICO, J. A. & A. G. ROBINSON, Eds. 1985. Oxytocin: Clinical and Laboratory Studies. Elsevier Science Publishers. Amsterdam.
5. DELANOY, R. L., A. J. DUNN & R. TINTNER. 1978. Behavioral responses to intracerebroventricularly administered neurohypophseal peptides in mice. Horm. Behav. 11: 348–362.
6. MEISENBERG, G. 1982. Short-term behavioral effects of neurohypophyseal hormones: Pharmacological characteristics. Neuropharmacology 21: 309–316.
7. DRAGO, F., C. A. PEDERSEN, J. D. CALDWELL & A. J. PRANGE, JR. 1986. Oxytocin potently enhances novelty-induced grooming behavior in the rat. Brain Res. 368: 287–295.
8. DRAGO, F., J. D. CALDWELL, C. A. PEDERSEN, G. CONTINELLA, U. SCAPAGNINI & A. J. PRANGE, JR. 1986. Dopamine neurotransmission in the nucleus accumbens may be involved in oxytocin-enhanced grooming behavior of the rat. Pharmacol. Biochem. Behav. 24: 1185–1188.
9. CALDWELL, J. D., F. DRAGO, A. J. PRANGE, JR. & C. A. PEDERSEN. 1986. A comparison of grooming behavior potencies of neurohypophyseal nonapeptides. Regulatory Peptides 14: 261–271.
10. CALDWELL, J. D., V. J. HRUBY, P. HILL, A. J. PRANGE, JR. & C. A. PEDERSEN. 1986. Is oxytocin-induced grooming mediated by uterine-like receptors? Neuropeptides 8: 77–86.
11. PEDERSEN, C. A. & A. J. PRANGE, JR. 1979. Induction of maternal behavior in virgin rats after intracerebroventricular administration of oxytocin. Proc. Natl. Acad. Sci. U.S.A. 76: 6661–6665.
12. PEDERSEN, C. A., J. A. ASCHER, Y. L. MONROE & A. J. PRANGE, JR. 1982. Oxytocin induces maternal behavior in virgin female rats. Science 216: 648–649.
13. PEDERSEN, C. A., J. D. CALDWELL, M. F. JOHNSON, S. A. FORT & A. J. PRANGE, JR. 1985. Oxytocin antiserum delays onset of ovarian steroid-induced maternal behavior. Neuropeptides 6: 175–182.
14. CALDWELL, J. D., A. J. PRANGE, JR. & C. A. PEDERSEN. 1986. Oxytocin facilitates the sexual receptivity of estrogen-treated female rats. Neuropeptides 7: 175–189.
15. GISPEN, W. H., V. M. WIEGANT, H. M. GREVEN & D. DE WIED. 1975. The induction of excessive grooming in the rat by intraventricular application of peptides derived from ACTH: structure-activity studies. Life Sci. 17: 645–652.

16. FUCHS, A.-R. 1985. Oxytocin in animal parturition. *In* Oxytocin: Clinical and Laboratory Studies. J. A. Amico & A. G. Robinson, Eds.: 207-235. Elsevier Science Publishers. Amsterdam.

17. NEMEROFF, C. B., P. W. KALIVAR & A. J. PRANGE, JR. 1984. Interaction of neurotensin and dopamine in limbic structures. *In* Catecholamines. Neuropharmacology and Central Nervous System: Theoretical Aspects. Alan R. Liss, Inc. New York. pp. 199-206.

18. GISPEN, W. H. & R. L. ISSACSON. 1981. ACTH-induced excessive grooming in the rat. Pharmacol. Ther. **12:** 209-246.

19. DRAGO, F., P. L. CANONICO, R. BITETTI & U. SCAPAGNINI. 1980. Systemic and intraventricular prolactin induces excessive grooming. Eur. J. Pharmacol. **65:** 457-458.

20. GISPEN, W. H., V. M. WIEGANT, A. F. BRADBURY, E. C. HULME, D. G. SMYTH, C. R. SNELL & D. DE WIED. 1976. Induction of excessive grooming in the rat by fragments of lipotropin. Nature **264:** 794-795.

21. DRAGO, F., G. L. KOVACS, G. CONTINELLA & U. SCAPAGNINI. 1984. Dopamine in the nucleus accumbens: Its role in prolactin-enhanced grooming behavior of the rat. Biogen. Amines **1:** 53-59.

22. MEISENBERG, G. 1981. Short-term behavioral effects of posterior pituitary peptides in mice. Peptides **2:** 1-8.

23. MANNING, M. & W. H. SAWYER. 1984. Design and uses of selective agonistic and antagonistic analogs of the neuropeptides oxytocin and vasopressin. Trends Neurosci. **7:** 6-9.

24. MANNING, M., J. LOWBRIDGE, J. HALDAR & W. H. SAWYER. 1977. Design of neurohypophyseal peptides that exhibit selective agonistic and antagonistic properties. Fed. Proc. **36:** 1848-1852.

25. MEISENBERG, G. & W. H. SIMMONS. 1982. Behavioral effects of intracerebroventricularly administered neurohypophyseal hormone analogs in mice. Pharmacol. Biochem. Behav. **16:** 819-825.

26. ACHER, R. 1974. Chemistry of the neurohypophyseal hormones: An example of molecular evolution. *In* Handbook of Physiology, Sect. 7, Vol. 4. S. R. Geiger, Ed.: 119-130. Waverly Press. Baltimore, MD.

27. BOLLES, R. C. & P. J. WOODS. 1964. The ontogeny of behaviour in the albino rat. Anim. Behav. **12:** 427-441.

28. RICHMOND, G. & B. D. SACHS. 1980. Grooming in Norway rats: The development and adult expression of a complex motor pattern. Behaviour **75:** 82-95.

29. SPEAR, L. P. & L. A. RISTINE. 1982. Suckling behavior in neonatal rats: Psychopharmacological investigations. J. Comp. Physiol. Psychol. **96:** 244-255.

30. RISTINE, L. A. & L. P. SPEAR. 1984. Effects of serotonergic and cholinergic antagonists on suckling behavior of neonatal, infant and weanling rat pups. Behav. Neural Biol. **41:** 99-126.

31. RISTINE, L. A. & L. P. SPEAR. 1985. Is there a "serotonergic syndrome" in neonatal rat pups. Pharmacol. Biochem. Behav. **22:** 265-269.

32. GUILLET, R. & S. M. MICHAELSON. 1978. Corticotropin responsiveness in the neonatal rat. Neuroendocrinology **27:** 119-125.

33. GUILLET, R., M. SAFFRAN & S. M. MICHAELSON. 1980. Pituitary-adrenal response in neonatal rats. Endocrinology **106:** 991-994.

34. LANG, R. E., J. W. E. HEIL, O. GANTEN, K. HERMANN, W. RASCHER, T. UNGER & W. RASCHER. 1983. Oxytocin unlike vasopressin is a stress hormone in the rat. Neuroendocrinology **37:** 314-316.

35. GIBBS, D. M. 1986. Stress-specific modulation of ACTH secretion by oxytocin. Neuroendocrinology **42:** 456-458.

36. MEYERSON, B. J., C. O. MALMNAS & B. J. EVERETT. 1985. Neuropharmacology, neurotransmitters and sexual behavior in mammals. *In* Handbook of Behavioral Neurobiology, Vol. 7. N. Adler, D. Pfaff & R. W. Goy, Eds.: 495-536. Plenum. New York.

37. ROTH, L. R. & J. S. ROSENBLATT. 1967. Changes in self-licking during pregnancy in the rat. J. Comp. Physiol. Psychol. **63:** 397-400.

38. FUCHS, A.-R., L. CUBILL & M. Y. DAWOOD. 1981. Effects of mating on levels of oxytocin and prolactin in the plasma of male and female rabbits. J. Endocrinol. **90:** 245-253.

Vasopressin-induced Grooming and Scratching Behavior in Mice

GERHARD MEISENBERG

Department of Biochemistry
Ross University School of Medicine
Dominica (West Indies)

Vasopressin and oxytocin, as well as their biosynthetic congeners, the neurophysins, have been described in fiber tracts originating from several hypothalamic and limbic structures.[1,2,3] Target areas of these fiber systems appear to be located in the lower brain stem, including the nucleus of the solitary tract and the dorsal column nuclei;[4,5] the intermediolateral column of the spinal cord;[6] and limbic forebrain structures including the lateral septum, the medial nucleus of the amygdala, and the lateral habenular nucleus.[7-9] Cerebrospinal fluid (CSF) vasopressin levels, although influenced by osmotic stimuli and hemorrhage, do not always parallel the vasopressin concentrations in plasma, suggesting that CSF vasopressin is derived from central release rather than the bloodstream or posterior pituitary gland.[10,11]

Behavioral and autonomic changes described after administration of these peptides either intracerebroventricularly (i.c.v.) or locally into specific brain areas include alterations of learning and memory processes,[12-15] development and maintenance of tolerance to and dependence on ethanol and opiates,[16-19] hypothermia and antipyresis,[20-22] analgesia,[23] changes of cardiovascular function,[24-26] an influence on intracranial self-stimulation,[27] and spontaneous behavioral changes.[28-32] The induction of grooming and scratching behavior in mice after acute i.c.v. injection was first described by Delanoy *et al.* in 1978.[28] The present report describes studies on the behavioral characteristics, pharmacology, structure-activity relationship, and possible sites of action for grooming and scratching behavior induced by vasopressin and oxytocin.

METHODS

Male albino mice of either the Swiss-Webster or the CF1/W68 strain, weighing 25–35 g, were used throughout. These two strains showed similar responsiveness to the acute behavioral effects of vasopressin and oxytocin. The peptide was injected intracerebroventricularly (i.c.v.) under light ether anesthesia in a volume of 20 µl saline or artificial cerebrospinal fluid. No difference was observed between the two vehicles on behavioral examination.

"Spontaneous" behavior was observed in a rectangular transparent plastic cage, 28.5×42 cm. Unless otherwise indicated, behavior was monitored 6–11 minutes postinjection. Immobility, grooming, and scratching were recorded in seconds per test session; locomotor activity was determined by the crossing of three imaginary lines in the cage. In the "tube test" the mouse was inserted in a transparent plastic tube, 30×2.6 cm and the tube was placed upright. Starting 2 minutes after insertion of the

257

mouse into the tube, immobility, scratching, grooming and escape behaviors were observed for 4 min. Details of the procedures have been described elsewhere.[29]

GENERAL EFFECTS

The injection procedure per se induced significant changes of "spontaneous" behavior as observed 6–11 minutes after the injection: a significant increase in immobility compared to noninjected controls with concomitant reductions of other behaviors was the most common finding. This "stress-induced" increase of immobility was due mostly to the penetration of the skull by the injection needle and is likely to be of nociceptive origin. Light ether anesthesia alone caused increases in immobility of a lesser magnitude, whereas an effect of the injection volume could not be observed (TABLE 1). A similar effect of the injection procedure on behavior in the tube test could not be observed. In the tube test, animals typically spent 180–225 out of 240 seconds immobile.

When mice were observed in the plastic cage, the total time spent immobile by vehicle-injected mice was variable between different experiments and ranged from 90 to 220 seconds per 5-min test session. The amount of grooming behavior was variable as well, and ranged from 10 to 40 seconds in different experiments. Scratching was rarely observed either in vehicle-injected or uninjected mice.

Vasopressin and oxytocin induced excessive grooming and scratching behavior over a wide dose range, with concomitant reductions of the time spent immobile.[29] [Arg-8] vasopressin (AVP), [Lys-8]vasopressin (LVP), and [Arg-8]vasotocin (AVT) were the most potent peptides. The minimal active dose for the induction of excessive scratching by these peptides was close to 0.5 ng.[31] Twenty µg, the highest dose of AVP used in the experiments,[29] did not cause overt signs of toxicity and did not result in any deaths. The extent to which grooming behavior was stimulated showed considerable variation between different experiments. Excessive scratching, accompanied by signs of general behavioral excitation after higher doses, could easily be recognized and was highly reproducible.

DURATION OF ACTION

In the experiment shown in FIGURE 1, mice were placed in a plastic cage immediately after injection of AVP or artificial CSF and their behavior was recorded for successive 5-minute periods, starting 2.5 minutes after the injection. The time course proved

TABLE 1. Effect of the Injection Procedure

Treatment[a]	Immobility (s)
H	16.2^b
H + E	87.6^c
H + E + P	173.7^d
H + E + P + I	172.8^d

[a] H = handling only; E = ether anesthesia; P = penetration of the skull with the injection needle; I = injection of 20 µl saline.
[b] $p < 0.1$, determined by ANOVA and Duncan's multiple-range test.
[c] $p < 0.01$.
[d] $p < 0.001$.

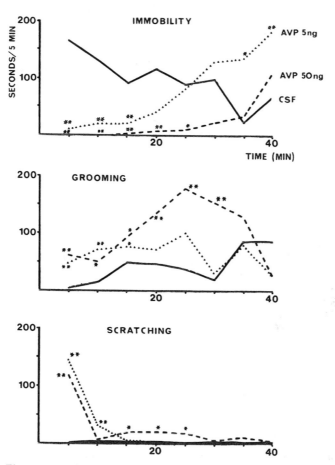

FIGURE 1. Time courses of immobility, grooming, and scratching after administration of 5 ng AVP, 50 ng AVP, or vehicle; **$p < 0.05$, **$p < 0.01$, determined by ANOVA and Duncan's multiple-range test. Behavior was recorded in successive 5-minute periods, starting 2.5 minutes after the injection.

to be different for different behavioral actions: stereotyped reciprocal hindlimb scratching, the most conspicuous effect of vasopressin, was usually observed immediately after awakening from the ether anesthesia.

After the higher dose of AVP (50 ng), signs of general motor excitation were present and mice often continued scratching while walking. Stereotyped scratching subsided within about 10 minutes, independent of the dosage.

Excessive grooming behavior occurred, during the first 10–15 minutes, in the form of short bouts often interspersed with stereotyped scratching. After the lower dose of AVP (5 ng), grooming returned to control levels within about 20 minutes after the injection. After the higher dose, however, grooming behavior tended to level off after about 10 minutes but recurred in the form of extended bouts (sometimes lasting more than 300 seconds) that were most prominent 20–30 minutes after the injection. This

TABLE 2. Spontaneous Behavior

Treatment		Immobility	Grooming	Scratching	Locomotion	Rearing
NaCl		129.7	16.3	1.2	29.8	9.4
AVP	(10 ng)	23.5^c	85.7^b	92.5^c	38.0	9.7
ACTH	(0.2 IU)	152.8	18.8	0.1	19.9	6.1
ACTH	(2.0 IU)	112.4	69.1^b	13.5^c	11.9^a	3.0^b
Subst. P	(1 µg)	119.5	25.7	2.0	30.5	12.0
Subst. P	(5 µg)	111.0	52.5^b	3.6	29.7	9.5
Bombesin	(4 ng)	157.8	23.2	0.7	29.9	9.3
Bombesin	(20 ng)	186.2	40.7	3.2	13.4	2.4^b
Bombesin	(100 ng)	19.0^c	129.6^c	117.8^c	17.5	1.0^c
Bombesin	(500 ng)	0.5^c	127.4^c	146.9^c	14.3	0.6^c
Mescaline	(4 µg)	98.4	3.9^a	70.6^c	33.6	2.0^b
Mescaline	(20 µg)	195.3	0.1^b	53.0^c	18.4	0.7^c

^{a} $p < 0.1$, determined by ANOVA and Duncan's multiple-range test.
^{b} $p < 0.01$.
^{c} $p < 0.001$.

included some scratching behavior that was not stereotyped and involved only one hind leg. Both grooming and scratching behavior appeared to be essentially normal after 35–40 minutes. A decrease of immobility appeared dose-related, both in magnitude and duration.

These short durations of action are of the same order of magnitude as those previously reported for behavioral changes in the tube test.[47,48] Since the time courses are different for the different behavioral parameters, it is unclear which effect most closely parallels the concentration of the peptide at its site of action.

COMPARISON WITH OTHER GROOMING-
AND SCRATCHING-INDUCING AGENTS

TABLE 2 shows a comparison of vasopressin-induced changes of "spontaneous" behavior with those induced by substance P, bombesin, ACTH, and mescaline. AVP induced a decrease of immobility and increases of grooming and scratching behavior with little change of locomotion and rearing. Bombesin induced phenotypically similar changes, but with a tendency for reduced locomotion and rearing. Mescaline induced scratching behavior with simultaneous reductions in grooming and, at the higher dose, decreased rearing and a tendency for increased immobility. Substance P and ACTH increased grooming with little concomitant scratching behavior.

In order to differentiate the effects of AVP from those of the other agents, behavioral effects were compared in the tube test (TABLE 3). While AVP induced a decrease of immobility and increase of escape-directed activity with little increase in grooming and scratching, bombesin induced scratching and grooming behavior with proportional reductions of immobility. Mescaline, like AVP, did not induce excessive grooming or scratching in this environment; a decrease of immobility or increase of escape-directed activity was, however, not evident. Substance P caused slight increases of both grooming and escape activity, while ACTH induced a slight increase of grooming behavior only.

TABLE 3. Behavior in the Tube Test

Treatment		Immobility	Grooming	Scratching	Escape Activity
NaCl		229.9	2.1	0.0	0.1
AVT	(50 ng)	50.9^c	3.0	0.0	93.4^c
Bombesin	(100 ng)	130.4	80.6	11.0^b	0.5
Bombesin	(500 ng)	47.2	132.0^c	47.2^c	5.3^a
Mescaline	(20 μg)	231.7	0.3	0.0	4.8
ACTH	(0.2 IU)	218.7	7.3	0.0	0.2
ACTH	(2.0 IU)	221.7	1.1	0.0	0.1
Substance P	(2 μg)	181.5^b	3.3	0.0	33.1^c
Substance P	(20 μg)	143.1^c	22.3^b	0.0	9.1^a

^{a} $p < 0.1$, determined by ANOVA and Duncan's multiple-range test.
^{b} $p < 0.01$.
^{c} $p < 0.001$.

These experiments show that vasopressin-induced grooming and scratching behavior is a behaviorally distinct phenomenon that can readily be distinguished from similar behaviors induced by other agents.

STRUCTURE-ACTIVITY RELATIONSHIPS

An activity unit was defined based on the ability of a substance to induce excessive scratching behavior 2–5 minutes after i.c.v. injection under light ether anesthesia in a volume of 20 μl saline.[31] One unit is 10,000 times the minimal active dose required to induce excessive scratching behavior. The most potent peptide, AVP, was found to have a mouse behavior potency of 220 units per mg, corresponding to a minimal active dose of about 0.45 ng.

Approximately 30 synthetic analogues of vasopressin and oxytocin were tested for their potencies in inducing excessive scratching behavior. The structure-activity relationships showed an excellent correlation of the behavioral potency with the potency in the rat pressor assay (FIG. 2), but not with antidiuretic, uterotonic, or avian depressor activities.[31]

This suggests that the behavioral effect is mediated by a V-2-type receptor. Oxytocin was about 30-fold less potent than vasopressin. Fragments of vasopressin or oxytocin that may be generated endogenously by partial proteolysis of the parent peptides, such as [Arg-8, Desglycinamide-9] vasopressin and prolyl-leucyl-glycinamide, were essentially inactive. Potencies in the induction of excessive scratching paralleled the potencies in the induction of reduced immobility and increased escape activity in the tube test.[31]

V-2 antagonists were able to antagonize the behavioral effects of vasopressin. These antagonists did not modify behavior when injected by themselves.[31,33] A vasopressin antagonist [1-(β-mercapto-β, β-cyclopentamethylene propionic acid), 2-(o-methyl)tyrosine] arginine-8 vasopressin (P-AVP), was not able to suppress grooming or scratching behaviors induced by mescaline, bombesin, or substance P (TABLE 4). This demonstrates that the behavioral syndromes induced by these agents are distinct not only behaviorally but also pharmacologically.

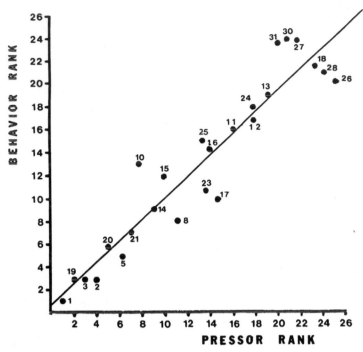

FIGURE 2. Relationship between potency for inducing stereotyped scratching behavior in mice and rat pressor potency. A variety of neurohypophyseal hormones and analogues tested for both behavior-altering and rat pressor activities were ranked according to their potency in each assay. See ref. 31 for a complete listing of peptides used in this study.

PHARMACOLOGICAL INTERACTIONS

In these experiments, 5 ng of AVP or the roughly equipotent peptide AVT were coinjected i.c.v. with various drugs, peptides, or neurotransmitters. A profound antagonism of AVP-AVT–induced grooming and scratching behavior was induced by morphine (1

TABLE 4. P-AVP Coinjected with Mescaline, Substance P, and Bombesin

Injection		Immobility	Grooming	Scratching
CSF		72.5	26.5	1.6
Mescaline	(2.0 µg)	17.4	24.9	38.4
Mescaline	(2.0 µg) + P-AVP	14.9	21.6	44.6
Substance P	(20 µg)	8.6	56.0	2.8
Substance P	(20 µg) + P-AVP	11.0	99.1	3.6
Bombesin	(100 ng)	17.6	159.5	29.0
Bombesin	(100 ng) + P-AVP	0.4	195.5	75.5

Values for antagonist-treated animals were not significantly different from their respective controls ($p > 0.05$).

TABLE 5. Antagonism of AVP-induced Behaviors by i.c.v. Coinjection of AVP with DADL, Morphiceptin, or TRH

Treatment		Immobility	Grooming	Scratching
CSF		94.5	49.6	1.0
AVP	5 ng	19.6	107.6	69.5
DADL	50 ng	51.0	19.7	1.5
DADL	50 ng + AVP	251.3^b	0.0^b	1.0^b
DADL	250 ng	293.0	0.0	0.0
DADL	250 ng + AVP	86.1^a	7.0^b	0.1^b
Morphiceptin	5 μg	152.5	0.9	0.1
Morphiceptin	5 μg + AVP	130.0^b	15.6^b	20.4^b
Morphiceptin	20 μg	162.2	13.5	0.1
Morphiceptin	20 μg + AVP	174.8^b	0.7^b	0.0^b
TRH	10 μg	101.6	14.5	0.2
TRH	10 μg + AVP	88.9	26.1^b	3.1^b

^{a} $p < 0.05$, compared with AVP-injected mice (ANOVA and Duncan's multiple-range test).
^{b} $p < 0.01$.

μg) and [D-Ala-2] Met-enkephalinamide (1 μg) (data not shown). AVP-AVT-induced grooming and scratching were also antagonized by morphiceptin and [D-Ala-2, D-Leu-5] enkephalin (DADL), which preferentially activate μ and α opioid receptors, respectively (TABLE 5). A less profound suppression of these behaviors, which was, however, often confounded by possible toxic and incapacitating effects of the drugs, was found after propranolol (59 μg, but not 10 μg), atropine (25 μg, but not 5 μg), muscimol (50 ng), D,L-3,4-methylenedioxy-amphetamine (MDA, 6 μg, 30 μg, 150 μg), thyrotropin-releasing hormone (TRH, 10 μg), and neurotensin (1 μg). Phentolamine (5 μg, 25 μg), chlorpromazine (40 μg), hexamethonium (10 μg), ergotamine (5 μg), naloxone (10 μg), glycine (25 μg), and bradykinin (10 μg) did not significantly or specifically suppress these behaviors.[30] In particular, naloxone had no effect either on spontaneous or AVT-induced behavior.

TOLERANCE FORMATION

After repeated twice daily injection with high doses of the active peptides, tolerance to the behavioral effects of a low challenge dose developed rapidly. This tolerance was reversible after cessation of the twice-daily pretreatment regimen (FIG. 3). There was complete cross-tolerance between vasopressin and oxytocin. Short-term desensitization (tachyphylaxis) to the behavioral effects of LVP was seen only after administration of very large doses of oxytocin (100 ng) (TABLE 6). All components of the vasopressin behavioral syndrome, including grooming and scratching behavior in unrestrained mice, immobility both in unrestrained animals and in the tube test (data not shown), and escape activity in the tube test (data not shown) were affected.

These results are in contrast to the induction of "barrel rotation" in rats, which shows a kindling phenomenon rather than tolerance formation.[34] This difference suggests that these two behavioral effects are mechanistically distinct although they may be mediated by the same receptor type.[35] We have not performed studies on cross-tolerance between neurohypophyseal hormones and other grooming- and scratching-inducing agents. Studies of this kind may be valuable in studying the interrelation-

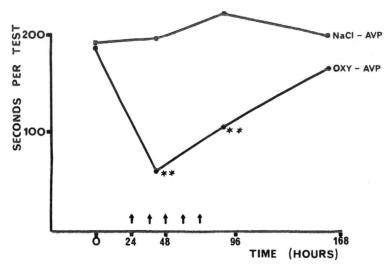

FIGURE 3. Formation of tolerance to the grooming and scratching behavior induced by a dose of 5 ng of AVP. *Arrows* represent the time at which the mice were subjected to repeated pretreatment injections with a high dose (8 μg) of oxytocin (*OXY-AVP*) or vehicle (*NaCl-AVP*). Tolerance to the AVP effect was readily induced by pretreatment with oxytocin, showing cross tolerance between the two peptides; **$p < 0.01$, ANOVA and Duncan's multiple-range test. This tolerance formation was reversible.

ships between the different behavioral syndromes (see chapters by Drago and Pedersen *et al.*, this volume).

SPECIES SPECIFICITY

Although different injection procedures were used, increased grooming behavior has been observed after the administration of vasopressin and oxytocin into the lateral ventricle of rats.[30,32] This effect is, however, of far smaller magnitude than the corresponding effects in mice. To date, stimulation of scratching behavior by vasopressin has been observed in only one rat study.[32] Thus there appears to be a considerable degree of species specificity in the acute behavioral effects of vasopressin and oxytocin.

TABLE 6. Tachyphylaxis

Pretreatment[a]	Test Injection	Immobility	Grooming and Scratching
NaCl	NaCl	148.4	28.5
NaCl	OXY (100 ng)	50.3	154.4
LVP (2 × 2 μg)	OXY (100 ng)	97.0[b]	96.1[b]

[a] Administered i.c.v. 155 and 95 min before test injection.
[b] $p < 0.1$, determined by ANOVA and Duncan's multiple-range test.

TABLE 7. Subcutaneous and Intracerebroventricular Administration

Injection		Behavior	
s.c.[a]	i.c.v.[b]	Immobility	Grooming and Scratching
NaCl		36.8	33.1
OXY		282.4[e]	6.0[c]
LVP		288.7[e]	0.0[c]
NaCl	NaCl	219.3	14.8
OXY	NaCl	267.5[d]	14.5
LVP	NaCl	270.7[d]	0.4[c]
LVP	LVP	10.8	260.5

[a] s.c. = subcutaneously; doses: lysine vasopressin (LVP) 5.5 µg; oxytocin (OXY) 160 µg.
[b] i.c.v. = intracerebroventricularly; LVP dose: 5 ng.
[c] $p < 0.1$, determined by ANOVA and Duncan's multiple-range test.
[d] $p < 0.01$.
[e] $p < 0.001$.

Difficulty in demonstrating enhanced grooming or scratching behavior after high doses of centrally administered vasopressin in rats may be related to toxic and particularly convulsant properties of the peptide in this species. Vasopressin and, occasionally, oxytocin induce "barrel rotation" in rats which, if it occurs, precludes any meaningful behavioral testing. The doses necessary to induce this seizure-like state vary greatly in different studies.[30,32,34,35] Epileptiform activity induced by subconvulsant doses of these peptides may interfere with the expression of grooming and scratching behaviors without producing overt seizures.

Preliminary observations (Meisenberg and Simmons, unpublished) suggest that i.c.v.-administered vasopressin stimulates grooming but not scratching behavior in hamsters, while overt behavioral changes could not be observed in guinea pigs.

SITE OF ACTION

TABLE 7 compares the effectiveness of lysine vasopressin (LVP) and oxytocin after both i.c.v. and subcutaneous (s.c.) injection in mice. This experiment shows that s.c. injection of even large doses of the peptides does not induce the typical grooming and scratching behavior, but rather an increase in immobility accompanied by reductions of grooming and scratching. This strongly suggests a central rather than peripheral site of action. Moreover, the ability of i.c.v.-administered LVP to induce grooming and scratching even in the presence of s.c.-administered peptide strongly suggests that failure of the peripherally administered peptide to induce these behaviors is due to an inability of the peptide to reach its central nervous system site of action rather than a suppression of the centrally mediated effect by its actions in the periphery. Thus, an effective blood-brain barrier seems to exist for these peptides. This is in agreement with previous findings about a poor penetration of neurohypophyseal peptides across the blood-brain barrier.[10,36]

Possible sites of action in the brain were studied in preliminary experiments using intracranial microinjections of 1.25 ng AVP in 1 µl 1% Evans blue solution. The dye did not appear to affect the behavioral response to AVP. The occurrence of excessive

TABLE 8. Possible Sites of AVP Action[a]

Location of Dye	
Cerebro-cerebellar cistern	5/13
Other cortical areas	0/2
Surfaces of the olfactory tubule	1/6
Hypothalmus .	0/4
Pons	0/5
Medulla	0/1
Olfactory tubercle	0/1
Anterior hypothalamus/thalamus	1/6
Lateral hypothalamus/thalamus	3/11
Midbrain tegmentum	1/4
Dorsal surface of the midbrain	1/2
Pons	0/1
Lateral ventricle	5/13
Third ventricle	2/5
Aqueduct	3/6
Fourth ventricle	5/8

[a] Number of mice exhibiting excessive scratching 2–5 min after intracranial injection of 1.25 ng AVP in 1 µl of 1% Evans blue solution per number of mice injected at each site.

scratching was observed 2–5 minutes after the injection. The animals were sacrificed 5–10 minutes after the injection, and the location of the dye was determined macroscopically. TABLE 8 lists sites in the mouse brain at which microinjection of vasopressin elicited or failed to elicit excessive scratching behavior. Injections onto the ventral or lateral surfaces of the medulla, pons, or basal forebrain, or into the cerebro-cerebellar cistern, lateral hypothalamus, or thalamus were mostly ineffective. Injections in the lateral ventricle were effective in less than half of the animals, and a positive response was most commonly associated with spread of the dye into the fourth ventricle.

These results suggest that the behavioral response is not mediated by a site in the hypothalamus, thalamus ventral midbrain, or near the ventral or lateral surface of the brainstem. The relevant site can be reached from the ventricular system. Since less than half of the lateral ventricle injections were effective, the site of action may be further downstream, possibly bordering the fourth ventricle. The anterior hypothalamus, which has been reported to mediate a vasopressin-induced flank-marking response in hamsters,[37–40] does not seem to be involved although the flank-marking response includes grooming behavior and appears to be mediated by a V-2-type receptor as well.

CONCLUSION

Grooming and scratching behavior induced by oxytocin, vasopressin, and structurally related peptides is a phenomenon distinct from behavioral changes induced by other peptides and drugs. Bombesin-induced grooming and scratching behavior, which greatly resembles the vasopressin effect (TABLE 2), may be a sign of direct nociceptive stimulation because of its persistence in a stressful environment (TABLE 3), its relative lack of species specificity,[30] and the aversive nature of the doses of bombesin required to induce grooming and scratching in rats.[41] Vasopressin-induced behavior, in contrast, is highly dependent on the environmental situation (TABLES 2 and 3). This suggests that the effect is not due to a direct stimulation of motor pathways or the stimulation of a nociceptive reflex.

The possibility that sensory systems are involved cannot, however, be excluded. In particular, the dorsal column nuclei, which are thought to receive vasopressinergic or oxytocinergic innervation,[45] may be sites for such an action. This would be compatible with the limited data available about the site of action of vasopressin-induced scratching (TABLE 8).

An action on higher integrative functions rather than specific sensory or motor pathways seems, however, more likely. This possibility is supported by the extreme variability of the "vasopressin behavioral syndrome" in different environments (TABLES 2 and 3). Specifically, the results suggest that vasopressin may generally stimulate active responses to stressful stimuli, including scratching in response to a painful injection (see TABLE 1), grooming during recovery from a distressing experience (FIG. 1), and escape activity in response to confinement (TABLE 3). Corresponding reductions of stress-induced immobility are generally observed. The previously reported[29] effectiveness of oxytocin and vasopressin in the "behavioral despair" test of Porsolt et al.[49] points in the same direction and suggests, in addition, that these peptides may act as "endogenous antidepressants." The frequently observed lack of a clear-cut and profound stimulation of grooming and scratching behavior in rats may be related to toxic and convulsant effects of vasopressin in this species,[34,35] which in turn may mask some of the more subtle behavioral alterations.

Although the structure-activity data show that the effect of vasopressin on grooming and scratching behavior, unlike its effect on learning and memory processes,[12-15] is mediated by a V-2-type receptor, the effect does not appear to be due to a stimulation of peripheral receptors. The possibility that the observed behaviors are secondary to ischemia caused by constriction of cerebral blood vessels is unlikely because of the highly specific nature of the behavioral changes, the lack of toxicity, and the known inability of vasopressin to constrict pial arterioles[42] or influence cerebral blood flow after central administration.[43] Also, the possibility that vasopressin induces grooming and scratching behavior at a brain site lacking an effective blood-brain barrier, such as the subfornical organ, the median eminence, or the area postrema, can virtually be excluded (TABLE 7). A role for these sites appears possible, however, in the much publicized actions of neurohypophyseal hormones on learning and memory processes that are observed after both central and peripheral peptide administration.[12-14]

Structure-activity relationships also suggest that fragments of vasopressin or oxytocin formed from the native peptides by partial proteolysis in the brain are unlikely to contribute to the grooming and scratching effect. This is compatible with the observed short duration of action of these peptides (FIG. 1). However, multiple receptor types for neurohypophyseal hormones have been described in the central nervous system,[44,45] and effects other than grooming and scratching, such as the modulation of memory processes[12-15] or an influence on ethanol or morphine tolerance and dependence,[16-19] may be related to the formation of biologically active fragments.[46] The short duration of vasopressin-induced escape activity in the tube test, which is of the same order of magnitude as the duration of AVP-induced grooming behavior, has tentatively been attributed to proteolytic inactivation of the peptide, probably by aminopeptidases.[47,48]

Vasopressin appears to be the most potent grooming and scratching-inducing peptide so far described. Nevertheless, the normal concentrations of vasopressin and oxytocin in the cerebrospinal fluid are on the order of a few picograms per ml,[10,11] while the minimal active dose of AVP for the induction of excessive scratching behavior is 0.45 ng.[31] Even if spread through the ventricular system, subarachnoid space, and brain tissue, as well as rapid inactivation by brain proteases are taken into consideration, the grooming and scratching-inducing concentrations at relevant receptor sites may be significantly higher than the physiological extracellular fluid concentration.

Also, repeated administration of high doses (but not repeated challenge with low doses, see FIG. 3) induced clear-cut tolerance. This suggests that the observed response may not mimic a physiological effect of normal, steady-state levels of the peptides in cerebrospinal fluid or brain extracellular space. It may, however, be relevant if it is induced at peptidergic synapses, where the peptides are released episodically and high extracellular concentrations may temporarily be reached. Release at relevant peptidergic synapses may not be continuous since this could lead to desensitization of the postsynaptic response.

Therefore, future studies should focus on establishing what relationships exist between the vasopressin-induced behavioral syndrome and the terminal fields of vasopressin- and oxytocin-containing nerve fibers. This should be a major step towards an understanding of the roles of vasopressin- and oxytocin-containing neuronal systems in the brain. The species specificity of the behavioral reponses should also be investigated in more detail. In particular, studies in higher mammals may yield valuable information about a possible role of these peptides in normal and abnormal human behavior.

REFERENCES

1. SOFRONIEW, M. V. 1985. Neuroscience **15:** 347–358.
2. CAFFE, A. R. & F. W. VAN LEEUWEN. 1983. Cell Tissue Res. **233:** 23–33.
3. VAN LEEUWEN, F. & A. R. CAFFE. 1983. Cell Tissue Res. **228:** 525–534.
4. STERBA, G., E. HOFFMANN, R. SOLECKI, W. NAUMANN, G. HOHEISEL & F. SCHOBER. 1979. Cell Tissue Res. **196:** 321–336.
5. STERBA, G., W. NAUMANN & G. HOHEISEL. 1980. Prog. Brain Res. **53:** 141–158.
6. SWANSON, L. W. & S. McKELLAR. 1979. J. Comp. Neurol. **188:** 87–106.
7. DOGTEROM, J. & R. M. BUIJS. 1980. Vasopressin and oxytocin distribution in rat brain. *In* Neuropeptides and Neural Transmission. C. Ajmone Marson & W. Z. Traczyk, Eds. IBRO Monogr. Ser. **7:** 307–314. Raven. New York, NY.
8. BUIJS, R. M. 1980. J. Histochem. Cytochem. **28:**357–360.
9. BUIJS, R. M. & D. F. SWAAB. 1979. Cell Tissue Res. **204:** 355–365.
10. WANG, B. C., L. SHARE & K. L. GOETZ. 1985. Fed. Proc. **44:** 72–77.
11. MORRIS, M., R. R. BARNARD, JR. & L. E. SAIN. 1984. Neuroendocrinology **39:** 377–383.
12. DE WIED, D. & D. H. G. VERSTEEG. 1979. Fed. Proc. **38:** 2348–2354.
13. DE WIED, D. 1977. Life Sci. **20:** 195–204.
14. RIGTER, H. & J. C. CRABBE. 1979. Vitam. Horm. **37:** 153–241.
15. VAN WIMERSMA GREIDANUS, TJ. B., B. BOHUS, G. L. KOVACS, D. H. G. VERSTEEG, J. P. H. BURBACH & D. DE WIED. 1983. Neurosci. Biobehav. Rev. **7:** 453–463.
16. HOFFMAN, P. L. & B. TABAKOFF. 1981. Centrally acting peptides and tolerance to ethanol. *In* Currents in Alcoholism, Vol. 8. M. Galanter, Ed.: 359–378. Grune & Stratton. New York.
17. HOFFMAN, P. L., R. F. RITZMANN, R. WALTER & B. TABAKOFF. 1978. Nature **276:** 614–616.
18. KRIVOY, W. A., E. ZIMMERMANN & S. LANDE. 1974. Proc. Natl. Acad. Sci. U.S.A. **71:** 1852–1856.
19. RITZMANN, R. F., K. A. STEECE, J. M. LEE & F. A. DELEON-JONES. 1985. Neuropeptides **6:** 255–258.
20. MEISENBERG, G. & W. H. SIMMONS. 1984. Neuropharmacology **23:** 1195–1200.
21. VEALE, W. L., N. W. KASTING & K. E. COOPER. 1981. Fed. Proc. **40:** 2750–2753.
22. BANET, M. & U. E. WIELAND. 1985. Brain Res. Bull. **14:** 113–116.
23. KORDOWER, J. H. & R. J. BODNAR. 1984. Peptides **5:** 747–756.

24. VARMA, S., B. P. JAJU & K. P. BHARGAVA. 1969. Circ. Res. **24:** 787–792.
25. ZERBE, R. L. 1985. Peptides **6:** 65–68.
26. RIPHAGEN, C. L. & Q. J. PITTMAN. 1985. Regul. Peptides **10:** 293–298.
27. SCHWARZBERG, H., K. BETSCHEN & H. UNGER. 1980. Action of neuropeptides on self-stimulation behavior in rats. *In* Neuropeptides and Neural Transmission. C. Ajmone Marsan & W. Z. Traczyk, Eds.: 339–341. Raven. New York.
28. DELANOY, R. L., A. J. DUNN & R. TINTNER. 1978. Horm. Behav. **11:** 348–362.
29. MEISENBERG, G. 1981. Peptides **2:** 1–8.
30. MEISENBERG, G. 1982. Neuropharmacology **21:** 309–316.
31. MEISENBERG, G. & W. H. SIMMONS. 1982. Pharmacol. Biochem. Behav. **16:** 819–825.
32. CALDWELL, J. D., F. DRAGO, A. J. PRANGE & C. A. PEDERSEN. 1986. Regul. Peptides **14:** 261–271.
33. MEISENBERG, G. & W. H. SIMMONS. 1987. Neuropharmacology **1:** 79–83.
34. KASTING, N. W., W. L. VEALE & K. E. COOPER. 1980. Can. J. Physiol. Pharmacol. **58:** 316–319.
35. KRUSE, H., TJ. B. VAN WIMERSMA GREIDANUS & D. DE WIED. 1977. Pharmacol. Biochem. Behav. **7:** 311–313.
36. CORNFORD, E. M., L. D. BRAUN, P. D. CRANE & W. H. OLDENDORF. 1978. Endocrinology **103:** 1297–1303.
37. FERRIS, C. F., H. E. ALBERS, S. M. WESOLOWSKI, B. D. GOLDMAN & S. E. LEEMAN. 1984. Science **224:** 521–523.
38. ALBERS, H. E., J. POLLOCK, W. H. SIMMONS & C. F. FERRIS. J. Neurosci. In press.
39. ALBERS, H. E. & C. F. FERRIS. 1985. Regul. Peptides **12:** 257–260.
40. FERRIS, C. F., J. POLLOCK, H. E. ALBERS & S. E. LEEMAN. 1985. Neurosci. Lett. **55:** 239–343.
41. MEISENBERG, G., W. H. SIMMONS & S. A. LORENS. Pharmacol. Biochem. Behav. In press.
42. LASSOFF, S. & B. M. ALTURA. 1980. Brain Res. **196:** 266–269.
43. RAICHLE, M. E. & R. L. GRUBB, JR. 1978. Brain Res. **143:** 191–194.
44. MEISENBERG, G. & W. H. SIMMONS. 1983. Neurosci. Biobehav. Rev. **7:** 263–280.
45. MUHLETHALER, M., W. H. SAWYER, M. M. MANNING & J. J. DREIFUSS. 1983. Proc. Natl. Acad. Sci. U.S.A. **80:** 6713–6717.
46. BURBACH, J. P. H. & J. L. M. LEBOUILLE. 1983. J. Biol. Chem. **258:** 1487–1494.
47. MEISENBERG, G. & W. H. SIMMONS. 1984. Life Sci. **34:** 1231–1240.
48. MEISENBERG, G. & W. H. SIMMONS. 1984. Peptides **5:** 535–539.
49. PORSOLT, R. D., A. BERTIN & M. JALFRE. 1977. Arch. Intern. Pharmacodyn. **229:** 327–336.

Actions of Psychoactive Drugs on ACTH- and Novelty-induced Behavior in the Rat

JÖRG TRABER,[a] DAVID G. SPENCER, JR.,[a]
THOMAS GLASER,[a] AND WILLEM HENDRIK GISPEN[b]

[a]*Neurobiology Department*
Troponwerke
D-5000 Cologne 80, Federal Republic of Germany

[b]*Division of Molecular Neurobiology*
Rudolf Magnus Institute for Pharmacology
and
Institute of Molecular Biology and Medical Biotechnology
University of Utrecht
Utrecht, The Netherlands

INTRODUCTION

"Grooming" or "maintenance" behavior is a common species-specific fixed action pattern with readily definable components. Both peripheral and central inputs govern the display of this "care of the body surface" behavior.[1-3] It is likely that functions other than body maintenance are also served by this behavior. Since grooming in rodents is often especially likely to occur after extremely arousing conditions, it has been suggested that the grooming response may play a deactivating role.[3-6] Grooming behavior as discussed in this paper consists of the following elements: forepaw vibration, face washing, body grooming, anogenital grooming, scratching, paw licking, head and body shaking, and tail preening. A time sampling technique was used for response quantification, as described by Gispen *et al.*[7]

Several groups have documented that upon placement into confined novel observation boxes, rats initially display grooming behavior invariably followed by sleep.[8-10] Although the initial frequency of novelty-induced grooming is higher than baseline levels of home-cage grooming, the structure of each grooming bout is the same.

Intracranial (but not systemic) treatment with ACTH and congeners of a variety of animal species such as dogs, cats, rabbits, and mice induces a behavioral response known as the stretching and yawning syndrome.[11] The onset of this syndrome in rodents is preceded by the display of excessive grooming behavior.[7,11,12] Spruijt and Gispen[13] found that ACTH enhanced the display of grooming behavior without changing the composition of the behavioral response relative to saline-injected control rats. Since under certain conditions intraventricularly injected anti-ACTH antibodies block novelty-induced grooming,[14] it may be argued that the neural structures underlying ACTH- and novelty-induced grooming are closely related.

The rather simple observational analysis technique has proven to be very useful in the study of this neural substrate.[4,5] Although central dopaminergic and GABAergic pathways are important components of this neural substrate,[15-17] we will present data indicating the relative importance of additional neurotransmitter systems. A variety of psychoactive drugs with completely different mechanisms of action have been previously shown to suppress the grooming response to ACTH.[10] In the present paper, the suspected involvement of serotoninergic neurotransmission was investigated in more detail. By using selective ligands, it was shown that 5-HT_{1A} receptors in particular may be part of the neural substrate underlying excessive grooming induced in the rat by ACTH and novelty.

THE NEURAL SUBSTRATE UNDERLYING ACTH-INDUCED EXCESSIVE GROOMING

Several authors have suggested that the grooming response depends on the integrity of neural pathways with dopamine and/or opiate receptors.[5] Such studies pointed to an important role of dopaminergic systems in general, and the substantia nigra in particular, in ACTH-induced excessive grooming in the rat.[18-22] In a series of experiments, Spruijt *et al.*[15-17] investigated the role that the GABAergic nigro-collicular-periaqueductal gray pathway plays in the expression of ACTH-induced grooming. These authors concluded that most likely the periaqueductal gray is the primary site of action of ACTH and that dopaminergic and GABAergic structures modulate the behavioral output of this structure, *i.e.*, the grooming response. Previously, Jacquet[23] implicated ACTH-sensitive sites in the periaqueductal gray in the induction of explosive motor behavior; Gmerek and Cowan[24] also indicated the importance of this structure in excessive grooming behavior in the rat. However, although these results certainly implicate dopamine and GABA in the modulation of grooming behavior in the rat, they should not be taken as evidence for an exclusive role for these transmitters.

There is increasing evidence that other transmitter systems may feed into the final common pathway, for example, noradrenergic and cholinergic manipulation may interfere with ACTH-induced excessive grooming as well.[5,25] Previous studies from our laboratories have also provided the basis for a more complex modulation of ACTH-induced grooming.[10] The effects of psychoactive drugs on ACTH-induced excessive grooming were examined in these studies. The neuroleptics haloperidol and clozapine, as well as the antidepressants amitriptyline, nomifensine and mianserin, all suppressed the grooming response in a dose-dependent manner. Interestingly, the neuroleptic drug sulpiride, which hardly crosses the blood-brain barrier,[26] was inactive even at high doses. A vital aspect of these findings was, however, that the suppression of the peptide-induced behavior was not the result of competing behavioral activity elicited by the drug itself. This was established by the additional examination of drug effects on novelty-induced behavior. At high doses, these drugs may have suppressed grooming by the initiation of competing behaviors (*e.g.*, motor activity in the case of nomifensine). However, a more specific interaction seemed to be responsible for the inhibition when doses at the ED_{50} level of grooming inhibition were used. Drug effects at these lower doses were rarely accompanied by drug-specific induction of competing responses.

In view of the complexity of the mechanisms of action of the drugs used, it was argued that the substrate underlying ACTH-induced grooming behavior could be modulated by several neurotransmitter systems. Certainly the effectiveness of the antidepressants mentioned above opened up the possibility that serotoninergic mechanisms might be involved in this modulation. In this context it seemed reasonable to investigate further the serotoninergic component in ACTH-induced grooming.

5-HT RECEPTORS

Since the pioneering work of Gaddum and Picarelli,[27] it has become clear that receptors for 5-hydroxytryptamine (5-HT) in the peripheral nervous system are heterogeneous. Using the guinea pig ileum preparation, these authors convincingly showed the existence of two pharmacologically distinct types of 5-HT receptors, namely the "D" and the "M" receptors. The former could be blocked by dibenzyline (phenoxybenzamine) and mediates contraction of the smooth muscle. The latter was indirectly blockable by morphine and mediates depolarization of the cholinergic nerves. Further studies during recent years have now indicated an even higher degree of receptor heterogeneity.[28,29]

As for the central nervous system (CNS), it was not until 1979 that Peroutka and Snyder[30] were able to provide evidence, by radioligand binding studies, for two distinct 5-HT binding sites. The sites that displayed low affinity for antagonists such as spiperone were named 5-HT_1 sites, whereas the sites with the opposite characteristics, $i.e.$, high affinity for spiperone and low affinity for agonists, were called 5-HT_2 sites.[30] In the meantime, it has been clearly shown that the affinities of a variety of 5-HT antagonists for the 5-HT_2 site correlate well with their IC_{50} values at the "D" receptors in functional studies using vascular and gastrointestinal smooth muscle. 5-HT_2 receptors are distinctly localized in certain parts of the rat brain such as the frontal cortex, the olfactory system, and the basal ganglia.[31]

Ketanserin[32] and ritanserin[33] are two antagonists that bind with high selectivity and high affinity to 5-HT_2 receptors. There are a number of compounds with purported 5-HT_2 agonistic activity, such as the hallucinogen 1-(2,5-dimethoxy-4-methylphenyl)-2-aminopropane (DOM)[34] and quipazine.[34,35] The degree of selectivity of these drugs, however, is less than that of the antagonists described above.

Less is known about the 5-HT_1 site, despite the fact that 5-HT has nanomolar affinity for this binding site. The functional role of the 5-HT_1 site is still under debate, largely due to the lack of specific 5-HT_1 antagonists.[36,37] The very recent appearance and pharmacological evaluation of several specific 5-HT_1 agonists, however, has provided strong evidence not only for further subtypes of 5-HT_1 binding sites in the brain, but also for a functional role of such sites. On the basis of receptor binding studies with spiperone[38] or 8-hydroxy-2-(di-n-propylamino)tetralin (8-OH-DPAT),[39] 5-HT_{1A} and 5-HT_{1B} subtypes were postulated. These findings were corroborated and extended by results from autoradiographical studies[40-42] showing that the two subtypes have differential distributions in the brain. Recently a third 5-HT_1 subtype site (5-HT_{1C}), present in the choroid plexus, has been described.[43]

New information is accumulating on the functional role of 5-HT_1 sites, especially the 5-HT_{1A} site, the site to which the most specific agonists have been characterized. For example, the 5-HT_{1A} agonist 8-OH-DPAT causes a marked stimulation of the sexual behavior of male rats,[44] lowers 5-HT turnover in rat brain,[45] and depresses firing of 5-HT neurons in the raphe nuclei.[46] BAY R 1531 (6-methoxy-4-(di-n-propylamino)-1,3,4,5-tetrahydrobenz(c,d)indole hydrochloride) is a more rigid structural analogue of 8-OH-DPAT and has been shown to be even more potent in a variety of test systems.[47,48] The pyrimidinylpiperazine derivative ipsapirone (TVX Q 7821), although structurally not related to 8-OH-DPAT or BAY R 1531, also displays high affinity and selectivity for the 5-HT_{1A} binding site[49,50] and depresses firing of 5-HT neurons in the raphe nuclei.[51] The behavioral profile, however, is quite different from that of the two 5-HT_{1A} ligands mentioned above. Ipsapirone shows marked anxiolytic and antiaggressive effects in animals[52] and stimulates rat sexual behavior only very weakly at doses much higher than those necessary to induce the antianxiety effects.[48]

There is much less information on the role of 5-HT$_{1B}$ binding sites because of the lack of appropriately selective compounds. The 5-HT$_{1B}$ sites seem to be located on presynaptic sites and to be involved in the regulation of neurotransmitter release.[53,54] Moreover, in contrast to the 5-HT$_{1A}$ sites, which are primarily located in the limbic system,[40-42] 5-HT$_{1B}$ sites are largely distributed in various sectors of the extrapyramidal motor system of the rat brain.[55] Currently, compounds such as RU 24969 (5-methoxy-3-(1,2,3,6-tetrahydro-4-pyridinyl)-1H-indole)[56] or TFMPP (trifluoromethyl-phenylpiperazine)[57] are in use as 5-HT$_{1B}$ agonists.[58]

The goal of the present series of experiments was to probe the involvement of 5-HT receptor subtypes in the grooming response. We have therefore used the more or less specific compounds for these receptors as tools.

MATERIALS AND METHODS

Serotonin Receptor Binding

The binding characteristics of the drugs used were determined as described by Glaser and Traber.[50] The selectivity of the drugs for the various 5-HT receptor subtypes was measured using the following labeled ligands: ^{3}H–5-HT, ^{3}H–ipsapirone, and ^{3}H–ketanserin.

ACTH- and Novelty-induced Grooming

The experimental design of the grooming test was identical to that described by Traber *et al.*[10] In short, male rats of an inbred Wistar strain bearing an intraventricular (i.c.v.) polypropylene cannula were injected with synthetic ACTH$_{1-24}$ (0.3 µg/3 µl saline) immediately followed by the i.p. injection of vehicle or the drug to be studied. Subsequently, they were placed individually in novel glass observation boxes and their display of grooming behavior was scored once every 15 s beginning 15 min after the i.c.v. injection and lasting for 50 min.[7] Novelty-induced behaviors were quantified as previously described[10]: every 15 s for a total of 50 min behavior of six control and six experimental rats was classified as sitting, sleeping, grooming, licking-sniffing, rearing, or locomotion. Moreover, 5-HT-related behaviors (the so-called 5-HT syndrome[59]) elicited by drug treatment itself were simultaneously assessed. The categories used for this purpose were head twitching, forepaw treading, and flat body posture.

The following drugs were selected for the present study: 8-OH-DPAT, BAY R 1531, and ipsapirone as 5-HT$_{1A}$ agonists, TFMPP as a 5-HT$_{1B}$ agonist, quipazine as a 5-HT$_2$ agonist, and ritanserin as a 5-HT$_2$ antagonist. The compounds came from the following sources: 8-OH-DPAT, BAY R 1531 and ipsapirone, Chemistry Department, Bayer AG, Wuppertal, FRG; TFMPP, Aldrich, Steinheim, FRG; quipazine, Miles, Munich, FRG; ritanserin, gift from Janssen, Beerse, Belgium.

RESULTS AND DISCUSSION

Serotonin Receptor Binding

In TABLE 1 the inhibition constants (K_i values) of the various 5-HT agonists and antagonists used in the present study are listed. 8-OH-DPAT, BAY R 1531, and ipsapi-

TABLE 1. Inhibition Constants K_i (nmol/1)

Drug	Receptor Type			
	5-HT[a]	5-HT$_{1A}$[b]	5-HT$_{1B}$[c]	5-HT$_2$[d]
5-HT	1.6 ± 0.1	5 ± 2	4 ± 1	1840 ± 350
8-OH-DPAT	2.0 ± 0.5	0.7 ± 0.1	6350 ± 1850	6230 ± 1050
BAY R 1531	0.6 ± 0.2	0.4 ± 0.1	2730 ± 220	730 ± 115
Ipsapirone	15 ± 2	2.6 ± 0.5	>10000	2940 ± 240
TFMPP	190 ± 40	nt[e]	10 ± 4	445 ± 125
Quipazine	1660 ± 160	nt	nt	810 ± 80
Ritanserin	510 ± 10	nt	nt	0.3 ± 0.2

[a] Ligand ^{3}H–5-HT, calf hippocampal membranes.
[b] Ligand ^{3}H–ipsapirone, rat hippocampal membranes.
[c] Ligand ^{3}H–5-HT in the presence of 10 µmol/1 8-OH-DPAT, rat cerebral cortical membranes.
[d] Ligand ^{3}H–ketanserin, rat prefrontal cortical membranes.
[e] nt, not tested.

rone display high affinity and selectivity for the 5-HT$_{1A}$ receptor subtype. TFMPP shows preference for the 5-HT$_{1B}$ site, which is in accordance with Hamon et al.[58] Quipazine is not markedly selective, but preferentially interacts with 5-HT$_2$ sites. Furthermore, the data confirm that the 5-HT$_2$ antagonist ritanserin is selective for the 5-HT$_2$ subtype.

ACTH- and Novelty-induced Grooming

As can be seen in FIGURE 1, the 5-HT$_{1A}$ agonists BAY R 1531 and 8-OH-DPAT potently suppressed ACTH-induced grooming in a dose-dependent manner. Although ipsapirone shows a subtype selectivity similar to that of BAY R 1531 and 8-OH-DPAT, the reduction of ACTH-induced grooming was only apparent at rather high dose levels. Similarly, the presumed 5-HT$_{1B}$ agonist TFMPP and the 5-HT$_2$-selective drugs quipazine and ritanserin were active (if at all) only at the highest dose tested (10 mg/kg; FIG. 1).

In order to establish whether the observed drug-induced inhibition of ACTH-related grooming was behaviorally specific, an analysis of the behavioral response to a novel confined chamber in the absence and presence of the drugs was performed. Furthermore, the drugs were studied for their ability to induce the 5-HT-syndrome. Induction by these drugs of active behaviors, including those of the 5-HT syndrome, was taken as evidence for the production of responses that could conceivably compete in an unspecific way with the grooming response.

TABLE 2A shows the percentage of time that saline-treated controls spent at the various behavioral elements or categories in the novel confined observation chambers used in the grooming test. After an initial phase of activity (rearing, grooming, licking-sniffing, locomotion), the animals sit and eventually fall asleep. In TABLE 2B, the effects of the drugs on the occurrence of components of the 5-HT syndrome are shown, as well as the statistically significant (Mann-Whitney U-test, $p \leqslant 0.05$) changes seen in the general behavioral response to the novel chamber. At the low doses where substantial inhibition of ACTH-induced grooming was seen, the compounds BAY R 1531 (0.1 mg/kg i.p.) and 8-OH-DPAT (0.3 mg/kg i.p.) show little competing 5-HT-related behavior. Novelty-related grooming was, however, strongly suppressed. At higher doses

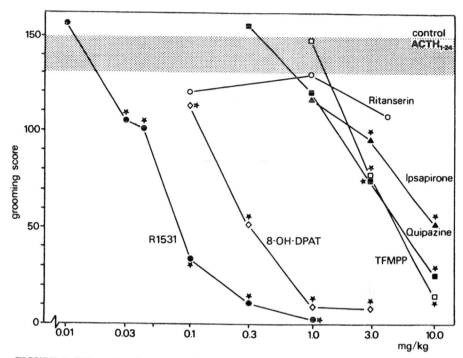

FIGURE 1. Effect of various serotoninergic agents on ACTH-induced excessive grooming in the rat. *Stippled bar* indicates mean grooming score $\pm$ SEM of control rats, calculated as $ACTH_{1-24}$ grooming score with 0.3 µg/3 µl (i.c.v.) − grooming score with vehicle 0.5 ml/100 g (i.p.). *Asterisks* denote significant difference in group mean value compared to control ($p \leqslant 0.05$, ANOVA followed by supplemental t-test; n = four to six rats).

the general behavioral response to the novel chamber was severely altered and signs of the 5-HT syndrome became visible. In contrast, treatment with ipsapirone did not induce the 5-HT syndrome. However, the general response to the novel chamber was also affected. Again, suppression of novelty-related grooming was observed, but always in the presence of other behavioral alterations. TFMPP treatment over the dose range tested did not induce signs of the 5-HT syndrome but did alter the novelty-related behaviors; grooming was not significantly affected. Quipazine treatment from 5 mg/kg i.p. onwards induced the 5-HT syndrome and significantly reduced sitting and sleeping. Novelty-induced grooming was reduced only at the highest dose tested (10 mg/kg i.p.).

From these data the following picture emerges. At low doses, the 5-HT$_{1A}$ agonists 8-OH-DPAT and BAY R 1531 specifically inhibited both ACTH- and novelty-related grooming. The presumed 5-HT$_{1B}$ agonist TFMPP and the 5-HT$_2$ agonist quipazine in higher doses suppressed ACTH-induced grooming in a nonspecific manner. Competing behavioral activities elicited by the drugs seem to be responsible for this inhibition. For the 5-HT$_{1A}$ agonist ipsapirone, the situation is more complex. Although the drug has been shown to be highly selective for the 5-HT$_{1A}$ site,[42,49,50] it was much less effective in blocking the ACTH-induced behavioral response as compared to 8-OH-DPAT and BAY R 1531. This may be due to the lower affinity of ipsapirone for the

TABLE 2A. Percentage of 50-Min Session during Which Saline-treated Rats Engaged in Specific Behaviors in a Novel, Confined Chamber (Typical Control Session, $n = 6$)

Behavior	%
Sitting	49
Grooming	16.4
Sleeping	11.5
Rearing	11.1
Lick and sniff	10.7
Locomotion	1.3

TABLE 2B. Effects of Serotoninergic Drugs Binding to Different 5-HT Receptor Subtypes on Serotonin Syndrome Elements and on General Behavioral Profile of Rats in a Novel, Confined Chamber ($n = 6$ per Drug Dose)

Substance	Dose (mg/kg i.p.)	Flat Body	Forepaw Tread	Head Twitch	Statistically Significant Alteration of Other Behaviors (% of Control)	
8-OH-DPAT	0.32	0	0	0	grooming	− 89
	0.64	0	0	0	grooming	− 52
					sitting	+ 44
	1.25	8.3	8.2	0.2	rearing	− 74
					grooming	− 67
					locomotion	+ 848
	2.5	12.5	22.2	0	sleeping	− 87
					grooming	− 85
					sitting	− 26
					locomotion	+ 1483
					lick and sniff	+ 98
BAY R 1531	0.05	0	0	0	sitting	+ 12
	0.1	0	0	0	grooming	− 63
	0.3	2.5	0	0	locomotion	+ 514
	1.0	7.7	28.0	0	locomotion	+ 11,400
Ipsapirone	5.0	0	0	0	grooming	− 88
					lick and sniff	+ 163
	10	0	0	0	grooming	− 89
					rearing	− 60
					lick and sniff	+ 96
					sitting	+ 40
	20	0	0	0	rearing	− 97
					grooming	− 91
					lick and sniff	+ 94
					sitting	+ 48
TFMPP	1.0	0	0	0	sitting	− 13
					lick and sniff	+ 53
	3.0	0	0	0	lick and sniff	+ 156
	10.0	0	0	0	sitting	− 24
					lick and sniff	+ 279
Quipazine	1.25	0	0	0		
	2.5	0	0	0	sitting	− 29
	5.0	1.8	6.0	6.7	sleeping	− 93
	10.0	3.3	17.3	8.8	sleeping	− 100
					grooming	− 74

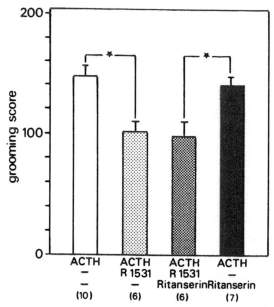

FIGURE 2. Failure of ritanserin to counteract the inhibition by BAY R 1531 of ACTH-induced grooming in the rat. Doses were as follows: $ACTH_{1-24}$, 0.3 µg/3 µl (i.c.v.); BAY R 1531, 0.05 mg/kg (i.p.); ritanserin, 1 mg/kg (i.p.). Ritanserin was injected 10 min prior to i.c.v. injection of $ACTH_{1-24}$ and i.p. injection of vehicle or BAY R 1531. (Numbers in parentheses indicate number of animals.) *Dashes* denote vehicle. *Bars* indicate mean grooming score ± SEM. *Asterisks* denote significance between relevant groups ($p \leqslant 0.05$, ANOVA followed by supplemental *t*-test).

$5HT_{1A}$ receptor (see TABLE 1) and/or a lower intrinsic activity. Indeed, it has been reported that in addition to being an agonist at the 5-HT autoreceptors of raphe nuclei neurons,[51] ipsapirone may also be an antagonist in other systems.[60,61] In the high doses necessary to block ACTH-induced grooming, the drug not only inhibits novelty-induced grooming behavior, but at the same time also considerably alters the general behavioral response. It is therefore suggested that the reduction of ACTH-induced grooming by ipsapirone is nonspecific.

These data point to a serotoninergic component in the expression or modulation of both ACTH- and novelty-induced excessive grooming. It is not surprising to find common pathways in the neural substrate underlying novelty- and ACTH-induced grooming, as others also have shown that both stimuli employ closely related mechanisms.[5] The data presented in this paper so far suggest that the serotoninergic modulation is brought about by interference with 5-HT_{1A} receptors. This is further supported by the fact that ritanserin, a selective 5-HT_2 antagonist[33] (TABLE 1) itself, had little if any effect on ACTH-induced grooming (FIG. 1) and was unable to block the inhibitory influences of the 5-HT_{1A} agonist BAY R 1531 (FIG. 2).

Taken together, the data confirm and extend the notion that the neural substrate underlying ACTH-induced excessive grooming is complex and contains other pathways in addition to the dopaminergic and GABAergic systems described by Spruijt *et al.*[15-17] It is known from electrophysiological studies that the autoreceptors on raphe nuclei serotoninergic neurons are of the 5-HT$_{1A}$ type.[51] 5-HT$_{1A}$ agonists mimic the effect of 5-HT and potently block the spontaneous firing. The substantia nigra is an area with a high density of innervation from the raphe nuclei.[62] Since this brain region forms an important part of the neural substrate of the excessive grooming response,[6] it is tempting to assume that 5-HT$_{1A}$ agonists exert their modulatory effect by influencing the pathway between raphe nuclei and substantia nigra. Studies on the effects of 5,7-dihydroxytryptamine lesions of the raphe nuclei could be important in establishing this link.

REFERENCES

1. FENTRESS, J. C. 1977. The tonic hypothesis and the patterning of behavior. Ann. N. Y. Acad. Sci. **290:** 370–395.
2. BORCHELT, P. L. 1980. Care of the body surface. *In* Comparative Psychology: An Evolutionary Analysis of Animal Behavior. M. Ray Denny, Ed.: 362–381. John Wiley & Sons. New York.
3. SPRUIJT, B. M. & W. H. GISPEN. 1983. ACTH and grooming behaviour in the rat. *In* Hormones and Behaviour in Higher Vertebrates. J. Balthazart, E. Pröve & R. Gilles, Eds.: 118–136. Springer-Verlag. Berlin.
4. GISPEN, W. H. & R. L. ISAACSON. 1981. ACTH-induced excessive grooming in the rat. Pharmacol. Ther. **12:** 209–216.
5. GISPEN, W. H. & R. L. ISAACSON. 1986. Excessive grooming in response to ACTH. *In* Neuropeptides and Behavior, Vol. 1. D. de Wied, W. H. Gispen & T. J. B. van Wimersma Greidanus, Eds.: 273–312. Pergamon Press. Oxford.
6. SPRUIJT, B. M. & W. H. GISPEN. 1986. ACTH and grooming. *In* Central Actions of ACTH and Related Peptides, Vol. 1. D. de Wied & W. Ferrari, Eds.: 179–187. Fidia Res. Ser. Symp. Neurosci. Liviana Press. Padua.
7. GISPEN, W. H., V. M. WIEGANT, H. M. GREVEN & D. DE WIED. 1975. The induction of excessive grooming in the rat by intraventricular application of peptides derived from ACTH: Structure activity studies. Life Sci. **17:** 645–652.
8. COLBERN, D. L., R. L. ISAACSON, E. J. GREEN & W. H. GISPEN. 1978. Repeated intraventricular injections of ACTH$_{1-24}$. The effects of home or novel environments on excessive grooming. Behav. Biol. **23:** 381–387.
9. JOLLES, J., J. ROMPA-BARENDREGT & W. H. GISPEN. 1979. Novelty and grooming behavior in the rat. Behav. Neural. Biol. **25:** 563–572.
10. TRABER, J., H. R. KLEIN & W. H. GISPEN. 1982. Actions of antidepressant and neuroleptic drugs on ACTH- and novelty-induced behavior in the rat. Eur. J. Pharmacol. **80:** 407–414.
11. FERRARI, W., G. L. GESSA & L. VARGIU. 1963. Behavioral effects induced by intracisternally injected ACTH and MSH. Ann. N. Y. Acad. Sci. **104:** 330–345.
12. IZUMI, K., J. DONALDSON & A. BARBEAU. 1973. Yawning and stretching in rats induced by intraventricularly administered zinc. Life Sci. **12:** 203–210.
13. SPRUIJT, B. M. & W. H.. GISPEN. 1984. Behavioral sequences as an easily quantifiable parameter in experimental studies. Physiol. Behav. **32:** 707–710.
14. DUNN, A. J., E. J. GREEN & R. L. ISAACSON. 1979. Intracerebral adrenocorticotropic hormone mediates novelty-induced grooming in the rat. Science **203:** 281–283.
15. SPRUIJT, B. M., A. R. COOLS, B. A. ELLERBROEK & W. H. GISPEN. 1986. The colliculus superior modulates ACTH-induced excessive grooming. Life Sci. **39:** 461–470.
16. SPRUIJT, B. M., A. R. COOLS, B. A. ELLERBROEK & W. H. GISPEN. 1986. Dopaminergic modulation of ACTH-induced grooming. Eur. J. Pharmacol. **120:** 249–256.
17. SPRUIJT, B. M., A. R. COOLS & W. H. GISPEN. 1986. The periaqueductal gray: A prerequisite for ACTH-induced excessive grooming. Behav. Brain Res. **20:** 19–25.
18. WIEGANT, V. M., A. R. COOLS & W. H. GISPEN. 1977. ACTH-induced excessive grooming involves brain dopamine. Eur. J. Pharmacol. **41:** 343–345.
19. COOLS, A. R., V. M. WIEGANT & W. H. GISPEN. 1978. Distinct dopaminergic systems in ACTH-induced grooming. Eur. J. Pharmacol. **50:** 265–268.

20. DRAGO, F., B. BOHUS, P. L. CANONICO & U. SCAPAGNINI. 1981. Prolactin induces grooming in the rat: Possible involvement of nigrostriatal dopaminergic system. Pharmacol. Biochem. Behav. **15**: 61–63.

21. CHESHER, G. B. & D. M. JACKSON. 1981. Swim-induced grooming in mice is mediated by a dopaminergic substrate. J. Neural Transm. **50**: 47–55.

22. ISAACSON, R. L., J. H. HANNIGAN, J. SPRINGER, J. RYAN & A. POPLAWSKY. 1983. Limbic and neurohormonal influences modulate the basal ganglia and behavior. *In* Integrative Neurohumoral Mechanisms. E. Endröczi, D. de Wied, L. Angelucci & U. Scapagnini, Eds.: 23–31. Elsevier Biomedical Press. Amsterdam.

23. JACQUET, Y. F. 1979. β-Endorphin and ACTH-opiate peptides with coordinated roles in the regulation of behavior. Trends Neurosci. **2**: 140–143.

24. GMEREK, D. E. & A. COWAN. 1983. $ACTH_{1-24}$ and RX 336-M induce excessive grooming in rats through different mechanisms. Eur. J. Pharmacol. **88**: 339–346.

25. DUNN, A. J. & G. VIGLE. 1985. Grooming behavior induced by ACTH involves cerebral cholinergic neurons and muscarinic receptors. Neuropharmacology **24**: 329–331.

26. BENAKIS, A., J. P. H. BROWN & P. BENARD. 1984. Autoradiographic study of ^{14}C-sulpiride in monkey. Eur. J. Drug Metab. Pharmacokinet. **9**: 365–370.

27. GADDUM, J. H. & Z. P. PICARELLI. 1957. Two kinds of tryptamine receptor. Br. J. Pharmacol. Chemother. **12**: 323–328.

28. HUMPHREY, P. P. A. 1983. Pharmacological characterisation of cardiovascular 5-hydroxytryptamine receptors. *In* Proceedings of IVth Vascular Neuroeffector Mechanisms Symposium. J. Bevan *et al.*, Eds.: 237–242. Raven Press. New York, NY.

29. HUMPHREY, P. P. A. 1984. Peripheral 5-hydroxytryptamine receptors and their classification. Neuropharmacology **23**: 1503–1510.

30. PEROUTKA, S. J. & S. H. SNYDER. 1979. Multiple serotonin receptors: Differential binding of ^3H-5-hydroxytryptamine, ^3H-lysergic acid diethylamide and ^3H-spiroperidol. Mol. Pharmacol. **16**: 687–699.

31. PAZOS, A., R. CORTES & J. M. PALACIOS. 1985. Quantitative autoradiographic mapping of serotonin receptors in the rat brain. II. Serotonin-2 receptors. Brain Res. **346**: 231–249.

32. LEYSEN, J. E., C. J. E. NIEMEGEERS, J. M. VAN NUETEN & P. M. LADURON. 1982. ^3H-Ketanserin (R 41 468), a selective ^3H-ligand for serotonin$_2$ receptor binding sites. Mol. Pharmacol. **21**: 301–314.

33. LEYSEN, J. E. & W. GOMMEREN. 1986. Drug receptor dissociation time, new tool for drug research: Receptor binding affinity and drug-receptor dissociation profiles of serotonin-S_2, dopamine-D_2, histamine-H_1 antagonists, and opiates. Drug Dev. Res. **8**: 119–131.

34. GLENNON, R. A., R. YOUNG & J. A. ROSECRANS. 1983. Antagonism of the effects of the hallucinogen DOM and the purported 5-HT agonist quipazine by 5-HT$_2$ antagonists. Eur. J. Pharmacol. **91**: 189–196.

35. RODRIGUEZ, R., J. ROJAS-RAMERIZ & R. DRUCKER-COLIN. 1973. Serotonin like actions of quipazine on the central nervous system. Eur. J. Pharmacol. **24**: 164–171.

36. FOZARD, J. R. 1983. Functional correlates of 5-HT$_1$ recognition sites. Trends Pharmac. Sci. **4**: 288–289.

37. LEYSEN, J. 1984. Problems in *in vitro* receptor binding studies and identification and role of serotonin receptor sites. Neuropharmacology **23**: 247–254.

38. PEDIGO, N. N., H. J. YAMAMURA & D. L. NELSON. 1981. Discrimination of multiple ^3H-5-hydroxytryptamine binding sites by the neuroleptic spiperone in the rat brain. J. Neurochem. **36**: 220–226.

39. MIDDLEMISS, D. N. & J. R. FOZARD. 1983. 8-Hydroxy-2-(di-*n*-propylamino) tetralin discriminates between subtypes of the 5-HT$_1$ recognition site. Eur. J. Pharmacol. **90**: 151–153.

40. DESHMUKH, P.P., H. J. YAMAMURA, L. WOODS & D. L. NELSON. 1983. Computer-assisted autoradiographic localization of subtypes of serotonin receptors in rat brain. Brain Res. **288**: 338–343.

41. MARCINKIEWICZ, M., D. VERGE, H. GOZLAN, L. PICHAT & M. HAMON. 1984. Autoradiographic evidence for the heterogeneity of 5-HT$_1$ sites in the rat brain. Brain Res. **291**: 159–163.

42. GLASER, T., M. RATH, J. TRABER, K. ZILLES & A. SCHLEICHER. 1985. Autoradiographic identification and topographical analyses of high affinity serotonin receptor subtypes as a target for the novel putative anxiolytic TVX Q 7821. Brain Res. **358**: 129–136.

43. PAZOS, A., D. HOYER & J. M. PALACIOS. 1984. The binding of serotonergic ligands to the porcine choroid plexus: Characterization of a new type of serotonin recognition site. Eur. J. Pharmacol. **106:** 539–546.
44. AHLENIUS, S., K. LARSSON, L. SVENSSON, S. HJORTH & A. CARLSSON. 1981. Effects of a new type of 5-HT receptor agonist on male rat sexual behavior. Pharmacol. Biochem. Behav. **15:** 785–792.
45. HJORTH, S., A. CARLSSON, P. LINDBERG, D. SANCHEZ, H. WILKSTRÖM, L. E. ARVIDSSON, U. HACKSELL & J. L. G. NILSSON. 1982. 8-Hydroxy-2-(di-*n*-propylamino)tetralin, 8-OH-DPAT, a potent and selective simplified ergot congener with central 5-HT receptor stimulating activity. J. Neural Transm. **55:** 169–188.
46. DAVIES, M & M. H. T. ROBERT. 1985. The effects of the 5-HT$_{1A}$ ligand, 8-OH-DPAT, applied iontophoretically to rat brainstem neurons. Br. J. Pharmacol. **86:** 594P.
47. SPENCER, D. G., T. GLASER, T. SCHUURMAN & J. TRABER. 1984. Behavioral and neurochemical correlates of the 5-HT$_1$ receptor. Soc. Neurosci. Abstr. **10:** 1072.
48. GLASER, T., W. U. DOMPERT, T. SCHUURMAN, D. G. SPENCER & J. TRABER. Differential pharmacology of the novel 5-HT$_{1A}$ receptor ligands 8-OH-DPAT, BAY R 1531 and ipsapirone. *In* Brain 5-HT$_{1A}$ Receptors: Behavioural and Neurochemical Pharmacology. C. T. Dourish, S. Ahlenius & P. Hutson, Eds. Ellis Horwood. Chichester. In press.
49. DOMPERT, W. U., T. GLASER & J. TRABER. 1985. ³H-TVX Q 7821: Identification of 5-HT$_1$ binding sites as target for a novel putative anxiolytic. Naunyn-Schmiedeberg's Arch. Pharmacol. **328:** 467–470.
50. GLASER, T. & J. TRABER. 1985. Binding of the putative anxiolytic TVX Q 7821 to hippocampal 5-hydroxytryptamine (5-HT) recognition sites. Naunyn-Schmiedeberg's Arch. Pharmacol. **329:** 211–215.
51. SPROUSE, J. S. & G. K. AGHAJANIAN. 1985. Serotonergic dorsal raphe neurons: Electrophysiological responses in rats to 5-HT$_{1A}$ and 5-HT$_{1B}$ receptor subtype ligands. Soc. Neurosci. Abstr. **11:** 47.
52. TRABER, J., M. A. DAVIES, W. U. DOMPERT, T. GLASER, T. SCHUURMAN & P.-R. SEIDEL. 1984. Brain serotonin receptors as a target for the putative anxiolytic TVX Q 7821. Brain Res. Bull. **12:** 741–744.
53. HOYER, D., G. ENGEL & H. O. KALKMAN. 1985. Characterization of the 5-HT$_{1B}$ recognition site in rat brain: Binding studies with (−)(^{125}I)iodocyanopindolol. Eur. J. Pharmacol. **118:** 1–12.
54. ENGEL, G., M. GÖTHERT, D. HOYER, E. SCHLICKER & K. HILLENBRAND. 1986. Identity of inhibitory presynaptic 5-hydroxytryptamine (5-HT) autoreceptors in the rat brain cortex with 5-HT$_{1B}$ binding sites. Naunyn-Schmiedeberg's Arch. Pharmacol. **332:** 1–7.
55. PAZOS, A. & J. M. PALACIOS. 1985. Quantitative autoradiographic mapping of serotonin receptors in the rat brain. I. Serotonin-1 receptors. Brain Res. **346:** 205–230.
56. EUVRARD, C & J. R. BOISSIER. 1980. Biochemical assessment of the central 5-HT agonist activity of RU 24969 (a piperidinyl indole). Eur. J. Pharmacol. **63:** 65–72.
57. FULLER, R. W., H. D. SNODDY, N. R. MASON, S. K. HEMRICK-LUECKE & J. A. CLEMENS. 1981. Substituted piperazines as central serotonin agonists: Comparative specificity of the postsynaptic actions of quipazine and *m*-trifluoromethylphenylpiperazine. J. Pharmacol. Exp. Ther. **218:** 636–641.
58. HAMON, M., J. M. COSSERY, U. SPAMPINATO & H. GOZLAN. 1986. Are there selective ligands for 5-HT$_{1A}$ and 5-HT$_{1B}$ receptor binding sites in brain? Trends Pharmac. Sci. **7:** 336–338.
59. ORTMANN, R. 1984. The 5-HT syndrome in rats as tool for the screening of psychoactive drugs. Drug Dev. Res. **4:** 593–606.
60. SMITH, L. M. & S. J. PEROUTKA. 1986. Differential effects of 5-hydroxytryptamine$_{1A}$ selective drugs on the 5-HT behavioral syndrome. Pharmacol. Biochem. Behav. **24:** 1513–1519.
61. GOODWIN, G. M., R. J. DE SOUZA & A. R. GREEN. 1986. The effects of a 5-HT$_1$ receptor ligand isapirone (TVX Q 7821) on 5-HT synthesis and the behavioural effects of 5-HT agonists in mice and rats. Psychopharmacology **89:** 382–387.
62. MOORE, R. Y. 1981. The anatomy of central serotonin neuron systems in rat brain. *In* Serotonin Neurotransmission and Behavior. B. L. Jacobs & A. Gelperin, Eds.: 35–71. MIT Press. Cambridge, MA.

The Effects of Anxiolytics and Other Agents on Rat Grooming Behavior[a]

T. W. MOODY,[b] Z. MERALI,[c] AND J. N. CRAWLEY[d]

[b]Department of Biochemistry
The George Washington University
School of Medicine and Health Sciences
Washington, D. C. 20037

[c]School of Psychology
Department of Pharmacology
University of Ottawa
Ottawa, Ontario K1N 6N5

[d]Clinical Neurosciences Branch
The National Institute of Mental Health
Bethesda, Maryland 20892

INTRODUCTION

Numerous peptides are known to alter grooming behavior in mammals. These include ACTH, β-endorphin, and α-MSH, which are derived from the pituitary prohormone pro-opiomelanocortin (POMC) as well as dynorphin.[1-7] In particular, i.c.v. injection of $ACTH_{1-24}$ (2 μg) increases all phases of grooming activity including head washing, body grooming, sexual (anogenital) grooming, paw licking, and scratching.[8] Similar results were observed for α-MSH and β-MSH. Because there is cross tolerance to grooming induced by $ACTH_{1-24}$, dynorphin, and β-endorphin, these peptides may induce grooming by a similar mechanism.[8,9] The neurohypophyseal peptides vasopressin and oxytocin increase grooming activity.[7,10] Intraventricular administration of vasopressin produces a behavioral syndrome that includes hyperactivity, foraging, and increased grooming. Recently, other categories of neuropeptides such as substance P[11] and bombesin (BN) were demonstrated to increase grooming.[12-18] In particular, BN (100 ng, i.c.v.) increased head washing, paw licking, and especially scratching but not body grooming or sexual grooming.[8] Furthermore, there was no cross tolerance between the grooming induced by BN and ACTH. Thus the grooming induced by ACTH, BN, and vasopressin may occur through different neural mechanisms.

It is possible that each of these peptides administered centrally increases stress in animals. Grooming is considered to be a specific rodent behavioral response to stressful situations.[19] While the grooming response varies with the degree of stress, even mild stress, such as placing a rat in a novel environment, causes increased grooming.[3,20,21] The increase in grooming caused by placement in an unfamiliar cage is less than the excessive grooming caused by $ACTH_{1-24}$ or BN.[12] We recently found that the increased

[a] This research was supported by NSF grant no. BNS-8500552 to TWM and Natural Sciences and Engineering Research Council of Canada and Medical Research Council grants to ZM.

grooming caused by a novel environment, $ACTH_{1-24}$, or BN is blocked by anxiolytics such as diazepam.[12] Here we further explore other anxiolytic agents that block the increased grooming caused by BN.

EFFECTS OF ANXIOLYTICS ON GROOMING BEHAVIOR

Previously, Dunn *et al.* reported that diazepam reduced the increased grooming in mice caused by a novel environment.[21] We investigated the effects of anxiolytics on rat grooming activity.[12] Drug injections and behavioral testing began 1 week after implantation of a stainless steel guide cannulae into the left lateral ventricle. Peptides were injected intraventricularly in 5 µl of saline 15 min before the start of behavioral testing. Antianxiety agents were injected i.p. 30 min prior to behavioral testing using a vehicle of 2% ethanol, 2% propylene glycol, and 96% physiological saline. Grooming behavior was scored every 15 s for a 30-min period. All forms of grooming were considered as equivalent, and because each rat was scored as either grooming or not grooming, a maximum score of 120 was possible. Each animal was used up to four times in a modified Latin Squares design. At least 2 days intervened between each treatment. TABLE 1 shows that the basal level of grooming was 11 ± 5. A novel environment, $ACTH_{1-24}$ (1 µg, i.c.v.), and BN (1 µg, i.c.v.) increased the grooming score to 21, 50, and 40, respectively. Diazepam (1 mg/kg, i.p.) significantly reversed the increased grooming caused by each of the three treatments, whereas it did not significantly affect the saline control. Also, diazepam did not significantly affect locomotor or exploratory behavior as quantitated on a video-tracking computer-assisted behavior monitor.[22] Thus the effect of diazepam does not appear to be due to sedation of the animals.

The effects of other antianxiety agents were investigated. TABLE 1 shows that meprobamate (50 mg/kg, i.p.) significantly reduced the increased grooming caused by ACTH (Newman Keuls $p < 0.01$). In addition, FIGURE 1 shows that BN significantly increased the grooming score (54 ± 12) relative to the saline control (8 ± 3). Chlordiazepoxide (5 mg/kg, i.p.) significantly blocked the BN-induced grooming (ANOVA $F4,29 = 6.6$, $p < 0.01$). Rats treated with chlordiazepoxide + BN had grooming scores significantly lower than rats treated with BN (Newman Keuls $p < 0.01$). Rats treated with chlordiazepoxide + BN had grooming scores that were not significantly different from the

TABLE 1. Effects of Anxiolytics on Grooming

Agent	Grooming Score
Saline, home cage	11 ± 5
Saline, unfamiliar cage	21 ± 5
Diazepam, unfamiliar cage	3 ± 2 *
$ACTH_{1-24}$	50 ± 10*
$ACTH_{1-24}$ + diazepam	22 ± 5 *
$ACTH_{1-24}$ + meprobamate	3 ± 1 *
BN	40 ± 13*
BN + diazepam	6 ± 4 *

The mean value ± SE of eight determinations is indicated. Saline or peptides were injected i.c.v. into rats pretreated with diazepam (1 mg/kg, i.p.), meprobamate (50 mg/kg), or vehicle. The number of total grooming episodes in 30 min is indicated. $ACTH_{1-24}$ (1 µg) and BN (1 µg) significantly increased grooming relative to the saline control. Diazepam and meprobamate significantly reduced the increased grooming caused by $ACTH_{1-24}$, BN, or placement in an unfamiliar cage (*$p < 0.05$).

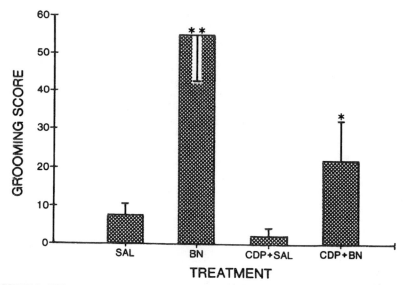

FIGURE 1. Effect of chlordiazepoxide (CDP) on grooming activity. The effects of chlordiazepoxide (5 mg/kg, i.p.) on basal and BN-stimulated grooming activity are shown. The mean value + SE of eight determinations is indicated. BN significantly increased the grooming score relative to the saline (SAL) control (**$p < 0.01$). Chlordiazepoxide significantly decreased the BN-induced grooming (*$p < 0.05$).

grooming scores of rats treated with saline control. In rats treated at the same doses and placed in an Omnitech optical activity monitor, chlordiazepoxide did not produce a significant effect on general locomotor activity as compared to vehicle treatment ($t = 2.13$, $df = 6$, ns): vehicle = 854 ± 196, chlordiazepoxide = 467 ± 84. This supported previous findings that anxiolytics do not appear to reduce grooming activity because of sedation. Thus diazepam, meprobamate, and chlordiazepoxide reverse the increased grooming caused by placement in an unfamiliar cage, BN, and/or ACTH.

The effects of antianxiety agents are well established in several conflict tests,[23-27] social interactions,[28] and exploratory behaviors.[29,30] Anxiety and stress appear to be related phenomena on many measures of biological response.[31,32] Our results suggest

TABLE 2. Pharmacology of [³H]Flunitrazepam Binding

Additions	cpm Bound
None	$13,776 \pm 261$
Diazepam (1 μM)	6660 ± 173
Chlordiazepoxide (1 μM)	6393 ± 162
Bombesin (1 μM)	$14,313 \pm 492$
[D-Phe-12]BN (10 μM)	$14,757 \pm 260$
Spantide (10 μM)	$13,286 \pm 409$

The mean value ± SD of three determinations is indicated. Rat brain homogenate (1 mg protein) was incubated with [³H]flunitrazepam (2.4 nM) for 60 min at 4°C in 50 mM Tris HCl (pH 7.4). Bound peptide was separated from free by filtration as described previously.[33]

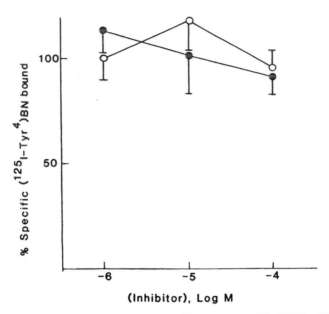

FIGURE 2. Effect of anxiolytics on BN binding. The percent specific [^{125}I-Tyr-4]BN bound is indicated as a function of increasing concentrations of diazepam (●) and chlordiazepoxide (O). The mean value + SE of three determinations, each performed in triplicate, is indicated.

that antianxiety compounds can reverse the behavioral actions of peptides that may be stress-related.

It is possible that peptides such as BN and drugs such as diazepam are antagonistic to one another. Therefore we investigated if peptides such as BN inhibit binding to the central benzodiazepine receptors. TABLE 2 shows that [^{3}H]flunitrazepam bound with high affinity to rat brain homogenate. In the absence of excess competitor, 13,776 cpm total bound to the brain membranes, whereas using excess competitor (1 μM diazepam or chlordiazepoxide) 6660 and 6393 cpm, respectively, bound nonspecifically to the brain membranes. The difference between the two represents specific binding to rat brain membranes. Also, TABLE 2 shows that BN (1 μM) did not significantly affect binding to the homogenate. Similarly, [D-Phe-12]BN and spantide (two pep-

TABLE 3. Regional Distribution of Dopamine following Unilateral MFB 6-OHDA Lesion

Region	Control Side (% of Control)	Lesioned Side (% of Control)
Nucleus accumbens	100 ± 19	25 ± 6
Caudate putamen	100 ± 14	17 ± 10
Central amygdaloid nucleus	100 ± 29	38 ± 8
Olfactory tubercle	100 ± 12	15 ± 3

The mean value ± SD of five determinations is indicated. The brain regions were dissected and extracted and the dopamine assayed using the electrochemical detector after fractionation by high-pressure liquid chromatography.

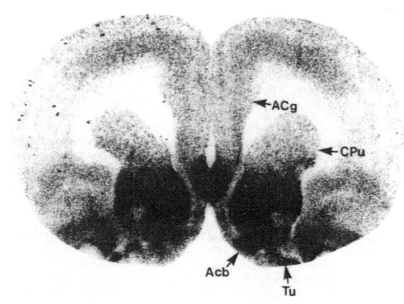

FIGURE 3. Distribution of receptors for BN-like peptides as determined by using *in vitro* autoradiographic techniques. Fresh frozen rat brain was dissected and coronal slices derived from Paxinos and Watson[44] coordinates at 11.2 mm were thaw mounted onto slides. The slides were incubated with [^{125}I-Tyr-4]BN (0.1 nM) in assay buffer for 45 min at 25°C as described previously.[45] Free peptide was removed by two consecutive washes in buffer at 4°C. The tissue-mounted slides containing bound peptide were then exposed to LKB Ultrafilm for 1 week. The dark areas indicate high receptor density. The left side was lesioned with 6-OHDA unilaterally. Densitometry analyses indicated that the receptor density was decreased approximately 5% on the left relative to the right side in the nucleus accumbens. Abbreviations: *Acb*, nucleus accumbens; *ACg*, anterior cingulate cortex; *CPu*, caudate putamen; *Tu*, olfactory tubercle.

tides that antagonize BN-induced grooming) at a 10 μM dose did not significantly affect the [^{3}H]flunitrazepam binding. Therefore, BN does not exert its effect at the level of the antianxiety binding site.

Conversely, we investigated if antianxiety drugs act directly on the BN receptor. Previously, we determined that [^{125}I-Tyr-4]BN, a potent BN analogue, served as a good receptor probe.[34] Specifically, [^{125}I-Tyr-4]BN binds with high affinity ($K_d = 4$ nM) to a single class of sites (80 fmol/mg protein) using rat brain homogenate. FIGURE 2 shows that diazepam or chlordiazepoxide do not affect specific binding of radiolabeled [Tyr-4]BN even at high concentrations of ligand such as 10^{-4} M. Therefore antianxiety drugs do not appear to exert their effect at the level of the receptor for BN-like peptides.

INTERACTION OF BN WITH DOPAMINE

It is generally accepted that grooming behavior may be mediated by certain neurotransmitters. Because ACTH-induced grooming is blocked by haloperidol and fluphenazine, ACTH may mediate its grooming effect through dopamine-containing neurons.[35,36]

Also, novelty-induced grooming is blocked by dopamine receptor antagonists as well as by clonidine and amitriptyline, suggesting involvement of both dopamine and noradrenergic systems.[3] Similarly, because novelty-induced grooming as well as the excessive grooming induced by $ACTH_{1-24}$ is blocked by naloxone or naloxozone, opiate receptors may be involved.[35,37,38] Therefore ACTH- and novelty-induced grooming may be mediated through dopamine, norepinephrine, and/or opiate receptor ligands.

BN-induced grooming is reversed by benzomorphans,[39] diazepam,[12] and recently described BN-receptor antagonists such as spantide.[40] In contrast, BN-induced grooming was not affected by atropine, diphenylhydramine, morphine, naloxone, or neurotensin.[14,41] Recently, the BN-induced grooming was associated with dopamine-containing neurons. Fluphenazine (0.1 mg/kg) and haloperidol (0.1 mg/kg) significantly reduced BN-induced grooming but did not significantly affect basal grooming activity.[42] At higher doses of dopamine receptor antagonist, however, such as 1 mg/kg, basal grooming activity was also affected. Because haloperidol (1 μM) does not affect binding to central receptors for BN-like peptides,[34] BN may exert its action on grooming utilizing traditional dopamine-containing neurons.

To investigate the role of dopamine on grooming, dopamine-containing neurons were lesioned using 6-OHDA. The medial forebrain bundle in rats was injected with 6-OHDA (8 μg/2 μl in 0.1% ascorbic acid, 30 min after pretreatment with pargyline, 50 mg/kg, i.p.) and desipramine (25 mg/kg, i.p.). After 12 days, the 6-OHDA-treated animals were compared to sham-operated controls for locomotion, floor, and rearing activity. BN (0.1 μg, i.c.v.) significantly increased locomotion, floor, and rearing activity in sham- but not 6-OHDA-lesioned animals.[43] Thus, for BN to cause its effect on grooming, an intact dopamine system may be required.

An important question concerns what brain regions may be required for BN to cause increased grooming activity. TABLE 3 shows that dopamine levels were reduced in numerous brain regions such as the nucleus accumbens, caudate putamen, central amygdaloid nucleus, and olfactory tubercle by approximately 75% after 6-OHDA injection into the medial forebrain bundle. The receptor density was then determined using *in vitro* autoradiographic techniques. FIGURE 3 shows that there is a high density in receptors for BN-like peptides in the nucleus accumbens as well as the olfactory tubercle and a moderate density in the cingulate and rhinal cortex as well as the ventral caudate putamen. The density of receptors for BN-like peptides was quantitated using a densitometer and found to be reduced by approximately 5% in the left (lesioned) relative to right (control) nucleus accumbens even though there was a 75% decrease in dopamine in the left relative to right nucleus accumbens. Similarly, the density of receptors for BN-like peptides was not affected in the olfactory tubercle and caudate putamen even though the dopamine levels decreased by approximatley 85% in these brain regions.

These data indicate that the majority of receptors for BN-like peptides are not present on dopamine-containing neurons in the nucleus accumbens, caudate putamen, or olfactory tubercle. This is surprising, as BN causes potent behavioral activation after direct injection into the nucleus accumbens. Also, BN causes an increase in the dopamine metabolites homovanillic acid (HVA) and 3,4-dihydroxyphenylacetic acid (DOPAC) in the hypothalamus, striatum, and olfactory tubercle. FIGURE 4 shows that there is also a high density of receptors for BN-like peptides in the hippocampus, rhomboid thalamic nucleus, amygdala, and central medial thalamic nucleus and a moderate density in the parietal cortex, neocortex, and paraventricular hypothalamic nucleus. Surprisingly, the density of receptors for BN-like peptides significantly decreased, by approximately 20%, in the central amygdaloid nucleus where the dopamine levels decreased by approximately 60%. Thus an appreciable number of BN receptors are present on

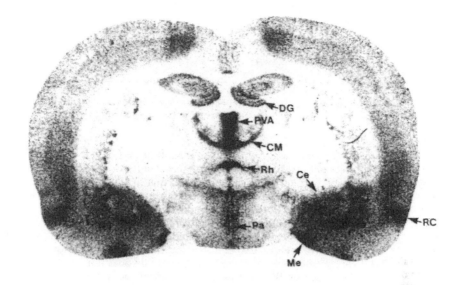

FIGURE 4. Distribution of receptors for BN-like peptides. The coronal section was derived from Paxinos and Watson[44] coordinates at 7.2 mm. The receptor density decreased approximately 20% on the 6-OHDA lesioned side (left) relative to the control side in the central amygdaloid nuclei. Abbreviations: *Ce*, central amygdaloid nuclei; *Pa*, paraventricular hypothalamic nuclei; *CM*, central medial thalamic nuclei; *DG*, dentate gyrus; *Me*, medial amygdaloid nuclei; *PVA*, paraventricular thalamic nuclei; *RC*, rhinal cortex; and *Rh*, rhomboid thalamic nuclei.

dopamine-containing neurons in the central amygdaloid nucleus. These BN receptors may be important in mediating the increased grooming caused by BN after injection into the lateral ventricle. Because BN has a blocked NH_2- and COOH-terminal and is not readily degraded by brain exopeptidases, it may diffuse and activate receptors far distant from the lateral ventricle where it is injected.

EFFECTS OF BN RECEPTOR ANTAGONISTS ON GROOMING

The physiological importance of a neurotransmitter or neuromodulator can be evaluated with the development of specific receptor antagonists. Putative substance P antagonists such as [D-Arg-1, D-Pro-2, D-Trp-7,9, Leu-11] substance P and [D-Arg-1, D-Trp-7,9, Leu-11] substance P (spantide) reverse the BN-stimulated amylase secretion from dispersed pancreatic acini derived from the guinea pig.[46] Also, these substance P analogues inhibit the binding of radiolabeled [Tyr-4]BN to exocrine pancreas receptors.[47] Here the effects of spantide on rat grooming behavior were investigated.

TABLE 4 shows that spantide reversed the increase in grooming behavior caused by i.c.v. injection of BN. In contrast, spantide (2 µg) did not significantly affect the basal grooming activity (7 ± 5). TABLE 5 shows that spantide also inhibited binding to central receptors for BN-like peptides (IC_{50} = 2 µM). Because BN had an IC_{50} of

TABLE 4. Effect of BN-Receptor Antagonists on Grooming Behavior

Agent	Grooming Score
None	8 ± 2
BN (1 µg)	75 ± 8**
BN (1 µg) + spantide (2 µg)	14 ± 4*
BN (1 µg) + [D-Phe-12]BN (10 µg)	28 ± 8*

The mean value ± SE of eight determinations is shown (**$p < 0.01$ relative to the saline control). Spantide (2 µg, i.c.v.) and [D-Phe-12]BN (10 µg, i.c.v.) significantly inhibited the BN-induced grooming (*$p < 0.05$), but had no effect on the basal grooming rate.

20 nM, spantide was approximately two orders of magnitude less potent than was BN. Thus spantide inhibits binding to central BN receptors and grooming activity, even though it does not have a primary sequence similar to that of BN. Recently [D-Arg-1, D-Pro-2, D-Trp-7,9, Leu-11] substance P was also demonstrated to inhibit [^{125}I-Tyr-4]BN binding to Swiss 3T3 cells and BN-stimulated uptake of [^{3}H]thymidine.[48,49]

Care must be taken when using spantide *in vivo*, however, as it may cause barrel rotations and death.[50] Because peptides such as spantide are highly enriched in D-amino acids, they cannot readily be metabolized and removed from the body. Thus they may accumulate in the body, causing toxic effects. Also spantide does not antagonize all biological responses caused by BN. In this regard, spantide does not affect the decrease in gastric acid secretion in rats caused by BN.[51] For this reason we developed more specific BN receptor antagonists.

TABLE 4 shows that [D-Phe-12]BN also inhibits the increase in grooming caused by BN. The grooming score for [D-Phe-12]BN (10 µg) alone was 4 ± 4, thus this peptide had no significant effect on the basal grooming activity. No toxic behavioral effects were observed using this compound at a 10 µg dose. Also, TABLE 5 shows that [D-Phe-12]BN inhibited binding of [^{125}I-Tyr-4]BN with an IC$_{50}$ of 2 µM. Thus [D-Phe-12]BN may function as a central BN receptor antagonist. Similarly, [D-Phe-12]BN inhibits binding of radiolabeled [Tyr-4]BN to and BN-stimulated amylase secretion from dispersed pancreatic acini.[52]

SUMMARY

BN and other peptides increase grooming activity in rodents. This increase in grooming activity caused by BN can be reversed by numerous agents. The most direct of these are BN receptor antagonists such as spantide and [D-Phe-12]BN. These agents reverse the increase in grooming activity caused by BN and inhibit binding to central BN receptors.

TABLE 5. Effect of BN-Receptor Antagonists on Receptor Binding

Agent	IC$_{50}$, µM
Spantide	2
[D-Phe-12]BN	2

[^{125}I-Tyr-4]BN (0.5 nM) was bound to rat brain homogenate at 4°C for 20 min in 50 mM Tris HCl, pH 7.0, that contained 0.5% BSA and 200 µg/ml bacitracin in the presence or absence of varying concentrations of competitor. Bound peptides were separated from free using the filtration techniques described previously.[34]

The increase in grooming activity caused by BN may be mediated by other neu-rotransmitters, such as dopamine. In this regard the increase in grooming activity caused by BN is reversed by fluphenazine and haloperidol. These agents do not affect binding to central BN receptors. Similarly, the increase in grooming caused by BN is absent if dopamine-containing neurons are lesioned with 6-OHDA. These lesions cause a sig-nificant reduction in central BN receptors, especially in the central amygdaloid nu-cleus. Thus some of the BN receptors may be present on dopamine-containing neurons.

The increase in grooming activity caused by BN was also reversed by diazepam and chlordiazepoxide. Neither of these agents affected binding to central BN receptors, and similarly, BN did not affect binding to central benzodiazepine receptors. Thus anxiolytics may act at a site anatomically downstream from the BN receptor and re-duce the apparent stress caused by central injection of BN.

ACKNOWLEDGMENTS

The authors thank Drs. R. Jensen for helpful discussions and C. Merchant for technical assistance.

REFERENCES

1. DUNN, A. J., S. R. CHILDERS, N. R. KRAMARCY & J. L. VILLEGAR. 1981. Behav. Neural Biol. **31:** 105–109.
2. DUNN, A. J., E. J. GREEN & R. L. ISAACSON. 1979. Science **203:** 281–283.
3. GISPEN, W. H. & R. L. ISAACSON. 1981. Pharmacol. Ther. **12:** 209–246.
4. GISPEN, W. H., V. M. WIEGANT, A. F. BRADBURY, E. C. HULME, D. G. SMYTH, C. R. SNELL & D. DE WIED. 1976. Nature **264:** 794–795.
5. GISPEN, W. H., V. M. WIEGANT, H. M. GREVEN & D. DE WIED. 1975. Life Sci. **17:** 645–652.
6. O'DONOHUE, T. L., G. E. HANDELMANN, Y. P. LOH, D. S. OLTON, J. LEIBOWITZ & D. M. JACOBOWITZ. 1981. Peptides **2:** 101–104.
7. REES, H. D., A. J. DUNN & P. M. IUVONE. 1976. Life Sci. **18:** 1333–1340.
8. VAN WIMERSMA GREIDANUS, TJ. B., D. K. DONKER, F. F. M. VAN ZINNIC BERGMANN, R. BEKENKAMP, C. MAIGRET & B. SPRUIJT. 1985. Peptides **6:** 369–372.
9. ALOYO, V. J., B. SPRUIJT, H. ZWIERS & W. H. GISPEN. 1983. Peptides **4:** 1–4.
10. DELANOY, R. L., A. J. DUNN & R. TINTNER. 1978. Horm. Behav. **11:** 348–363.
11. KATZ, R. 1979. Neuropharmacology **19:** 143–146.
12. CRAWLEY, J. N. & T. W. MOODY. 1983. Brain Res. Bull. **10:** 399–401.
13. GIRARD, F., C. AUBE, S. ST.-PIERRE & F. B. JOLICOEUR. 1983. Neuropeptides **3:** 443–452.
14. GMEREK, D. E. & A. COWAN. 1983. Peptides. **4:** 907–913.
15. KATZ, R. 1980. Neuropharmacology **19:** 143–146.
16. KULKOSKY, P. J., J. GIBBS & G. P. SMITH. 1982. Physiol. Behav. **28:** 505–512
17. KULKOSKY, P. J., J. GIBBS & G. P. SMITH. 1982. Brain Res. **242:** 194–196.
18. MERALI, Z., S. JOHNSTON & S. ZALCMAN. 1983. Peptides **4:** 693–697.
19. EISENBERG, J. F. A. 1967. Proc. U.S. Natl. Mus. **122:** 1–51.
20. COLBERN, D. L., R. L. ISAACSON, E. J. GREEN & W. H. GISPEN. 1978. Behav. Biol. **23:** 381–387.
21. DUNN, A. J., A. L. GUILD, N. R. KRAMARCY & M. D. WARE. 1981. Pharmacol. Biochem. Behav. **15:** 605–608.
22. CRAWLEY, N. H., S. SZARA, G. T. PRYOR, C. R. CREVELING & B. K. BERNARD. 1982. J. Neu-rosci. Methods **5:** 235–247.
23. GELLER, I. & J. SEIFTER. 1960. Psychopharmacology (Berlin) **1:** 482–493.
24. LIPPA, A. S., C. A. KEPNER, M. C. YUNGER, W. V. SANO, W. F. SMITH & B. BEER. 1978. Pharmacol. Biochem. Behav. **9:** 853–856.

25. MARGULES, D. L. & L. STEIN. 1968. Psychopharmacology (Berlin) 13: 74–80.
26. PATEL, J. B. & J. B. MALICK. 1981. Biochem. Behav. 12: 819–821.
27. VOGEL, J. R., B. BEER & D. E. CLODY. 1971. Psychopharmacology (Berlin) 21: 1–7.
28. FILE, S. E. & J. R. G. HYDE. 1979. Pharmacol. Biochem. Behav. 11: 65–69.
29. CRAWLEY, J. N. 1981. Pharmacol. Biochem. Behav. 15: 695–699.
30. NOLAN, N. A. & M. W. PARKES. 1973. Psychopharmacology (Berlin) 29: 277–288.
31. HENNESSY, J. W. & S. LEVIN. 1979. In Progress in Psychology and Physiological Psychology. Academic Press. New York. p. 133.
32. FILE, S. E. & J. B. PEARSE. 1981. Br. J. Pharmacol. 74: 593–599.
33. SKOLNICK, P., M. SCHWERI, E. KUTTER, E. WILLIAMS & S. PAUL. 1982. J. Neurochem. 39: 1142–1146.
34. MOODY, T. W., C. B. PERT, J. RIVIER & M. R. BROWN. 1978. Proc. Natl. Acad. Sci. U.S.A. 75: 5372–5376.
35. GREEN, E. J., R. L. ISAACSON, A. J. DUNN & T. H. LANTHORN. 1979. Behav. Neural Biol. 27: 546–551.
36. WIEGANT, V. M., A. R. COOLS & W. H. GISPEN. 1977. Eur. J. Pharmacol. 41: 343–345.
37. DUNN, A. J., S. R. CHILDERS, N. R. KRAMARCY & J. W. VILLIGER. 1981. Behav. Neural Biol. 31: 105–109.
38. GREEN, E. J., R. L. ISAACSON, A. J.. DUNN & T. H. LANTHORN. 1979. Behav. Neural Biol. 27: 546–551.
39. GMEREK, D. E. & A. COWAN. 1982. Life Sci. 31: 2229–2235.
40. YACNHIS, A. T., J. N. CRAWLEY, R. T. JENSEN, M. M. McGRANE & T. W. MOODY. 1984. Life Sci. 35: 1963–1969.
41. SCHULZ, D. W., P. W. KALIVAS, C. B. NEMEROFF & A. J. PRANGE, JR. 1984. Brain Res. 304: 377–382.
42. JOHNSTON, S., P. PARMASHWAR & Z. MERALI. 1986. In Neural and Endocrine Peptides and Receptors. T. W. Moody, Ed.: 389–402. Plenum. New York.
43. MERALI, Z., S. JOHNSTON & J. SISTEK. 1985. Pharmacol. Biochem. Behav. 23: 243–248.
44. PAXINOS, G. & C. WATSON. 1982. The Rat Brain in Stereotaxic Coordinates. Academic Press. New York.
45. ZARBIN, M. A., M. J. KUHAR, T. L. O'DONOHUE, S. S. WOLF & T. W. MOODY, 1985. J. Neurosci. 5: 429–437.
46. JENSEN, R. T., S. W. JONES, Y. A. LU, S. C. XU, K. FOLKERS & J. D. GARDNER. 1984. Biochem. Biophys. Acta 804: 181–191.
47. JENSEN, R. T., S. W. JONES, K. FOLKERS & J. D. GARDNER. 1984. Nature 309: 61–63.
48. ZACHARY, I. & E. ROZENGURT. 1985. Proc. Natl. Acad. Sci. U.S.A. 82: 7616–7620.
49. CORPS, A. N., L. H. REES & K. D. BROWN. 1985. Biochem. J. 231: 781–784.
50. COWAN, A., P. KHUNAWAT, X. Z. ZHU & D. E. GMEREK. 1985. Life Sci. 37: 135–144.
51. TACHE, Y. & M. GUNION. 1985. Life Sci. 37: 115–123.
52. MOODY, T. W., C. A. MERCHANT, R. T. JENSEN, P. HEINZ-ERIAN, D. H. COY, V. LANCE, M. TAMURA & J. N. CRAWLEY. 1986. Soc. Neurosci. Abstr. 12: 372.

Role of Opioid Receptors in Bombesin-induced Grooming[a]

DEBRA E. GMEREK[b] AND A. COWAN[c]

[b]Department of Pharmacology
University of Michigan Medical School
Ann Arbor, Michigan 48109-0010

[c]Department of Pharmacology
Temple University School of Medicine
Philadelphia, Pennsylvania 19140

INTRODUCTION

Bombesin, a tetradecapeptide originally isolated from the skin of the frog *Bombina bombina*,[1] elicits dose-related excessive grooming when injected centrally in rats[2-4] as well as other species.[5] The grooming induced by intracerebroventricular (i.c.v.) bombesin in rats consists of hindlimb scratching directed primarily at the head and neck, although facial grooming, body washing, nail licking and biting, forepaw tremors, wet-dog shakes, and stretching also occur. Bombesin-induced grooming behavior can be observed within a few minutes of i.c.v. injection and continues for 30–45 minutes.

In contrast to the grooming induced by a variety of other peptides such as ACTH$_{1-24}$, thyrotropin-releasing hormone, substance P, β-endorphin, and nonpeptides such as RX 336-M (7,8-dihydro-5',6'-dimethylcyclohex-5'-eno-1',2',8',14 codeinone), bombesin-induced grooming is not inhibited by nonsedative doses of morphine.[6-11]

In keeping with multiple types of cholinergic, adrenergic, histaminic, and dopaminergic receptors, it is also well established that there are multiple types of opioid receptors. Based on observations in dogs, the existence of three types of opioid receptors was postulated: mu (for which morphine was the prototype), kappa (for which ketocyclazocine was the prototype) and sigma (for which *N*-allyl-normetazocine was the prototype).[12,13] Since then, mu and kappa types of opioid receptors have been differentiated by biochemical studies using receptor binding and visualization techniques (see, for example, refs. 14 and 15) and by pharmacological studies in which differences in acute effects, differences in the ease of antagonism by the opioid-specific antagonist naloxone, and the selective development of tolerance and dependence were identified (see refs. 16–18). Whereas sigma receptors are no longer considered to be opioid (*i.e.*, naloxone is not an antagonist of effects mediated by the "sigma" receptor), a third type of opioid receptor has been identified primarily through *in vitro* studies, the delta receptor.[19] The endogenous enkephalins have been proposed as natural ligands for the delta receptor.

In light of the multiple classes of opioids, we decided to examine opioids other than morphine for their ability to inhibit bombesin-induced grooming in rats. In our

[a] This work was supported by NIH grant no. BRSG S07 RR05417 and USPHS grant nos. DA 03681 and DA 00254.

preliminary studies,[8,9,20] it became evident that ethylketazocine (a compound structurally and pharmacologically related to ketazocine) could antagonize bombesin-induced grooming. It was therefore decided to study further the role of opioid receptors in the modulation of bombesin-induced grooming.

Ethylketocyclazocine, and more recently tifluadom and U-50,488 (*trans*-3,4-dichloro-*N*-methyl-*N*-[2-(pyrrolidinyl)cyclohexyl-1]benzeneacetamide), have been used as prototypical kappa agonists. Ethylketazocine is a benzomorphan showing selectivity for kappa receptors in binding assays as well as in most tests *in vitro* and *in vivo*. It can act through mu receptors, however, to produce hot water tail-withdrawal analgesia and inhibit gastrointestinal transit in rats.[21] Tifluadom is a very interesting compound in that it has the structure of a benzodiazepine but little or no activity at benzodiazepine receptors; it has been found to be very selective for the kappa type of opioid receptor, however.[22,23] U-50,488 is a compound of novel structure for opioids.[24] In binding studies, U-50,488 has a selectivity of approximately 250-fold for kappa over mu or delta binding sites in quinea pig brain.[25] It has a strictly kappa profile of activity both *in vitro* and *in vivo* (see, for example, refs. 18, 24, 26, and 27).

In this work, the role of opioid receptors in bombesin-induced grooming was investigated by means of classical methods in opioid pharmacology: by classification of opioid-like compounds on the basis of their ability to inhibit the grooming elicited by a standard submaximal dose of bombesin; by examination of the stereospecificity of the interaction between the opioid and bombesin; by examination of the sensitivity of the opioids to naloxone; and through tolerance and cross tolerance studies.

METHODS

Animals and Surgery

Male Sprague-Dawley albino rats weighing 180–200 g were each implanted with a stainless steel cannula (Plastic Products Co., Roanoke, VA) directed towards the lateral cerebral ventricle as described previously.[4] The rats were housed individually at 23 ± 1°C with free access to food and water for at least 4–7 days before testing. They were exposed to a 12-h timer-regulated light period from 7 A.M. to 7 P.M.

General Procedure

Experiments were carried out in the afternoons. At least 15 min before testing, the rats were transfered from their home cages and placed individually in Plexiglas observation boxes (22 cm long; 18 cm wide; 25 cm high). Four rats were monitored at a time with the aid of a portable microcomputer as described previously.[11,28] Test compounds, with the exception of dynorphin A, were injected 15 min prior to a standard submaximal groom-inducing dose of bombesin[4] (0.1 µg, i.c.v.). Dynorphin A and [D-Pen-2,D-Pen-5]enkephalin were given i.c.v. 5 min before bombesin (0.1 µg, i.c.v.).

Quantification of Behavior

Beginning immediately after the last bombesin or i.c.v. saline control injection, each rat was observed for 5 s out of every 20 s for a total of 30 min. A positive grooming score was given if the rat demonstrated any type of grooming behavior (*e.g.*, scratching, washing, licking, biting, etc.) during the 5-s observation period. Consequently, there was a total of 90 grooming episodes scored.

Results are presented as the %MGE (maximum number of grooming episodes scored as positive) ± SEM. The %MGE was calculated for each rat (*i.e.*, number of positive grooming scores times 100, divided by 90). Groups of at least five rats were used to determine each data point. Intracerebroventricular saline control animals had %MGE scores of 6 ± 1. If, following the pretreatment with a test compound and bombesin administration, the majority of rats had %MGE of less than 6, their grooming behavior was said to be suppressed. Compounds were said to have the ability to antagonize bombesin-induced grooming only if they inhibited the scratching behavior in a dose-related manner at doses that did not suppress grooming behavior below i.c.v. saline controls.

Absolute grooming scores were analyzed statistically by means of ANOVA followed by Dunnett's test or the Mann-Whitney U-test, as appropriate.[29] Percent inhibition of bombesin-induced grooming was calculated as follows:

$$1 - \frac{(\text{bombesin grooming score in opioid-pretreated rats } - \text{ saline controls})}{(\text{bombesin grooming score in saline-pretreated rats } - \text{ saline controls})} \times 100.$$

A_{50} values (*i.e.*, doses at which test compounds antagonize an effect by 50%) were determined by linear regression analysis[30] of absolute data from which saline control values had been subtracted. Confidence limits (95%) of A_{50} values were also determined.

Compounds and Injections

The following compounds were used: azidomorphine bitartrate (J. Knoll, Semmelweise University of Medicine, Budapest, Hungary), bombesin (Sigma and Boehringer Mannheim, Indianapolis, IN), bremazocine HCl (D. Romer, Sandoz, Basel, Switzerland), buprenorphine HCl (National Institute on Drug Abuse, Rockville, MD), codeine phosphate (Merck Sharp & Dohme, West Point, PA), cyclazocine (Sterling-Winthrop, Rensselaer, NY), dextrorphan (NIDA), dynorphin A (Penninsula Laboratories, Belmont, Calif.), β-endorphin (A. A. Manian, National Institute of Mental Health, Bethesda, MD), [D-Pen-2,D-Pen-5]enkephalin (DPDPE, Penninsula), ethylketazocine methanesulfonate (EKC, Sterling-Winthrop), d-EKC and l-EKC (Sterling-Winthrop), ethylmorphine HCl (Merck Sharp & Dohme), heroin HCl (NIDA), ketocyclazocine methanesulfonate (Sterling-Winthrop), levorphanol tartrate (Hoffmann-La Roche, Nutley, NJ), meperidine HCl (Sterling-Winthrop), methadone HCl (NIDA), metkephamid (R. C. A. Frederickson, Eli Lilly Co., Indianapolis, IN), morphine sulfate (Mallinckrodt Inc., St. Louis, MO), nalbuphine HCl (Endo Laboratories, Inc., Garden City, NY), nalorphine HCl (Sigma), d-naloxone HCl and l-naloxone HCl (A. Jacobson, National Institutes of Health, Bethesda, MD), naloxone HCl (Endo), l-pentazocine (Sterling-Winthrop), phenazocine HBr (Smith Kline & French, Philadelphia, PA), quadazocine methanesulfonate (also known as Win 44,441-3; Sterling-Winthrop), thebaine (NIDA), racemic tifluadom, (+)- and (−)-tifluadom (referring to the optical rotation as measured in ethanol; Kali-Chemi AG, Hannover, FRG), U-50,488 (*trans*-3,4-dichloro-*N*-methyl-*N*-[2-(pyrolidinyl)cyclohexyl-1]benzeneacetamide) (J. Collins, Upjohn Co., Kalamazoo, MI), xorphanol mesylate (H. G. Pars, Pharmaceutical Laboratories, Inc., Cambridge, MA), and zomepirac (McNeil Laboratories, Fort Washington, PA).

The alkaloids were dissolved in a minimal volume of glacial acetic acid, if necessary, and diluted with saline. Tifluadom (racemic and optical isomers) were dissolved in a solution of 40% propylene glycol, 10% ethanol, 5% sodium benzoate and ben-zoic acid, and 15% benzyl alcohol by sonication and heat. Injections were given s.c. in volumes of 1 ml/kg with doses calculated as the base or salt as indicated above; the dose of *d*- and *l*-naloxone was calculated as the free base. Aliquots of bombesin, DPDPE, dynorphin A, and β-endorphin were each dissolved in saline daily as needed. Bombesin, DPDPE, dynorphin A and β-endorphin were injected i.c.v. to hand-held, conscious rats in volumes of 3–4 μl followed by 0.5–1 μl saline wash.

RESULTS

Opioid Classification

The following compounds given as 15 min s.c. pretreatments to a standard 0.1 μg dose of i.c.v. bombesin did not attenuate bombesin-induced grooming at nonsedative doses, *i.e.*, doses that did not suppress grooming behavior below i.c.v. saline controls (highest nonsedative dose used in mg/kg, s.c., in parentheses): azidomorphine (0.05), buprenorphine (0.5), codeine (40), dextrorphan (30), ethylmorphine (100), heroin (0.5), levorphanol (1), meperidine (25), methadone (1), metkephamid (30), morphine (10), nalbuphine (10), naloxone (10), and thebaine (25). β-Endorphin (10 μg, i.c.v.) given

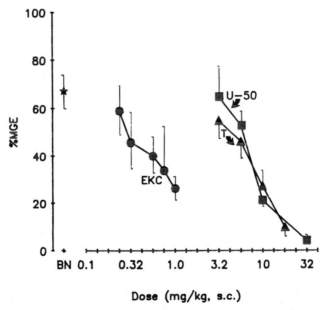

FIGURE 1. The effect of 15-min pretreatment with ethylketazocine methanesulfonate (*EKC*), U-50,488 (*U-50*), or tifluadom (*T*) on the grooming induced by bombesin (0.1 μg, i.c.v.). %*MGE* is the percentage of the maximum number of grooming episodes scored as positive during the 30-min observation session. Data points indicate the mean ± SE for 5–8 rats. *BN*: %MGE following saline (1 ml/kg, s.c.) pretreatment.

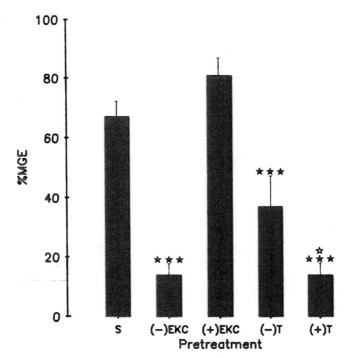

FIGURE 2. The effect of 15-min pretreatment with saline (1 ml/kg, s.c.), the stereoisomers of EKC (both 0.5 mg/kg, s.c.), or the optical isomers of tifluadom (T, both 10 mg/kg, s.c.) on the grooming induced by bombesin (0.1 µg, i.c.v.). %MGE is the percentage of the maximum number of grooming episodes scored as positive over the 30-min observation session. *Bars* indicate the mean ± SE for 5–8 rats; ***$p < 0.001$, ANOVA and Mann-Whitney U-test compared to saline + bombesin controls; ☆$p = 0.02$, ANOVA and Mann-Whitney U-test compared to (−)-tifluadom + bombesin.

15 min before bombesin or DPDPE (1 and 5 µg, i.c.v.) given 5 min before bombesin also did not attenuate bombesin-induced grooming at doses that did not produce competing behaviors (catalepsy and rearing, respectively). Zomepirac (5 mg/kg, s.c.) had no marked effect on bombesin-induced grooming.

Bremazocine (10 mg/kg), cyclazocine (5 mg/kg), ketazocine (1 mg/kg), *l*-pentazocine (20 mg/kg), and phenazocine (0.5 mg/kg) attenuated bombesin-induced grooming significantly compared to saline + bombesin controls ($p < 0.001$; ANOVA followed by Dunnett's test for multiple comparisons) at nonsedative doses (%MGE scores remained greater than i.c.v. saline controls). Xorphanol attenuated bombesin-induced grooming in a dose-related manner with an A_{50} of 0.71 (0.41–1.21) mg/kg, s.c. Dynorphin A also antagonized bombesin-induced grooming in a dose-related manner with an A_{50} of 4.1 (2.8–5.9) µg, i.c.v.

FIGURE 1 shows the dose-related decrease in bombesin-induced grooming produced by EKC, U-50,488, and tifluadom. The A_{50}'s for EKC, tifluadom, and U-50,488 to antagonize bombesin-induced grooming were 0.36 (0.33–0.40), 6.61 (5.37–8.13), and 6.80 (5.52–8.36) mg/kg, respectively.

TABLE 1. Naloxone Antagonism of the Effect of s.c. EKC (0.5 mg/kg), U-50,488 (10 mg/kg), and Tifluadom (10 mg/kg) on Bombesin-induced Grooming (0.1 µg, i.c.v.).

Opioid Agonist	Naloxone A_{50} (95% C.L.[a]), mg/kg, s.c.
EKC	0.076 (0.068–0.086)
Tifluadom	0.12 (0.08–0.20)
U-50,488	0.43 (0.30–0.60)

[a] Confidence limit.

Stereoselectivity of EKC and Tifluadom

FIGURE 2 shows the stereoselectivity of EKC and tifluadom in their ability to antagonize bombesin-induced grooming. The levorotatory enantiomer, but not the dextrorotatory enantiomer, of EKC inhibited bombesin-induced grooming. Whereas both isomers of tifluadom antagonized bombesin-induced grooming, the (+)-optical isomer of tifluadom was significantly more effective than the (−)-isomer.

Sensitivity to Naloxone

Naloxone itself, up to 10 mg/kg, had no effect on bombesin-induced grooming. Naloxone, however, reversed the effects of EKC, tifluadom, and U-50,488 (doses that inhibited bombesin-induced grooming by approximately 80%) in a dose-related manner. TABLE 1 shows the potency of naloxone against the effects of the three kappa agonists on bombesin-induced grooming. This action of naloxone was stereospecific in that l-naloxone (0.08 mg/kg) prevented the action of 0.5 mg/kg EKC on bombesin-induced grooming significantly ($p < 0.01$), whereas d-naloxone (0.1 mg/kg) had no effect on the ability of the same dose of EKC to attenuate bombesin-induced grooming (data not shown).

TABLE 2. Effect of Multiple Injections of EKC or Morphine on the Ability of EKC to Attenuate Bombesin-induced Grooming

Treatment[a] (8×)	Challenge[b]	Bombesin-induced Grooming (%MGE ± SEM)
saline	saline	68 ± 3
saline	EKC	20 ± 4[c]
EKC	saline	73 ± 6
EKC	EKC	85 ± 4[d]
morphine	saline	62 ± 6
morphine	EKC	23 ± 4

[a] Eight injections over consecutive days as described in RESULTS.

[b] Challenge with saline (1 ml/kg, s.c.) or EKC (0.5 mg/kg, s.c.) was given 19–20 h after the eighth injection of saline, EKC, or morphine, and 15 min before bombesin (0.1 µg, i.c.v.).

[c] $p < 0.001$, ANOVA followed by the Mann-Whitney U-test compared to saline + saline controls.

[d] $p < 0.001$, ANOVA followed by the Mann-Whitney U-test compared to saline + EKC controls.

Tolerance Studies

Morphine was given at 5 P.M. and 8 A.M. at doses of 10, 10, 30, 30, 100, 100, 100, and 100 mg per kg per injection for a total of eight injections over consecutive days to two groups of eight rats. EKC was given eight times over consecutive days at the doses of 5, 5, 10, 10, 20, 20, 20, and 20 mg per kg per injection to two groups of rats. Another two groups of rats received twice-daily injections of saline (1 ml/kg, s.c.) for four consecutive days. Beginning 19–20 h after the last of the multiple morphine, EKC, or saline injections, the rats were challenged with saline (1 ml/kg, s.c.) or EKC (0.5 mg/kg, s.c.) and followed 15 min later with the standard dose of bombesin (0.1 µg, i.c.v.) as indicated in TABLE 2. Multiple treatments with saline, morphine, or EKC had no effect on bombesin-induced grooming. Multiple injections of morphine also did not affect the ability of EKC to attenuate bombesin-induced grooming. However, multiple injections of EKC decreased significantly ($p < 0.001$) the ability of EKC (0.5 mg/kg) to inhibit bombesin-induced grooming; i.e., tolerance developed to EKC.

DISCUSSION

The excessive grooming (scratching) behavior elicited by a standard submaximal dose of bombesin was attenuated in a dose-related and stereoselective manner by opioids showing activity at the kappa type of receptor. These included benzomorphan analgesics; the endogenous kappa-selective opioid peptide, dynorphin A; a morphinan mixed agonist-antagonist analgesic of the kappa type, xorphanol; and the most kappa-receptor-selective opioid available, U-50,488. It is important to note that the inhibition of bombesin-induced grooming by these opioids occurred in a dose-related fashion and at doses that did not obliterate natural grooming behavior.

In contrast, compounds whose agonist effects are thought to be mediated by the mu or delta types of opioid receptors were unable to affect bombesin-induced scratching at doses that did not completely suppress grooming behavior. Most of these compounds eventually prevented bombesin-induced grooming, but it was not a dose-related effect, and grooming behavior was prevented altogether: %MGE scores were below those of i.c.v. saline controls. These included the prototype mu-receptor agonist, morphine; the prototype delta-receptor agonist, DPDPE; and a mixed agonist-antagonist analgesic of the mu-type, nalbuphine. In addition, the nonopioid analgesic, zomepirac, had no marked effect on bombesin-induced grooming.

Ethylketazocine, tifluadom, and U-50,488 were examined more thoroughly as prototype kappa opioid agonists in their ability to affect bombesin-induced grooming. All three compounds inhibited bombesin-induced grooming in a dose-related and parallel manner. Ethylketazocine was approximately 18 times as potent as U-50,488 and tifluadom, which were equipotent. A similar potency relationship has been observed among EKC and U-50,488 and tifluadom in the rat tail-immersion test.[24,31] U-50,488 was found to be 10 times less potent than EKC in the rat tail-flick test.[24]

Ethylketazocine was stereospecific in its ability to affect bombesin-induced grooming in that the (+)-enantiomer was ineffective at doses at which the (−)-enantiomer had a significant effect. Tifluadom showed stereoselectivity, but not to as great an extent. Thus, 10 mg/kg of both the (+)- and (−)-isomers did suppress bombesin-induced grooming. However, (−)-tifluadom was significantly less effective than the (+)-isomer. The stereoselectivity of the (+)-optical isomer of tifluadom (determined in ethanol) for opioid receptors has been observed previously in mice[32] and monkeys.[33] The (−)-isomer of tifluadom has some affinity for the benzodiazepine receptor[32,34] and may inhibit bombesin through this receptor; diazepam also inhibits bombesin-induced grooming at nonsedative doses (unpublished observation). This could be tested easily by examining the ability of naloxone to antagonize the actions of ($-$)-tifluadom on bombesin-induced grooming.

The ease at with which naloxone reversed the actions of EKC, U-50,488, and tifluadom on bombesin-induced grooming and the stereospecificity of naloxone in having this effect verifies that opioid receptors mediate their action. Naloxone is less potent at kappa than mu receptors. Its decreasing potency in antagonizing EKC, tifluadom, and U-50,488 may therefore correspond to the increasing relative affinity of the three agonists for kappa over mu receptors.[26]

The development of tolerance to the ability of EKC to inhibit bombesin-induced grooming, taken together with the stereoselectivity, reversibility by naloxone, and the lack of effect of the nonopioid analgesic zomepirac indicate that opioid receptors mediate the attenuation of bombesin-induced scratching by analgesics. It becomes clear, however, based on tests of 35 opioid-like compounds with a large variety of chemical and pharmacological properties, that the kappa type of receptor is the only known opioid receptor linked to bombesin-induced grooming. Thus, mu opioid receptor agonists (morphine, codeine, levorphanol, heroin), mixed agonist-antagonists of the mu type (buprenorphine, nalbuphine), opioid antagonists (naloxone), peptides with mu- and delta-receptor activity (β-endorphin, metkephamid, DPDPE), and chemically related nonopioids (dextrophan, thebaine) had no effect at nonsedative doses in this test. However, all the compounds tested that act at the kappa-type of opioid receptor were effective. Furthermore, the lack of tolerance to EKC in rats that received multiple injections of morphine is in agreement with the kappa-receptor specificity of opioid link to bombesin-induced scratching.

Dynorphin A was proposed as an endogenous ligand for the kappa receptor.[35] However, its kappa receptor activity *in vivo* has been questioned. The dose-related antagonism of bombesin-induced grooming may be used as new evidence of a kappa-receptor-mediated effect of supraspinally administered dynorphin A. The inhibition of bombesin-induced grooming occurred at doses similar to those at which dynorphin A increased feeding in rats[36] (another effect thought to be mediated by kappa receptors). Similarly, the attenuation of bombesin-induced grooming by xorphanol may be used as further evidence of the kappa agonist nature of this compound.[38]

Having established that opioids act through kappa receptors to affect bombesin-induced scratching, we can speculate on the role of bombesin and opioids in scratching behavior. It is known that low doses of opioids acting at mu receptors (*e.g.*, morphine and levorphanol) often cause scratching behavior in animals, and itching in humans. Bombesin given i.c.v. to morphine-dependent rhesus monkeys produces scratching behavior that is similar to, but more intense, than that observed following low doses of morphine in drug-naive monkeys (ref. 5 and unpublished observations). The itching produced by low doses of morphine is thought to be due to the release of histamine; however, it is not clear that histamine is the sole cause of this effect. In contrast, the acute administration of kappa opioids does not elicit scratching behavior. However, abstinence or naloxone-induced withdrawal behavior following the chronic administration of kappa opioids includes scratching as a prominent response.[18,38] This behavior, and the possible role of bombesin in its production, needs to be studied further. Similarly, the role of bombesin in different types of pruritus and a possible link to kappa opioids should be investigated further. For example, opioids with kappa-receptor selectivity might be a drug class of choice for the relief of the pain and itching caused by small cell carcinoma of the lung, which is known to produce large amounts of bombesin-like immunoreactivity.[39]

ACKNOWLEDGMENTS

We thank the various individuals and pharmaceutical companies (listed in METHODS) who donated samples of the compounds used in these studies.

REFERENCES

1. ANASTASI, A., V. ERSPAMER & M. BUCCI. 1971. Isolation and structure of bombesin and alytesin, two analogous active peptides from the skin of the European amphibians *Bombesin* and *Alytes*. Experientia 27: 166–167.
2. BROWN, M. R., J. RIVIER & W. VALE. 1977. Bombesin: Potent effects on thermoregulation in the rat. Science 196: 998–1000.
3. GMEREK, D. E. & A. COWAN. 1981. Tolerance and cross-tolerance studies on the grooming and shaking effects of ACTH (1–24), bombesin and RX 336-M. Fed. Proc. 40: 274.
4. GMEREK, D. E. & A. COWAN. 1983. Studies on bombesin-induced grooming in rats. Peptides 4: 907–913.
5. COWAN, A., P. KHUNAWAT, X. ZUZHU & D. E. GMEREK. 1985. Effects of bombesin on behavior. Life Sci. 37: 135–145.
6. COWAN, A. & D. E. GMEREK. 1981. Studies on the antinocisponsive action of neurotensin. Fed. Proc. 40: 273.
7. DOBRY, P. J. K., M. R. PIERCEY & L. A. SCHROEDER. 1981. Pharmacological characterization of scratching behaviour induced by intracranial injection of substance P and somatostatin. Neuropharmacology 20: 267–272.
8. GMEREK, D. E. & A. COWAN. 1982. Bombesin-induced grooming is antagonized by benzomorphan analgesics but not by morphine. Fed. Proc. 31: 1301.
9. GMEREK, D. E. & A. COWAN. 1982. Classification of opioids on the basis of their ability to antagonize bombesin-induced grooming in rats. Life Sci. 31: 2229–2232.
10. GMEREK, D. E. & A. COWAN. 1983. ACTH (1–24) and RX 336-M induced excessive grooming in rats through different mechanisms. Eur. J. Pharmacol. 88: 339–346.
11. GMEREK, D. E. & A. COWAN. 1982. A study of the shaking and grooming induced by RX 336-M in rats. Pharmacol. Biochem. Behav. 16: 929–932.
12. MARTIN, W. R., C. G. EADES, J. A. THOMPSON, R. E. HUPPLER & P. E. GILBERT. 1976. The effects of morphine- and nalorphine-like drugs in the nondependent and morphine-dependent chronic spinal dog. J. Pharmacol. Exp. Ther. 197: 517–532.
13. GILBERT, P. E. & W. R. MARTIN. 1976. The effects of morphine- and nalorphine-like drugs in the nondependent, morphine-dependent and cyclazocine-dependent chronic spinal dog. J. Pharmacol. Exp. Ther. 198: 66–82.
14. HUTCHINSON, M., H. W. KOSTERLITZ, F. M. LESLIE, A. A. WATERFIELD & L. TERENIUS. 1975. Assessment of the guinea pig ileum and mouse vas deferens of benzomorphans which have strong antinociceptive activity but do not substitute for morphine in the dependent monkey. Br. J. Pharmacol. 55: 541–546.
15. KOSTERLITZ, H. W., S. J. PATERSON & L. E. ROBSON. 1981. Characterization of the kappa subtype of opiate receptor in guinea-pig brain. Br. J. Pharmacol. 73: 939–949.
16. WOODS, J. H., C. B. SMITH, F. MEDZIHRADSKY & H. H. SWAIN. 1979. Preclinical testing of new analgesic drugs. *In* Mechanisms of Pain and Analgesic Compounds. R. F. Beers, Jr. & E. G. Bassett, Eds.: 429–445. Raven. New York.
17. HERLING, S. & J. H. WOODS. 1981. Discriminative stimulus effects of narcotics: evidence for multiple receptor-mediated actions. Life Sci. 28: 1571–1584.
18. GMEREK, D. E., L. A. DYKSTRA & J. H. WOODS. 1987. Kappa opioids in rhesus monkeys: III. Dependence association with chronic administration. J. Pharmacol. Exp. Ther. 242: 428–436.
19. LORD, J. A. H., A. A. WATERFIELD, J. HUGHES & H. W. KOSTERLITZ. 1977. Endogenous opioid peptides: multiple agonists and receptors. Nature 267: 495–499.
20. GMEREK, D. E. & A. COWAN. 1984. *In vivo* evidence for benzomorphan-selective receptors in rats. J. Pharmacol. Exp. Ther. 230: 110–115.

21. PORRECA, F., R. B. RAFFA, A. COWAN & R. J. TALLARIDA. 1982. A comparison of the receptor constants of morphine and ethylketocyclazocine for analgesia and inhibition of gastrointestinal transit in the rat. Life Sci. **31:** 1955-1961.

22. ROMER, D., H. H.. BUSCHER, R. C. HILL, R. MAURER, T. J. PETCHER, H. ZEUGNER, W. BENSON, E. FINNER, W. MILKOWSKI & P. W. THIES. 1982. An opioid benzodiazepine. Nature **298:** 759-760.

23. ROMER, D., H. H. BUSCHER, R. C. HILL, R. MAURER, T. J. PETCHER, H. ZEUGNER, W. BENSON, E. FINNER, W. MILKOWSKI & P. W. THIES. 1982. Unexpected opioid activity in a known class of drug. Life Sci. **31:** 1217-1220.

24. VONVOIGTLANDER, P. F., R. A. LAHTI & J. H. LUDENS. 1983. U-50,488: A selective and structurally novel non-mu (kappa) opioid agonist. J. Pharmacol. Exp. Ther. **224:** 7-12.

25. LAHTI, R. A., M. M. MICKELSON, J. M. MCCALL & P. F. VONVOIGTLANDER. 1985. [^{3}H]U-69593, a highly selective ligand for the opioid kappa receptor. Eur. J. Pharmacol. **109:** 281-284.

26. JAMES, E. F. & A. GOLDSTEIN. 1984. Site-directed alkylation of multiple opioid receptors. I. Binding selectivity. Mol. Pharmacol. **25:** 337-342.

27. TANG, A. H. & R. J. COLLINS. 1985. Behavioral effects of a novel kappa opioid analgesic, U-50,488, in rats and rhesus monkeys. Psychopharmacology **85:** 309-314.

28. MURRAY, R. B., D. E. GMEREK, A. COWAN & R. J. TALLARIDA. 1981. Use of a programmable protocol timer and data logger in the monitoring of animal behavior. Pharmacol. Biochem. Behav. **15:** 135-140.

29. ALDER, H. L. & E. B. ROESSLER. 1977. Introduction to Probability and Statistics, 6th edit. W. H. Freeman. San Francisco, CA. p. 337.

30. TALLARIDA, R. J. & R. B. MURRAY. 1981. Manual of Pharmacologic Calculations with Computer Programs. Springer-Verlag. New York.

31. UPTON, N., J. P. GONZALEZ & R. D. E. SEWELL. 1983. Characterization of a kappa-agonist-like antinociception action of tifluadom. Neuropharmacology **22:** 1241-1242.

32. RUHLAND, M & H. ZEUGNER. 1983. Effects of the opioid benzodiazepine tifluadom and its optical isomers on spontaneous locomotor activity of mice. Life Sci. **33**(Supp.I): 631-634.

33. DYKSTRA, L. A., D. E. GMEREK, G. WINGER & J. H. WOODS. 1987. Kappa opioids in rhesus monkeys. I. Diuresis, sedation, analgesia and discriminative stimulus effects. J. Pharmacol. Exp. Ther. **242:** 413-420.

34. KLEY, H., U. SCHEIDEMANTEL, B. BERLING & W. E. MULLER. 1983. Reverse stereoselectivity of opiate and benzodiazepine receptors for the opioid benzodiazepine tifluadom. Eur. J. Pharmacol. **87:** 503-504.

35. CHAVKIN, C., I. F. JAMES & A. GOLDSTEIN. 1982. Dynorphin is a specific endogenous ligand of the kappa opioid receptor. Science **215:** 79-81.

36. MORLEY, J. E. & A. S. LEVINE. 1983. Involvement of dynorphin and the kappa opioid receptor in feeding. Peptides **4:** 797-800.

37. HOWES, J. F., J. E. VILLARREAL, L. S. HARRIS, E. M. ESSIGMANN & A. COWAN. 1985. Xorphanol. Drug Alc. Dependence **14:** 373-380.

38. WOODS, J. H. & D. E. GMEREK. 1985. Substitution and primary dependence studies in animals. Drug. Alc. Dependence **14:** 233-247.

39. MOODY, T. W., C. B. PERT, P. I. F. GAZDAR, D. N. CARNEY & J. D. MINNA. 1981. High levels of intracellular bombesin characterize human small-cell lung carcinoma. Science **214:** 1246-1248.

Grooming Reflexes in the Cat: Endocrine and Pharmacological Studies[a]

WALTER RANDALL

Department of Psychology
University of Iowa
Iowa City, Iowa 52242

Goltz and Gergens[1] may have been the first to describe the grooming reflexes induced in dogs by lesions of the central nervous system. Gergens,[2] who showed that both cerebral lesions and transections of the spinal cord were effective, established that these reflexes were lesion induced by carefully examining the intact dogs at Goltz's institute. He found the reflexes in attenuated form in only a few animals. In a lengthy description of the behavior of one decerebrated dog, Goltz[3] described the "new and peculiar" reflexes induced by decerebration. These reflexes are components of normal grooming behavior and were peculiar because they were elicited by light tactile stimuli. For example, the shake reflex, the familiar shaking movement used by dogs to remove water from their fur, was elicited by light tactile stimulation when there was no water on the dog. The grooming scratch, lick, and bite reflexes also were described. Foster[4] described the lesion-induced scratch reflex in the dog in his *Textbook of Physiology* as the "remarkable scratching movements." Sherrington used this text as a student in Foster's laboratory and subsequently did his own review of Goltz's research,[5] attributing the induction of the scratch reflex to the destruction of the pyramidal tract, an "inhibito-motor path." Sherrington stated that the "ablation of the cerebrum removes a nervous centre which under normal conditions constantly restrains these spinal reflexes." Sherrington and Laslett[6] found that transection of the lateral column of the spinal cord (the area of the cord where the pyramidal axons are found) induced the scratch reflex and again suggested that the pyramidal tract may be the critical pathway whose destruction induces the scratch reflex in Goltz's lesions of the "cortex cerebri." The scratch reflex is thus a neurological sign, comparable to Babinski's reflex in the human.

Sherrington[7-9] did intensive research on the scratch reflex induced by spinal transection, and the scratch reflex is considered to be the most complex reflex that he investigated. Granit[10] considered the scratch reflex "too complex for the terms in which we think." In a nonexhaustive survey, Sherrington[9] found 39 muscles involved in the execution of the scratch reflex. The complex spatial and temporal pattern exhibited by the scratch reflex is seen in our cinematographic analysis (FIG. 1), which shows the successive positions of the tip of the foot every 20.8 ms for one complete beat of the scratch. The complex spatial and temporal pattern consists of a fast and a slow component. The low velocity component occurs near and on the body surface. When contact with the body surface is lost, the foot displays its high velocity as it moves around

[a] This research was supported by the National Institutes of Health (MH-15402 and MH-15773) and by a grant from the Epilepsy Foundation of America.

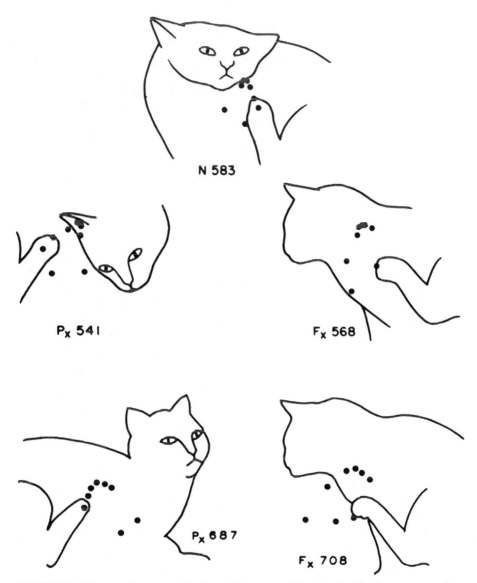

FIGURE 1. The complex temporal and spatial pattern of the scratch. The dots represent the position of the tip of the foot at 20.8-ms intervals traced from the successive frames of a movie obtained with a Hy Cam 16 mm camera with an attached timing light that marked the film every 100 ms. One beat of the scratch is shown for each of five cats. The abbreviations are: *N*, normal; *Px*, pontile lesion; *Fx*, frontal neocortical lesion. (Based on Randall *et al.*[61])

for the next contact. The low- and high-velocity phases are of approximately equal duration, each lasting about 80 ms (four dots each in FIG. 1). FIGURE 1 shows the normal scratching behavior in three cats: a normal cat (N583), a cat with a pontile lesion (Px 541), and a cat with a frontal neocortical lesion (Fx 568). The scratch reflex in two

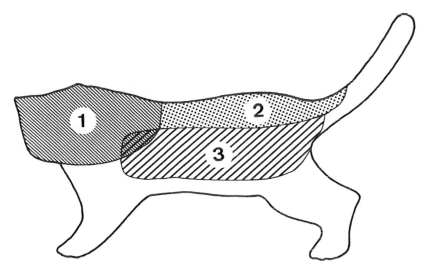

FIGURE 2. The receptive field for the grooming reflexes in a cat with a pontile lesion. Light tactile stimulation of areas 1, 2, and 3 elicits the scratch, bite, and lick reflexes, respectively. Areas 1 and 3 overlap and the reflex that is elicited depends on the stimulus, whether it is like a scratch (in which case a scratch reflex is elicited) or like a lick (in which case a lick reflex is elicited). Typically, both the bite and lick reflexes may be obtained from area 2, depending on the type of cutaneous stimulus that is applied. (From Randall et al.[61] Copyright 1985 by the American Psychological Association. Reprinted by permission of the publisher.)

cats with lesions also is illustrated (Px 687 and Fx 708, the two cats at the bottom of the figure). Some distortion of the low-velocity component is typical for the scratch reflex, but the low- and high-velocity components are always present, and because of this typical topography the movement was categorized as a scratch. Skin irritants elicit the scratching *behavior* in the upper three cats in the figure, whereas light tactile stimulation (stimulation that is without effect in normal cats) elicits the scratch *reflex* in the two cats with lesions at the bottom of the figure. The scratch reflexes occur in midair, without making contact with the body surface; a hand lightly touching the head of the two cats is not shown.

The receptive field for the grooming reflexes (that area on the body surface where stimulation elicits the reflex) can be reliably determined and sketched on an outline drawing of a cat (FIG. 2). The area on the outline drawing is estimated, and the percentage of the total area covered by the sketch of the receptive fields is our quantitative estimate of the size of the receptive field for the grooming reflexes. The receptive fields shown in FIGURE 2 are those of a cat with a pontile lesion and represent a near maximum extent. Cutaneous stimulation elicited the scratch reflex from area 1 (FIG. 2), the lick reflex from area 3, and both the lick and bite reflexes from area 2. Typically, scratch and lick receptive fields overlap on the shoulder (FIG. 2). The three different reflexes are elicited by different kinds of cutaneous stimulation, namely, the kind of stimulation normally delivered by the scratching foot or licking tongue or scratching bite to parts of the body surfaces indicated in the figure. The receptive fields change in size synchronously among cats as a function of season[11] (see below).

In 50 pages of detailed analysis, Sherrington[8] demonstrated that the scratch reflex was an innate movement generated by the spinal cord. He eliminated, by careful ex-

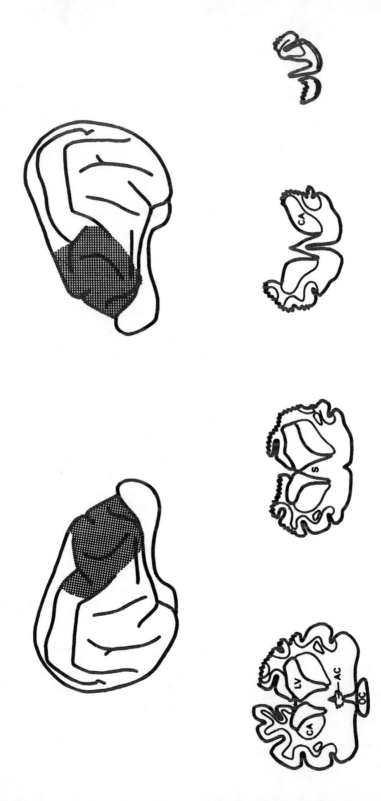

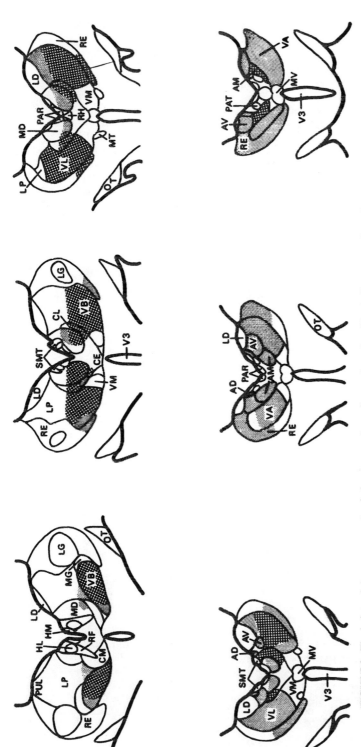

FIGURE 3. The lesion of the neocortex that induces grooming reflexes. *Cross-hatching* indicates partial absence of neurons, and *stippling* indicates complete absence of neurons. The abbreviations are: *AC,* anterior commissure; *AD,* anterodorsal nucleus; *AM,* anteromedial nucleus; *AV,* anteroventral nucleus; *CA,* caudate nucleus; *CE,* central medial nucleus; *CL,* central lateral nucleus; *CM,* nucleus centrum medianum; *HL,* lateral habenular nucleus; *HM,* medial habenular nucleus; *LD,* lateral dorsal nucleus; *LG,* lateral geniculate body; *LP,* lateral posterior complex; *LV,* lateral ventricle; *MD,* mediodorsal nucleus; *MG,* medial geniculate body; *MT,* mamillothalamic tract; *MV,* medioventral nucleus; *OC,* optic chiasm; *OT,* optic tract; *PAR,* paraventricular complex of the thalamus; *PAT,* parataenial nucleus; *PUL,* pulvinar; *RE,* reticular complex of the thalamus; *RF,* retroflex bundle; *RH,* rhomboid nucleus; *S,* septal nuclei; *SMT,* stria medullaris thalami; *V3,* third ventricle; *VA,* ventroanterior nucleus; *VB,* ventrobasal complex; *VL,* ventrolateral complex; *VM,* ventromedial complex. (From Elbin and Randall.⁴¹ Copyright 1977 by the American Psychological Association. Reprinted by permission of the publisher.)

periment and analysis, the input path, the motor unit, and the proprioceptive feedback from the muscles of the leg, leaving only the spinal cord as the source of the complex movement. The proprioceptive input from the leg muscles was abolished, and the typical complex spatial and temporal pattern of the scratch reflex was not disrupted. The typical frequency of the scratch reflex existed independently of the frequency of the stimulus, i.e., the stimulus frequency was varied and did not determine or influence the innate frequency of the scratch reflex. The motor unit was known to be capable of responding over a wide range of frequencies, and thus was not considered as a source of the innate frequency of the scratch reflex. It was obvious that the motor unit was not responding passively to stimulus input because, when an additional stimulus was applied at a second position on the skin, no change in the frequency of the scratch reflex occurred. Also, the presence of homogeneous stimulus summation indicated the redundancy and irrelevancy of a specific input path. Thus a scratch "black box" was in the central nervous system, called "the internuncial path" in Sherrington's intentionally simple model. Sherrington's concept of the integrative action of the nervous system[12] is the black box "welding together . . . [the] metamers into the unity of an animal individual." This unity is obtained by innate, preformed patterns of neural output from the central nervous system that are endogenously generated, of considerable spatial and temporal complexity, and triggered by "stimuli."

Sherrington's[12] concept of behavior involved purpose as an important property of the reflex. The reflex was the unit of behavior, and the reflex exhibited purpose: "the reflex reaction cannot be really intelligible to the physiologist until he knows its aim" and "the skeletal musculature is treated practically as a whole and in a manner suitably anticipatory of a later event." Stimuli act as "releasing forces," with the "change in the environment (acting) as a releasing force upon the living machinery of the organism." Thus Sherrington provided important insights about behavior that became traditional in neurophysiology: innate patterns of behavior are released by stimuli, with these innate patterns exhibiting the important property of purposiveness.

Subsequent research on the grooming reflexes induced by lesions of the cortex cerebri[13-15] indicated that the neocortex was involved and that ablation of the frontal third of the neocortex was sufficient for inducing the grooming reflexes. A reconstruction of a lesion of the frontal neocrotex that induces the grooming scratch, lick, and bite reflexes in the cat is shown in FIGURE 3. The neocortical lesions include all the cortex rostral to the ansate and anterior suprasylvian sulci, extending ventrally to the rhinal sulcus and medially to include the superior frontal and rectus gyri. The top row shows the lesion on the lateral surfaces of the left and right hemispheres. In the second row, tranverse sections are shown through the extent of the primary lesion. The bottom two rows show the retrograde changes in the dorsal thalamus, which typically involve the anterior and ventral groups of nuclei, as well as portions of the medial dorsal and parafascicular nuclei. Large portions of the internal capsule and cerebral peduncles (the "pyramidal tract," not indicated in the figure) exhibited anterograde degeneration because the primary lesion included the "motor" cortex. The anterior group projects Papez's circuit onto the cingulate gyrus, and the ventral group projects cerebellar input to the frontal neocortex (VA and VL) and somatic input to the somatic cortex (VB). The medial dorsal nucleus receives input from the hypothalamus and from other dorsal thalamic nuclei and projects to the frontal neocortex. All these projections to the cortex are reciprocated. Any one or more of these complex systems of interconnections may be involved as the critical system whose destruction induces grooming reflexes. The cerebellar connection is of interest because of the research of Berntson and his colleagues.[16,17]

Grooming reflexes also are induced by pontile lesions (FIG. 4), which are stereo-

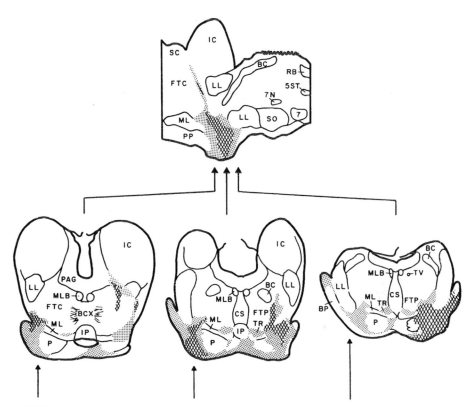

FIGURE 4. The pontile lesion that induces grooming reflexes. The upper half of the figure is a reconstruction on the parasagittal plane indicated by the three arrows at the bottom of the figure. In the bottom half, three transverse sections are shown, with their caudal-rostral position indicated by the arrows in the middle of the figure. *Cross-hatching* indicates complete absence of neurons, and *stippling* indicates partial absence. The abbreviations are: *BC*, brachium conjunctivum; *BCX*, decussation of the brachium conjunctivum; *BP*, brachium pontis; *CS*, superior central nucleus; *FTC*, central tegmental field; *FTP* paralemniscal tegmental field; *IC*, inferior colliculus; *IP*, interpeduncular nucleus; *LL*, lateral lemniscus; *ML*, medial lemniscus; *MLB*, medial longitudinal bundle; *P*, pyramidal tract; *PAG*, periaqueductal gray; *PP*, pes pedunculi; *RB*, restiform body; *SC*, superior colliculus; *SO*, superior olive; *TR*, tegmental reticular nucleus; *TV*, ventral tegmental nucleus.

taxic, anodal, DC lesions made with stainless steel Kirschner pins. These lesions are centered in the lateral and ventral portions of the central and paralemniscal tegmental fields, extending ventrally into the pons and to the ventral and lateral boundary of the brain stem at this level. Thus the primary lesion involves the pontile nuclei and the brachia ponti as well as portions of the medial lemnisci, tectopontile, corticopontile, corticotegmental, and corticospinal tracts. Again, as in the lesion of the frontal neocortex, both cerebellar and corticofugal systems are involved.

Pharmacological manipulations that abolish grooming reflexes enhance the serotoninergic system (TABLE 1). Serotoninergic function is enhanced by 5-hydro-

TABLE 1. A Summary of the Effective and Noneffective Pharmaca for Reducing the Size of the Receptive Field for Grooming Reflexes in Cats with Pontile Lesions[a]

	Effective				No Effect	
Pharmaca	Dosage	No. of Doses	% Reduction	Pharmaca	Dosage	No. of Doses
L-5HTP	5 mg·kg⁻¹ body wt	1	50	D-5HTP	50 mg·kg⁻¹ body wt	1
hydrocortisone	25 mg·kg⁻¹ body wt·day⁻¹	12	90	L-tryptophan	200 mg·kg⁻¹ body wt·day⁻¹	12
MAOI (tranylcypromine)	2.5 mg·kg⁻¹ body wt	1	45	deoxycorticosterone	7.5 mg·kg⁻¹ body wt·day⁻¹	7
MAOI and tryptophan	2.5 & 25 mg·kg⁻¹ body wt	1	85	adrenaline	.15 mg per cat·day⁻¹	4
methysergide maleate	10 mg·kg⁻¹ body wt	1	100	diphenylhydantoin	10 mg·kg⁻¹ body wt·day⁻¹	5
morphine sulfate	.25 mg·kg⁻¹ body wt	1	65	phenobarbital	4 mg·kg⁻¹ body wt·day⁻¹	26
diazepam	10 mg·kg⁻¹ body wt	1	65	L-DOPA	100 mg·kg⁻¹ body wt	1

[a] Abbreviations are: 5HTP, 5-hydroxytryptophan; DOPA, dihydroxyphenylalanine; MAOI, monoamine oxidase inhibitor.

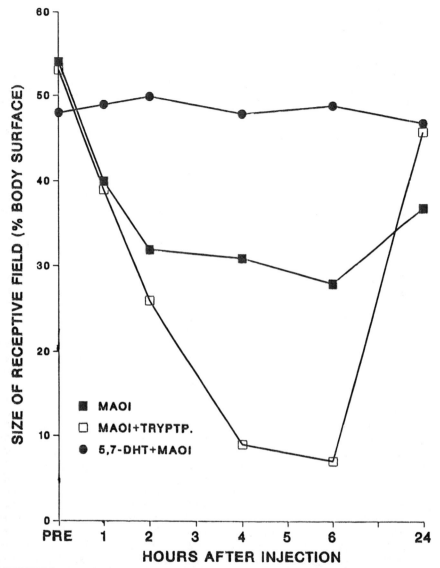

FIGURE 5. Cats with frontal neocortical lesions were injected with a monoamine oxidase inhibitor (tranylcypromine, 2.5 mg/kg body weight, intraperitoneal) at time zero. Over a month before the study, one group of eight cats with frontal neocortical lesions was pretreated (5,7-DHT + MAOI) with the serotoninergic neurotoxin 5, 7-DHT. Two mg of 5, 7-DHT creatinine sulfate were injected into the third ventricle; to prevent damage to catecholamine neurons, the MAOI, desmethylimipramine HCl (15 mg/kg body weight, intraperitoneal) was injected 1 hour before the 5, 7-DHT injection. Another group of five cats with frontal neocortical lesions was injected with the MAOI on one occasion and with the MAOI combined with tryptophan (100 mg/kg body weight, intramuscular) on another occasion. Vehicle injections into the third ventricle and pretreatment with desmethylimipramine HCl was without effect (not shown, *i.e.*, the curve was like the MAOI group). The abbreviations are: MAOI, monoamine oxidase inhibitor; TRYPTP, tryptophan; 5, 7-DHT, 5, 7-dihydroxytryptamine. (Based on Elbin.[55])

xytryptophan (5-HTP),[18] glucocorticoids,[19,20] monoamine oxidase inhibitors (MAOI),[21] tryptophan,[22] methysergide,[23,24] morphine,[25] and diazepam.[26] Tryptophan, which has no effect on the grooming reflexes when administered alone, is a precursor of serotonin and increases serotonin concentration in neurons, but in other studies tryptophan has a behavioral effect only in the presence of a monoamine oxidase inhibitor.[27-29] Increased synthesis of serotonin may merely increase the storage of serotonin within the neuron and have nothing to do with function. Function of serotoninergic neurons appears to involve autoregulation, reuptake, MAO activity, and number and density of receptors, rather than rate of synthesis.[30-32] In our study with tryptophan, the lack of effect of tryptophan was obtained with the cats on a special diet so that 90% of their ingested amino acids was tryptophan, therefore decreasing the competition from other neutral amino acids for uptake into the brain.

Although the data in TABLE 1 were obtained on cats with pontile lesions, many of the studies also have been conducted on other groups of cats with grooming reflexes. For example, the outcomes listed in TABLE 1 for tryptophan, L-5-HTP, L-DOPA, hydrocortisone, monoamine oxidase inhibitor, and the combined treatment with tryptophan and a monoamine oxidase inhibitor, also have been obtained in two other groups with grooming reflexes, cats with frontal neocortical lesions,[33] and thyroidectomized cats.[34]

The important dimensions of time and dose are not included in TABLE 1, and FIGURES 5 and 6 provide examples of the type of studies that TABLE 1 attempts to summarize. In FIGURE 5, time-response curves with a monamine oxidase inhibitor and tryptophan are shown for cats with frontal neocortical lesions.[33] One group was pretreated with 5, 7-dihydroxytryptamine (5, 7-DHT), a serotoninergic neurotoxin, several weeks before the study indicated in FIGURE 5. FIGURE 5 demonstrates that 5, 7-DHT blocks the effect of the MAOI of abolishing the receptive fields and that tryptophan enhances this effect. The log dose-time-response curves for morphine (FIG. 6) indicate that the grooming reflexes are abolished completely with the higher dosages (the dosages are in terms of morphine sulfate)[35]; for example, at 1 hour, 1 mg/kg morphine sulfate reduced the receptive fields to 94% of their original size.

Although the administration of substances that increase serotoninergic function abolishes grooming reflexes (or decreases the size of the receptive field), procedures that depress serotoninergic function do not induce grooming reflexes in intact cats. For example, the administration of para-chlorophenylalanine (PCPA; 120 mg per kg body wt per day for 7 to 11 days in 22 normal cats) does not induce grooming reflexes, even though this treatment has been shown to reduce midbrain serotonin in the cat by over 90%.[36] McGeer et al.[37] found that a combined treatment with reserpine and DOPA reduced the serotonin in the tectum of a cat to zero, but this combination of treatments does not induce grooming reflexes. These and other methods that are known to deplete serotonin do not induce grooming reflexes in normal cats.[38]

A temporal dysfunction in the annual glucocorticoid rhythm in cats with pontile lesions suggested that an independent glucocorticoid defect might exist.[39] Thus normal cats were adrenalectomized, but no grooming reflexes appeared. However, grooming reflexes did appear when para-chlorophenylalanine was administered to the adrenalectomized cats.[40] This study has been replicated.[41] These grooming reflexes were abolished by intramuscular injections of 5-HTP or glucocorticoid hormones. Thus the induction of grooming reflexes by lesions may be mediated by a combined effect of a hormonal and a neurotransmitter change.

The combined hormonal and neurotransmitter hypothesis was tested with lesions of the raphe nuclei and with 5, 7-DHT injections.[42] Neither raphe lesions nor 5, 7-DHT induced grooming reflexes in normal cats, but both were effective in cats that

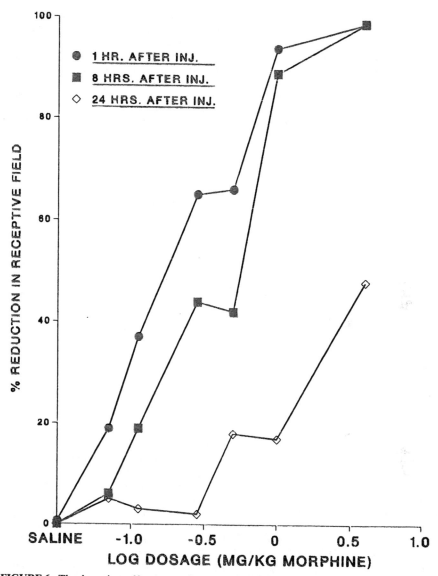

FIGURE 6. The dose-time effect curves for mg/kg body weight of morphine sulfate (intramuscular) on the size of the receptive field for grooming reflexes in a group of six cats with pontile lesions. The same group of six cats was retested with the different doses of morphine, with the doses chosen randomly and with at least 5 days between tests. (Based on Cunningham.[35])

were adrenalectomized. The grooming reflexes were abolished in both groups by the administration of either glucocorticoid hormones or 5-HTP.

Turnover measures of iodine in the thyroid gland of one group of cats with pontile lesions indicated that a hypothyroid condition existed.[43] Thus, a hypothesis for the induction of grooming reflexes was that the lesions of the central nervous system depressed thyroid function, and the depressed thyroid function effected the glucocorticoid and serotoninergic changes. Thyroid hormones have an effect on serotoninergic function [44] and on glucocorticoid levels.[45,46] Thus normal cats were thyroidectomized, and grooming reflexes appeared and were abolished by intramuscular injections of either thyroid hormones, 5-HTP, or glucocorticoid hormones. However, thyroid hormone administration was without effect in cats with pontile lesions, and cats with pontile lesions did not have a thyroid deficiency when tested more systematically and with more specific methods.[47] Furthermore, a longitudinal study for a period of a year with normal cats and cats with pontile lesions[11] failed to detect any differences in the level, amplitude, or pattern of the seasonal rhythms. A further indication that the hypophyseal-thyroid system was not involved is that thyroidectomized cats and cats with pontile lesions that are receiving thyroid injections both exhibit grooming reflexes; thus, grooming reflexes are present regardless of the levels of thyroid hormones and thyroid stimulating hormone. Thus the hypothesis that the lesions of the central nervous system induce a thyroid dysfunction was incorrect. But thyroid hormone administration elevated glucocorticoid levels in the thyroidectomized cats, glucocorticoid levels that were abnormally low before thyroid hormone replacement was initiated.[47] The pharmacological studies implicate both glucocorticoid hormones and serotonin in the induction of grooming reflexes by thyroidectomy, so that it appears that thyroidectomy may induce independent effects on both these systems, effects that then interact to induce the grooming reflexes by the same mechanism as in the cats with lesions of the central nervous system.

In the study with 5, 7-DHT and raphe lesions combined with adrenalectomy,[42] the 5, 7-DHT injections were made locally into the superior colliculus. The superior colliculus has been implicated previously in the mediation of the grooming reflexes by an ablation study. Because cutaneous stimuli elicit grooming reflexes, and because the superior colliculus receives cutaneous input, the superior colliculus was removed in cats with pontile lesions and the grooming reflexes were abolished. Lesions of the somatic neocortex were without effect. Subsequent neurochemical and microinjection studies on cats with pontile lesions and on cats with lesions of the frontal neocortex established that tryptophan hydroxylase activity and serotonin content of the superior colliculus (but not the dorsal thalamus, hypothalamus, cerebellar neocortex, or cerebral neocortex) was decreased,[38,48,49] and that microinjections of serotonin or 5-hydroxytryptophan (but not other substances) into the superior colliculus (but not the ventricle or dorsal thalamus or midbrain tegmentum) abolished the grooming reflexes.[41,48] A log dose-response curve for microinjections of 5-hydroxytryptophan into the superior colliculus (FIG. 7) shows that the lowest dose that gave a prominent effect was .34 µg of L-5-hydroxytryptophan (1.5 nmol) which reduced the size of the receptive field by 28%. Two different postsynaptic receptors for serotonin have been described that differ in their affinity for serotonin and in their drug profile.[32] Because as little as 5 mg/kg body weight of L-5-HTP injected intramuscularly will abolish grooming reflexes in cats with pontile lesions,[50] and because of the nanomole dose-response curve, the high-affinity receptor (5-HT$_1$ receptor) may be involved in the mechanism of 5-HTP reversal of the grooming reflexes. The effect of methysergide on the grooming reflexes, an effect like 5-HTP, is consistent with the possibility of 5-HT$_1$ involvement. Twitches are induced in guinea pigs[51] and rats[52] with 500 mg/kg body weight of 5-HTP, and a U-shaped dose-effect curve for the effects of 5-HTP on pentylenetetrazol-induced seizures has been described[53] with low doses inhibiting the seizures and high doses enhancing the seizures.

In summary, grooming reflexes are induced by three different methods of depleting serotonin (PCPA, 5, 7-DHT, raphe lesions) when combined with adrenalectomy. None

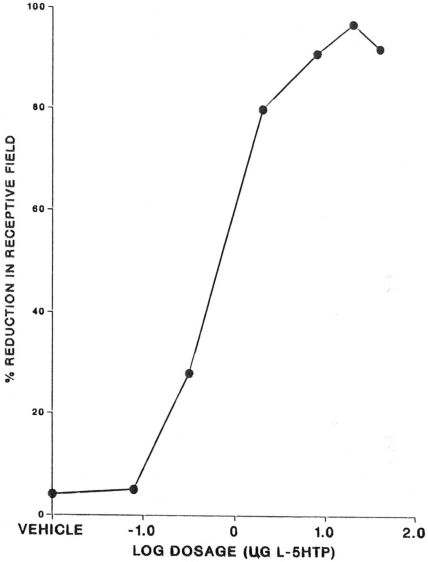

FIGURE 7. Microinjections of L-5-hydroxytryptophan into the superior colliculus of a cat with a frontal neocortical lesion affects the size of the receptive field for grooming reflexes in a dose-dependent manner; 1.2 microliters of solution were injected gradually over a 15-second interval. The same cat was injected with different concentrations of 5-HTP in seven sessions separated by at least 2 weeks. (Based on Elbin and Randall.[41])

of 23 cats that were only adrenalectomized developed grooming reflexes. Thus two independent changes must occur in order to induce grooming reflexes, a depression of serotoninergic function, and a depression of glucocorticoid hormone levels. Enhancing either the function of the serotoninergic system or increasing the systemic levels of glucocorticoid hormones abolishes the grooming reflexes, regardless of how the grooming reflexes have been induced. In cats with pontile lesions and frontal neocortical lesions, the serotoninergic deficit appears to be confined to the superior colliculus: tryptophan hydroxylase activity and serotonin content is decreased only in the superior colliculus, and serotonin abolishes grooming reflexes only when injected into the superior colliculus.

In cats with grooming reflexes induced by lesions of the central nervous system, the depletion of serotonin in the superior colliculus is at most 50%. Additional depletion of serotonin to 70% with 5, 7-DHT does not increase the size of the receptive field for grooming reflexes. Adrenalectomy does, however; when cats with pontile or frontal neocortical lesions are adrenalectomized, large increases in the size of the receptive field occur.[54,55] Thus the serotoninergic deficit is permissive and the level of glucocorticoid hormone controls the size of the receptive field. Several kinds of evidence indicate that glucocorticoid hormones control the size of the receptive field: (1) Longitudinal studies of the size of the receptive field and glucocorticoid secretion indicated seasonal rhythms in both variables with a statistically significant and negative correlation.[39] (2) The seasonal variations in the size of the receptive field are blocked by adrenalectomy; when cats with pontile lesions are adrenalectomized and maintained on a constant level of glucocorticoid hormones by daily intramuscular injections, no variation in the size of the receptive field occurs.[54] (3) Increases and decreases in the size of the receptive field are obtained in adrenalectomized cats by decreasing and increasing the level of glucocorticoid replacement, respectively.[54]

During the longitudinal study of glucocorticoid hormones and the receptive fields, the cats with pontile lesions exhibited above-normal levels of these hormones for most times of the year (FIG. 8). In FIGURE 8 all values of "11OH" (the 11-hydroxycorticoids that are measured) that are above 0.0 are above normal; the values that are plotted are the differences between the group with lesions and the control group. Except for September and October, the glucocorticoid scores for the group with pontile lesions were either higher than normal or very close to normal. Thus, because depressed levels of glucocorticoids are involved in inducing grooming reflexes in adrenalectomized cats, and because glucocorticoid hormones are elevated in cats with pontile lesions, an unusual equivalence between adrenalectomy and pontile lesion is indicated. The lesion may have removed important sites of action of the hormones and increased the threshold for glucocorticoid action. Glucocorticoid receptors are widely distributed throughout the central nervous system[58] and manipulations of glucocorticoid hormone levels affect single unit neuronal activity in hypothalamus, hippocampus, midbrain, amygdala, and septum.[19]

Chronic implants of bipolar recording electrodes in the superior colliculus of cats were used to evaluate the effects of pontile lesions and of pharmacological manipulations on the averaged evoked potentials to cutaneous stimulation.[59] In a group of five cats, pontile lesions decreased the amplitude of the four main peaks of the collicular evoked potentials. No significant changes were detected in the somatic neocortex, and the decreases in amplitude were not found in a group of cats with sham lesions. The intramuscular injection of 33 mg/kg body weight of hydrocortisone induced statistically significant increases in the first, third, and fourth peaks of the averaged evoked potentials of the cats with pontile lesions. The glucocorticoid also reduced the size of the receptive field for the grooming reflexes. Thus electrophysiological indices of function implicate the superior colliculus and demonstrate that the lesion-induced changes in the superior colliculus are reversed with pharmacologically high dosages of glucocorticoid hormones. Glucocorticoid hormones regulate the number of 5HT receptors in the superior colliculus.[20]

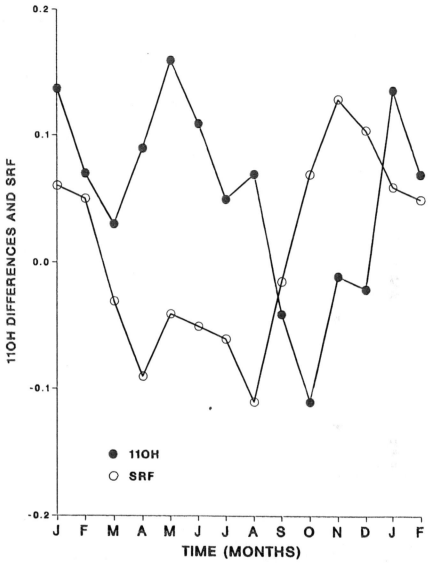

FIGURE 8. The relationship between the glucocorticoid abnormalities and the size of the receptive field (SRF) for the grooming reflexes in a group of six cats with pontile lesions. The total duration of the study was 12 months with the data for 2 months repeated. The glucocorticoid values (*11OH*) that are plotted were obtained by subtracting the monthly values for a group of eight normal cats from the monthly values for the group with pontile lesions (thus a zero score was obtained whenever the cats with pontile lesions exhibited normal values). The data on the size of the receptive field were plotted on an arbitrary scale in order to see the relationship with the temporal abnormality in glucocorticoid hormones. (Based on Randall and Parsons.[39])

Glucocorticoid hormones have two roles in cats with grooming reflexes: (1) their seasonal variation controls the seasonal changes in the size of the receptive field for grooming reflexes, and (2) their depressed function (either by adrenalectomy or by lesions of the central nervous system) is essential (but not sufficient) for induction of the grooming reflexes. Decreases of glucocorticoid hormone levels in humans depress thresholds and discrimination abilities of sensory stimuli of all modalities.[60]

In about 15% of cats with pontile lesions, the grooming reflexes gradually increase in vigor over a period of a year or longer to progress into complete seizures with tonic and clonic components.[61] This type of seizure was first described by Brown-Sequard[62] in the guinea pig and is referred to in the literature as "Brown-Sequard epilepsy." Graham Brown, a colleague of Sherrington, applied the then famous and well-known kymograph methods to Brown-Sequard's guinea pig "preparations,"[63-67] and established that the early myoclonic component which progressively developed into the complete tonic and clonic seizures was the scratch reflex. Brown-Sequard's "epileptogenic zone" was Sherrington's "receptive field." That the scratch reflex is a stimulus-induced myoclonia is consistent with the fact that it is lesion induced. In addition, nothing is "adequate" about the "adequate stimulus" that elicits a grooming reflex. Dog owners who are familiar with the scratch reflex have an important insight about the adequate stimulus: to elicit the scratch reflex, you imitate the scratching movements of the animal's leg with respect to intensity, excursion, frequency, and locus; you do not imitate a flea or embedded foreign matter. Sherrington applied two criteria in naming a reflex: one was the topography of the response ("the time relations and the spatial form") and the other was the "mode of the adequate stimulus."[12] No studies of the temporal and spatial pattern of the scratch reflex were made by Sherrington, but Sherrington did systematic studies on the adequate stimulus, finding that the best stimuli for eliciting the scratch reflex were "dragging along the surface a scratching point,"[7] cathodal electrical stimulation of the skin at the frequency of the scratch reflex (about 5 per second), and the dog scratching his own receptive field,[8] i.e., stimuli that match the self-imposed stimulation in terms of frequency, intensity, excursion, and locus. The existence of a best frequency for elicitation of the scratch reflex that matches the frequency of the movement clearly documents the peculiarity of the adequate stimulus.

Sherrington used spinal animals, where the final common path of the grooming lick and grooming bite reflexes is not connected to their receptive fields. Thus the scratch reflex is the only grooming reflex that was available for study. In cats with frontal neocortical lesions or with pontile lesions, the lick and bite reflexes may be used to advantage to obtain information on the adequate stimulus. In most cases, the receptive fields for the lick and bite reflexes correspond: the bite reflex is a shallow scratching bite that is directed to those parts of the body surface that the scratching foot cannot reach. Thus the caudal two-thirds of the cat's body surface is groomed with licks and bites, and in cats with lesions, both reflexes may be elicited from the same area depending on whether the stimulus is like a lick or like a bite. Because the lick and bite reflexes are directed toward any object placed near the cat's face, the cat can be used to groom another cat, and a crucial experiment on the adequate stimulus may be performed. The licks and bites of the first cat, elicited by the experimenter, are directed toward the body surface of the second cat. When this is done, the second cat emits licks when it is licked and emits bites when it receives bites, with both the licks and the bites occurring in midair and in synchrony with the licks and bites of the first cat.

Thus a "new and peculiar" adequate stimulus has appeared. The stimulus that elicits the scratch reflex and other grooming reflexes is not like the cutaneous stimuli that cause grooming behavior (e.g., parasites, foreign matter stuck in the fur, etc.) but like the cutaneous stimuli caused by grooming behavior (i.e., the foot scratching, the teeth biting, the tongue licking). Sherrington and his colleagues referred to the electric stimulus that was delivered to the skin of the spinal dog as an "electric flea"[68,69] when in reality it was an electric foot. The frequency of the best stimulus for eliciting the scratch reflex is the innate frequency of the scratch reflex.

Another abnormality of the grooming reflexes is orientation. Neither the lick, bite, nor scratch reflexes are properly directed to the body surface, but occur in midair or are run off on any object which happens to be in contact with them. The lick and bite reflexes were used by Bard and Rioch[14] as a method of force-feeding their cats in the immediate post-operative period; they elicited the grooming lick and bite reflexes with the cat's head held in the food. The food was ingested, just like the hair and any other materials that are groomed off the body surface in normal grooming. Different "tonic" positions of the leg in the scratch reflex may be obtained as a function of the position of the stimulus on the body surface, but the orientation to the body surface is never complete.[8] Thus the scratch reflex and the other grooming reflexes have no function, a finding that is consistent with their absence in the majority of adult animals. Normal grooming behavior is purposive, consistent with Sherrington's concept of behavior, because it continues until a clean body surface has been obtained. But the termination of grooming reflexes has nothing to do with the condition of the body surface, but only with the absence of its peculiar adequate stimulus. The otherwise astute analyses of Sherrington and his colleagues therefore failed in this one regard: the recognition of the inadequacies of the movement that was labeled the scratch reflex. The misinterpretation by such an authoritative, influential, and venerable group was not without its effect. The scratch reflex is still typically viewed as a normal behavior, with the lesion merely "enhancing its excitability."[70]

Graham Brown added more confusion about the scratch reflex with inconsistent interpretations of his studies on Brown-Sequard epilepsy. On most occasions Graham Brown drew the proper conclusion from his data, stating, for example, that the scratching *movement* in Brown-Sequard epilepsy "appears to be a scratch reflex."[64] But on other occasions he asserted that Brown-Sequard *epilepsy* was "nothing more or less than a specific instance of the scratch reflex"[67] or "a special instance of the scratch reflex."[63] What makes Brown-Sequard epilepsy special, of course, is that it is a stimulus-induced seizure that begins with the scratch reflex and ends with the animal falling, rolling, and thrashing on the floor, urinating, defecating, ejaculating, and exhibiting a complete tonic and clonic seizure (as observed by Brown-Sequard, Graham Brown, Vulpian, Westphal, and Horsley among others, and described as "wonderfully like that of ordinary epilepsy" by *The Medical Times and Gazette*[71]).

The "authoritative" reviewers of the field of epilepsy accepted Graham Brown's occasional overstatement and ignored his data. Penfield and Erickson,[72] for example, asserted that Graham Brown had demonstrated that Brown-Sequard epilepsy "involved an erroneous interpretation of the scratch reflex". In a later review by Penfield and Jasper,[73] Brown-Sequard's work was relegated to a mere footnote, which stated that the seizures were "probably no more than a scratch reflex." In a systematic review on epilepsy by O'Leary and Goldring,[74] Graham Brown's research is seen as "a crushing blow" to Brown-Sequard's work on this animal model of epilepsy. The now traditional and standard misinterpretation is included in Olmsted's biography[75] of Brown-Sequard. Thus Brown-Sequard epilepsy is currently identified as a scratch reflex rather than the scratch reflex being identified as a stimulus-induced myoclonia. And the guinea pig, apparently a highly epilepsy-prone species, has been mostly ignored as an animal model for this disease.

But a knowledge of the basic neural mechanisms of the scratch reflex and of its relationship to behavior has relevance to important problems, not only in the neurology of epilepsy, but more generally in conceptualizing the functions of the central nervous system. The important properties lacking in the scratch reflex that make it less than a behavior are properties that are imposed by the brain superimposed on the spinal cord mechanisms. Brain calls forth the movement, brain directs and persists, brain provides the adaptive and purposive triggering and maintenance of the movement neurologically mediated totally by cord. Cord integrates the metamers into a complex spatial and temporal pattern that is the scratch reflex, but brain welds the integrated metamers to purpose. In Sherrington's words,[12] "The motile and consolidated individual is driven, guided, and controlled by above all organs, its cerebrum." The disintegration of cord and "cerebrum" in cats with pontile or frontal neocortical lesions and in thyroidectomized cats appears to involve independent deficits in serotoninergic neuronal function and in glucocorticoid function. The hierarchically organized system of controls and regulators of grooming behaviors involve in part a tectal influence on the endogenous pattern of the spinal cord.

ACKNOWLEDGMENTS

The author wishes to thank Virginia Parsons, Walter Rogers, Michael Trulson, Richard Swenson, John Elbin, Brian Cooper, and Ralph Johnson for their intelligent devotion to the problems presented in this paper.

REFERENCES

1. GOLTZ, F. & E. GERGENS. 1876. Pfleugers Arch. **13:** 1–44.
2. GERGENS, E. 1877. Pfleugers Arch. **14:** 340–344.
3. GOLTZ, F. 1892. Pfluegers Arch. **51:** 570–614.
4. FOSTER, M. 1879. A Textbook of Physiology, Vol. 3. 3rd edit. Macmillan. New York.
5. SHERRINGTON, C. 1900. *In* Text-book of Physiology, Vol. 2. Macmillan. New York. pp. 793–883.
6. SHERRINGTON, C. & E. LASLETT. 1903. J. Physiol. **29:** 58–96.
7. SHERRINGTON, C. 1903. J. Physiol. (London) **30:** 39–46.
8. SHERRINGTON, C. 1906. J. Physiol. (London) **34:** 1–50.
9. SHERRINGTON, C. 1910. J. Exp. Physiol. **3:** 213–220.
10. GRANIT, R. 1967. Charles Scott Sherrington. Doubleday. Garden City, NY.
11. ROGERS, W., V. PARSONS & W. RANDALL. Psychon. Sci. **23:** 375–376.
12. SHERRINGTON, C. 1906. The Integrative Action of the Nervous System. Yale Univ. Press. New Haven, CT.
13. SCHALTENBRAND, G. & S. COBB. 1930. Brain **53:** 449–488.
14. BARD, P. & D. RIOCH. 1937. Johns Hopkins Med. J. **60:** 73–148.
15. BRADFORD, F. 1939. J. Neurophysiol. **2:** 192–201.
16. BERNSTON, G. & H. HUGHE. 1974. Exp. Neurol. **44:** 255–265.
17. BERNSTON, G. 1988. This volume.
18. SHORE, P. 1972. Ann. Rev. Pharmacol. **12:** 209–222.
19. MCEWEN, B. B. 1979. *In* Glucocorticoid Hormone Action. Springer-Verlag. New York. pp. 467–492.
20. DEKLOET, R., H. SYBESMAN & H. REUL. 1986. Neuroendocrinology **42:** 513–521.
21. GOODRICH, C. 1969. Brit. J. Pharmacol. **37:** 87–93.
22. MOIR, A. & D. ECCLESTON. 1969. J. Neurochem. **15:** 1093–1108.
23. PEROUTKA, S. & S. SNYDER. 1981. Brain Res. **208:** 339–347.
24. PEROUTKA, S. & S. SNYDER. 1982. *In* Biology of Serotonergic Transmission. N. Osborne, Ed.: 279–298. John Wiley & Sons. New York.
25. SNELGAR, R. & M. VOGT. 1980. J. Physiol. (London) **314:** 395–410.

26. SOUBRIE, P. & J. GLOWINSKI. 1984. *In* Progress in Tryptophan and Serotonin Research. H. Schlossberger *et al.*, Eds.: 217-230. Walter de Gruyter. Berlin.
27. HESS, S. & W. DOEPFNER. 1961. Arch. Int. Pharmacodyn. Ther. **134:** 89-99.
28. MARSDEN, C., J. CONTI, E. STROPE, G. CURZON & R. ADAMS. 1979. Brain Res. **171:** 85-99.
29. ELKS, M. W., W. YOUNGBLOOD & J. KISER. 1979. Brain Res. **172:** 471-486.
30. GREEN, A. & D. GRAHAM-SMITH. 1975. *In* Handbook of Psychopharmacology, Vol 3. L. Iversen *et al.*, Eds.: 162-245. Plenum. New York.
31. SANDERS-BUSH, E. & L. MARTIN. 1982. *In* Biology of Serotonin Transmission. N. Osborne, Ed.: 95-118. John Wiley & Sons. New York.
32. BAUMGARTEN, H. & H. SCHLOSSBERGER. 1984. *In* Progress in Tryptophan and Serotonin Research. H. Schlossberger *et al.*, Eds.: 173-188. Walter de Gruyter. New York.
33. ELBIN, J. 1980. Ph.D. dissertation, University of Iowa, Iowa City, IA.
34. TRULSON, M. & W. RANDALL. 1976. J. Comp. Physiol. Psychol. **90:** 917-924.
35. CUNNINGHAM, J. 1984. M. A. thesis, University of Iowa, Iowa City, IA.
36. ZITRIN, A. F., F. BEACH, J. BARCHAS & W. DEMENT. 1970. Science **170:** 868-869.
37. MCGEER, P., E. MCGEER & J. WADA. 1963. Arch. Neurol. **9:** 91-99.
38. TRULSON, M., J. NICOLAY & W. RANDALL. 1975. Pharmac. Biochem. Behav. **3:** 87-94.
39. RANDALL, W. & V. PARSONS. 1971. J. Interdiscip. Cycle Res. **1:** 3-24.
40. RANDALL, W., J. ELBIN & R. SWENSON. 1974. J. Comp. Physiol. Psychol. **86:** 747-750.
41. ELBIN, J. & W. RANDALL. 1977. J. Comp. Physiol. Psychol. **91:** 300-312.
42. SWENSON, R. & W. RANDALL. 1980. J. Comp. Physiol. Psychol. **94:** 353-364.
43. RANDALL, W. & J. LIITTSCHWAGER. 1967. J. Psychiatr. Res. **5:** 39-58.
44. SHOPSIN, B., L. SHENKMAN, I. SANGHVI & C. HOLLANDER. 1974. *In* Advances in Biochemical Psychopharmacology, Vol. 10. E. Costa *et al.*, Eds.: Raven. New York.
45. BAJUSZ, E. 1969. *In* Physiology and Pathology of Adaptation Mechanisms. E. Bajusz, Ed.: 89-145. Pergamon. New York.
46. MARTIAL, J., P. SEEBURG, D. MATULICH, H. GOODMAN & J. BASTER. 1979. *In* Glucocorticoid Hormone Action. J. Baster *et al.*, Ed.: 279-289. Springer-Verlag. New York.
47. RANDALL, W., M. TRULSON & V. PARSONS. 1976. J. Comp. Physiol. Psychol. **90:** 231-243.
48. TRULSON, M. & W. RANDALL. 1973. J. Comp. Physiol. Psychol. **85:** 1-10.
49. RANDALL, W. & M. TRULSON. 1974. Pharmacol. Biochem. Behav. **2:** 355-360.
50. JOHNSON, R. & W. RANDALL. 1983. Behav. Neurosci. **97:** 195-209.
51. KLAWANS, H, C. GOETZ & W. WEINER. 1973. Neurology **23:** 1234-1240.
52. STEWARD, R. & J. GROWDON. 1976. Neuropharmacology **15:** 449-455.
53. HEHMAN, K. & A. VONDERAHE. 1961. Neurology **11:** 1011-1016.
54. SWENSON, R. 1978. Ph.D. thesis, University of Iowa, Iowa City, IA.
55. ELBIN, J. 1980. Ph.D. thesis, University of Iowa, Iowa City, IA.
56. ENDROCZI, E. & K. LISSAK. 1964. Acta Physiol. **24:** 65-77.
57. LINDQUIST, F. 1956. Design and Analysis of Experiments in Psychology and Education. Houghton Mifflin. Boston, MA.
58. STUMP, W. & M. SAR. 1981. *In* Steroid Hormone Regulation of the Brain. K. Fuxe *et al.*, Eds.: 41-50. Pergamon. New York.
59. COOPER, B. 1980. Ph.D. thesis, University of Iowa, Iowa City, IA.
60. HENKIN, R. 1970. Prog. Brain Res. **32:** 270-293.
61. RANDALL, W., S. RANDALL & R. JOHNSON. 1985. Behav. Neurosci. **99:** 109-121.
62. BROWN-SEQUARD, C. 1850. C. R. Soc. Biol. (Paris) **2:** 105-106.
63. GRAHAM BROWN, T. 1909. Q. J. Exp. Physiol. **2:** 243-275.
64. GRAHAM BROWN, T. 1910. Q. J. Exp. Physiol. **3:** 21-52.
65. GRAHAM BROWN, T. 1910. Q. J. Exp. Physiol. **3:** 139-170.
66. GRAHAM BROWN, T. 1910. Q. J. Exp. Physiol. **3:** 319-353.
67. GRAHAM BROWN, T. 1912. Proc. R. Soc. London, Ser. B **84:** 555-579.
68. SHERRINGTON, C. 1931. Brain **54:** 1-28.
69. LIDDELL, E. 1960. The Discovery of Reflexes. Clarendon Press. Oxford.
70. SWAZEY, J. 1969. Reflexes and Motor Integration: Sherrington's Concept of Integrative Action. Harvard Univ. Press. Cambridge, MA.
71. MEDICAL TIMES AND GAZETTE (LONDON). 1863. **1:** 324-326.

72. PENFIELD, W. & T. ERICKSON. 1941. Epilepsy and Cerebral Localization. Thomas. Spring-field, IL.
73. PENFIELD, W. & H. JASPER. 1954. Epilepsy and the Functional Anatomy of the Human Brain. Little, Brown. Boston.
74. O'LEARY, J. & S. GOLDRING. 1976. Science and Epilepsy. Raven. New York.
75. OLMSTED, J. 1946. Charles-Edouard Brown-Sequard, a Nineteenth Century Neurologist and Endrocrinologist. Johns Hopkins Univ. Press. Baltimore.

Excessive Grooming in Response to Opiate Drugs

The Ontogeny of Responsiveness

ROBERT L. ISAACSON

Center for Neurobehavioral Sciences
Department of Psychology
University Center at Binghamton
Binghamton, New York 13901

Most interesting scientific stories have a history, a chain of events or experiments and human interactions that have led up to the current projects. Such is the case for the work I am about to describe.

The story goes back to a type of research in which I was involved some years ago: developmental pharmacology. I was stimulated to undertake some new studies in this field when I read articles by Dr. Alan Cowan and his associates[1,2] in which it was reported that the peripheral injection of a novel opioid drug, the experimental codeione RX 336-M, induced excessive grooming in rats. My colleagues and I have been studying excessive grooming in the rat for about the past 10 years, but usually following the central injection of the active agents, most often neuropeptides related to some portion of the ACTH molecule (for a review see Gispen and Isaacson.[3]) We had *not* been successful in inducing excessive grooming through the peripheral administration of any of the agents with which we had worked.

I called Dr. Cowan and he was kind enough to send me some of the experimental drug, which we injected into some of the rats from our colony. To our surprise there was no observable effect of the drug. I called Dr. Cowan to ask why he had played a trick on me by sending me an inactive compound. He said it was no trick and asked the weights of the animals we used. When I told him that the animals had weighed between 250 and 300 g, he informed me that animals that large would not respond to systemic administration of the drug.

Because under usual conditions a rat's weight is closely related to its age, it seemed that we had stumbled onto an interesting age-related drug-responsiveness paradigm.

I therefore undertook, with John Hannigan and Cheryl-Ann Hardy, a study to clearly define the age-related responsiveness of young animals to the codeione.[4] The overall relationships we discovered are shown in FIGURE 1.

These results show a decreasing effectiveness of the drug in inducing excessive grooming as the animal grows older. The administration of 6 mg/kg of RX 336-M produces remarkably high levels of grooming through the first 40 days after birth, declines between 40 and 50 days, but does not reach control levels until the animals are 60 days of age or older. Female animals are less responsive overall, and show a more rapid decline in grooming with age. There is also a substantial reduction in the females' responsiveness at 35–40 days of age. The reduction in female responsiveness to the drug can be restored at that age by the administration of testosterone.

To further understand the nature of the grooming induced by RX 336-M, we exam-

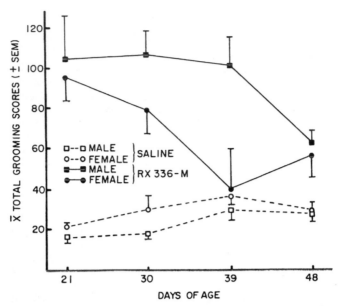

FIGURE 1. Mean total grooming scores (± SEM) of male and female rats of four different ages given 6 mg/kg RX 336-M in saline or saline alone. (From Isaacson *et al.*[4])

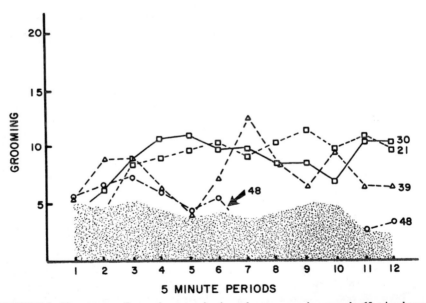

FIGURE 2. Time course of excessive grooming in male rats occurring over the 60-min observation period. The numbers to the right of each set of symbols, or behind the arrow, indicate the age at which the group of animals was tested. The *y*-axis indicates the mean number of grooming scores achieved in each 5-min period, using the 15-s time-sampling technique developed by Gispen *et al.*[5] (maximum for 5 min: 20). The *shaded area* on the lower portion of the figure is the area below the highest score of any of the saline-treated control groups. (From Isaacson *et al.*[4])

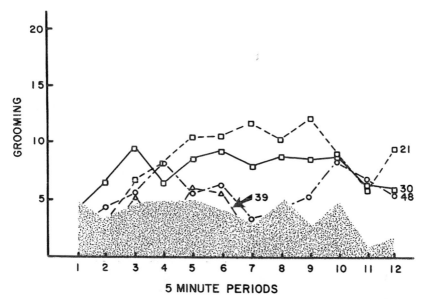

FIGURE 3. The time course of excessive grooming in female rats occurring during the 60-min observation period. As in FIGURE 2, the numbers indicate the ages of the animals at the time of testing. The other aspects of the figure are as in FIGURE 2. (From Isaacson *et al.*[4])

ined the time course of the effects in males and in females over the hour-long observation period. The data from the males and females are shown in FIGURES 2 and 3. The performance of the saline control groups at different ages were intermixed with each other and not significantly different from one another. Therefore, I have combined them all into the stippled area at the bottom of the two figures. The top of the stippled area represents the highest mean value of any of the saline-injected control groups at the specified test age.

The shape of this stippled area contains information of value. It shows that the excessive grooming found about 15–25 min after an animal is handled, transported, and placed into a novel environment does not occur in animals less than 40 days of age. The bottom curve of FIGURE 4 shows the nature of this "novelty-induced" grooming found in older animals.

It should be noted that the effect of the experimental codeione is detected soon after injection and lasts over the hour-long observation period. This was a bit of a surprise, since given our previous knowledge of morphine-induced grooming I had presumed that the effectiveness of codeine (and the experimental codeione) was due to its demethylation to morphine, probably soon after the drug entered the bloodstream.

Accordingly, we tested the effects of peripherally administered morphine in young animals and did find a small increase in excessive grooming. However, the magnitude of the effect was much less than that of the codeione at any dose and at any age studied. We then went on to test animals with codeine itself in order to compare its effects with those of morphine.

A substantial amount of information on the effects of codeine on grooming in young animals has recently been collected in Utrecht with the enthusiastic cooperation of Jan Brakkee. Before describing this new information, however, let me remind you that there was prior evidence that morphine only began to induce excessive grooming

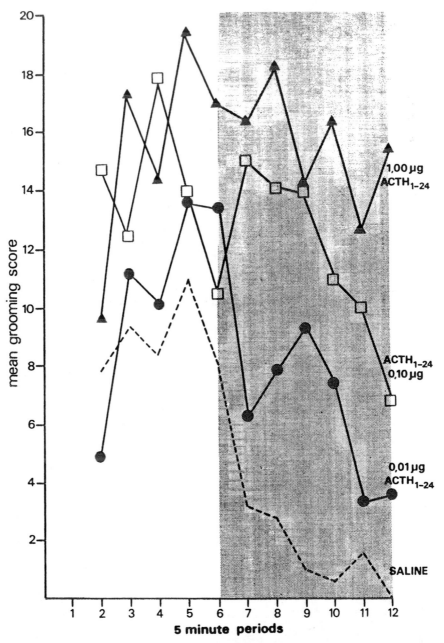

FIGURE 4. A time-course analysis of excessive grooming induced by the i.c.v. injection of different doses of ACTH$_{1-24}$. The time of onset of a possible second, later period of excessive grooming is indicated by the *shaded area*. (From Isaacson *et al.*[6] Reprinted with permission from *Brain Research Bulletin*.)

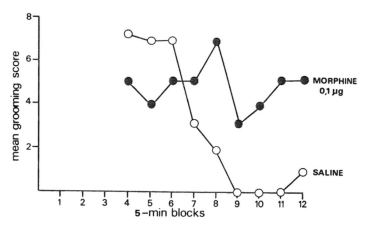

FIGURE 5. Time course of excessive grooming induced by i.c.v. saline or by 0.1 µg of morphine. Note that there was a 15-min delay between the time at which the animals were placed in the test chambers and the commencement of scoring.

40 minutes or more after i.c.v. administration and only produced modest increases in the response. In adult animals given morphine i.c.v., the grooming produced by morphine never produced anything like the amount of grooming induced by even very low doses of one of the active neuropeptides. We had discovered that the reason it produced excessive grooming was because "novelty grooming" did not decline as it usually did after i.c.v. saline injection 40 and more min after administration. Rather, it continued at a moderately high level for the next hour or so, as shown in FIGURE 5. In further support of the delayed effect of morphine activation, when naloxone is given peripherally to animals that have received i.c.v. ACTH, grooming is only reduced in the second half of the observation period.

It therefore seemed safe to conclude that morphine's effects occur late in the usual observation period in adult animals. Would this be true for animals with peripheral injections early in life?

Much of our recent data are presented in FIGURE 6. The curves M_1 and M_2 represent groups of 12 animals, 23–25 days of age. The data points on the right side with stars above them, however, represent data points with only six animals in them. Morphine was given subcutaneously at a dose of 3.0 mg/kg, free base. The dose of codeine used was equimolar to that of the morphine.

The effects of morphine appeared only late in the observation period, just as is found in adult animals given the drug intracerebroventricularly. The group M_1 subjects were tested on the day they were separated from their mothers and siblings. It was the first time they had been so isolated. The M_2 group was tested 2 days later after their separation from their mother and siblings. It seems likely that the difference between the M_1 and the M_2 groups is due to a suppressive effect of isolation stress on the morphine-induced excessive grooming.

The codeine animals, group C, in this figure were a bit older than the animals of the M_1 and the M_2 groups. They were about 45–50 days of age at testing. Like the RX 336-M animals, the excessive grooming began early in the observation period, in sharp contrast to the effects of morphine. However, precisely the same codeine curve is found in animals tested at 25 days after birth.

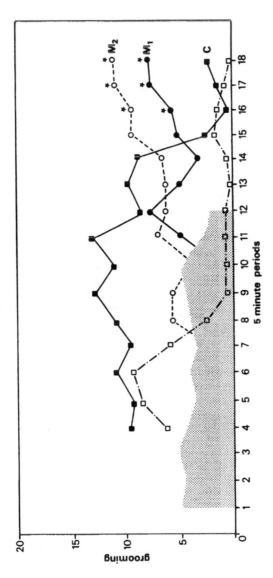

FIGURE 6. Time course of excessive grooming of animals receiving subcutaneous injections of 3.0 mg/kg morphine (expressed as the free base) at one of the two time periods after birth, either at 23 or 25 days of age (M_1 and M_2 respectively). The animals of group C received a dose of codeine equimolar to that of 3.0 mg/kg morphine at 45–50 days of age, although similar effects are found in younger animals (*i.e.,* 25 days of age). The *open squares* represent the grooming of animals given subcutaneous saline when 50 days old. The *shaded area* represents the scores of saline-treated male animals at earlier ages, as in FIGURE 2. Note the extension of number of the observation periods in the figure.

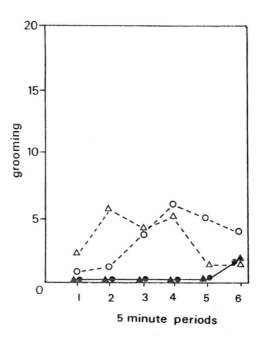

FIGURE 7. Suppression of grooming by subcutaneous injection of 3.0 mg/kg morphine in both male (*triangles*) and female (*circles*) rats 25 days after birth in the first half hour of the observation period. *Open symbols* represent saline-treated animals, *filled symbols* morphine-treated animals.

As can be seen in the figure, the observation period has been extended from the usual 50 or 60 min to 90 min because we noted two unexpected phenomena that needed to be documented. The first is the unexpected sudden drop in excessive grooming that occurred about 70 min after the administration of codeine. According to most published studies, this effect cannot be correlated with a corresponding rapid decline in the level of codeine in the blood. Second, we noted that morphine-induced grooming began to increase greatly after 70–80 min, an effect that obviously cannot be seen when the observation period lasts only 60 min.

Briefly, I would like to point out two other results of the study. First, morphine actually suppresses the amount of excessive grooming exhibited during the first 30 min of the test period. This is shown in FIGURE 7. During this period, the animals are alert, sit quietly, are capable of movement, but do not groom. On the other hand, a few minutes after codeine, many animals exhibit brief bouts of jumping or hopping that often precede bouts of grooming. However, such reactions are not necessary precursors of grooming. Often, in fact, they interrupt bouts of grooming.

Grooming induced by codeine can be blocked by 1 mg/kg naloxone, as is shown in FIGURE 8. In the original work of Cowan and his associates, it was reported that grooming produced by RX 336-M could not be blocked by this opiate antagonist. It will be of interest to determine if we can block the effects of the RX compound on grooming by naloxone under the conditions used in our laboratory which are quite different from those in Cowan's studies.

Our results have generated more questions than answers. At the very least, they show that the effects of morphine and codeine are quite dissimilar and the drugs may not act on the same receptor systems. The fact that only about 10% of codeine becomes demethylated into morphine in the blood supports this possibility. An important study to be undertaken is the determination of whether i.c.v. codeine in adult animals

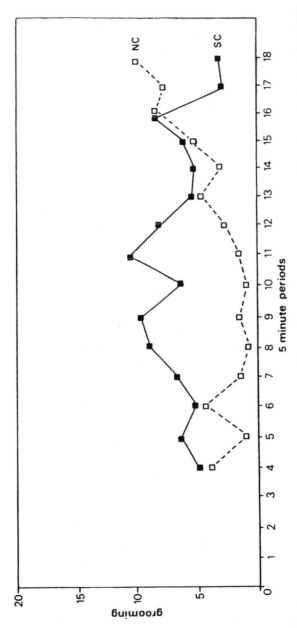

FIGURE 8. Suppression of subcutaneously administered codeine-induced excessive grooming in 22-day-old female rats by pretreatment with 1.0 mg/kg naloxone or saline. *NC*, group given naloxone before codeine; *SC*, group given saline before codeine. Note again the extended observation periods.

mimics the curves found after systemic administration in young animals or the curves produced by i.c.v. morphine in adults.

Our morphine results are also surprising. When given intravenously, morphine has poor penetration into the brain but great analgesic effectiveness when it gets there, Codeine penetrates much more readily but has much less analgesic effect. Furthermore, the effectiveness of codeine in inducing excessive grooming is inversely related to the numbers of high affinity mu receptors at different ages. Since these "analgesic receptors" develop quite differently from the low affinity types to which codeine has substantial affinity, the relationship between these low affinity opiate systems and excessive grooming should be pursued.

In any case, in trying to further our understanding of the developmental effects of the opiates on behavior, we seem to have opened some new directions of research that need to be pursued.

[NOTE ADDED IN PROOF: Since the preparation of this manuscript, a more extensive report of studies with codeine and morphine has been submitted for publication (Isaacson, Danks, Brakkee, Schefman, and Gispen. Behav. Neur. Biol. Submitted). In addition, recent work by Spector and his associates (Donnerer, J. *et al.* 1986. Proc. Nat. Acad. Sci. **83**: 4566–4567; Donnerer, J. *et al.* 1987. J. Pharm. Exp. Therap. In press) has found evidence for the presence of endogenous codeine and morphine in the central nervous system and in the rat, and demonstrated the potential for their alteration through physiological interventions.]

ACKNOWLEDGMENTS

The assistance of Ms. Lia Claessens and others on the staff of the Institute of Molecular Biology and Medical Biotechnology, Division of Molecular Neurobiology, University of Utrecht, in the preparation of this manuscript is greatly appreciated.

REFERENCES

1. COWAN, A. 1981. RX 336-M, a new chemical tool in the analysis of the quasi-morphine withdrawal syndrome. Fed. Proc. **40**: 1497–1501.
2. GMEREK, D. E. & A. COWAN. 1982. A study of shaking and grooming induced by RX 336-M in rats. Pharmacol. Biochem. Behav. **16**: 929–932.
3. GISPEN, W. H. & R. L. ISAACSON. 1985. Excessive grooming in response to ACTH. *In* CNS Effects of ACTH, MSH, and Opioid Peptides. International Encyclopedia of Pharmacology and Therapeutics, Vol. 1. D. de Wied, W. H. Gispen, and Tj. B. van Wimersma Greidanus, Eds.: 273–312. Pergamon. Oxford.
4. ISAACSON, R. L., C.-A. HARDY & J. H. HANNIGAN, JR. 1987. Age-and sex-related induction of excessive grooming and "wet-dog" shakes in the rat. Behav. Neural Biol. **47**: 250–261.
5. GISPEN, W. H., V. M. WIEGANT, H. M. GREVEN & D. DE WIED. 1975. The induction of excessive grooming in the rat induced by intracerebroventricular application of peptides derived from ACTH: Structure-activity studies. Life Sci. **17**: 645–652.
6. ISAACSON, R. L., J. H. HANNIGAN, JR., J. BRAKKEE & W. H. GISPEN. 1983. The time course of excessive grooming. Brain Res. Bull. **11**: 289–293.

Yawning and Penile Erection: Central Dopamine-Oxytocin-Adrenocorticotropin Connection[a]

ANTONIO ARGIOLAS, MARIA R. MELIS,
AND GIAN L. GESSA

Institute of Pharmacology
University of Cagliari
09100 Cagliari, Italy

Repeated episodes of yawning and penile erection can be induced in rats and other experimental animals by the systemic administration of low doses of dopamine (DA) agonists, such as apomorphine (see ref. 1 for review), by the central administration of adrenocorticotropin (ACTH), α-melanocyte-stimulating hormone (α-MSH) and related peptides (see ref. 2 for review), and by the intracerebroventricular (i.c.v.) injection of oxytocin.[3] While the importance of penile erection in reproduction does not need to be further stressed, it is pertinent to recall that yawning, alone or associated with stretching, is considered to be an ancestral vestige surviving throughout evolution, that subserves the purpose of arousal. In particular, the role of yawning could be that of increasing attention when sleep is imminent due to fatigue or boredom but cannot be engaged in, as in face of a danger or social circumstances (for a review on the physiological significance of yawning, see ref. 4).

The capability of the above unrelated substances to induce such similar symptomatology raises the possibility that a neuronal link exists among DA, oxytocin, and ACTH in the central nervous system. The results of the experiments presented below, which were performed with the aim of demonstrating the existence of such a link and to clarify the neural mechanisms underlying the expression of yawning and penile erection as well, provide evidence for the first time that DA, oxytocin, and ACTH act in the hypothalamus in a sequence to induce these behavioral responses.

Effect of DA Antagonists on Yawning and Penile Erection Induced by Apomorphine, Oxytocin, and ACTH$_{1-24}$

The first group of experiments was aimed at the identification of possible interactions among DA agonists, oxytocin, and ACTH-MSH peptides in the induction of yawning and penile erection, namely to clarify whether oxytocin and/or ACTH induce the above responses by releasing DA in brain or vice versa. Male Sprague-Dawley rats were used in all the experiments. For i.c.v. injections, chronic guide cannulae aimed at one lateral ventricle were stereotaxically implanted under chloral hydrate anaesthesia 5 days before the experiments, as previously described.[3] Peptides were injected into a lateral ventricle by means of an internal cannula connected by a polyethylene tubing

[a] This work was supported by CNR grant no. 85.00605.56.

TABLE 1. Effect of the DA Receptor Blockers Haloperidol and (−)Sulpiride on Yawning and Penile Erection Induced by Apomorphine, Oxytocin, and $ACTH_{1-24}$

Pretreatment	mg/kg	Treatment	Yawns/Rat (Mean ± SEM)	Penile Erections/Rat (Mean ± SEM)
saline	—	saline	2.0 ± 0.5	0.3 ± 0.2
saline	—	apomorphine	16.5 ± 1.8*	4.0 ± 0.5*
haloperidol	0.2	apomorphine	3.5 ± 0.6+	0.5 ± 0.2+
(−)sulpiride	10	apomorphine	3.0 ± 0.4+	0.6 ± 0.3+
saline	—	oxytocin	18.0 ± 1.6*	3.9 ± 0.6*
haloperidol	2	oxytocin	15.6 ± 1.0*	3.0 ± 0.5*
(−)sulpiride	50	oxytocin	17.8 ± 2.0*	3.6 ± 0.7*
saline	—	$ACTH_{1-24}$	16.5 ± 1.3*	3.8 ± 0.8*
haloperidol	2	$ACTH_{1-24}$	15.6 ± 1.4*	3.4 ± 1.0*
(−)sulpiride	50	$ACTH_{1-24}$	17.0 ± 1.2*	3.5 ± 0.6*

Haloperidol and sulpiride were injected intraperitoneally in 1 ml of saline per rat at a pH of 4.5–5. Saline was injected in controls. Apomorphine-HCl was dissolved in saline and injected in a volume of 200 μl in the back of the neck, 30 min after haloperidol and sulpiride. Oxytocin (30 ng) and $ACTH_{1-24}$ (5 μg) were dissolved in saline and injected i.c.v. in a volume of 10 μl, 30 min after neuroleptics. The same volume of saline was injected i.c.v. in controls. After treatments, animals were placed individually in Plexiglas cages (30 × 30 × 25 cm) and observed for 60 min during which yawning and penile erection episodes were counted. Each value is the mean ± SEM of three experiments (15 rats per group); * $p < 0.001$ with respect to saline-treated rats; + $p < 0.001$ with respect to the corresponding group pretreated with saline (Duncan's new multiple range test).

to a 10-μl Hamilton syringe driven by a micrometric screw. Other drugs were given systemically as reported in the legends of tables and figures. TABLE 1 shows the effect of haloperidol and (−)-sulpiride, two specific DA receptor blockers, on yawning and penile erection induced by apomorphine, oxytocin, and $ACTH_{1-24}$. In agreement with previous studies,[1-3] apomorphine (80 μg/kg s.c.), oxytocin (30 ng i.c.v.), and $ACTH_{1-24}$ (5 μg i.c.v.) induced repeated episodes of yawning and penile erection. Haloperidol (0.2 mg/kg i.p.) and (−)-sulpiride (10 mg/kg i.p.) administered 30 min beforehand completely suppressed apomorphine-induced response; these receptor blockers, however, were ineffective against yawning and penile erection induced by oxytocin and $ACTH_{1-24}$ even at doses of 2 mg/kg and 50 mg/kg, respectively.

The failure of haloperidol and sulpiride to antagonize yawning and penile erection induced by oxytocin and $ACTH_{1-24}$ suggests that the two peptides do not induce the above responses by releasing DA in the central nervous system, but act downstream from DA receptors.

Effect of the Oxytocin Antagonist [d(CH₂)₅,Tyr (Me),Orn-8]Vasotocin on Yawning and Penile Erection Induced by Apomorphine, Oxytocin and $ACTH_{1-24}$

The availability of a potent oxytocin antagonist [d(CH₂)₅,Tyr(Me),Orn-8]vasotocin (kindly provided by Dr. M. Manning, University of Toledo)[5] gave us the possibility to verify if DA agonists and/or ACTH induce yawning and penile erection by releasing oxytocin in brain. The effect of the i.c.v. administration of [d(CH₂) ₅,Tyr(Me), Orn-8] vasotocin on yawning and penile erection induced by apomorphine, oxytocin, and $ACTH_{1-24}$ is shown in FIGURE 1. The oxytocin antagonist injected i.c.v. 15 min before the administration of the above substances not only antagonized in a dose-dependent

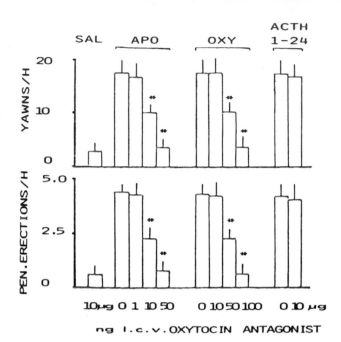

FIGURE 1. Effect of [dCH₂)₅,Tyr(Me), Orn-8]vasotocin on penile erection and yawning induced by apomorphine, oxytocin, and ACTH$_{1-24}$. Pretreatment i.c.v. with saline or [dCH₂)₅, Tyr(Me), Orn-8]vasotocin was performed 15 min before i.c.v. oxytocin (30 ng), ACTH$_{1-24}$ (5 μg), or apomorphine (80 μg/kg s.c.). After treatment the animals were placed individually into Plexiglas cages and observed for 60 min. Penile erection and yawning episodes were counted; $**p < 0.001$ with respect to the corresponding group not receiving [dCH₂)₅,Tyr(Me), Orn-8]vasotocin (Duncan's new multiple range test).

TABLE 2. Yawning and Penile Erection in MSG-treated Rats: Effect of Apomorphine, Oxytocin, and ACTH$_{1-24}$

| | | Neonatal Treatment | | | |
| | | Yawns/Rat (Mean ± SEM) | | Penile Erections/Rat (Mean ± SEM) | |
Treatment	Dose	Saline	MSG	Saline	MSG
Saline	1 ml i.p.	2.9 ± 0.3	2.0 ± 0.5	0.3 ± 0.2	0.5 ± 0.2
Apomorphine	80 μg/kg s.c.	18.5 ± 1.0*	17.2 ± 0.9*	3.6 ± 0.5*	3.6 ± 0.3*
Oxytocin	30 ng i.c.v.	18.0 ± 1.3*	18.2 ± 1.8*	3.9 ± 0.6*	3.5 ± 0.5*
ACTH$_{1-24}$	1 μg i.c.v.	3.0 ± 0.6	2.7 ± 0.5	0.6 ± 0.1	0.9 ± 0.3
ACTH$_{1-24}$	10 μg i.c.v.	19.0 ± 3.0*	18.5 ± 2.0*	3.9 ± 0.5*	3.2 ± 0.4*

Neonatal saline and MSG treatment was performed in pups in the first, third, fifth, seventh, and ninth day of life by administering 4 g/kg of MSG or the same volume of saline. The experiments were performed when the rats were 4 months old. At this age, rats were chronically implanted with guide cannulae aimed at one lateral ventricle 5 days before the experiments as described in the test. Other conditions for apomorphine, oxytocin, and ACTH$_{1-24}$ are the same described in the legend of TABLE 1. Yawning and penile erection episodes were counted for 60 min after treatment. Each value is the mean ± SEM of two experiments (10 rats per group); $* p < 0.001$ with respect to saline-treated rats (Duncan's new multiple range test).

manner oxytocin-induced response, as was expected, but also even more effectively antagonized the response induced by apomorphine. A 50% inhibition of the apomorphine and oxytocin effect was already obtained with 10 ng and 50 ng of the oxytocin antagonist, respectively. A complete suppression of either the apomorphine or oxytocin effect was obtained with 100 ng of the peptide. On the contrary, a dose as high as up to 10 μg of the oxytocin analogue was unable to antagonize yawning and penile erection induced by 5 μg of i.c.v. $ACTH_{1-24}$. It is noteworthy that doses of the oxytocin antagonist that suppressed apomorphine-induced yawning and penile erection were totally ineffective in antagonizing stereotypy and hypermotility induced by 1 mg/kg of apomorphine administered subcutaneously (results not shown).

The results obtained with [d(CH_2) $_5$,Tyr(Me), Orn-8]vasotocin suggest that apomorphine and other DA agonists, but not ACTH-MSH peptides, induce yawning and penile erection by releasing oxytocin in some brain area, and that ACTH-MSH peptides act downstream from DA receptors as well as oxytocin to induce the above responses.

Effect of Neonatal Monosodium Glutamate (MSG) Treatment on Yawning and Penile Erection Induced by Apomorphine, Oxytocin, and $ACTH_{1-24}$

The results obtained with DA and oxytocin antagonists suggest that DA, oxytocin, and ACTH act in sequence to induce yawning and penile erection. However, the possibility that DA and/or oxytocin induce the above responses by releasing an ACTH-derived peptide from the recently discovered opiomelanotropinergic neurons in the hypothalamus (for a review, see ref. 6) remains to be verified. Since specific antagonists of ACTH-MSH peptides capable of antagonizing their central effects are not available at present, we have attempted to verify the above possibility by studying the effect of apomorphine, oxytocin, and $ACTH_{1-24}$ on yawning and penile erection in rats neonatally treated with monosodium glutamate (MSG). Such treatment has been found to cause the almost complete depletion of brain ACTH-, α-MSH-, and endorphin-like peptides without altering their pituitary and circulating concentrations.[7,8] Under our conditions, neonatal MSG treatment caused both the expected marked reduction in growth, secondary to the destruction of hypothalamic growth-hormone-releasing hormone,[9] and a decrease of about 90% in the hypothalamic concentration of ACTH and α-MSH as measured by specific radioimmunoassays.[10,11] The results obtained with neonatally MSG-treated rats are shown in TABLE 2. Surprisingly, the depletion of ACTH-MSH-like peptides from the hypothalamus was completely ineffective in modifying yawning and penile erection induced not only by apomorphine and oxytocin but also by $ACTH_{1-24}$.

The ineffectiveness of the hypothalamic depletion of ACTH-MSH peptides to modify the behavioral effects of the DA agonist and oxytocin effect suggests that oxytocin and DA agonists do not induce yawning and penile erection by releasing an ACTH-derived peptide in brain, although it is possible that the small amount of ACTH and α-MSH remaining in the hypothalamus still might be sufficient to mediate the DA and/or oxytocin effect. This possibility is unlikely, however, since doses of ACTH or α-MSH much higher than those of DA agonists or oxytocin are needed to induce yawning and penile erection, and no supersensitivity to ACTH or α-MSH was found in neonatally MSG-treated rats not only with regard to yawning and penile erection (present results) but also to other behavioral and biochemical responses.[12,13]

Another possibility that cannot be completely ruled out by the results obtained with MSG-treated rats is that ACTH or α-MSH are released by DA agonists or oxytocin in some brain area where they are not depleted by MSG treatment, since opiomelanotropinergic neurons have been identified also in extrahypothalamic brain areas.[14]

However, previous studies have shown that the most sensitive brain areas for the induction of yawning and penile erection by ACTH-MSH peptides are localized in the hypothalamus, and are those surrounding the third ventricle.[2] Hence, it is unlikely that ACTH-MSH peptides act to induce yawning and penile erection in an area other than the hypothalamus.

Oxytocin- and Apomorphine-induced Yawning and Penile Erection: Site of Action in Brain

Besides the studies cited above showing that ACTH-MSH peptides induce yawning and penile erection by acting in the hypothalamic regions surrounding the third ventricle,[2] no information was available so far about the brain areas where DA agonists and/or oxytocin act in order to induce such responses. In an attempt to identify these brain areas we have microinjected apomorphine and/or oxytocin in discrete brain regions through chronic guide cannulae, stereotaxically implanted under chloral hydrate anaesthesia 5 days before the experiments, according to a stereotaxic atlas of the rat brain.[15] Apomorphine and oxytocin were dissolved in saline and injected into the various nuclei in a volume of 0.3 µl per site by means of an internal cannula connected by polyethylene tubing to a 10-µl Hamilton syringe driven by a Stoelting microinfusion pump. The length of the internal cannula was adjusted according to the position nuclei to be injected. The correct position of the cannula tip was verified at the end of the experiments by histological analysis as previously described.[16] The brain areas that were microinjected with saline, apomorphine (1 µg), and oxytocin (30 ng) are listed in TABLE 3. The paraventricular nucleus of the hypothalamus (PVN) was found to be the only area where microinjections of apomorphine and oxytocin induced yawning and penile erection. Surprisingly, no effect was observed when apomorphine was microinjected in areas very rich in DA and DA receptors, such as the striatum, the nucleus accumbens, or the substantia nigra. Microinjections of apomorphine or oxytocin in other hypothalamic nuclei very close to the PVN, such as the ventromedial and dorsomedial nuclei and the preoptic area were also ineffective. The effect of apomor-

TABLE 3. Yawning and Penile Erection by Apomorphine and Oxytocin Microinjections in Different Brain Areas

Brain Area	Apomorphine	Oxytocin
Striatum	none	none
Nucleus accumbens	none	none
Substantia nigra	none	none
Paraventricular nucleus	yawning, penile erection	yawning, penile erection
Dorsomedial nucleus	none	none
Ventromedial nucleus	none	none
Preoptic area	none	none

1 µg of apomorphine or 10 ng of oxytocin were injected in each site through bilateral chronic guide cannulae by means of an internal cannula connected by polyethylene tubing to a 10-µl Hamilton syringe driven by a Stoelting microinfusion pump, except for the paraventricular and dorsomedial nucleus where 1 µg of apomorphine and 10 ng of oxytocin were injected unilaterally. The injection volume was 0.3 µl per site. After microinjections, rats were placed individually in Plexiglas cages and observed for 60 min during which yawning and penile erection episodes were counted. Ten rats per group were used: four were injected with saline alone (controls) and the other six with apomorphine or oxytocin. The correct position of the cannula tip in the various nuclei was verified by histological analysis.[16]

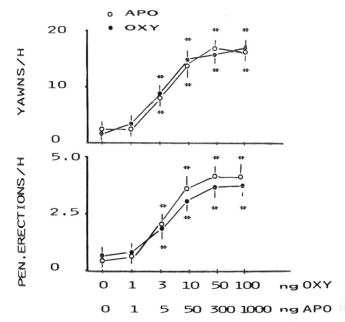

FIGURE 2. Yawning and penile erection induced by apomorphine and oxytocin microinjections into the paraventricular nucleus of the hypothalamus. Apomorphine and oxytocin were microinjected unilaterally into the paraventricular nucleus in a volume of 0.3 μl. Saline was injected in control rats. Microinjections were performed through chronic guide cannulae aimed at the hypothalamic nucleus by means of an internal cannula connected by polyethylene tubing to a 10 μl Hamilton syringe driven by a Stoelting microinfusion pump. After microinjections, the animals were placed individually in Plexiglas cages and observed for 60 min, during which penile erection and yawning episodes were counted; $**p < 0.001$ with respect to saline treated rats (Duncan's multiple range test).

phine and oxytocin microinjections into the PVN was then studied in detail. As shown in FIGURE 2, yawning and penile erection were induced in a dose-dependent manner by both substances. The minimal effective dose of oxytocin and apomorphine was 3 ng and 5 ng, respectively, which induced the response in about 60% of the treated animals. The yawning and penile erections induced by intracranial microinjection of apomorphine or oxytocin microinjection was similar to that observed after systemic apomorphine or i.c.v. oxytocin, except the response started within 5 min after the microinjections. Even at the highest dose tested (1 μg), apomorphine failed to induce stereotypy and hypermotility.

The above results show that both apomorphine and oxytocin act in the hypothalamic PVN to induce yawning and penile erection. The potency of the two substances and the fact that both are present in the PVN suggest that DA and oxytocin might have a physiological role in the control of the nucleus responses. Indeed, PVN contains the cell bodies of at least two types of oxytocinergic neurons: the magnocellular neurons, projecting to the neurohypophysis, from which oxytocin is released into the circulation to exert its hormonal role in parturition and lactation (for a review see ref. 17), and the parvocellular neurons, many of which send their projections to several extrahypothalamic brain areas.[18,19] In addition to oxytocinergic cell bodies, the PVN also contains the cell bodies of dopaminergic neurons of the group A14,[20] which together with those of the groups A11 and A13 constitute the so-called incertohypothalamic system.[21] Our results suggest that DA agonists interact with DA receptors in the PVN or surrounding structures to stimulate the activity of oxytocinergic neurons, which in turn mediate the appearance of yawning and penile erection. In support of this hypothesis, immunocytochemical studies have shown that DA neurons in the PVN are mainly located in the proximity of oxytocinergic neurons.[22] As for the kind of DA receptors mediating yawning and penile erection, previous studies have shown that they belong to the D2 type, although whether they are DA autoreceptors (a special kind of inhibitory receptor located in the nerve terminal and cell body of the neuron itself) or postsynaptic DA receptors is still controversial (for a review on this subject, see ref. 23).

As to the mechanism by which oxytocin acts in the PVN to induce yawning and penile erection, only some speculation is possible at present. A possible explanation is that oxytocin activates oxytocinergic neurons. Supporting this hypothesis, oxytocinergic receptors have been identified in the rat PVN,[24] and exogenous oxytocin has been found to increase *in vivo* the activity of oxytocinergic neurons[25] and to stimulate *in vitro* the release of endogenous oxytocin.[26] Furthermore, oxytocinergic synapses have been found to impinge on oxytocinergic neurons in hypothalamic nuclei.[27]

As previously mentioned the hypothalamic regions surrounding the third ventricle, comprising the PVN, were found to be the most sensitive for the induction of yawning, stretching, and penile erection by ACTH-MSH peptides.[2] ACTH fibers have been identified in the PVN, but they seem not to contact oxytocinergic or DA neurons.[28]

Conclusions

These results show for the first time that DA agonists induce yawning and penile erection by releasing oxytocin in the central nervous system. The brain area where DA agonists apparently act in order to induce the release of oxytocin release seems to be the hypothalamic PVN. This finding suggests a direct involvement of the incertohypothalamic dopaminergic system in the expression of yawning and penile erection.

On the other hand, DA agonists and oxytocin do not appear to induce yawning and penile erection by releasing an ACTH-derived peptide from hypothalamic opiomelanotropinergic neurons. However, our results suggest that ACTH-MSH peptides induce their effect by acting at sites localized downstream from DA receptors and oxytocin. Finally, the finding that oxytocin is implicated in the expression of penile erection opens new clinical perspectives, raising the possibility that abnormalities in the central oxytocinergic function might be responsible for penile erection disturbances.

REFERENCES

1. SERRA, G., M. COLLU, G. L. GESSA & W. FERRARI. 1986. Melanocortins and dopamine link in yawning behaviour. *In* Central Actions of ACTH and Related Peptides. D. de Wied & W. Ferrari, Eds.: 163–178. Liviana Press. Padua.
2. BERTOLINI, A. & G. L. GESSA. 1981. Behavioural effects of ACTH-MSH peptides. J. Endocrinol. Inv. **4:** 241–251.
3. ARGIOLAS, A., M. R. MELIS & G. L. GESSA. 1986. Oxytocin: an extremely potent inducer of penile erection and yawning. Eur. J. Pharmacol. **130:** 265–272.
4. FERRARI, W., G. L. GESSA & L. VARGIU. 1963. Behavioural effects induced by intracisternally injected ACTH and MSH. Ann. N.Y. Acad. Sci. **104:** 330–345.
5. BANKOWSKI, K., M. MANNING, J. SETO, J. HALDER & W. H. SAWYER. 1980. Design and synthesis of potent *in vivo* antagonists of oxytocin. Int. J. Pept. Protein Res. **16:** 382–391.

6. O'DONOHUE, T. L. & D. M. DORSA. 1982. The opiomelanotropinergic neuronal and endocrine system. Peptides 3: 353-395.

7. ESKAY, R. L, M. J. BROWNSTEIN & R. T. LONG. 1979. α-Melanocyte stimulating hormone: reduction in adult rat brain after monosodium glutamate treatment in neonates. Science 205: 827-828.

8. KRIEGER, D. T., A. S. LIOTTA, G. NICHOLSON & J. S. KIZER. 1979. Brain ACTH and endorphin reduced in rats with monosodium glutamate-induced arcuate nucleus lesions. Nature 278: 562-563.

9. BLOCH, B., N. LING, R. BENOIT, W. B. WEHRENBERG & R. GUILLEMIN. 1984. Specific depletion of immunoreactive growth hormone-releasing factor by monosodium glutamate in rat median eminence. Nature 307: 272-273.

10. ARGIOLAS, A., M. R. MELIS, W. FRATTA & G. L. GESSA. 1986. Existence of different forms of adrenocorticotropin in rat hypothalamus. Peptides 7: 591-596.

11. FRATTA, W., A. ARGIOLAS, F. FADDA, G. L. GESSA & W. FERRARI. 1986. Involvement of ACTH-MSH peptides in opiate withdrawal syndrome. In Central Actions of ACTH and Related Peptides. D. de Wied & W. Ferrari, Eds.: 223-230. Liviana Press. Padua.

12. DUNN, A. J., E. L. WEBSTER & C. B. NEMEROFF. 1985. Neonatal treatment with monosodium glutamate does not alter grooming behaviour induced by novelty or adrenocorticotropic hormone. Behav. Neur. Biol. 44: 80-89.

13. YOUNG, E., J. OLNEY & H. AKIL. 1983. Selective alteration of opiate receptor subtypes in monosodium glutamate-treated rats. J. Neurochem. 40: 1558-1564.

14. FINLEY, J. C. W., P. LINDSTROM & P. PETRUSZ. 1981. Immunocytochemical localization of β-endorphin-containing neurons in the rat brain. Neuroendocrinology 33: 28-42.

15. PELLEGRINO, L. J. & A. J. CUSHMAN. 1971. A Stereotaxic Atlas of the Rat Brain. Meredith. New York.

16. MELIS, M. R., A. ARGIOLAS & G. L. GESSA. 1986. Oxytocin-induced penile erection and yawning: site of action in brain. Brain Res. 398: 259-265.

17. POULAIN, D. A. & J. B. WAKERLY. 1982. Electrophysiology of hypothalamic magnocellular neurones secreting oxytocin and vasopressin. Neuroscience 7: 773-808.

18. BUIJS, R. M. 1978. Intra-and extrahypothalamic vasopressin and oxytocin pathways in the rat. Cell Tissue Res. 192: 423-435.

19. SOFRONIEW, M. W. 1983. Vasopressin and oxytocin in the mammalian brain and spinal cord. Trends Neurosci. 6: 467-472.

20. DAHLSTROM, A. & K. FUXE. 1964. Evidence for the existence of monoamine-containing neurons in the central nervous system. I: Demonstration of monoamines in the cell bodies of brainstem neurons. Acta Physiol. Scand. 62 (Suppl. 232): 1-55.

21. LINDVALL, O. & A. BJORKLUND. 1978. Anatomy of the dopaminergic neuron systems in the rat brain. In Advances in Biochemical Psychopharmacology, Vol. 19. P. J. Roberts et al., Eds.: 1-23. Raven. New York.

22. SWANSON, L. W. & P. E. SAWCHENKO. 1983. Hypothalamic integration: Organization of the paraventricular and supraoptic nuclei. Ann. Rev. Neurosci. 6: 269-324.

23. SERRA, G., M. COLLU & G. L. GESSA. 1986. Dopamine receptors mediating yawning: Are they autoreceptors? Eur. J. Pharmcol. 120: 187-192.

24. BRINTON, R. E., J. K. WAMSLEY, K. W. GEE, P. WAN YEIH & H. I. YAMAMURA. 1984. ^{3}H-oxytocin binding sites in the rat brain demonstrated by quantitative light microscopic autoradiography. Eur. J. Pharmacol. 102: 365-367.

25. FREUND-MERCIER, M. J. & P. RICHARD. 1981. Excitatory effects of intraventricular injections of oxytocin on the milk ejection reflex in the rat. Neurosci. Lett. 23: 193-198.

26. MOOS, F., M. J. MERCIER, Y. GUERNE, J. M. GUERNE, M. E. STOECKEL & P. RICHARD. 1984. Release of oxytocin and vasopressin by magnocellular nuclei in vitro: specific facilitatory effect of oxytocin on its own release. J. Endocrinol. 102: 63-72.

27. THEODOSIS, D. T. 1985. Oxytocin-immunoreactive terminals synapse on oxytocinergic neurones in the supraoptic nuclei. Nature 313: 682-684.

28. KNIGGE, K. M. & S. A. JOSEPH. 1982. Relationship of the central ACTH-immunoreactive opiocortin system to the supraoptic and paraventricular nuclei of the hypothalamus of the rat. Brain Res. 237: 655-658.

Role of Central Dopamine in ACTH-induced Grooming Behavior in Rats

ALEXANDER R. COOLS, BERRY M. SPRUIJT,[a] AND
BART A. ELLENBROEK

Psychoneuropharmacological Research Unit
Faculty of Medicine
University of Nijmegen
Nijmegen, The Netherlands
and
[a]*Division of Molecular Neurobiology*
Institute of Molecular Biology
Rudolf Magnus Institute for Pharmacology
State University of Utrecht
3521 CG Utrecht, The Netherlands

INTRODUCTION

About 10 years ago it was discovered that intraperitoneal injections of dopaminergic antagonists such as haloperidol attenuate excessive grooming behavior induced by intracerebroventricular (i.c.v.) administration of adrenocorticotropic hormone fragment (ACTH$_{1-24}$) in rats.[1] The original study of Wiegant *et al.*[1] has now been replicated and extended in several laboratories.[2-6] Apart from sulpiride, which poorly penetrates the central nervous system, almost all the tested dopaminergic antagonists attenuate this peptide-induced behavior in rats.[3-6] The degree to which these effects are dopamine specific is not yet proven. However, studies on the effects of intracerebrally administered agents are unequivocal in that respect[7]: when intracerebrally injected, dopaminergic agonists potently counteract the attenuating effects of dopaminergic antagonists upon the ACTH-induced grooming behavior[7] (FIGURES 1–6).

The attenuating effects of dopaminergic antagonists are not due to drug-induced behavior that prevents the display of ACTH-induced grooming behavior. First, compounds such as haloperidol impair ACTH-induced grooming behavior at doses which do not affect the behavior of rats in a so-called novel box test.[6] Second, doses of dopaminergic antagonists that attenuate the ACTH-induced grooming behavior do not yet attenuate novelty-induced grooming behavior.[6] Finally, a particular dose of neostriatally administered haloperidol that attenuates grooming behavior induced by i.c.v. administration of ACTH[8] enhances grooming behavior induced by intranigral administration of ACTH,[7] implying that the muscular rigidity induced by neostriatally administered haloperidol[9] does not prevent the display of ACTH-induced grooming behavior. Furthermore, the attenuating effects of dopaminergic antagonists can not be ascribed to drug-induced effects upon the brain's handling of external stimuli, since the ACTH-induced grooming behavior is almost completely independent of changes in these stimuli.[10]

338

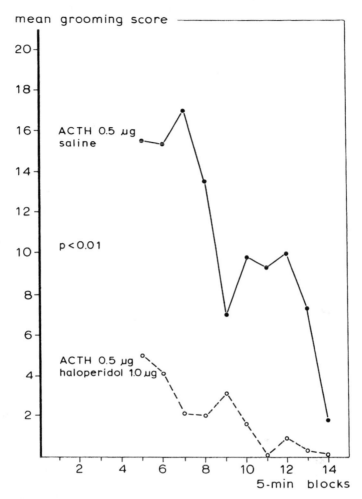

FIGURE 1. Attenuating effects of neostriatally administered haloperidol (0.5 μl per side) upon excessive grooming behavior induced by i.c.v. administration of ACTH$_{1-24}$ (3 μl).[1] Haloperidol ($n = 7$) and saline (control, $n = 6$) were given 5 min prior to the peptide. Intracerebroventricular and neostriatal injections were carried out and verified according to the method described previously by Brakkee *et al.*[28] and Cools *et al.*,[8] respectively. The behavior was studied according to the time-sampling method described by Gispen *et al.*[29]

Given the selectivity and dopamine specificity of the attenuating effects of dopaminergic antagonists, it follows that these drugs exert their effects via dopaminergic receptors. From this point of view it is evident that the affected dopaminergic receptors form part and parcel of the network in which the neural substrate underlying the ACTH-induced grooming behavior is embedded.

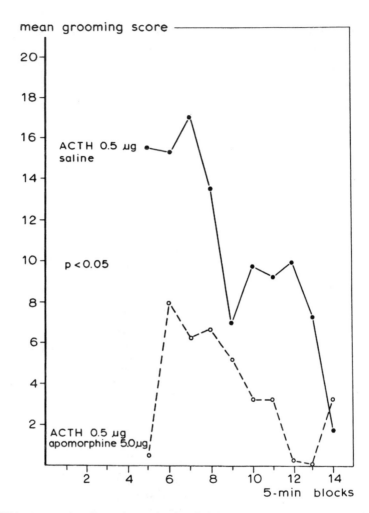

FIGURE 2. Attenuating effects of neostriatally administered apomorphine (0.5 µl per side) upon excessive grooming behavior induced by i.c.v. administration of ACTH$_{1-24}$ (3 µl).[8] Apomorphine ($n = 9$) and saline (control, $n = 6$) were given 5 min prior to the peptide (see also legend to FIGURE 1).

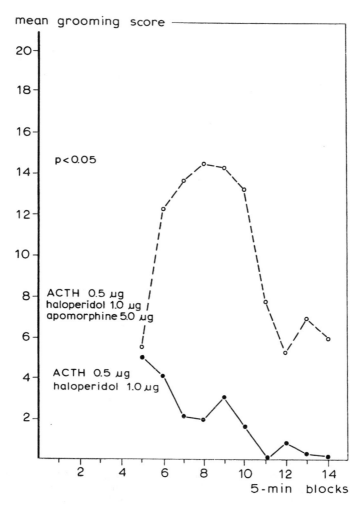

FIGURE 3. Counteracting effects of neostriatally administered apomorphine (0.5 μl per side) upon the haloperidol-induced attenuation of excessive grooming behavior induced by i.c.v. administration of ACTH$_{1-24}$ (3 μl). Apomorphine together with haloperidol ($n = 7$) and haloperidol (control, $n = 7$) were given 5 min prior to the peptide (see also legend to FIG. 1).

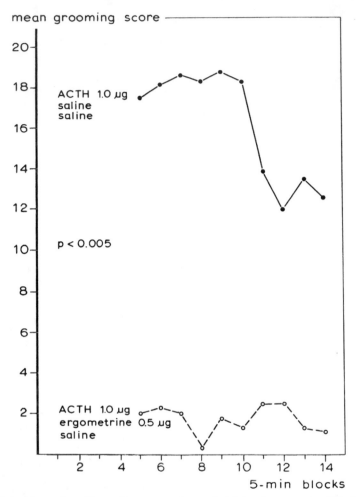

FIGURE 4. Attenuating effects of intra-accumbens-administered ergometrine (0.5 μl per side) upon excessive grooming behavior induced by i.c.v. administration of ACTH$_{1-24}$ (3 μl).[7,8] Ergometrine ($n = 10$) and saline (control, $n = 6$) were given 60 min prior to the peptide; an additional saline injection was given 5 min prior to the peptide in 5 animals (see also legend to FIG. 1).

INVOLVEMENT OF CENTRAL DOPAMINE IN ACTH-INDUCED GROOMING BEHAVIOR

Studies using the intracerebral injection technique have shown that both stimulation of neostriatal dopaminergic receptors by means of the dopaminergic agonist apomorphine and inhibition of these receptors by means of the dopaminergic antagonist haloperidol attenuate the ACTH-induced grooming behavior in a dopamine-specific manner[1,7,8]: the apomorphine-induced effect is counteracted by haloperidol, and vice

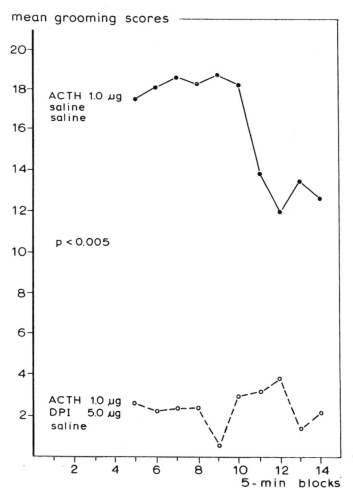

FIGURE 5. Attenuating effects of intra-accumbens-administered DPI (0.5 μl per side) upon excessive grooming behavior induced by i.c.v. administration of $ACTH_{1-24}$ (3 μl).[7,8] DPI (3,4-dihydroxyphenylamino-2-imidazoline, $n=9$) and saline (control, $n=6$) were given 5 min prior to the peptide; an additional saline injection was given 60 min prior to the peptide in four animals (see also legend to FIG. 1).

versa (FIGS. 1–3). The hypothesis that the attenuating effects of neostriatally administered apomorphine are due to stimulation of presynaptic dopaminergic receptors can be rejected, since enhancement of the firing rate of nigrostriatal dopaminergic neurons by means of intranigral administration of haloperidol also attenuates the grooming behavior induced by i.c.v. administration of ACTH.[1] Studies on the nucleus accumbens[7,8] have produced similar results when apomorphine and haloperidol are used (unpublished data). Since the nucleus accumbens is known to contain both classic neostriatal

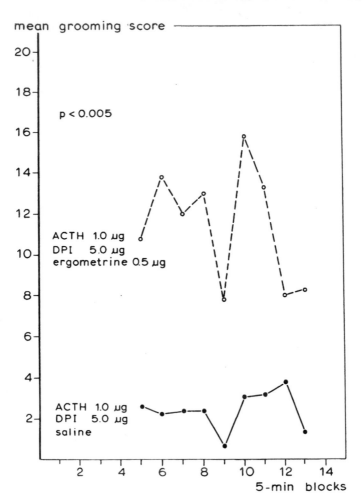

FIGURE 6. Counteracting effects of intra-accumbens-administered DPI (0.5 μl per side) upon the ergometrine-induced attenuation of excessive grooming behavior induced by i.c.v. administration of ACTH$_{1-24}$ (3 μl).[7] Ergometrine was given 60 min and DPI ($n = 7$) or saline (control, $n = 5$) 5 min prior to the peptide (see also legend to FIG. 1).

dopaminergic receptors that are stimulated by apomorphine and inhibited by haloperidol, and atypical mesolimbic dopaminergic receptors stimulated by 3,4-dihydroxyphenylamino-2-imidazoline (DPI) and inhibited by ergometrine,[11,12] it is useful to recall the outcome of studies with the latter compounds as well. Both stimulation of these mesolimbic receptors by means of DPI and inhibition of these receptors by means of ergometrine attenuate the ACTH-induced grooming behavior in a dopamine-specific manner[7]: the attenuating effect of DPI is counteracted by ergometrine, and vice versa (FIGURES 4-6); moreover, noradrenergic agents remain ineffective in that

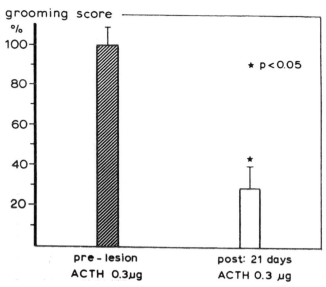

FIGURE 7. Suppressing effect of a lesion in the periaqueductal gray (PAG) upon excessive grooming behavior induced by i.c.v. administration of $ACTH_{1-24}$ (3 μl) in 9 rats.[14] The lesion was produced by 0.25 μg kainic acid dissolved in phosphate buffer.

respect.[7] When the nucleus accumbens of rats is lesioned by injection of 6-hydroxydopamine, the ACTH-induced grooming behavior is also suppressed,[13] underlining the important role of this nucleus in the ACTH-induced grooming behavior. These data have led to the suggestion that only an imbalance between classic nigrostriatal dopaminergic receptors and atypical mesolimbic dopaminergic receptors is responsible for the attenuating effects of dopaminergic antagonists.[8] This suggestion is supported by the observation that the attenuating effect of the mesolimbic dopaminergic antagonist ergometrine is at least partly counteracted by the neostriatal dopaminergic antagonist haloperidol.[7]

The above-mentioned data indicate that there must exist a particular relationship between the dopaminergic neurons and the neural substrate underlying the ACTH-induced grooming behavior. Below we summarize evidence in favor of the hypothesis that dopaminergic neurons direct the neural activity of the neural substrate underlying the ACTH-induced grooming behavior.

DOPAMINERGIC NEURONS AND NEURAL NETWORK OF ACTH-INDUCED GROOMING BEHAVIOR

For a long time the target sites for grooming behavior following the i.c.v. administration of $ACTH_{1-24}$ were unknown. Recently, however, evidence has been provided that the periaqueductal gray (PAG) is at least one of the brain structures indispensable for the display of ACTH-induced grooming behavior.[14] Lesioning of this area significantly suppresses this peptide-induced behavior (FIG. 7), whereas local administration of

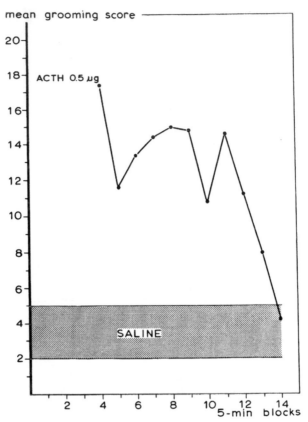

FIGURE 8. Excessive grooming behavior induced by administration of ACTH$_{1-24}$ into the periaqueductal gray.[14]

ACTH$_{1-24}$ into the PAG elicits grooming behavior (FIG. 8). This finding makes it possible to understand the complex role of central dopamine in the ACTH-induced grooming behavior. First, the PAG is inter alia innervated by both the substantia nigra, pars reticulata,[15,16] *i.e.*, one of the main output stations of the nigrostriatal dopaminergic neurons,[17] and the substantia innominata,[15,16] *i.e.*, one of the main output stations of the tegmento-accumbens dopaminergic neurons.[18] From this point of view it will be evident that changing the dopaminergic activity within either the neostriatum or the nucleus accumbens affects the display of behavior elicited by stimulation of ACTH-sensitive sites within the PAG. Second, the PAG is innervated by the deeper layers of the superior colliculus.[15,16] The latter layers are known to alter grooming behavior induced by i.c.v. administration of ACTH$_{1-24}$. Intracollicular administration of the GABAergic agonist muscimol enhances this peptide-induced behavior, whereas intracollicular administration of the indirectly acting GABAergic antagonist picrotoxin attenuates the ACTH-induced behavior.[19] Since lesioning of the deeper layers of the superior colliculus does not prevent the display of grooming behavior induced by i.c.v. administration of ACTH (FIG. 9), these data imply that the deeper layers of the supe-

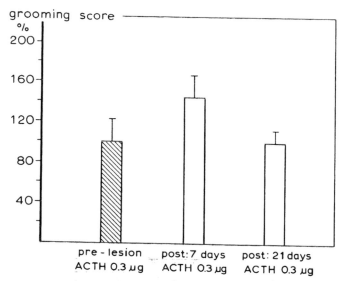

FIGURE 9. The absence of any significant effect of a lesion of the superior colliculus upon excessive grooming behavior induced by i.c.v. administration of ACTH$_{1-24}$. The lesion was induced by suction.[19]

rior colliculus control the neural activity at ACTH-sensitive sites lying outside these deeper layers. Given this knowledge it is evident that brain structures directing the overall activity in the deeper layers of the superior colliculus will affect the grooming behavior induced by i.c.v. administration of ACTH. The latter layers receive inter alia information both from the neostriatum via the striatonigrocollicular pathway[20] and from the nucleus accumbens via the accumbens-substantia, innominata-collicular pathway (unpublished data). Thus, the superior colliculus has to integrate information from both sources into a single efferent command (FIG. 10). From this point of view it becomes possible to understand why both stimulation and inhibition of dopaminergic receptors belonging to the same system produce similar effects: they simply disrupt the integration of the incoming signals at the level of the deeper layers of the superior colliculus with the consequence that the PAG receives faulty information in both cases.

CONCLUSIONS

Grooming behavior induced by i.c.v. administration of ACTH$_{1-24}$ requires at least an intact periaqueductal gray. The neural activity of the periaqueductal gray is inter alia dictated by the activity at the level of the deeper layers of the superior colliculus. The information leaving the deeper layers of the superior colliculus is the integrative product of information received from the substantia nigra, pars reticulata, *i.e.*, one of the main output stations of the nigrostriatal dopaminergic neurons, and the substantia innominata, *i.e.*, one of the main output stations of the tegmento-accumbens dopaminergic neurons. It is this network that underlies the ability of dopaminergic agents to attenuate

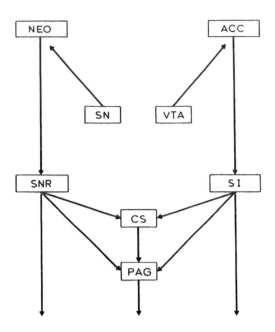

FIGURE 10. Oversimplified diagram of the network in which at least one of the neural sub-strates underlying the excessive grooming induced by i.c.v. administration of ACTH$_{1-24}$ (*i.e.*, PAG) is embedded: *SN*, substantia nigra, pars compacta containing cell bodies of the nigrostriatal dopaminergic neurons; *NEO*, neostriatum containing classic neostriatal dopaminergic receptors belonging to the nigrostriatal neurons[11]; *SNR*, substantia nigra, pars reticulata; *CS*, superior colliculus, deeper layers; *PAG*, periaqueductal gray; *VTA*, ventral tegmental area of Tsai containing cell bodies of the tegmento-accumbens dopaminergic neurons; *ACC*, nucleus accumbens containing primarily atypical mesolimbic dopaminergic receptors belonging to tegmento-accumbens neurons[11,12]; *SI*, substantia innominata.

the grooming behavior induced by i.c.v. administration of ACTH$_{1-24}$. As mentioned, both classic neostriatal dopaminergic receptors and atypical mesolimbic dopaminergic receptors play a role in the effects of dopaminergic agents. The involvement of the described network in the ACTH-induced grooming behavior implies that all the drugs that affect the neural activity of one of the intercalated stations will also affect this peptide-induced behavior. Against this background it becomes understandable why the ACTH-induced behavior can also be attenuated by antidepressants, benzodiazepines, etc.[6,21]

Biochemical studies on dopaminergic receptors have documented the presence of two additional subclasses of dopaminergic receptors [22]: D$_2$ receptors, which are stimulated by LY 17 1555 and inhibited by metoclopramide, and D$_1$ receptors, which are stimulated by SKF-38393 and inhibited by SCH-23390. Since the above-mentioned insight into the role of central dopamine in ACTH-induced grooming behavior derives from studies with dopaminergic agents with either a pure D$_2$ profile (sulpiride and metoclopramide) or a mixed D$_1$/D$_2$ profile (apomorphine, etc.), the involvement of D$_1$ and D$_2$ receptors awaits further clarification. Still, it is worthwhile to note that the D$_1$ agonist SKF-38393 per se induces grooming behavior in rodents.[23-27] In contrast to the ACTH-induced grooming behavior, which is almost completely independent of environmental conditions,[10] the SKF-induced grooming behavior is only induced in habituated rodents and not in rodents introduced in a new environment.[23-27] These findings underline the notion that only nonhabituated animals should be used in the ACTH paradigm in order to prevent unwanted confounding effects of the dopaminergic agents themselves.

REFERENCES

1. WIEGANT, V. M., A. R. COOLS & W. H. GISPEN. 1977. Eur. J. Pharmacol. **41:** 343–345.
2. GUILD, A. L. & A. J. DUNN. 1982. Pharmacol. Biochem. Behav. **17:** 31–36.
3. GREEN, E. J., R. L. ISAACSON, A. J. DUNN & T. H. LANTHORN. 1979. Behav. Neural. Biol. **27:** 546–551.
4. ISAACSON, R. L., J. H. HANNIGAN, J. H. BRAKKEE & W. H. GISPEN. 1983. Brain Res. Bull. **11:** 289–293.
5. GMEREK, D. E. & A. COWAN. 1983. Eur. J. Pharmacol. **88:** 339–346.
6. TRABER, J., H. R. KLEIN & W. H. GISPEN. 1982. Eur. J. Pharmacol. **80:** 407–414.
7. SPRUYT, B. M., A. R. COOLS, B. A. ELLENBROEK & W. H. GISPEN. 1986. Eur. J. Pharmacol. **120:** 249–256.
8. COOLS, A. R., V. M. WIEGANT & W. H. GISPEN. 1978. Eur. J. Pharmacol. **50:** 265–268.
9. ELLENBROEK, B. A., M. SCHWARZ, K.-H. SONTAG, R. JASPERS & A. R. COOLS. 1985. Brain Res. **345:** 132–140.
10. JOLLES, J., J. ROMPA-BARENDREGT & W. H. GISPEN. 1979. Horm. Behav. **12:** 60–72.
11. COOLS, A. R. 1977. Adv. Biochem. Pharmacol. **16:** 215–225.
12. COOLS, A. R. & S. K. OOSTERLOO. 1983. J. Neural Transmission Suppl. **18:** 181–188.
13. SPRINGER, J. E., R. L. ISAACSON, J. P. RYAN & J. H. HANNIGAN. 1983. Life Sci. **33:** 207–211.
14. SPRUYT, B. M., A. R. COOLS & W. H. GISPEN. 1986. Behav. Brain Res. **20:** 19–25.
15. BEITZ, A. J. 1982. Neuroscience **7(1):** 133–159.
16. MANTYH, P. W. 1982. J. Comp. Neurol. **206:** 146–158.
17. GRAYBIEL, A. M. & C. W. RAGSDALE. 1979. *In* Development and Chemical Specificity of Neurons. M. Cuenod, G. Kreutzberg & F. E. Bloom, Eds.: 239–283. Elsevier. Amsterdam.
18. NAUTA, W. J. H., G. P. SMITH, R. L. M. FAULL & V. B. DOMESICK. 1978. Neuroscience **3:** 385–401.
19. SPRUYT, B. M., B. A. ELLENBROEK, A. R. COOLS & W. H. GISPEN. 1986. Life Sci. **39:** 461–470.
20. COOLS, A. R., J. M. M. COOLEN, J. C. A. SMIT & B. A. ELLENBROEK. 1984. Eur. J. Pharmacol. **100:** 71–77.
21. CRAWLEY, J. N. & T. W. MOODY. 1983. Brain Res. Bull. **10:** 399–401.
22. STOOF, J. C. & J. W. KEBABIAN. 1984. Life Sci. **35:** 2281–2284.
23. MOLLOY, A. G. & J. L. WADDINGTON. 1984. Psychopharmacol. **82:** 409–410.
24. MOLLOY, A. G. & J. L. WADDINGTON. 1985. Eur. J. Pharmacol. **108:** 305–308.
25. MOLLOY, A. G. & K. M. O'BOYLE, M. T. PUCH & J. L. WADDINGTON. 1986. Pharmac. Biochem. Behav. **25:** 249–253.
26. STARR, B. S. & M. S. STARR. 1986. Pharmacol. Biochem. Behav. **24:** 837–839.
27. STARR, B. S. & M. S. STARR. 1986. Neuropharmacology **25(5):** 455–463.
28. BRAKKEE, J. H., V. M. WIEGANT & W. H. GISPEN. 1979. Lab. Animal Sci. **29:** 78–81.
29. GISPEN, W. H., V. M. WIEGANT, H. J. GREVEN & D. DE WIED. 1975. Life Sci. **17:** 645–652.

Brainstem Systems and Grooming Behaviors[a]

GARY G. BERNTSON, JAYE F. JANG, AND
APRIL E. RONCA

Departments of Psychology and Pediatrics
Ohio State University
Columbus, Ohio 43210

John Hughlings Jackson believed that with the phylogenetic and ontogenetic development of the nervous system, higher neural systems re-represent and elaborate on lower brain mechanisms.[1] According to this view, the functional contributions of lower mechanisms are not bypassed or replaced by higher systems. Rather, lower mechanisms continue to participate in adaptive functions, serving as the basic substrates through which newer re-representative systems achieve expression. Consistent with this suggestion is the fact that decerebrate animals retain many behavioral capacities, including righting, locomotion, eating, grooming, sexual responses, and escape and aggressive behaviors.[2-4] In addition, decerebrate animals have demonstrated both habituation and associative learning.[5-7] Similarly, decerebrate humans also retain many of the adaptive capacities of the normal neonate. In the absence of the cerebral hemispheres, anencephalic and hydranencephalic infants demonstrate typical infantile reflexes, pleasure and aversive reactions, and ingestive behaviors, including the stereotyped oro-facial acceptance and rejection responses characteristic of normal infants.[8-10] Consistent with animal studies, habituation and associative learning have also been reported in the decerebrate human infant.[11-13] Taken together, these findings document the remarkable behavioral capacities of brainstem circuits.

The appearance of grooming behaviors in the decerebrate animal clearly indicates that many of the neural substrates for grooming responses are organized at lower levels of the neuraxis.[2] Although the grooming behaviors of the decerebrate have not been systematically characterized, these responses often appear somewhat fragmentary or poorly directed.[3] One significant consequence of decerebration, however, is the appearance of extensor rigidity that can severely interfere with the expression of adaptive behaviors. This and other acute effects of decerebration may obscure the adaptive potential of brainstem circuits and confound interpretation of the decerebrate state. Since decerebrate rigidity partially resolves with time, one approach to this problem has been to maintain decerebrated animals beyond the immediate postsurgical period. Significantly, the chronic decerebrate has been reported to show a dramatically enhanced behavioral repertoire relative to the acute decerebrate state. An additional approach has been to employ immature animals in decerebration studies, since denervation phenomena and decerebrate rigidity are much diminished in this preparation.[14-15] Consequently, the behavioral features of the immature decerebrate may provide a less confounded view of brainstem functions.[2,7]

[a] This work was supported in part by USPHS grant no. MH 25630 to G. G. Berntson.

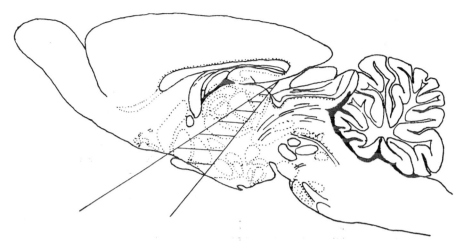

FIGURE 1. Neuraxial level of decerebration. Brain transections for all animals were located within the *shaded* zone. (Brain diagram drawn from sagittal plane, 0.2 mm, of the 39-day-old rat brain, based on Sherwood, N. M. & P. S. Timiras. 1970. A Stereotaxic Atlas of the Developing Rat Brain. Univ. Calif. Press. Berkeley, CA.)

STUDIES OF DECEREBRATE ANIMALS

To further explore the intrinsic organizational capacities of the brainstem, we systematically examined grooming behaviors in rats who were decerebrated at an early age (21 days). Eleven hooded rats served as sham-operated controls, while an equal number of experimental animals received a surgical transection of the neuraxis at the level of the diencephalon (FIG. 1). Experimental animals also received bilateral parasagittal transections along the lateral aspect of the diencephalon in order to eliminate lateral connections between remaining diencephalic structures and more rostral systems (basal and posterior hypothalamic areas were left intact to preserve basic regulatory functions). As illustrated in FIGURE 1, the decerebration procedure completely isolated lower levels of the neuraxis from the cerebral hemispheres, dorsal diencephalon, basal ganglia, and septal area. The mesencephalon was largely intact in all animals, as was the posterior and ventrobasal hypothalamus. In three cases, minimal remnants of ventral tissue bridged the hypothalamus and the septal-preoptic area. Behavioral data from these animals did not appreciably differ from animals with complete decerebrations, suggesting that the residual tissue contributed minimally to the results.

Within 24 hours of surgery, decerebrate subjects would readily ingest food introduced into the mouth, often grasping the tip of the food tube with the forepaws. Chewing was vigorous and prolonged, relative to the amount of food consumed. While decerebrate animals demonstrated robust feeding responses following surgery, they did not actively seek out food. Consequently, caloric intake and water balance were maintained by hand feeding of a suspension of laboratory chow in evaporated milk (eight times daily; control animals were also handled daily to equate the groups as much as possible). In spite of this feeding regimen, decerebrate animals failed to gain weight as fast as the control pups, and were consistently lighter than their sham-operated littermates. At the initiation of testing (day 26) the mean weight of the decerebrate animals was approximately 75% that of the normal subjects.

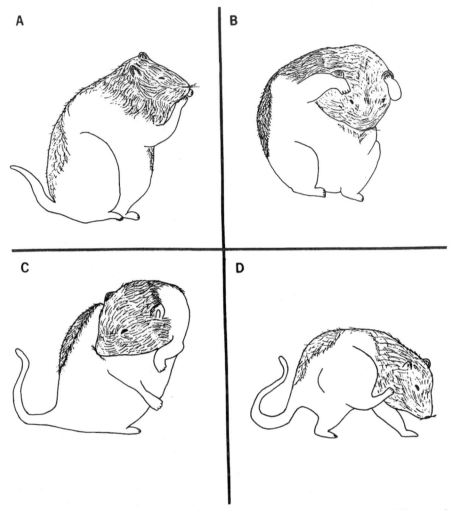

FIGURE 2. Illustrations of four general categories of grooming behaviors observed in normal and decerebrate animals. **(A)** Face washing, forepaw licking. **(B)** Face washing, repetitive head stroking. **(C)** Fur licking. **(D)** Scratching.

No gross differences in posture or locomotion were apparent between decerebrated and normal subjects within the first few days of surgery. The decerebrate animals were highly sensitive to environmental stimuli, however, and mild perturbations often elicited a brief but vigorous bout of motor activity. Over the course of postoperative recovery (from days 21 to 35), the decerebrated subjects progressively adopted a characteristic hunchbacked "tiptoe" posture typical of the thalamic animal.[3] In addition, while normal subjects invested a considerable amount of time in social play behaviors, decerebrate animals did not actively interact with littermates. Rather, the decerebrate subjects ex-

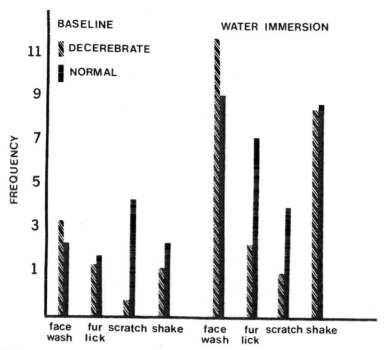

FIGURE 3. Frequency of occurrence of grooming responses in normal and decerebrated subjects under baseline conditions and after water immersion.

hibited long bouts of apparently aimless locomotor activity, interspersed with brief periods of relative quiescence. Responses to tactile stimuli, including painful stimuli, were similar to those of normal subjects, as were righting reflexes and responses to acoustic startle stimuli.

Decerebrated subjects maintained their fur poorly in the early weeks after surgery, although grooming was one of the predominant activities displayed by these subjects. Spontaneous grooming behaviors of all animals were formally examined in 5-min sessions on days 26, 27, 32, and 33. For each test, cage mates were removed from the home cage, and the target animal was given a 5-min adaptation period. Specific grooming behaviors were then recorded for a 5-min test period. A sample of tests was also videotaped for further analysis. Two additional tests were given on days 29 and 34 to explore the effects of water immersion. The general procedures for these tests were as described, except that subjects were immersed in water (33°C) immediately after the 5-min adaptation period.

Largely consistent with previous studies of the development of grooming in the rat,[16] four general categories of grooming behavior were observed in the control animals. These included (1) face washing, (2) fur licking, (3) scratching, and (4) "wet-dog" shaking. Face washing entailed licking of the forepaws and repetitive stroking of the face and head with the upper limbs as illustrated in FIG. 2A and 2B. Fur licking (FIG. 2C) consisted of repetitive licking of the body surface, generally directed to the flank or belly, but occasionally also to the tail or hind limbs. Scratching entailed the use of a hind limb to rhythmically scratch the body surface, generally the flank area (FIG. 2D). Fi-

TABLE 1. Bout Durations for Grooming Responses of Normal and Decerebrate Pups

	Normal		Decerebrate	
	Median	Interquartile Range	Median	Interquartile Range
Face washing	6.0	3.8–7.9	5.8	0.7–7.5
Fur washing	9.2	4.0–15.5	12.5	11.5–14.3
Scratching	5	1.4–15.5	4	2.0–4.8
Shaking	<1	<1–<1	<1	<1–<1

nally, "wet dog" shakes consisted of brief, high-frequency shaking and axial rotation of the head and torso. No significant differences or systematic trends were apparent in grooming responses over test days for either the control or decerebrate subjects. Consequently, results under given conditions were averaged over sessions for illustration and analysis.

No significant group differences were apparent in the total number of grooming bouts per session (mean for control animals, 16.5; for decerebrates, 15.2; $F(1,20) = 0.82$, $p > 0.05$). Moreover, as illustrated in FIGURE 3, water immersion resulted in a significant overall increase in grooming responses for all animals ($F(3,60) = 35.6$, $p < 0.001$) with no significant group interaction. All categories of grooming responses observed in the normal animals were also apparent in the decerebrate subjects (FIG. 3), and the bout durations of grooming components were comparable for the two groups (TABLE 1). In addition, similarities were observed in the effects of water immersion, which resulted in significant increases in face washing and wet-dog shakes in both control and decerebrate animals. Fur licking also increased somewhat, but to a greater extent in control animals. Scratching responses were largely unaffected by water immersion.

In spite of the similarities in the grooming responses of control and decerebrate animals, notable differences were also apparent. Decerebrate subjects demonstrated significantly fewer scratching responses than did control animals, and differences were apparent in the serial organization of grooming components. In control animals, grooming typically began with paw licking and face washing, followed by fur licking (which was initially directed anteriorly but progressively shifted to a more posterior body focus). In contrast, scratching and wet-dog shakes were not temporally integrated with other grooming components, but tended to appear in isolation. The progressive pattern of face washing to fur licking observed in the control animals was not generally observed in the decerebrates, which tended to show a more disjointed appearance of all grooming components. Some of these group differences may have been attributable to postural instability in the decerebrate subjects. Thus, while hind-limb scratching responses were rare in the decerebrates, these subjects tended to lose their balance when scratching responses were evidenced. Consequently, the infrequent occurrence of scratching may have been associated with the impairment in postural control. This is a less likely explanation, however, of the loss of serial integration between face washing and fur licking in the decerebrate subjects.

Additional studies further argue against a simple postural explanation of differences in grooming between normal subjects and decerebrate subjects. Three decerebated subjects were chronically maintained to adulthood. The motor patterns of the chronic decerebrates were highly coordinated, and their postural control appeared comparable to that of normal animals. Indeed, it was often difficult to discriminate these animals from the normal controls. Unlike the younger decerebrates, the chronic animals maintained their fur in a clean and healthy condition, and they ate and drank without assistance if food and water were distributed throughout the cage. In spite of their good locomotor and postural abilities, these animals still showed little hind-limb scratching in formal tests as described above. Moreover, they continued to demonstrate disturbances in the serial progression of face washing to fur licking.

In summary, decerebrate rats demonstrated all major components of grooming observed in normal animals, although some elements were less frequently seen (fur licking and scratching). In spite of reduced fur licking, chronic decerebrate rats adequately maintained their coat. These results are in accord with the retention of effective grooming responses following decerebration in young cats.[14] The most notable disturbance in the grooming behavior of the decerebrate was in the serial organization of discrete components, which did not appear to be secondary to motor impairment.

To further explore the potential brainstem substrates that serve to maintain grooming in the decerebrate animal, we have employed chronic brain stimulation techniques to examine specific brainstem areas in several species.

STIMULATION STUDIES

W. R. Hess pioneered the development of techniques for focal stimulation of subcortical brain structures in free-moving animals. He was awarded the Nobel Prize in 1949, in part for his application of these methods to the study of the functional organization of the diencephalon.[17] These early studies revealed that sequentially organized behaviors, including grooming-like responses, could be induced by stimulation of localized zones of the diencephalon in the cat. These findings were soon replicated in a wide variety of species.[18-19] In a series of studies, our laboratory has extended Hess's findings to the lower reaches of the brainstem in the cat, rat, and opossum.[20-25] Since the most thorough mapping has been completed in the cat, we summarize here the results of both published and unpublished stimulation studies of a total of 741 brainstem loci in 76 animals of this species. Two to four chronic electrodes with multiple stimulation tips were stereotaxically implanted into brainstem sites and secured to the skull by standard procedures. Following a minimum of 1 week of recovery, each electrode was tested for the elicitation of grooming responses. For each electrode site, 30–60 s trains of monophasic cathodal pulses (100–200 μsec; 50–100 Hz) were repeatedly delivered at a number of intensity levels. For each session, stimulation was begun at a subthreshold current level, and was then progressively increased in small steps over the course of testing. Behavioral protocols were taken at each stimulation level, and results were confirmed in a minimum of two separate sessions.

Stimulation of 103 electrode sites was found to consistently induce robust and stimulation-bound grooming behaviors, including fur licking and groom-biting. The latter response consisted of repetitive nibbling and pulling at the fur with the front teeth. This response was associated with only minimal jaw opening, and differed considerably from the deep biting associated with predatory attack or eating. FIGURE 4 illustrates the distribution of effective and ineffective brainstem loci for the elicitation of grooming behaviors. In general, effective electrodes produced both licking and groom-biting, although groom-biting frequently showed a higher threshold intensity than did licking (mean threshold for licking, 4.8 μA rms; groom-biting, 5.1). As is apparent from the figure, most effective electrodes were distributed in three general areas; (1) the paleocerebellum (anterior cerebellar vermis and fastigial nucleus), (2) the parabrachial region surrounding the superior cerebellar peduncle, and (3) the ventrolateral medulla. In addition, a few points were located in the dorsal reticular forma-

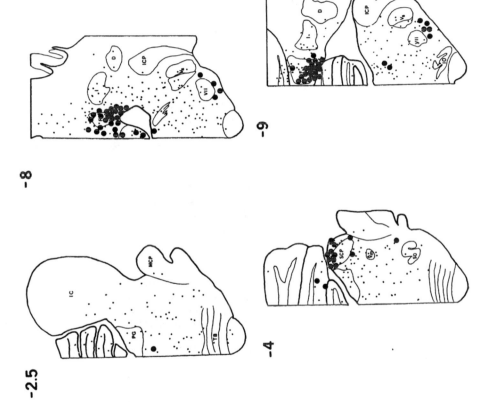

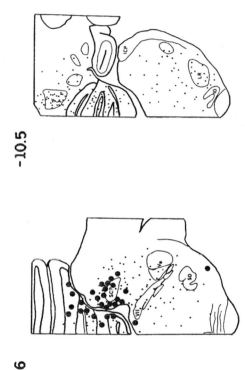

FIGURE 4. Distribution of effective (*large dots*) and ineffective (*small dots*) electrode sites for the elicitation of grooming behaviors in cats. Numerical designations indicate the frontal plane, relative to the interaural line. Letter designations are as follows: *D*, nucleus dentatus; *F*, nucleus fastigii; *I*, nucleus interpositus; *IC*, inferior colliculus; *ICP*, inferior cerebellar peduncle; *IO*, inferior olivary nucleus; *LR*, lateral reticular nucleus; *MCP*, middle cerebellar peduncle; *PG*, periacqueductal gray; *SCP*, superior cerebellar peduncle; *SO*, superior olivary nucleus; *TB*, trapezoid body; *Vm*, trigeminal motor nucleus; *VII*, nucleus and tractus facialis. (Brain drawings are based on atlas plates of Snider, R. J. & W. T. Neimer. 1961. A Stereotaxic Atlas of the Cat Brain. Univ. Chicago Press. Chicago, IL.)

tion. These areas are comparable to those from which grooming responses can be elicited in the rat[20] and the opossum.[26]

Characteristics of Stimulation-induced Grooming Responses

Grooming behaviors elicited by brain stimulation did not appear to be merely motor automatisms, but rather showed a high degree of directedness and serial organization. These grooming responses appeared to be subject to normal sensory controls, as evidenced by their preferential directedness to foreign objects embedded in the fur.[21,25] While the majority of grooming was directed to the cats' own fur, many animals would also groom nearby objects. The amount of grooming of environmental goal objects was dependent on their stimulus features. An anesthetized rat received significantly more grooming than did a similarly sized furry block, while wood blocks were virtually never groomed.[21-23] Thus, the stimulus features of the goal object served as important determinants of the directedness of the elicited grooming behavior. Moreover, it appears that the sensory features of the goal object can control not only the directedness of the grooming responses, but their topography as well. The proportion of fur-licking and groom-biting responses was examined as a function of the surface features of the goal object.[23] Two anesthetized rats were used as grooming objects; one had clean fur, and the other had bits of foreign material (dental acrylic) embedded within the fur. All testing was accomplished at a fixed stimulus intensity. When the cats were grooming the rat with clean fur, the predominant component was licking, and relatively little groom-biting was evidenced. In contrast, when grooming fur having surface irregularities, the relative proportion of licking and groom-biting responses changed significantly. Under these conditions, groom-biting was the most prevalent component.[23] In summary, both the directedness and topography of stimulation-induced grooming appear to be under the control of the stimulus features of the environment.

Nature of Elicited Grooming Responses

The serial organization and environmental sensitivity of stimulation-induced grooming rules out the possibility that such responses reflect simple motor automatisms. Furthermore, the directed grooming of environmental goal objects argues against the possibility that stimulation merely induces an aversion sensation to which grooming responses are directed. This latter interpretation is also inconsistent with the finding that electrode sites which yield grooming responses can also support self-stimulation.[20,27] Rather, brainstem stimulation appears to facilitate or sensitize the sensorimotor mechanisms that normally support grooming behaviors. This is consistent with the findings outlined above, and is further supported by the fact that stimulation of electrode sites that induce grooming behaviors also facilitate specific licking and biting reflexes. In studies with cats, Flynn[28] has reported that stimulation of hypothalamic sites which induce predatory-like biting attack on a rat also sensitizes perioral sensory fields from which discrete tactile stimulation can elicit jaw opening and biting reflexes. These sensory fields expand with increasing stimulation intensity, and have a predominant distribution contralateral to the site of stimulation. Denervation of these sensory fields not only blocks biting reflexes but also eliminates biting attack of a rat. Consistent with these findings, we have observed that stimulation of grooming sites in the cat results in the lateralized facilitation of perioral sensory fields for licking and biting reflexes. These findings suggest that stimulation-bound grooming responses do not result from the direct activation of motor mechanisms but rather arise from the sensitization of behavioral systems to the sensory inputs that normally serve to control grooming behaviors.

Anatomical Substrates

The complexity and coordination of grooming behaviors evoked by brain stimulation suggests that the serial organization of these responses does not arise from the neurons at the stimulation site. Repetitive stimulation pulses induce a mindless synchrony of neuronal firing, overriding the normal influences associated with synaptic events. The informational content of this rhythmical output would be insufficient to specify the subtle patterning of neuronal activity necessary for the elaboration of coordinated behaviors. Rather, the outputs from the stimulated tissue appear to facilitate, either directly or indirectly, response substrates located elsewhere. A growing number of physiological studies indicate that neurological mechanisms for adaptive behaviors may be organized at brainstem levels, and central patterning mechanisms for biting, chewing, and swallowing have been identified within the reticular formation.[28-30] In view of these considerations, it is likely that grooming responses induced by brainstem stimulation result from the activation of intrinsic brainstem mechanisms. This possibility is further suggested by the finding that grooming responses elicited by brainstem stimulation in the opossum are not disrupted by decerebration at a thalamic level.[26]

The cerebellum has recently received increasing attention as an organ that contributes to behavioral functions.[31-34] The cerebellum receives afferent input from all levels of the neuraxis and exerts modulatory control over all major neural systems, including limbic networks.[35-36] In the studies outlined above, grooming responses were elicited by stimulation of anatomically interconnected regions of the anterior cerebellar vermis and the rostral fastigial nucleus. The specific cerebellar projections contributing to the elicitation of grooming responses remain unclear, however. The major efferents of the fastigial nucleus project via the uncinate tract of Russell (lying above the superior cerebellar peduncle) and the juxtarestiform body (in association with the inferior cerebellar peduncle) to widespread areas of the brainstem reticular formation.[35] It appears that the juxtarestiform body contains the critical efferents mediating grooming responses induced by cerebellar stimulation. While stimulation of the superior cerebellar peduncle can also induce grooming, this does not appear to reflect the activation of a cerebellar efferent pathway. Fastigial stimulation in the cat continues to induce grooming responses, at prelesion thresholds, following bilateral destruction of the superior cerebellar peduncle.[24] Rather, the elicitation of grooming from the parabrachial region may reflect (1) the excitation of intrinsic parabrachial nuclei which have been shown to have wide projections to limbic and brainstem areas,[37-39] or (2) the activation of cerebellar afferents (such as the catecholamine projections from the locus ceruleus to the anterior vermis[40]). In this regard, the locus ceruleus has frequently been implicated in behavioral processes,[41-42] and Micco[43] has reported grooming-like responses and self-stimulation from electrodes in the region of the locus ceruleus of the rat.

Especially puzzling are positive electrode sites that lie ventrolateral to the facial nucleus. This region primarily contains longitudinal fiber pathways projecting to and from the spinal cord and brainstem. If these fiber bundles were the relevant anatomical substrates for stimulation-induced grooming, however, one would also expect to see similar responses from the ventrolateral zone anterior and posterior to the facial nucleus. This was not the case (see FIG. 4). In view of their proximity to the facial nucleus, positive electrodes in this ventrolateral region may have directly activated this nucleus through current spread. It is unlikely that this could account for elicited grooming responses, however, since electrodes directly within the facial nucleus failed to induce such responses. Given the wide dendritic distribution patterns of reticular neurons,[44] stimulation techniques alone may not be sufficiently precise to further address this issue.

CONCLUSION

While additional work is clearly necessary to enumerate the precise anatomical systems involved, it appears that grooming behaviors are organized to a significant extent at brainstem levels. Moreover, it seems likely that paleocerebellar circuits may figure prominently in the neurological control of these mechanisms. The cerebellum is not an essential site for the generation or elaboration of grooming responses, since animals will groom following paleocerebellar lesions or in the total absence of the cerebellum.[45-46] Nevertheless, the paleocerebellum projects widely to brainstem, limbic, and telencephalic structures that have been implicated in behavioral processes, and exerts important modulatory control over these systems.[31,34,36] It is likely that cerebellar networks contribute significantly not only to motor control,[47-48] but also to the inhibition and selection of behavioral responses relative to existing sensory inputs and motivational dispositions.[31,34,36]

ACKNOWLEDGMENT

We thank Erin Cassidy-Schairer for assistance with testing.

REFERENCES

1. JACKSON, J. H. 1958. Evolution and dissolution of the nervous system. *In* Selected Writings of John Hughlings Jackson, Vol. 2. J. Taylor, Ed. Basic Books. New York.
2. BERNTSON, G. G. & D. J. MICCO. 1976. Organization of brainstem behavioral systems. Brain Res. Bull. **1:** 471–483.
3. GRILL, H. J. & R. NORGREN. 1978. Neurological tests and behavioral deficits in chronic thalamic and chronic decerebrate rats. Brain Res. **142:** 299–312.
4. BARD, P. & M. B. MACHT. 1958. The behavior of chronically decerebrate cats. *In* Ciba Foundation Symposium on the Neurological Basis of Behavior. G. E. W. Wohlstenholme & C. M. O'Connor, Eds. Churchill. London.
5. NORMAN, R. J., J. S. BUCHWALD & J. R. VILLABLANCA. 1977. Classical conditioning with auditory discrimination of the eyeblink in decerebrate cats. Science **196:** 551–553.
6. OAKLEY, D. A. 1983. Learning capacity outside neocortex in animals and man: Implications for therapy after brain injury. *In* Animal Models of Human Behavior. G. C. L. Davey, Ed. John Wiley & Sons. New York.
7. RONCA, A. E., G. G. BERNTSON & D. S. TUBER. 1986. Cardiac orienting and habituation to auditory and vibrotactile stimuli in the infant decerebrate rat. Dev. Psychobiol. **18:** 545–558.
8. HALSEY, J. H., N. ALLEN & H. R. CHAMBERLAIN. 1967. Chronic decerebrate state of infancy. Arch. Neurol. **19:** 339–344.
9. HALSEY, J. H., N. ALLEN & H. R. CHAMBERLAIN. 1971. The morphogenesis of hydranencephaly. J. Neurol. Sci. **12:** 187–217.

10. STEINER, J. E. 1973. The gustofacial response: observations on normal and anencephalic newborn infants. *In* Fourth Symposium on Development in the Human Infant: Oral Sensation and Behavior. J. F. Bosma, Ed. National Institutes of Health. Bethesda, MD.

11. BERNTSON, G. G., D. S. TUBER, A. E. RONCA & D. S. BACHMAN. 1983. The decerebrate human: Associative learning. Exper. Neurol. **81:** 77–88.

12. GRAHAM, F. K., L. A. LEAVITT, B. D. STROCK & J. W. BROWN. 1978. Precocious cardiac orienting in a human anencephalic infant. Science **199:** 322–324.

13. TUBER, D. S., G. G. BERNTSON, D. S. BACHMAN & J. N. ALLEN. 1980. Associative learning in premature hydranencephalic and normal twins. Science **210:** 1035–1037.

14. BIGNALL, K. E. & L. SCHRAMM. 1974. Behavior of chronically decerebrated kittens. Exper. Neurol. **42:** 519–531.

15. WEED, L. H. 1917. The reactions of kittens after decerebration. Amer. J. Physiol. **43:** 131–157.

16. BOLLES, R. C. & P. J. WOODS. 1964. The ontogeny of behavior in the albino rat. Anim. Behav. **12:** 427–441.

17. HESS, W. R. 1957. The Functional Organization of the Diencephalon. Grune & Stratton. New York, NY.

18. MOGENSON, G. J. & Y. H. HUANG. 1973. The neurobiology of motivated behavior. Prog. Neurobiol. **1:** 55–83.

19. ROBERTS, W. W. 1970. Hypothalamic mechanisms for motivational and species-typical behavior. *In* The Neural Control of Behavior. R. E. Whalen, R. F. Thompson, M. Verzeano & N. M. Weinberger, Eds. Academic Press. New York.

20. BALL, G. G., D. J. MICCO, & G. G. BERNTSON. 1974. Cerebellar stimulation in the rat: Complex stimulation-bound oral behaviors and self-stimulation. Physiol. Behav. **13:** 123–127.

21. BERNTSON, G. G. 1973. Attack, grooming, and threat elicited by stimulation of the pontine tegmentum in cats. Physiol. Behav. **11:** 81–87.

22. BERNTSON, G. G. & H. C. HUGHES. 1974. Medullary mechanisms for eating and grooming behaviors in the cat. Exper. Neurol. **44:** 255–265.

23. BERNTSON, G. G. & H. C. HUGHES. 1976. Behavioral characteristics of grooming induced by hindbrain stimulation in the cat. Physiol. Behav. **17:** 165–168.

24. BERNTSON, G. G. & T. S. PAULUCCI. 1979. Fastigial modulation of brainstem behavioral mechanisms. Brain Res. Bull. **4:** 549–552.

25. BERNTSON, G. G., S. J. POTOLICCHIO & N. E. MILLER. 1973. Evidence for higher functions of the cerebellum: Eating and grooming elicited by cerebellar stimulation in cats. Proc. Natl. Acad. Sci. **70:** 2497–2499.

26. BUCHHOLZ, D. 1975. Spontaneous and centrally induced behaviors in normal and thalamic opossums. J. Comp. Physiol. Psychol. **90:** 898–908.

27. WATSON, P. J. 1978. Behavior maintained by electrical stimulation of the rat cerebellum. Physiol. Behav. **21:** 749–755.

28. FLYNN, J. B. 1972. Patterning mechanisms, patterned reflexes and attack behavior in cats. *In* Nebraska Symposium on Motivation. J. K. Cole & D. D. Jenson, Eds. University of Nebraska. Lincoln, NE.

29. DOTY, R. W. 1968. Neural organization of deglutition. *In* Handbook of Physiology, Vol. 4, Sec. 6. Alimentary canal. C. F. Code & W. Heidl, Eds. American Physiological Society. Washington, D. C.

30. DUBNER, R. & Y. KAWAMURA, EDS. 1971. Oral-facial Sensory and Motor Mechanisms. Appleton-Century-Crofts. New York.

31. BERNTSON, G. G. & M. W. TORELLO. 1982. The paleocerebellum and the integration of somatovisceral and behavioral function. Physiol. Psychol. **10:** 2–12.

32. MARTNER, J. 1975. Cerebellar influences of autonomic mechanisms. Acta Physiol. Scand. Suppl. **425:** 1–42 (and addenda).

33. LAVOND, D. G., D. A. McCORMICK & R. F. THOMPSON. 1984. A nonrecoverable learning deficit. Physiol. Psychol. **12:** 103–110.

34. WATSON, P. J. 1978. Nonmotor functions of the cerebellum. Psychol. Bull. **85:** 944–967.

35. BRODAL, A. 1981. Neurological Anatomy. Oxford Univ. Press. New York.

36. SNIDER, R. S. & A. MAITI. 1976. Cerebellar contributions to the Papez circuit. J. Neurosci. Res. **49:** 529–539.

37. SAPER, C. B. & A. D. LOEWY. 1980. Efferent connections of the parabrachial nucleus in the rat. Brain Res. **197:** 291–317.

38. TAKAGI, H., S. SHIOSAKA, M. TOHYAMA, E. SENBA & M. SAKANAKA. 1980. Ascending components of the medial forebrain bundle from the lower brainstem in the rat, with special reference to raphe and catecholamine cell groups. A study by the HRP method. Brain Res. **193:** 315–337.

39. TRAVERS, J. B. & R. NORGREN. 1983. Afferent projections to the oral motor nuclei in the rat. J. Comp. Neurol. **220:** 280–298.

40. CHU, N. S. & F. E. BLOOM. 1974. The catecholamine-containing neurons in the cat dorsolateral pontine tegmentum: Distribution of the cell bodies and some axonal projections. Brain Res. **66:** 1–22.

41. FOOTE, S. L., G. ASTON-JONES & F. E. BLOOM. 1980. Impulse activity of locus coeruleus neurons in awake rat and monkeys is a function of sensory stimulation and arousal. Proc. Natl. Acad. Sci. **77:** 3033–3037.

42. REDMOND, D. E. & Y. H. HUANG. 1979. New evidence for a locus coeruleus-norepinephrine connection with anxiety. Life Sci. **25:** 2149–2162.

43. MICCO, D. J. 1974. Complex behaviors elicited by stimulation of the dorsal pontine tegmentum in rats. Brain Res. **75:** 172–176.

44. SCHEIBEL, M. & A. SCHEIBEL. 1964. Structural substrates for integrative patterns in the brainstem reticular core. *In* Reticular Formation of the Brain. H. H. Jasper, L. D. Proctor, R. S. Knighton, W. S. Noshay & R. T. Costello, Eds. Little, Brown. Boston.

45. BERNTSON, G. G. & K. S. SCHUMACHER. 1980. Effects of cerebellar lesions on activity, social interactions, and other motivated behaviors in the rat. J. Comp. Physiol. Psychol. **94:** 706–717.

46. KARAMYAN, A. J. 1962. Evolution of the Function of the Cerebellum and Cerebral Hemispheres. Israel Program for Scientific Translations. Jerusalem.

47. LLINAS, R. & J. I. SIMPSON. 1981. Cellular control of movement. Handb. Behav. Neurobiol. **5:** 231–302.

48. MARR, D. 1969. A theory of cerebellar cortex. J. Physiol. **202:** 437–470.

Differential Thermosensor Control of Thermoregulatory Grooming, Locomotion, and Relaxed Postural Extension[a]

WARREN W. ROBERTS

Department of Psychology
University of Minnesota
Minneapolis, Minnesota 55455

The sensory signals on which regulation of body temperature depends originate from a large number of superficial sites in the skin and nasal cavity and deeper sites within the brain, spinal cord, esophagus, abdominal cavity, and skeletal muscles.[1-8] The traditional assumption has been that these multiple inputs are combined by a central integrator or controller to produce a unitary warm or cold error signal that in turn activates the full set of regulatory responses against hyperthermia or hypothermia.[9,10] Most neural and mathematical models of thermoregulation have been concerned with the manner in which thermosensory inputs are combined to yield these unitary signals.[2] However, in the last 10–12 years this assumption has come into question because of studies that have found that a number of behavioral and physiological thermoregulatory responses are differentially controlled by different central and peripheral thermosensors.[6,7,11-13] These relatively independent linkages between sensors and effectors point toward a more fragmentary parallel organization of thermal homeostasis that offers the possibility of specialized response patterns adapted to different kinds of thermal stress that produce different distributions of heat between different sets of thermosensors. This paper will describe some of the differences in sensory control of thermoregulatory grooming, locomotion, and relaxed postural extension in the rat and the adaptive functions of this mode of control in different kinds of heat stress that distribute hyperthermia differently between deep and superficial sensors.

RESPONSES OF RATS TO HYPERTHERMIA

Rats display three behavioral thermoregulatory responses in warm environments: increased grooming, increased locomotion, and assumption of a relaxed extended posture.[14,15] The grooming, in combination with increased salivation, wets the fur and thus increases heat dissipation by evaporation. At an ambient temperature of 40–41°C, which is close to the limit of the rat's ability to thermoregulate, evaporation of saliva accounts for as much as 60% of total heat dissipation.[16] The grooming typically begins with paw licking and face washing with the wet paws, followed during longer bouts

[a] This work was supported by National Institute of Mental Health grant no. MH 6901.

by licking of the abdomen, sides, legs, and tail. The movements do not differ qualitatively from those present at thermally neutral ambient temperatures, although the proportion of time spent grooming is about two to three times greater.

The second heat-induced behavior, locomotion, which is accompanied by an increase in the reward effect of cool air,[17] increases the likelihood of escape to a cooler environment and facilitates learning of escape routes for future use. In small observation cages (e.g., 27 cm in diameter), the locomotion is usually slow, consists of intermittent bursts of a few steps, and after deep temperatures have risen about 1°C, is often executed in a low extended posture.

The third response to heat, relaxed postural extension, reduces heat production by postural tonus and nonthermoregulatory activities and increases heat transfer to the air and substrate by increasing body surface area. It consists of relaxed elongation of the body on the cage floor in a prone or near-prone position with the normally convex spine straightened horizontally, forelegs placed under the neck, and hindlegs folded outward and toward the rear. At first the rat usually keeps the head elevated, but later lowers it to the floor and becomes drowsy, as evidenced by partial closure and retraction of the eyes.

These three behavioral responses to hyperthermia are employed by a wide variety of mammals, although there are appreciable differences between species in the relative amounts and detailed movement topography of the behaviors. In a comparison of rats with other rodents, Roberts, Mooney, and Martin[15] found that ambient heat induced increased grooming in gerbils and guinea pigs, increased locomotion in gerbils, guinea pigs, and hamsters, and relaxed prone or lateral postural extension in gerbils, guinea pigs, hamsters, and mice. Heat-induced grooming of saliva over the fur has also been observed in cats and Virginia opossums, increased locomotion in opossums, dogs, and rabbits, and relaxed postural extension or lying on the side in cats, opossums, dogs, and rabbits.[18-20] Some form of postural relaxation with drowsiness is probably the most widespread behavioral response of mammals to deep hyperthermia[20] and in extreme form becomes estivation in some species under conditions of prolonged heat.[21]

In addition to the behavioral responses, rats employ a number of physiological heat dissipation responses, including tail vasodilation, salivation, and in some circumstances urination, which wets the lower abdomen.[22-25] All of these increase the rate of heat loss, but do not deal directly with the basic problem of external heat as does escape to a cooler area or the problem of internal metabolic heat from exertion as does the relaxed postural extension response.

FRACTIONAL ORGANIZATION OF THERMOREGULATORY BEHAVIOR MECHANISMS

A necessary first step in the analysis of thermoregulatory mechanisms is identification of the sites of thermosensitivity and the responses controlled by the thermosensors at those sites. For this purpose, we have used implanted diathermy electrode arrays, resistance heaters, or a thermal floor through which controlled temperatures could be applied to the ventral surface.

Central Thermosensors

When we applied local diathermic warming (2 MHz) to the preoptic/anterior hypothalamic region (PO/AH) of rats, we were surprised to obtain consistently strong

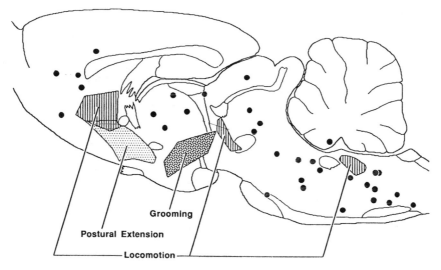

● = No behavioral response to local warming

FIGURE 1. Thermosensitive zones where localized diathermic warming in midline of rat brain elicited thermoregulatory behaviors: sagittal diagram, 6 mm from midline. *Black circles* indicate individual negative sites where thermoregulatory behaviors were not obtained. More detailed sagittal and frontal plots are found in an earlier report.[7]

and clear-cut postural extension but no grooming and only a small amount of high threshold locomotion in a small proportion of animals (FIG. 1).[7] Since we had earlier found some anatomical differentiation in the PO/AH zones where local warming induced grooming and sleep-like relaxation in the Virginia opossum,[19] we increased the anatomical scatter of our diathermy electrodes and discovered that vigorous grooming could be obtained from the posterior hypothalamus and anterior ventral midbrain without postural extension and with only occasional weak and high-threshold locomotion. Exploration of additional brain areas disclosed that strong and consistent locomotion without grooming or postural extension could be elicited by warming the septal area, the ventromedial midbrain, and the dorsomedial medulla. Tail vasodilation was obtained from all of the thermosensitive areas. From these findings, it was concluded that signals from central thermosensors in the rat are not combined to yield a unitary error signal that controls antihyperthermia behaviors generally, but instead, the signals from different groups of sensors take separate paths to the mechanisms for the different behaviors.

Similar evidence of preferential control of different thermoregulatory responses by different groups of central thermosensors was obtained by Carlisle and Ingram,[1] who found that cooling the anterior hypothalamus of pigs was more effective in eliciting bar pressing for radiant heat than cooling the cervical spinal cord, while cooling the cord was more effective in eliciting heat-conserving postural adjustments. Similar differential linkages have been reported in the guinea pig between anterior hypothalamic cold sensors and nonshivering (brown fat) thermogenesis and between cervical cord cold sensors and shivering thermogenesis.[26]

A number of reports that PO/AH or other localized brain lesions impair some thermoregulatory responses to hyper- or hypothermia while sparing other responses[27-33] are consistent with the concept of relatively separate sensor-to-effector pathways, but could alternatively be due to selective interruption of divergent output pathways from a general sensory integrator to separate response mechanisms.

Peripheral Thermosensors

When superficial or deep peripheral thermosensors were warmed by a temperature-controlled cage floor or by implanted resistance heaters in the subcutaneous fascia, nasal cavity, pharynx, esophagus, or abdominal cavity, temperatures below the pain threshold consistently elicited grooming, locomotion, and tail vasodilation, but postural extension was never observed[6] (although a very small but statistically significant amount was detected in later tests with a warm floor[34]).

From these studies of the behaviors induced by localized central and peripheral warming, it appears that grooming and locomotion are controlled by superficial thermosensors in the skin and possibly the anterior nasal passages and by deep thermosensors located in separate brain areas and scattered widely through the body tissues. Postural extension is almost entirely dependent on deep central thermosensors in the PO/AH area[7] and perhaps the medulla and spinal cord.[4,35]

INTERACTION OF THERMOSENSORY INPUTS

Although localized thermal stimulation is useful in determining the presence of thermosensors at specific sites and the responses they control, it leaves open the question of how inputs from different sites combine and interact under natural conditions when hyperthermia is more generally distributed in the body.

One of the most important functional characteristics of thermosensors is their depth beneath the body surface. Although this is a continuum, it is useful to distinguish between superficial and deep sensors. Superficial thermosensors are most directly and rapidly affected by changes in ambient temperature and infrared radiation and are thus in a position to elicit protective or escape responses before development of deep hyperthermia. Deep thermosensors, whose temperatures are more slowly altered by changes in the external thermal environment, signal the more critical state of deep hyperthermia and are the primary monitors when excessive heat is generated internally by exertion.

Superficial sensors are exposed to a considerably wider range of temperatures than deep sensors, and as a consequence, the functions relating the intensity of thermoregulatory responses to superficial temperatures are less steep than the functions relating them to deep temperatures. In goats, for example, the function relating metabolic heat production to skin temperature is about one-third as steep as the function relating it to core temperature,[36] and the functions relating panting and sweating to skin temperature are about one-fourth as steep as the functions relating them to core temperature.[37] For this reason, superficial and deep hyperthermia cannot validly be compared in terms of absolute number of degrees deviation from the normal local temperatures present in a thermally neutral environment, but must be corrected for differences in response-temperature functions, or as a first approximation, differences in the ranges of superficial and deep temperatures within the limits of thermoregulation.

A number of studies have found that the different distributions of heat between deep and superficial thermosensors that occur in different natural conditions of heat

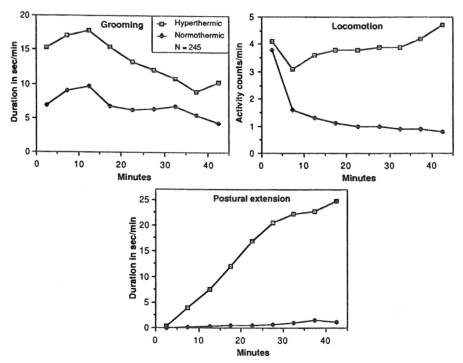

FIGURE 2. Differing temporal patterns of thermoregulatory behaviors during exposure to ambient heat (37–38°C) and thermoneutrality (25–27°C). Standard errors for grooming and locomotion were less than 0.77 and 0.33, respectively.

stress elicit different combinations of thermoregulatory responses.[15,34,37,38] The thermoregulatory behaviors of the rat furnish a particularly clear example of this form of adaptation.

Predominantly Superficial Hyperthermia

The effects of predominantly superficial hyperthermia can be observed during the initial 10–15 min of exposure to ambient heat before deep temperatures have risen much above normal. For example, when rats were shifted from 24° to 37°C for 45 min, 91% of the increase in the average of four skin temperatures occurred during the first 15 min, while 83% of the increase in rectal temperature occurred during the following 30 min.[38]

FIGURE 2 shows the temporal patterns of emergence of grooming, locomotion, and postural extension in 245 rats in a number of published[15,30] and unpublished studies. During the first 15 min of heat exposure, when the temperatures of superficial thermosensors were near-maximal for the session, but deep temperatures had just begun to rise, the rats evidenced large increases in grooming and locomotion above control levels but relatively little postural extension. Grooming increased immediately to a near-maximum during the first 5 min. The increase in locomotion above the control level was delayed by 5 min, probably because of the initial burst of exploratory locomotion in both conditions, a typical response of rats in a new environment. During the next two 5-min blocks, locomotion rose to about 75% of the session maximum, while extension increased at a considerably lower rate to about 30% of the session maximum. The early emergence of grooming and locomotion and the relatively small amount of postural extension during this initial phase indicate that the earlier finding that localized warming of the skin elicits grooming and locomotion but little or no extension holds for generalized predominantly superficial hyperthermia in the more natural condition of ambient heat exposure.

Increased grooming and locomotion would have high utility after a rise in ambient temperature caused by movement into a warmer environment or a change in cloud cover or the incidence of the sun. The probability of escape by locomotion to a burrow or other cool area is fairly high for rats and most other small animals across a broad range of such conditions, and the cooling effect of evaporation of the saliva spread on the fur by increased grooming would reduce the rate of net heat gain in the meantime. During this early stage of minimal deep hyperthermia, the need for reduction of internal heat production by the postural extension response would not be great enough to exceed the cost of interference with other adaptive behaviors, such as feeding, defending territory, and vigilance for predators.

Mixed Superficial and Deep Hyperthermia

A second pattern of heat distribution is the more balanced mixture of superficial and deep hyperthermia that occurs in ambient heat after deep temperatures have risen to a plateau, which in rats can be $1.5-2.5°C$ above normal core temperature in a thermally neutral environment. During the last 30 min of the 45 min test shown in FIGURE 2, when superficial temperatures were nearly constant, but deep temperatures were progressively rising, extension increased markedly by about 230%, locomotion increased by about 30%, and grooming declined by about 50% but remained significantly above the control level. All of these trends were statistically significant. The large increase in extension confirms the conclusion from the localized warming studies that extension is entirely or almost entirely controlled by deep thermosensors, most or all of which are in the brain and spinal cord. The additional increase in locomotion and decline in grooming during the latter part of the session occurred too late to be due to warming of superficial sensors, but are consistent with a study of the interactions between localized central and superficial warming in which warming of each of the brain areas controlling grooming, locomotion, and extension was combined with superficial warming of the ventral surface by a thermal floor.[34] In this study, locomotion induced by warming the thermal floor was facilitated by the addition of diathermic warming of any of the central thermosensitive areas, while grooming elicited by the warm floor was partially inhibited by warming of the PO/AH, septal, or midbrain areas where responses other than grooming were obtained. The posterior hypothalamic area where local warming elicits grooming evidently does not exert a strong-enough influence to overcome these inhibitory effects, but may still serve an adaptive function in protecting a moderate level of grooming against still greater suppression. Whether the reduction in grooming was due to direct inhibition within thermoregulatory mechanisms or to response interference at the motor level is unclear, but it demonstrates that the effect on responses of combining thermal signals of like sign is not always facilitation and that direct determination of interaction effects is essential.

Mixed superficial and deep hyperthermia, which develops after escape has failed for a considerable time, is the most life-threatening and the most likely to be prolonged of the three possible patterns of distribution of hyperthermia. When body heating has reached this stage, the need for escape to a cooler environment is even greater than during the preceding phase of predominantly superficial hyperthermia. Evaporative cooling continues to be useful, but it is only a relatively temporary palliative with a rising cost because of small animals' limited content of dispensable water and salt. The role of relaxed postural extension in reducing heat generation by nonthermoregulatory behaviors and postural tonus becomes increasingly important in this condition, partly because of the elevated core temperature, and partly because the probability of extended exposure to the heat is greater if locomotor escape attempts have failed for a long enough time (20–40 min) for appreciable deep hyperthermia to develop. To suppress nonthermoregulatory behaviors, extension must be relatively strongly facilitated. Simultaneous suppression of grooming and locomotion is reduced or prevented by the additional facilitation they receive from central and deep peripheral thermosensors, which is appreciably greater for locomotion, the more essential response.

Another species in which mixed deep and superficial hyperthermia elicits a different pattern of responses than predominantly superficial hyperthermia is the black Bedouin goat. When skin thermosensors are heated relatively more than deep tissues by strong solar radiation accompanied by a mild elevation of air temperature, as occurs at times in the desert, sweating is utilized more and panting less than when a more balanced mixture of superficial and deep hyperthermia is produced by a high air temperature without solar radiation, even when total evaporative heat loss is equated in the two conditions.[37,39]

Predominantly Deep Hyperthermia

A third pattern of heat distribution is predominantly deep hyperthermia, which occurs after escape from a warm area to a burrow or other thermally sheltered area or during and following vigorous physical exertion in a thermally neutral environment. In these conditions, thermoregulatory responses are mainly elicited by deep sensors in the CNS and other body tissues, whose temperatures are high within their normal range of variation, while superficial sensors, whose local temperatures may be raised almost as much in absolute degrees by heat conduction from deep tissues, exert a considerably weaker influence because of the lesser steepness of the functions relating thermoregulatory responses to superficial temperatures.

Our first attempt to induce predominantly deep hyperthermia by means of exercise failed because of the unwillingness of our middle-aged, somewhat heavy rats (6–8 months, 450–650 g) to run fast enough in a treadmill without supplementary motivation likely to confound the findings. We then turned to escape from a warm area (39°C) to a thermally neutral area (24°C) after a rise of about 2°C in core temperature to produce the condition of predominantly deep hyperthermia. The effects on grooming, locomotion, and postural extension of the exposure to ambient heat and the subsequent drop in ambient temperature to thermoneutrality are shown in FIGURE 3. There was a 5-min delay in the onset of thermoregulatory responses in these tests, probably because air flow through the animal cage was reduced and heat radiation from the thermal chamber walls was interrupted by white cardboard screens located 4 cm from the cage walls to block extra-cage visibility before and after escape. With this exception, the trends were generally similar to the earlier findings shown in FIGURE 2, including early emergence of heat-induced grooming and locomotion, followed by a partial decline in grooming, a further increase in locomotion, and a gradual but large increase

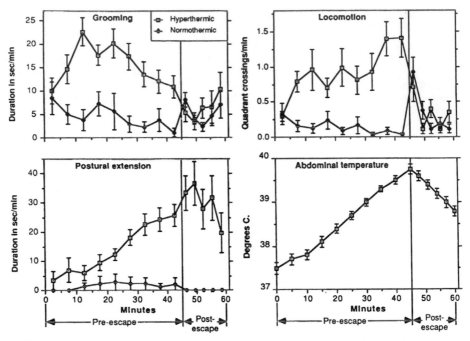

FIGURE 3. Effect of a shift from mixed superficial and deep hyperthermia to predominantly deep hyperthermia on thermoregulatory behaviors. After 45-min exposure to 39°C, during which abdominal temperature rose about 2°C, rats escaped to a thermally neutral area (24°C), where surface (but not deep) temperatures rapidly declined by about 80% of their increase in the heat. In normothermic control tests, ambient temperature was 24°C throughout. *Bars* indicate standard errors.

in extension. Following escape to thermoneutrality, both grooming and locomotion fell to control levels (although locomotion was somewhat inflated by exploration of the new cage during the first 3 min after escape in both hyperthermic and normothermic conditions). Postural extension increased about 30%, suggesting that it may have been reduced somewhat by interference from grooming and locomotion in the heat.

The rapid drop in grooming and locomotion and the reciprocal increase in postural extension indicate that extension, in addition to being wholly or almost wholly dependent on deep thermosensors as indicated by local warming studies,[6,7,34] is the paramount response when signals from deep sensors are dominant. This conclusion is reinforced by a second study in which predominantly deep hyperthermia was produced by intermittent exposure to microwave radiation in a thermally neutral environment. As shown in FIGURE 4, extension rose progressively with the increase in deep body temperature, while grooming and locomotion declined below control levels.

When predominantly deep hyperthermia is caused by vigorous exertion or by escape from a hot to a cool environment, the relaxation associated with postural extension would be maximally effective in minimizing further heat production by locomotion and other muscle activities, while core temperatures would return to normal within a reasonable time through efficient heat dissipation to the cooler air and ground sur-

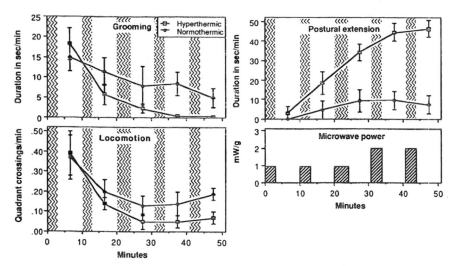

FIGURE 4. Increased postural extension and mild reduction in grooming and locomotion induced by predominantly deep hyperthermia created by five 3- or 4-min exposures to weak microwave radiation in a thermally neutral environment. Normothermic control tests for spontaneous behavior were given without microwave radiation. Behavior measures were taken during 7-min periods following each microwave exposure. Mean rectal temperature increased from 38.1°C at the beginning of the first microwave exposure to 39.8°C at the end of the last microwave exposure.

face with assistance from vasodilation in the tail and other vascularized exposed skin.[40] In this condition, locomotion would contribute nothing and grooming relatively little (especially in a humid burrow) to compensate for the added muscle heat production and water and salt loss they entail.[41-43]

Overall, it appears that the rat, in the course of evolution, has arrived at a balance between the relative strength of superficial and deep thermosensor control of the three behaviors that yields different optimal combinations of responses across a variety of heat stress conditions. Along similar lines, Taylor[44] has proposed that dissipation of heat loads from exertion and from ambient heat may be accomplished by different evaporative cooling mechanisms in some large species, although the sensors responsible for selecting among the mechanisms have not been specified.

Differentiated fractional control of thermoregulatory responses is especially likely to occur when responses are accompanied by appreciable physiological or behavioral costs and differ in effectiveness in different thermal stress conditions. These properties are most often possessed by behavioral responses, but are also associated with some physiological responses, such as evaporative cooling by sweating and panting.[44] Other physiological responses that are relatively free of such limitations on their net effectiveness may be more generally and uniformly activated by thermosensory input from different body sites, as was found for vasodilation in local warming studies in the rat.[6,7] Detection of differentiated fractional control of thermoregulatory responses requires techniques by which thermal stimuli can be applied to different groups of thermosensors together with measurement of a number of thermoregulatory responses to error signals of the same sign, a difficult combination of variables not often employed.

SUMMARY AND CONCLUSIONS

The rat displays three behaviors when surface and/or deep body temperatures rise above their normal levels: grooming, which spreads saliva over the fur to cool by evaporation; locomotion, which results in escape to a cooler environment; and relaxed postural extension, which reduces heat production by muscle activity and increases heat dissipation by increasing body surface area. Each of these behaviors can be elicited in pure or nearly pure form by localized warming of thermosensors in different brain areas. In addition, localized warming of superficial and deep peripheral thermosensors elicits grooming and locomotion, but little or no postural extension. Because of the marked differences in the behaviors induced by localized warming of different sets of thermosensors, it is concluded that sensors are linked to the effector mechanisms for these behaviors by relatively separate and independent pathways rather than a master controller through which sensory signals are integrated to produce a unitary error signal that drives all responses.

When more generalized hyperthermia in natural heat stress situations affects mainly superficial thermosensors, as during the initial phase of exposure to ambient heat, grooming and locomotion are predominant and extension is absent or weak. When deep hyperthermia is predominant and superficial hyperthermia is relatively weak, as occurs after escape from ambient heat to thermoneutrality or during and following exercise in a thermally neutral environment, the resulting predominant activation of deep sensors elicits marked postural extension, but little or no grooming or locomotion. When both superficial and deep sensors are hyperthermic, as in the later stages of exposure to ambient heat, all three behaviors are present, although locomotion is somewhat greater and grooming is somewhat less than during predominantly superficial hyperthermia, and extension is somewhat less than during predominantly deep hyperthermia. As a consequence, grooming is maximal during predominantly superficial hyperthermia, locomotion during mixed deep and superficial hyperthermia, and extension during predominantly deep hyperthermia.

It is concluded that the relatively fractionated control of the rat's thermoregulatory behaviors by different groups of central and peripheral thermosensors makes possible a differentiation of responses to different distributions of hyperthermia between superficial and deep sensors that optimizes cost-benefit trade-offs in different kinds of heat stress.

REFERENCES

1. CARLISLE, H. J. & D. L. INGRAM. 1973. The effects of heating and cooling the spinal cord and hypothalamus on thermoregulatory behavior in the pig. J. Physiol. 231: 353–364.
2. JESSEN, C. 1985. Thermal afferents in the control of body temperature. Pharmac. Ther. 28: 107–134.
3. JESSEN, C., G. FEISTKORN & A. NAGEL. 1983. Temperature sensitivity of skeletal muscle in the conscious goat. J. Appl. Physiol. 54: 880–886.
4. LIPTON, J. M. 1973. Thermosensitivity of medulla oblongata in control of body temperature. Am. J. Physiol. 224: 890–897.
5. RAWSON, R. D. & K. P. QUICK. 1972. Localization of intra-abdominal thermoreceptors in the ewe. J. Physiol. 222: 665–677.
6. ROBERTS, W. W. & J. R. MARTIN. 1974. Peripheral thermoreceptor control of thermoregulatory responses of the rat. J. Comp. Physiol. Psychol. 87: 1109–1118.
7. ROBERTS, W. W. & R. D. MOONEY. 1974. Brain areas controlling thermoregulatory grooming,

prone extension, locomotion, and tail vasodilation in rats. J. Comp. Physiol. Psychol. **86:** 470–480.

8. THAUER, R. 1970. Thermosensitivity of the spinal cord. *In* Physiological and Behavioral Temperature Regulation. J. D. Hardy, A. P. Gagge & J. A. J. Stolwijk, Eds.: 472–492. Charles C Thomas. Springfield, IL.

9. BRÜCK, K. & P. HINCKEL. 1982. Thermoafferent systems and their adaptive modifications. Pharmac. Ther. **17:** 357–381.

10. HAMMEL, H. T. 1968. Regulation of internal body temperature. Ann. Rev. Physiol. **30:** 641–710.

11. CHATONNET, J. 1983. Some general characteristics of temperature regulation. J. Therm. Biol. **8:** 33–36.

12. SATINOFF, E. 1978. Neural organization and evolution of thermal regulation in mammals. Science **201:** 16–22.

13. SATINOFF, E. 1983. A reevaluation of the concept of the homeostatic organization of temperature regulation. *In* Handbook of Behavioral Neurobiology, Vol. 6: Motivation. E. Satinoff & P. Teitelbaum, Eds.: 443–472. Plenum. New York.

14. HAINSWORTH, F. R. 1967. Saliva spreading, activity, and body temperature regulation in the rat. Am. J. Physiol. **212:** 1288–1292.

15. ROBERTS, W. W., R. D. MOONEY & J. R. MARTIN. 1974. Thermoregulatory behaviors of laboratory rodents. J. Comp. Physiol. Psychol. **86:** 693–699.

16. HAINSWORTH, F. R. & E. M. STRICKER. 1970. Salivary cooling by rats in the heat. *In* Physiological and Behavioral Temperature Regulation. J. D. Hardy, A. P. Gagge & J. A. J. Stolwijk, Eds.: 611–626. Charles C Thomas. Springfield, IL.

17. SATINOFF, E. & R. HENDERSEN. 1977. *In* Handbook of Operant Behavior. W. K. Honig & J. E. R. Staddon, Eds.: 153–173. Prentice-Hall. Englewood Cliffs, NJ.

18. HIGGINBOTHAM, A. C. & W. E. KOON. 1955. Temperature regulation in the Virginia opossum. Am. J. Physiol. **181:** 69–71.

19. ROBERTS, W. W., E. H. BERGQUIST & T. C. L. ROBINSON. 1969. Thermoregulatory grooming and sleep-like relaxation induced by local warming of preoptic area and anterior hypothalamus in opossum. J. Comp. Physiol. Psychol. **67:** 182–188.

20. ROBINSON, K. W. & D. H. K. LEE. 1946. Animal behaviour and heat regulation in hot atmospheres. Univ. Qld. Papers, Dept. Physiol. **1(9):** 1–8.

21. BERGER, R. J. 1984. Slow wave sleep, shallow torpor, and hibernation: Homologous states of diminished metabolism and body temperature. Biol. Psychol. **19:** 305–326.

22. ELMER, M. & P. OHLIN. 1971. Salivary secretion in the rat in a hot environment. Acta Physiol. Scand. **83:** 174–178.

23. HUBBARD, R. W., C. B. MATTHEW & R. FRANCESCONI. 1982. Heat-stressed rat: Effects of atropine, desalivation, or restraint. J. Appl. Physiol.: Respirat. Environ. Exercise Physiol. **53:** 1171–1174.

24. RAND, R. P., A. C. BURTON & T. ING. 1965. The tail of the rat in temperature regulation and acclimatization. Can. J. Physiol. Pharmacol. **43:** 257–267.

25. THOMPSON, G. E. & J. A. F. STEVENSON. 1965. The temperature response of the male rat to treadmill exercise, and the effect of anterior hypothalamic lesions. Can. J. Physiol. Pharmacol. **43:** 279–287.

26. BRÜCK, K. & W. WÜNNENBERG. 1970. Meshed control of two effector systems: Non-shivering and shivering thermogenesis. *In* Physiological and Behavioral Temperature Regulation. J. D. Hardy, A. P. Gagge & J. A. J. Stolwijk, Eds.: 562–580. Charles C Thomas. Springfield, IL.

27. CARLISLE, H. J. 1969. The effects of preoptic and anterior hypothalamic lesions on behavioral thermoregulation in the cold. J. Comp. Physiol. Psychol. **69:** 391–402.

28. GILBERT, T. M. & C. M. BLATTEIS. 1977. Hypothalamic thermoregulatory pathways in the rat. J. Appl. Physiol.: Respirat. Environ. Exercise Physiol. **43:** 770–777.

29. LIPTON, J. M. 1968. Effects of preoptic lesions on heat-escape responding and colonic temperature in the rat. Physiol. Behav. **3:** 165–169.

30. ROBERTS, W. W. & J. R. MARTIN. 1977. Effects of lesions in central thermosensitive areas on thermoregulatory responses in rat. Physiol. Behav. **19:** 503–511.

31. SATINOFF, E. & J. RUTSTEIN. 1970. Behavioral thermoregulation in rats with anterior hypothalamic lesions. J. Comp. Physiol. Psychol. **71:** 77–82.

32. SATINOFF, E. & S. SHAN. 1971. Loss of behavioral thermoregulation after lateral hypothalamic lesions in rats. J. Comp. Physiol. Psychol. **77:** 302–312.
33. VAN ZOEREN, J. G. & E. M. STRICKER. 1977. Effects of preoptic, lateral hypothalamic, or dopamine-depleting lesions on behavioral thermoregulation in rats exposed to the cold. J. Comp. Physiol. Psychol. **91:** 989–999.
34. ROBERTS, W. W. & A. FROL. 1979. Interaction of central and superficial peripheral thermosensors in control of thermoregulatory behaviors of the rat. Physiol. Behav. **23:** 503–512.
35. LIN, M. T., T. H. YIN & C. Y. CHAI. 1972. Effects of heating and cooling of spinal cord on CV and respiratory responses and food and water intake. Am. J. Physiol. **223:** 626–631.
36. JESSEN, C. 1981. Independent clamps of peripheral and central temperatures and their effects on heat production in the goat. J. Physiol. **311:** 11–22.
37. DMI'EL, R. & D. ROBERTSHAW. 1983. The control of panting and sweating in the black Bedouin goat: A comparison of two modes of imposing a heat load. Physiol. Zool. **56:** 404–411.
38. ROBERTS, W. W., J. A. NAGEL & S. A. FRUTIGER. Thermoregulatory behaviors of rats vary with distribution of hyperthermia between superficial and deep thermosensors. In preparation.
39. BORUT, A., R. DMI'EL & A. SHKOLNIK. 1979. Heat balance of resting and walking goats: Comparison of climatic chamber and exposure in the desert. Physiol. Zool. **52:** 105–112.
40. STRICKER, E. M. & F. R. HAINSWORTH. 1971. Evaporative cooling in the rat: Interaction with heat loss from the tail. Q. J. Exp. Physiol. **56:** 231–241.
41. MORRISON, S. D. 1968. The constancy of the energy expended by rats on spontaneous activity, and the distribution of activity between feeding and non-feeding. J. Physiol. **197:** 305–323.
42. HAINSWORTH, F. R., E. M. STRICKER & A. N. EPSTEIN. 1968. Water metabolism of rats in the heat: Dehydration and drinking. Am. J. Physiol. **214:** 983–989.
43. STRICKER, E. M. & F. R. HAINSWORTH. 1970. Evaporative cooling in the rat: Effects of dehydration. Can. J. Physiol. Pharmacol. **48:** 18–27.
44. TAYLOR, C. R. 1977. Exercise and environmental heat loads: Different mechanisms for solving different problems? In Environmental Physiology II. International Review of Physiology, Vol. 15. D. Robertshaw, Ed.: 119–146. University Park Press. Baltimore.

Behavioral and Electrophysiological Effects of Intracranially Applied Neuropeptides with Special Attention to DC Slow Potential Changes[a]

J. P. HUSTON AND M.-S. HOLZHÄUER

Institute of Psychology III
University of Düsseldorf
D-4000 Düsseldorf, Federal Republic of Germany

INTRODUCTION

Intracranial administration of certain neuropeptides can induce massive changes in EEG activity and the DC slow potential, referred to as "spreading depression." Spreading depression, which is difficult to recognize in EEG recordings, has various behavioral consequences. This report summarizes studies dealing with the elicitation of "spreading depression" by neuropeptides and the concomitant behavioral changes commonly (mistakenly) attributed to physiological effects of intracranially injected neuropeptides.

The application of various neuropeptides into the brain can elicit behaviors such as "wet-dog" shaking, grooming, circling, changes in locomotor activity, immobility, analgesia, and barrel rotations. Similar behaviors accompany spreading depression (SD) triggered by electrical, mechanical, or chemical stimulation in certain brain regions. Therefore, we hypothesized that certain behavioral changes that follow the intracranial administration of some neuropeptides are a direct consequence of SD.

Our first experiments demonstrated that the injection of certain neuropeptides into the brain can induce SD in the anesthetized animal. The application of nanogram amounts of vasopressin and microgram amounts of $ACTH_{1-24}$, Leu- and Met-enkephalin, and the Met-enkephalin analogue [D-Ala-2]Met-enkephalinamide (DAME) into the cortex and/or hippocampus of rats elicited SD.[1-3] In the case of the enkephalins and DAME, their SD- and seizure-inducing activity was blocked or partly antagonized by the opiate antagonist naloxone.

The finding that some neuropeptides are able to induce SD in the anesthetized rat led us to record EEG, DC slow potentials, and behavioral responses concurrently after hippocampal injection of $ACTH_{1-24}$ and DAME in the freely moving animal.[4] We wished to determine whether certain behavioral effects of intracranial injection of neuropeptides (*e.g.*, wet-dog shaking, hyperactivity or immobility, and analgesia for DAME; grooming for $ACTH_{1-24}$) are the behavioral expression of SD rather than peptide-specific responses.

[a] This work was supported by grant no. Hu 306/4-2 from the Deutsche Forschungsgemeinschaft.

375

GENERAL CHARACTERISTICS OF "SPREADING DEPRESSION" (SD)

Comprehensive reviews have been devoted to the phenomenon of SD, including its mechanisms, application, behavioral effects, and methods of registration.[5-9]

The electrophysiological manifestation of SD was probably first described by Leão in 1944[6] as a decrease of spontaneous and evoked EEG activity in the cortex of rabbits after local electrical stimulation. The depression of EEG activity spreads slowly with a velocity of 2–3 mm per min through the brain tissue, *e.g.*, neocortex, nucleus caudatus, hippocampus, and thalamus. Concomitant recording of the DC slow potentials reveals a DC slow potential change (DC-SPC) of up to 40 mV negativity for 1–2 min, followed by a lower but longer-lasting positivity. This sequence constitutes a "wave" of SD. During SD unit activity is depressed, since the neurons are depolarized. Increased neural activity often precedes and/or follows the onset of a wave of SD.

Three criteria that have been employed to verify the presence of SD are (1) the negative DC-SPC; (2) the depression of the amplitude of EEG activity during the DC-SPC; and (3) behavioral signs, such as the loss of posture and placing reactions. In the anesthetized animal both the EEG and DC-SPC can be registered easily and reliably. However, since in the awake animal the amplitude of EEG may not decrease significantly during SD, the DC-SPC is a more reliable indicator for its occurrence.

SD can be triggered by electrical, mechanical, thermal, and chemical stimulation of brain tissue. The most commonly used experimental technique is to apply potassium chloride (KCl; 1–25% solution). Single or multiple waves of SD are induced, depending on the method and magnitude of stimulation.

FIGURE 1 represents an example of EEG and DC-SPC characteristics of spreading depression as described above; single (FIG. 1A) and multiple (FIG. 1B) waves of SD were elicited by unilateral hippocampal injection of neuropeptides in the anesthetized rat.[2,3]

Although SD may be localized to the hemisphere and brain structure where it was elicited, it is often accompanied by changes of neural activity in other brain areas. For example, cortical SD decreases neural activity in the hypothalamus, caudate nucleus, and amygdala and increases neural activity in the reticular formation.[10-13]

CHEMICAL CHANGES IN RELATION TO SD

During SD the concentration of K^+ ions increases, whereas Na^+, Cl^- and Ca^{2+} concentrations decrease in the extracellular space.[14-16] Furthermore, SD influences the release of transmitters and hormones. For example, there is a release of ACh and GABA and a decrease in the release of glutamate[17,18] as well as the release of LH and prolactin[19] after cortical SD. In order to restore physiological conditions metabolic mechanisms (energy metabolism) are activated by this process.[20,21]

During cortical SD glucose utilization is markedly increased in the cortex, but reduced in many subcortical structures.[22,23] The onset of cortical SD is paralleled by a pronounced decrease in glucose content, high-energy phosphates, and tissue pH.[24,25] Studies on cerebral blood flow have shown vasoconstriction during SD followed by vasodilation and a subsequent decrease in blood flow.[26] These changes persist beyond the transient depolarization of cortical cells and the DC-SPC.[25]

The mechanisms for the initiation and propagation of SD are not thoroughly understood; however, a central role is ascribed to potassium (K^+)[27] and/or glutamate.[28,29] To trigger SD the accumulation of K^+ ions must exceed a threshold level of 10–12 mM.[30]

The adjacent neurons are then activated and in turn release additional K^+ ions. A close relationship between SD and convulsions was already recognized in the first reports on SD; SD can elicit or suppress epileptiform activity.[31] On the other hand, it can be triggered[32,33] or blocked[34,35] by epileptiform discharges.

Possible electrophysiological and chemical mechanisms responsible for the action of SD and its behavioral manifestation were summarized by Ueda and Bureš.[34] These include (1) the depression of inhibitory processes by interference with the presynaptic release of neurotransmitters or the blockade of postsynaptic receptor sites; (2) potentiation of excitatory processes by enhancing the release of neurotransmitters, e.g., blockade of presynaptic inhibition; and (3) direct influence on the membrane properties of the postsynaptic neuron, e.g., increase of membrane resistance or selective decrease of membrane conductance.

Considering the electrophysiological and chemical changes observed in relation to SD, the behavioral expression of SD must obviously reflect a complex interaction of neurochemical changes in many neuronal systems.

BEHAVIORAL EFFECTS OF SD

The following examples represent a short overview of behavioral concomitants of SD.[6] Multiple waves of unilateral or bilateral cortical SD produce an impairment of coordination and posture related reflexes, accompanied by a state of diminished sensibility and immobility. Behavioral effects of unilateral cortical SD, such as the loss of placing reactions, are observed on the side of the body contralateral to the induction of SD. Multiple waves of cortical SD can be accompanied by recurrent behavioral recovery between the waves of SD, depending on the interwave interval.[36] Changes in thermoregulation have also been reported after cortical SD.[37]

Multiple waves of bilateral SD in the cortex and hippocampus of rats were followed by immobility or increased locomotor activity.[6,38,39] Although the rats were able to move when prodded, their reactivity to footshock was diminished.[40] Unilateral cortical and caudate SD was reported to induce consumatory and circling behavior,[41-44] increased locomotion, grooming,[45,46] yawning, and sexual behavior.[47-49] In the hippocampus unilateral SD was accompanied by wet-dog shaking and increased locomotor activity interrupted by bouts of grooming and eating.[50] Wet-dog shaking was reported independent of whether the stimulus used to induce SD in the hippocampus was electrical, mechanical, or chemical. It was invariably linked to periods of paroxysmal activity, and coincided regularly with the short positive phase preceding and following the negative SPC.[50]

INTRACRANIALLY APPLIED NEUROPEPTIDES

Few studies dealing with intracranial effects of neuropeptides (and other compounds) have employed the combined registration of EEG and behavioral responses. Recording of DC-SPC, which best reflects the occurrence of SD, is seldom if ever used.

The following sections review behavioral and electrophysiological effects of ACTH, opioid peptides, vasopressin, and substance P after their intracranial administration. Judging from the range of "unusual" behavioral responses described in the literature after intracranial application of neuropeptides, it is likely that SD is the cause for behavioral effects in many cases.

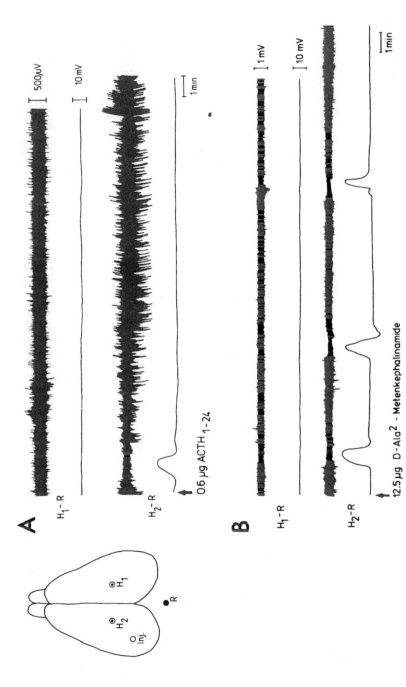

FIGURE 1. EEG (*upper trace*) and DC slow potential changes (*lower trace*) accompanying hippocampal spreading depression (SD) in the anesthetized rat. SD was elicited by unilateral injection of (**A**) 0.6 µg ACTH$_{1-24}$ and (**B**) 12.5 µg [D-Ala-2]Met-enkephalinamide in 1.0 µl volume into the dorsal hippocampus (*Inj.*). H_1 and H_2 represent the position of the recording electrodes, R the reference electrode. The *arrow* indicates the end of the 30-s duration of injection. The typical slow potential change consists of a negativity (upward) preceded and followed by a slight positivity. The amplitude of EEG shows a slight depression preceded by a burst of increased activity.

$ACTH_{1-24}$

The intracranial application of ACTH in mammals has long been known to induce stretching, yawning, and sexual behavior.[51-53] These behaviors were preceded by excessive grooming in rats and mice.[51,54] Grooming has been investigated extensively as an example of a peptide-modulated behavior, particularly by ACTH.[55-59]

The possibility that SD could be involved in ACTH-induced behavioral effects was suggested by the findings that cortical and hippocampal spreading depression, elicited by KCl and electrical stimulation, was accompanied by stretching and yawning, penile erection, and grooming.[46-50,60]

Few data are available on the electrophysiological effects of $ACTH_{1-24}$; the ionophoretic application of ACTH increases the firing rate of brainstem neurons,[61] has no influence on spontaneous activity of hippocampal neurons, but modulates transmitter-induced activity.[62] No changes in cortical and hippocampal EEG activity were observed after i.c.v. injection of 10 μg $ACTH_{1-24}$ in the rabbit.[63]

In a preliminary study using anesthetized rats, the injection of $ACTH_{1-24}$ into the cortex (2.5–15 μg) and hippocampus (0.24–2.88 μg; FIG. 1A) was found to elicit SD, as reflected by DC-SPC and depression of EEG activity.[2] Single and multiple waves of SD (one to eight waves) were observed. Spike discharges were recorded in 16% of trials following the wave of SD. However, the application of $ACTH_{1-24}$ unilaterally into the hippocampus of freely moving rats elicited SD in only 13% of 76 rats with an unusually long onset latency of about 3 min.[4] Behaviors and electrical brain activity were registered for 50 min after injection of ACTH or KCl. ACTH did not lead to gross changes of EEG activity, and epileptiform discharges neither preceded nor followed the wave of ACTH-induced SD. Control injections of KCl elicited SD in all trials and the depression of hippocampal EEG activity was preceded and followed by bursts of epileptiform discharges of 20–40 s duration (FIG. 2).

ACTH-induced grooming behavior was observed irrespective of the occurrence of SD, the strain of rats (Sprague-Dawley, Wistar), and the dosages used (0.5 and 10 μg $ACTH_{1-24}$ in 0.5-μl volume). However, SD induced by KCl resulted in excessive grooming comparable to that induced by ACTH (FIG. 3). One difference was seen in the time course of this behavioral response (FIG. 2 and FIG. 3A): whereas grooming began shortly after KCl-induced SD and diminished slowly after 25 min, ACTH-induced grooming reached its peak 15 to 20 min after injection. Grooming outlasted the observation period and was interrupted by stretching and yawning. Whenever ACTH elicited SD, grooming also commenced immediately after the wave of SD (FIG. 3B). Since grooming induced by ACTH occurred also in the absence of SD, it cannot be considered solely as a behavioral expression of SD.

Still unaccounted for is the discrepancy between the high incidence of SD initially reported in the anesthetized rat[2] after intrahippocampal application of $ACTH_{1-24}$ and the low incidence of SD found in the freely moving animal.[4]

OPIOID PEPTIDES

Numerous studies have been performed on endogenous opioid peptides since the discovery of the enkephalins in 1975.[64] Opioid peptides have been reported to affect neuroendocrine regulation, nociception, sexual and consumatory behavior, locomotion, learning and memory, and other functions.[65-67] Opioid peptides have a prominent inhibitory effect on neurons.[68] Their excitatory effect in the hippocampus is mediated by the inhibition of interneurons, which results in an increase in activity of the pyram-

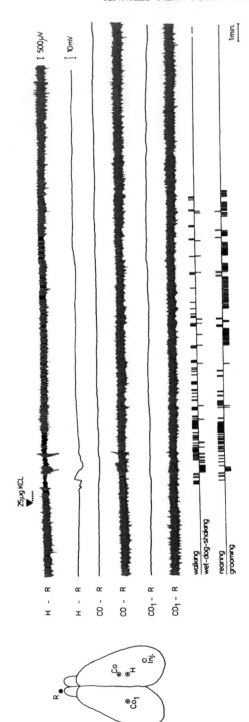

FIGURE 2. Sample of EEG and DC recordings after unilateral injection of 25 µg KCl (0.5 µl) into the hippocampus of a freely moving rat. *H* indicates the position of the hippocampal, *Co* and *Co₁* the cortical electrodes, *R* the reference electrode, and *Inj.* the site of injection. Behavioral activities were registered concurrently; walking, wet-dog shaking, rearing and grooming are indicated on the bottom two traces.

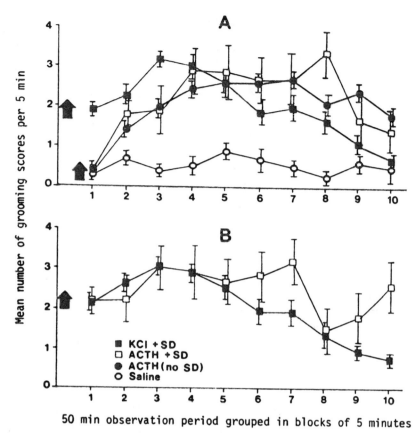

FIGURE 3. Incidence of grooming behavior sampled once per minute for 50 min after unilateral injection of ACTH$_{1-24}$ (0.5 or 10 µg), 25 µg KCl, or 0.5 µl saline into the hippocampus. In **A** the scores are grouped from the time of injection, independently of the onset of SD. In **B** below, the scores are matched in relation to the peak of the first wave of spreading depression (*SD*). The occurrence of SD is indicated by the *arrows*. Scores are mean ± SEM.

idal neurons.[69,70] This disinhibition is reflected by epileptiform discharges, which can also be registered in the cortex and amygdala after i.c.v. application of opioid peptides.[71,72]

Behavioral responses (*e.g.*, changes in locomotor activity, wet-dog shaking, analgesia) and changes in EEG activity (*e.g.*, decrease of the amplitude of the EEG preceded and followed by bursts of epileptiform activity) induced by the intracranial application of microgram amounts of opioid peptides seemed very similar to the effects of SD. Therefore, we decided to investigate whether opioid peptides can elicit SD in the anesthetized rat.

Injections of Leu- and Met-enkephalin and DAME into the neocortex and hippocampus elicited SD and epileptiform discharges.[3,73] As depicted in FIGURE 4 the threshold doses for the induction of SD were 6.25–12.5 µg for Leu-enkephalin and DAME, and 100 µg for Met-enkephalin. Naloxone completely or partly blocked the effects of the enkephalins, but had no influence on KCl-induced cortical SD.

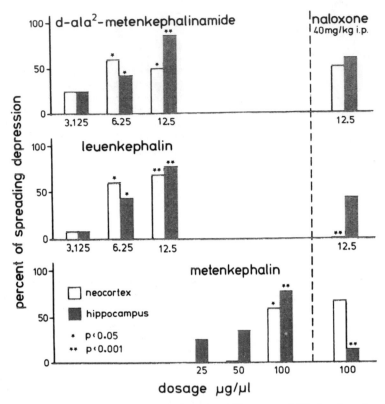

FIGURE 4. Incidence of cortical (*white columns*) and hippocampal (*black columns*) spreading depression expressed as percentage of 12 animals per group after unilateral injection of various doses of [D-Ala-2]Met-enkephalinamide, Leu-enkephalin, and Met-enkephalin in 1.0 µl volume, and the effect of i.p. naloxone pretreatment on enkephalin-induced SD in the anesthetized rat. The *p* values (compared to vehicle controls; for naloxone trials compared to the corresponding enkephalin group) based on Fisher's exact probability test are marked with *stars.*

The results of the experiments on opioid peptide-induced SD in the anesthetized animal led us to inject 10 µg DAME unilaterally into the dorsal hippocampus of freely moving rats and to concurrently monitor the EEG, DC-SPC and behavior for 50 min after injection.[4] Typical effects of KCl- and DAME-induced hippocampal SD are represented in FIGURES 2 and 5, respectively.

In 84% of 26 rats DAME elicited SD and epileptiform discharges in the hippocampus. The epileptiform activity usually outlasted the recording period. Recordings of EEG and DC-SPC showed that epileptiform activity was reflected in the cortical EEG traces, although the depression of EEG activity and the DC-SPC remained restricted to the hippocampus (FIG. 5).

Behavioral effects of DAME were closely related to the occurrence of SD. Wet-dog shaking occurred in all animals in which the application of DAME or KCl elicited SD. Whenever DAME elicited multiple waves of SD, multiple phases of wet-dog shakes occurred which were always restricted to the occurrence of SD. Hippocampal SD induced by DAME and KCl also influenced locomotor activity: SD was followed by an

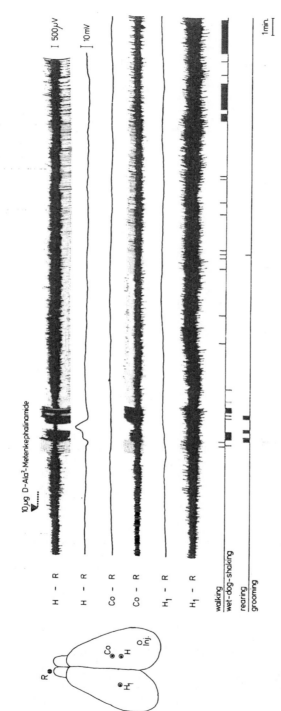

FIGURE 5. Sample of cortical and hippocampal EEG and DC recordings with concurrent registration of behavioral activities (bottom two traces) after injection of 10 μg [D-Ala-2]Met-enkephalinamide into the hippocampus. *H* and *H*, indicate the position of hippocampal recording electrodes; *Co*, the cortical electrode; *R*, the reference electrode; and *Inj.*, the site of injection. Note the wet-dog shaking behavior concurrent with the wave of SD and the increase in locomotor activity after a prolonged period of behavioral depression.

increase of rearing and walking behavior. In some animals DAME-induced SD was followed by periods of immobility (during which the animals did not react to tail pinch) that suddenly changed to activity (FIG. 5).

The effects of the opioid peptides and ACTH are similar in that both ACTH$_{1-24}$ (0.1–10 µg i.c.v.) and β-endorphin (0.03–3 µg i.c.v.) induce grooming, which can be blocked by the opiate antagonist naloxone.[74,75] The behavioral effects of KCl-induced SD (FIG. 2) were similar to β-endorphin-induced behavioral patterns described by Gispen et al.[74] After injection, the animals showed wet-dog shakes and their grooming behavior was interrupted by walking and rearing more often than in ACTH-treated rats; the latter were mostly engaged in grooming. Since opioid peptides were found to elicit SD when injected into the hippocampus and neocortex of rats,[3,4,73] it seems very likely that β-endorphin-induced grooming is a behavioral response to SD induced by this peptide.

Local DC slow potential changes have also been reported after intracranial application of opioid peptides (Leu-, Met-enkephalin, [D-Ala-2]Met-enkephalin) by Sadile and coworkers.[76-78] When these peptides were microinjected into the cortex and hippocampus (CA4 area dentata) of anesthetized rats, large changes in the DC slow potential were elicited. Differences were found in the threshold dose to elicit the DC-SPC depending on the genotype of rats, the brain structure injected, and the age of the animals. Surprisingly small dosages (nanogram to picogram amounts) of these opioid peptides induced a local DC-SPC, comparable in duration (1–2 min) and amplitude (up to −20 mV) to the DC-SPC measured during SD. Since these authors recorded the DC-SPC only close to the injection site, it is not known whether the DC slow potential change in their experiments reflects spreading depression. It is possible that the DC-SPC remained at the locus of initiation; SD involves the spread of a local DC-SPC. Even if the DC-SPC was a local effect, its relevance to the understanding of the action and behavioral effects of intracranially applied neuropeptides must be seriously considered.

SD is difficult to detect in AC recordings of EEG activity alone (see also FIG. 1), and has therefore not been detected in the many studies dealing with intracranial application of opioid peptides. Depression of EEG activity, preceded and followed by short lasting epileptiform discharges, has been registered after the i.c.v. application of β-endorphin, DAME, Met- and Leu-enkephalin (dosages: 10 to 250 µg) but were not recognized to be indices of SD.[71,72,79-84] The time course of the electrophysiological changes, as well as the occurrence of wet-dog shakes, corresponds well with our measures of opioid-induced SD. Since the dosage of DAME (17 nmol ≙ 10 µg) we used is well within the range commonly used in the other studies cited above, we can assume that the intracranial injection of opioid peptides and their analogues in these studies elicited SD. It follows that some of the effects of such injections were due to SD induced by the injected compounds.

We suggest that the electrophysiological and many behavioral effects of intracranial injection of opioid peptides that have been attributed to specific activation of certain opioid receptors are, in fact, artifacts of SD induced by these compounds. The fact that opiate antagonists, such as naloxone, can block these effects may merely indicate that the elicitation of SD has thereby been prevented.

VASOPRESSIN AND SUBSTANCE P

Similar behavioral responses have been elicited by the i.c.v. application of the neuropeptides vasopressin (LVP, AVP) and substance P. Behavioral changes that have been reported include, for example, increased or decreased locomotor activity,[85-87] scratching and grooming,[86,88,89] analgesia,[90-93] immobility or prostration, often followed by "barrel rotation."[94-97]

The changes of EEG activity in the hippocampus[98] and cortex[99] reported after i.c.v. injection of vasopressin are similar to SD-induced responses. The amplitude of EEG activity was decreased (described as "flattening of EEG activity which returned to baseline levels"[98]) during immobility and/or barrel rotation. Spike discharges were registered only in hippocampal recordings.

Ionophoretic application of vasopressin shows it to have an excitatory action on cortical and hippocampal neurons.[100,101] A dose-dependent increase or decrease of the firing rate of hypothalamic neurons has also been reported.[100]

Vasopressin (LVP) was found to induce SD in the hippocampus of anesthetized rats.[1] LVP (5-12 ng) elicited single as well as multiple waves of SD. The amplitude of EEG activity was decreased and followed by spike activity in 38% of trials. Changes of the DC slow potential comparable to SD were induced close to the injection site of vasopressin (nanogram to picogram amounts) in the cortex and hippocampus of anesthetized rats.[102-105]

Substance P (SP) is also known to excite central neurons.[106] Preliminary findings showed that a low dose of 100 ng injected into the hippocampus of anesthetized rats did not elicit SD. However, for lack of systematic studies, we cannot rule out that SP can elicit SD at other doses and in other brain structures.

The time course of behavioral responses to vasopressin and substance P corresponds well to that of behaviors measured after peptide-induced SD, which shows onset latencies of up to 6 min linked to the wave of SD.[1-4,73] Therefore, we must consider the possibility that SD is also the cause for some of the behavioral effects of these substances.

CONCLUDING REMARKS

We have summarized data showing that neuropeptides (opioid peptides, vasopressin, ACTH) are able to induce spreading depression when injected into certain brain areas. Some of the behavioral effects found after intracranial injection of these, and probably other, neuropeptides reflect the behavioral expression of SD rather than peptide-receptor-specific responses.

Even the possibility that some of the reported behavioral effects of neuropeptides are nonspecific concomitants of SD requires reinterpretation of the available experimental results and the use of appropriate controls in future research dealing with central effects of neuropeptides. It should be apparent that the sole use of AC recording of neural activity is inadequate to gauge the electrophysiological effects of substances injected intracranially. We propose that DC recordings, with the explicit purpose of monitoring possible DC slow potential changes as an index of spreading depression, must be measured routinely in behavioral and electrophysiological studies on intracranial effects of neuropeptides and other substances. This procedure will be necessary until the experimental substance in question has been investigated thoroughly in terms of its possible SD-inducing properties.

REFERENCES

1. HUSTON, J. P. & L. JAKOBARTL. 1977. Evidence for selective susceptibility of hippocampus to spreading depression induced by vasopressin. Neurosci. Lett. 6: 69-72.

2. JAKOBARTL, L. & J. P. HUSTON. 1977. Spreading depression in hippocampus and neocortex of rats induced by ACTH 1-24. Neurosci. Lett. **5:** 189-192.
3. SPRICK, U., M.-S. OITZL, K. ORNSTEIN & J. P. HUSTON. 1981. Spreading depression induced by microinjection of enkephalins into the hippocampus and neocortex. Brain Res. **210:** 243-252.
4. OITZL, M.-S. & J. P. HUSTON. 1984. Electroencephalographic spreading depression and concomitant behavioral changes induced by intrahippocampal injections of ACTH $_{1-24}$ and d-Ala²-metenkephalinamide in the rat. Brain Res. **308:** 33-42.
5. BUREŠ, J. & O. BUREŠOVÁ. 1972. Inducing cortical spreading depression. *In* Methods in Psychobiology. R. D. Myers, Ed.: 319-343. Academic Press. New York.
6. BUREŠ, J., O. BUREŠOVÁ & J. KŘIVÁNEK. 1974. The Mechanism and Application of Leão's Spreading Depression of Electroencephalographic Activity. Academic Press. New York.
7. SHIBATA, M., B. SIEGFRIED & J. P. HUSTON. 1977. Miniature calomel electrode for recording DC potential changes accompanying spreading depression in the freely moving rat. Physiol. Behav. **18:** 1171-1174.
8. BUREŠ, J., O. BUREŠOVÁ & J. P. HUSTON. 1983. Techniques and Basic Experiments for the Study of Brain and Behavior. Elsevier. Amsterdam.
9. SCHNEIDER, A. M. & P. E. SIMSON. 1983. Spreading depression: a behavioral analysis. *In* The Physiological Basis of Memory. J. A. Deutsch, Ed.: 121-138. Academic Press. New York.
10. FIFKOVÁ, E. & J. SYKA. 1964. Relationship between cortical and striatal spreading depression in rat. Exp. Neurol. **9:** 355-366.
11. TRACHTENBERG, M. C., C. D. HULL & N. A. BUCHWALD. 1970. Electrophysiological concomitants of spreading depression in caudate and thalamic nuclei of the cat. Brain Res. **20:** 219-231.
12. IRWIN, D. A., H. E. CRISWELL & J. W. KAKOLEWSKI. 1974. Prolonged slow potential variation associated with cortical spreading depression. Physiol. Behav. **13:** 377-380.
13. HORI, T., M. SHIBATA, T. KIYOHARA & T. NAKASHIMA. 1982. Effects of cortical spreading depression on hypothalamic thermosensitive neurons in the rat. Neurosci. Lett. **32:** 47-52.
14. HEINEMANN, U., H. D. LUX & M. J. GUTNICK. 1978. Changes in extracellular free calcium and potassium activity in the somatosensory cortex of cats. *In* Abnormal Neuronal Discharges. N. Chalazonitis & M. Boisson, Eds.: 329-345. Raven Press. New York.
15. KRAIG, R. P. & C. NICHOLSON. 1978. Extracellular ionic variations during spreading depression. Neuroscience **3:** 1045-1059.
16. PHILLIPS, J. M. & C. NICHOLSON. 1979. Anion permeability in spreading depression investigated with ion-sensitive microelectrodes. Brain Res. **173:** 567-571.
17. RODRIGUES, P. S. & H. MARTINS-FERREIRA. 1980. Cholinergic neurotransmission in retinal spreading depression. Exp. Brain Res. **38:** 229-236.
18. CLARK, R. M. & G. G. S. COLLINS. 1976. The release of endogenous amino acids from the rat visual cortex. J. Physiol. **262:** 383-400.
19. COLOMBO, J. A., C. A. BLAKE, R. J. LORENZ & C. H. SAWYER. 1973. Plasma prolactin changes following cortical spreading depression in female rats. Am. J. Physiol. **225:** 766-769.
20. ROSENTHAL, M. & D. L. MARTEL. 1979. Ischemia-induced alterations in oxidative "recovery" metabolism after spreading cortical depression *in situ*. Exp. Neurol. **63:** 367-378.
21. MAYEVSKY, A., N. ZARCHIN & C. M. FRIEDLI. 1982. Factors affecting the oxygen balance in the awake cerebral cortex exposed to spreading depression. Brain Res. **236:** 93-105.
22. GJEDDE, A., A. J. HANSEN & B. QUISTORFF. 1981. Blood brain glucose transfer in spreading depression. J. Neurochem. **37:** 807-812.
23. SHINOHARA, M., B. DOLLINGER, G. BROWN, S. RAPOPORT & L. SOKOLOFF. 1979. Cerebral glucose utilization: Local changes during and after recovery from spreading cortical depression. Science **203:** 188-190.
24. KŘIVÁNEK, J. 1961. Some metabolic changes accompanying Leão's spreading cortical depression in the rat. J. Neurochem. **6:** 183-189.
25. CSIBA, L., W. PASCHEN & G. MIES. 1985. Regional changes in tissue pH and glucose content during cortical spreading depression in rat brain. Brain Res. **336:** 167-170.
26. HANSEN, A. J., B. QUISTORFF & A. GJEDDE. 1980. Relationship between local changes

in cortical blood flow and extracellular K^+ during spreading depression. Acta Physiol. Scand. **109:** 1–6.

27. GRAFSTEIN, B. 1956. Mechanism of spreading cortical depression. J. Neurophysiol. **19:** 154–171.

28. VAN HARREVELD, A. & E. FIFKOVÁ. 1973. Mechanisms involved in spreading depression. J. Neurobiol. **4:** 375–387.

29. VAN HARREVELD, A. 1978. Two mechanisms for spreading depression in the chicken retina. J. Neurobiol. **9:** 419–431.

30. VYSKOČIL, F., N. KŘÍŽ & J. BUREŠ. 1972. Potassium-selective microelectrodes used for measuring the extracellular brain potassium during spreading depression and anoxic depolarization in rats. Brain Res. **39:** 255–259.

31. VAN HARREVELD, A. & J. S. STAMM. 1955. Cortical responses to metrazol and sensory stimulation in the rabbit. Electroencephalogr. Clin. Neurophysiol. **7:** 363–370.

32. MACHEK, J., E. UJEC, V. PAVLÍK & F. HORÁK. 1977. Interhemispheric field potentials, spreading depression and kindling. Neurosci. Lett. **4:** 337–341.

33. KOROLEVA, V. I. & J. BUREŠ. 1983. Cortical penicillin focus as a generator of repetitive spike-triggered waves of spreading depression in rats. Exp. Brain Res. **51:** 291–297.

34. UEDA, M. & J. BUREŠ. 1977. Differential effects of cortical spreading depression on epileptic foci induced by various convulsants. Electroencephalogr. Clin. Neurophysiol. **43:** 666–674.

35. KOROLEVA, V. I. & J. BUREŠ. 1982. Stimulation induced recurrent epileptiform discharges block cortical and subcortical spreading depression in rats. Physiol. Bohemoslov. **31:** 385–400.

36. FREEDMAN, N. L. 1969. Recurrent behavioral recovery during spreading depression. J. Comp. Physiol. Psychol. **68:** 210–214.

37. SHIBATA, M., T. HORI, T. KIYOHARA & T. NAKASHIMA. 1983. Facilitation of thermoregulatory heating behavior by single cortical spreading depression in the rat. Physiol. Behav. **31:** 651–656.

38. TESCHKE, E. J., J. D. MASER & G. G. GALLUP. 1975. Cortical involvement in tonic immobility ("animal hypnosis"): Effect of spreading cortical depression. Behav. Biol. **13:** 139–143.

39. GREENE, E. 1971. Comparison of hippocampal depression and hippocampal lesion. Exp. Neurol. **31:** 313–325.

40. BUREŠ, J., O. BUREŠOVÁ & T. WEISS. 1960. Functional consequences of hippocampal spreading depression. Physiol. Bohemoslov. **9:** 219–227.

41. HUSTON, J. P. & J. BUREŠ. 1970. Drinking and eating elicited by cortical spreading depression. Science **169:** 702–704.

42. JAKOBARTL, L. & J. P. HUSTON. 1977. Circling and consumatory behavior induced by striatal and neocortical spreading depression. Physiol. Behav. **19:** 673–677.

43. HUSTON, J. P., L. JAKOBARTL, G. PAPADOPOULOS & B. SIEGFRIED. 1978. Effects of nigrostriatal 6-OHDA lesions on turning elicited by cortical spreading depression. Pharmacol. Biochem. Behav. **9:** 837–843.

44. SIEGFRIED, B., F. HEFTI, W. LICHTENSTEIGER & J. P. HUSTON. 1979. Lateralized hunger: ipsilateral attenuation of cortical spreading depression-induced feeding after unilateral 6-OHDA injection into the substantia nigra. Brain Res. **160:** 327–340.

45. SHIBATA, M. 1978. Unilateral cortical spreading depression in the rat: effects on feeding, drinking and other behaviors. Brain Res. Bull. **3:** 395–400.

46. PERÉZ-SAAD, H. & J. BUREŠ. 1983. Cortical spreading depression blocks naloxone-induced escape behaviour in morphine pretreated mice. Pharmacol. Biochem. Behav. **18:** 145–147.

47. HUSTON, J. 1971. Yawning and penile erection induced in rats by cortical spreading depression. Nature **232:** 274–275.

48. ROSS, J. W. & R. A. GORSKI. 1973. Effects of potassium chloride on sexual behavior and the cortical EEG in the ovariectomized rat. Physiol. Behav. **10:** 643–646.

49. KURTZ, R. G. & R. SANTOS. 1979. Supraspinal influences on the penile reflexes of the male rat: a comparison of the effects of copulation, spinal transection, and cortical spreading depression. Horm. Behav. **12:** 73–94.

50. SIEGFRIED, B. & J. P. HUSTON. 1977. Properties of spreading depression-induced consumatory behavior in rats. Physiol. Behav. **18:** 841–851.

51. FERRARI, W. & L. M. VARGIU. 1956. Effetti dell' introduzione endocisternale di interme-
dina in differenti specie animali. Boll. Soc. Ital. Biol. Sper. **32:** 1368-1369.
52. GESSA, G. L., L. VARGIU & W. FERRARI. 1966. Stretchings and yawnings induced by
adrenocorticotrophic hormone. Nature **211:** 426-427.
53. BERTOLINI, A., W. VERGONI, G. L. GESSA & W. FERRARI. 1969. Induction of sexual excitement by the action of adrenocorticotrophic hormone in brain. Nature **221:** 667-669.
54. FERRARI, W., G. L. GESSA & L. VARGIU. 1963. Behavioral effects induced by intracisternally injected ACTH and MSH. Ann. N.Y. Acad. Sci. **104:** 330-345.
55. GISPEN, W. H., V. M. WIEGANT, H. M. GREVEN & D. DE WIED. 1975. The induction
of excessive grooming in the rat by intraventricular application of peptides derived from
ACTH: structure-activity studies. Life Sci. **17:** 645-652.
56. GISPEN, W. H. & R. L. ISAACSON. 1981. ACTH-induced excessive grooming in the rat.
Pharmac. Ther. **12:** 209-246.
57. COLBERN, D., R. L. ISAACSON, B. BOHUS & W. H. GISPEN. 1977. Limbic-midbrain lesions
and ACTH-induced excessive grooming. Life Sci. **21:** 393-402.
58. BÄR, P. R., W. H. GISPEN & R. L. ISAACSON. 1981. Behavioral and regional neurochemical sequelae of hippocampal destruction in the rat. Pharmac. Biochem. Behav. **14:**
305-312.
59. VAN WIMERSMA GREIDANUS, T. B., B. BOHUS, G. L. KOVÁCS, D. H. G. VERSTEEG, J. P. H.
BURBACH & D. DE WIED. 1983. Sites of behavioral and neurochemical action of ACTH-
like peptides and neurohypophyseal hormones. Neurosci. Biobehav. Rev. **7:** 453-463.
60. MACLEAN, P. D. 1957. Chemical and electrical stimulation of hippocampus in unrestrained
animals. II. Behavioral findings. Arch. Neurol. Psychiatry **78:** 128-142.
61. GENT, J. P. & J. R. NORMANTON. 1980. Antagonism of the excitatory action of ACTH
(1-24) by D-Phe⁷ ACTH (4-10) on single neurones in the rat medulla. Proc. B. P. S. 80P-81P.
62. SEGAL, M. 1976. Interactions of ACTH and norepinephrine on the activity of rat hippocampal cells. Neuropharmacology **15:** 329-333.
63. BALDWIN, D. M., C. K. HAUN & C. H. SAWYER. 1974. Effects of intraventricular infusions of ACTH 1-24 and ACTH 4-10 on LH release, ovulation and behavior in the
rabbit. Brain Res. **80:** 291-301.
64. HUGHES, J., T. W. SMITH, H. W. KOSTERLITZ, L. A. FOTHERGILL, B. A. MORGAN & H. R.
MORRIS. 1975. Identification of two related pentapeptides from the brain with potent
opiate agonist activity. Nature **258:** 577-579.
65. BROWNE, R. G. & D. S. SEGAL. 1980. Behavioral activating effects of opiates and opioid
peptides. Biol. Psychiatry **15:** 77-86.
66. FREDERICKSON, R. C. A. & L. E. GEARY. 1982. Endogenous opioid peptides: review of
physiological, pharmacological and clinical aspects. Prog. Neurobiol. **19:** 19-69.
67. IZQUIERDO, I. & C. A. NETTO. 1985. The brain β-endorphin system and behavior: the modulation of consecutively and simultaneously processed memories. Behav. Neural Biol.
44: 249-265.
68. NICOLL, R. A., G. R. SIGGINS, N. LING, F. E. BLOOM & R. GUILLEMIN. 1977. Neuronal
actions of endorphins and enkephalins among brain regions: a comparative microion-
tophoretic study. Proc. Natl. Acad. Sci. **74:** 2584-2588.
69. ZIEGLGÄNSBERGER, W., E. D. FRENCH, G. R. SIGGINS & F. E. BLOOM. 1979. Opioid peptides may excite hippocampal pyramidal neurons by inhibiting adjacent inhibitory interneurons. Science **205:** 415-417.
70. LEE, H. K., T. DUNWIDDIE & B. HOFFER. 1980. Electrophysiological interactions of
enkephalins with neuronal circuitry in the rat hippocampus. II. Effects on interneuron
excitability. Brain Res. **184:** 331-342.
71. HENRIKSEN, S. J., F. E. BLOOM, F. MCCOY, N. LING & R. GUILLEMIN. 1978. β-endorphin
induces nonconvulsive limbic seizures. Proc. Natl. Acad. Sci. **10:** 5221-5225.
72. FRENK, H., G. URCA & J. C. LIEBESKIND. 1978. Epileptic properties of leucine- and
methionine-enkephalin: comparison with morphine and reversibility by naloxone. Brain
Res. **147:** 327-337.
73. OITZL, M. S., V. I. KOROLEVA & J. BUREŠ. 1985. D-Ala²-metenkephalinamide blocks the
synaptically elicited cortical spreading depression in rats. Experientia **41:** 625-627.
74. GISPEN, W. H., V. M. WIEGANT, A. F. BRADBURY, E. C. HULME, D. G. SMYTH, C. R.

SNELL & D. DE WIED. 1976. Induction of excessive grooming in the rat by fragments of lipotropin. Nature **264:** 794-795.

75. GISPEN, W. H. 1982. Neuropeptides and behavior: ACTH. Scand. J. Psychol. Suppl. **1:** 16-25.

76. SADILE, A. G., A. CERBONE & L. A. CIOFFI. 1982. Genotype-dependent developmental patterns of multiple opioid receptors in the dorsal hippocampus and neocortex. An electrophysiological study in the Naples high (NHE) and low excitable (NLE) rat strains. Behav. Brain Res. **5:** 116-117.

77. SADILE, A. G. & A. CERBONE. 1984. High affinity *in vivo* systems for leu- and metenkephalin in CA4-area dentata subregion in the rat: differential maturation and plasticity with field slow depolarization as a functional probe. *In* Central and Peripheral Endorphins: Basic and Clinical Aspects. E. E. Müller & A. R. Genazzani, Eds.: 59-63. Raven. New York.

78. CERBONE, A. & A. G. SADILE. 1986. Differential functional maturation of "δ" and "μ" opioid-induced slow potentials in the rat hippocampus. *In* Ontogenesis of the Brain. S. Trojan & F. Stastny, Eds. Charles University Press. Prague. In press.

79. URCA, G., H. FRENK & J. C. LIEBESKIND. 1977. Morphine and enkephalin: analgesic and epileptic properties. Science **197:** 83-86.

80. NEAL, H. & P. E. KEANE. 1978. The effects of local micro injections of opiates and enkephalins into the forebrain on the electrocorticogram of the rat. Electroencephalogr. Clin. Neurophysiol. **45:** 655-665.

81. ELAZAR, Z., E. MOTLES, Y. ELY & R. SIMANTOV. 1979. Acute tolerance to the excitatory effect of enkephalin microinjections into hippocampus. Life Sci. **24:** 541-548.

82. FIREMARK, H. M. & R. E. WEITZMAN. 1979. Effects of β-endorphin, morphine and naloxone on arginine vasopressin secretion and the electroencephalogram. Neuroscience **4:** 1895-1902.

83. ALOISI, F., A. S. DE CAROLIS & V. LONGO. 1980. EEG and behavioral effects of morphine, enkephalins and derivatives administered into the lateral cerebral ventricles of rats and rabbits. Pharmacol. Res. Commun. **12:** 467-477.

84. ELAZAR, Z., R. SIMANTOV & E. MOTLES. 1982. Local electrographic effects of Leu-enkephalin microinjections into the brain. Electroencephalogr. Clin. Neurophysiol. **54:** 91-95.

85. MEISENBERG, G. 1982. Short-term behavioural effects of neurohypophyseal hormones: Pharmacological characteristics. Neuropharmacology **21:** 309-316.

86. MEISENBERG, G. & W. H. SIMMONS. 1982. Behavioral effects of intracerebroventricularly administered neurohypophyseal hormone analogs in mice. Pharmacol. Biochem. Behav. **16:** 819-825.

87. NARANJO, J. R. & J. DEL RIO. 1984. Locomotor activation induced in rodents by substance P and analogues. Neuropharmacology **23:** 1167-1171.

88. KATZ, R. J. 1980. Substance P elicited grooming in the mouse: behavioral and pharmacological characteristics. Int. J. Neurosci. **10:** 187-189.

89. HALL, M. E., P. A. GRANTHAM & J. M. STEWART. 1985. Age and strain differences in some behavioral effects of intracranial substance P. Peptides **6:** 363-368.

90. BERNTSON, G. G. & B. S. BERSON. 1980. Antinociceptive effects of intraventricular or systemic administration of vasopressin in the rat. Life Sci. **26:** 455-459.

91. KORDOWER, J. H., V. SIKORSZKY & R. J. BODNAR. 1982. Central antinociceptive effects of lysine-vasopressin and an analogue. Peptides **3:** 613-617.

92. FREDERICKSON, R. C. A., V. BURGIS, C. E. HARRELL & J. D. EDWARDS. 1978. Dual actions of substance P on nociception: possible role of endogenous opioids. Science **199:** 1359-1362.

93. NARANJO, J. R., M.-P FERNANDEZ-TOME & J. DEL RIO. 1984. Analgesic activity of substance P in rats: tolerance and cross-tolerance with enkephalin. Europ. J. Pharmacol. **98:** 133-136.

94. KRUSE, H., T. B. VAN WIMERSMA GREIDANUS & D. DE WIED. 1977. Barrel rotation induced by vasopressin and related peptides in rats. Pharmacol. Biochem. Behav. **7:** 311-313.

95. BOAKES, R. J., J. M. EDNIE, J. A. EDWARDSON, A. B. KEITH, A. SAHGAL & C. WRIGHT. 1985. Abnormal behavioural changes associated with vasopressin-induced barrel rotations. Brain Res. **326:** 65-70.

96. RONDEAU, D. B., F. B. JOLICOEUR, F. BELANGER & A. BARBEAU. 1978. Motor activity induced by substance P in rats. Pharmacol. Biochem. Behav. **9:** 769–775.
97. JAMES, T. A. & M. S. STARR. 1979. Effects of substance P injected into the substantia nigra. Br. J. Pharmacol. **65:** 423–429.
98. ABOOD, L. G., R. KNAPP, T. MITCHELL, H. BOOTH & L. SCHWAB. 1980. Chemical requirements of vasopressins for barrel rotation convulsions and reversal by oxytocin. J. Neurosci. Res. **5:** 191–199.
99. EHLERS, C. L., T. K. REED, M. WANG, C. J. LEBRUN & G. F. KOOB. 1985. EEG effects of subcutaneous and intracerebroventricular injections of arginine vasopressin in the rat. Psychopharmacology **87:** 430–433.
100. NICOLL, R. A. & J. L. BARKER. 1971. The pharmacology of recurrent inhibition in the supraoptic neurosecretory system. Brain Res. **35:** 501–511.
101. MÜHLETHALER, M., J. J. DREIFUSS & B. H. GÄHWILER. 1982. Vasopressin excites hippocampal neurones. Nature **296:** 749–751.
102. SADILE, A. G., A. CERBONE & B. DE LUCA. 1980. Vasopressin-induced slow potential change in the hippocampus and neocortex of two rat strains with different behavioural arousal, during development. Neurosci. Lett. Suppl. **5:** 210.
103. SADILE, A. G., A. CERBONE & L. A. CIOFFI. 1982. Vasopressinergic modulation of behavioural plasticity. Electrophysiological and behavioural study in the Naples high and low excitable rats. Behav. Brain Res. **5:** 117.
104. SADILE, A. G. 1982. Hippocampal vasopressinergic terminals are modulated by peripheral vasopressin through short-term negative and long-term positive feed back mechanisms. Soc. Neurosci. Abstr. **8:** 367.
105. CERBONE, A. & A. G. SADILE. 1986. Differential effect of vasopressin and oxytocin on hippocampal slow field depolarization in the rat. Soc. Physiol. Ital. **62:** 47–48.
106. NICOLL, R. A., C. SCHENKER & S. E. LEEMAN. 1980. Substance P as a transmitter candidate. Ann. Rev. Neurosci. **3:** 227–268.

Excessive Grooming Behavior in Rats and Mice Induced by Corticotropin-releasing Factor[a]

ADRIAN J. DUNN, CRAIG W. BERRIDGE, YEN I. LAI,
TAMMY L. YACHABACH, AND SANDRA E. FILE[b]

Department of Neuroscience
University of Florida
Gainesville, Florida 32610

[b] *MRC Neuropharmacology Research Group*
School of Pharmacy
University of London
London, United Kingdom

Previous reports have described increases in self-grooming behavior following intracerebroventricular (i.c.v.) administration of corticotropin-releasing factor (CRF) in rats[1-5] but not in mice.[6,7] These observations raise the possibility of species differences in the ability of i.c.v. CRF to elicit grooming behavior.

Male CD-1 mice (25–35 g) were obtained from Charles River (Wilmington, MA). Male Sprague–Dawley rats (200–250 g) were bred at the University of Florida. Rats and mice were implanted bilaterally with cannulae in the lateral ventricles as previously described.[8,9] Synthetic human/rat CRF was obtained from Peninsula Laboratories, and as a gift from Dr. Jean Rivier of the Salk Institute. ACTH$_{1-24}$ was a gift from Organon (Oss, The Netherlands). Peptides were dissolved in 0.14 M NaCl containing 0.001 M HCl and stored frozen until immediately before injection. Procedures for studying the grooming behavior were as previously described.[8,9] Briefly, the animals were injected with CRF or vehicle (2 µl in each cannula) and immediately placed in clean Plexiglas cages. Behavioral scores were recorded by a trained observer blind as to the treatment groups, at 15-s intervals for 60 min after injection using a time-sampling procedure.[10-13]

Intracerebroventricularly administered ACTH$_{1-24}$ (1 µg) markedly increased grooming behavior in both rats and mice (FIG. 1). CRF (1 µg) increased grooming in the rat (the material from the Salk Institute was significantly more effective than that obtained from Peninsula). CRF had no effect on grooming behavior in mice. These results confirm previous observations in rats[1-5,9,10,12,14] and our previous observations in mice.[6-8,11]

In FIGURE 2 the grooming response to different doses of CRF has been broken down into subcategories. The minimum effective dose for increasing grooming was 300 ng. This may be considered a relatively high dose because significant behavioral effects of CRF have been observed at lower doses.[1,4,5,7] CRF increased face washing

[a] This research was supported by the National Institute of Mental Health (grant no. MH25486).

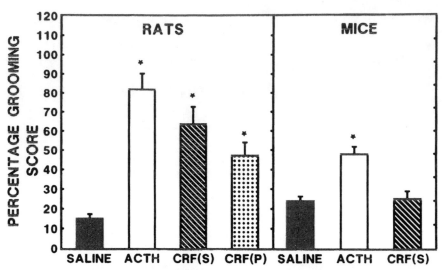

FIGURE 1. Grooming responses of rats and mice to i.c.v. injections of ACTH$_{1-24}$ or CRF. Rats ($n = 5$) and mice ($n = 6$) were injected i.c.v. with 1 µg of ACTH$_{1-24}$ (*ACTH*), 1 µg of Peninsula CRF (*CRF(P)*), 1 µg of Salk CRF (*CRF(S)*), or saline vehicle. Percentage grooming scores were increased in rats by ACTH or CRF, but in mice only by ACTH; *$p < 0.05$ (paired t-test).

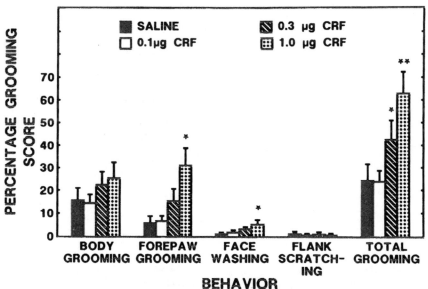

FIGURE 2. Grooming responses of rats ($n = 5$) to different doses of CRF. Grooming was scored in the categories: body grooming, forepaw grooming, face washing, and flank scratching. Salk CRF at 0.3 or 1 µg significantly increased percentage total grooming scores in rats, and 1 µg significantly increased forepaw grooming and face washing (Dunnett's test: *$p < 0.05$, **$p < 0.01$).

and forepaw grooming, but not flank scratching with the hind limb (the predominant form of grooming affected by vasopressin.[11,13]) By contrast, $ACTH_{1-24}$ in the same rats significantly increased all forms of grooming, including flank scratching. The increases in grooming were accompanied by small increases in locomotor activity, no change in rearing or nonlocomotor moving, and decreases in quiet and sleeping scores. The major effect of $ACTH_{1-24}$ was to increase the mean length of grooming bouts, with only a small effect on the number.[12,14] However, CRF significantly increased the number of grooming bouts relative to either saline or ACTH. Thus the pattern of the grooming response to CRF was similar but not identical to that induced by ACTH.

REFERENCES

1. SUTTON, R. E., G. F. KOOB, M. LE MOAL, J. RIVIER & W. VALE. 1982. Nature. 297: 331-333.
2. BRITTON, D. R., G. F. KOOB, J. RIVIER & W. VALE. 1982. Life Sci. 31: 363-367.
3. MORLEY, J. & A. S. LEVINE. 1982. Life Sci. 31: 1459-1464.
4. VELDHUIS, H. D. & D. DE WIED. 1984. Pharmacol. Biochem. Behav. 21: 707-713.
5. DUNN, A. J. & S. E. FILE. 1987. Horm. Behav. 21: 193-202.
6. DUNN, A. J. & C. W. BERRIDGE. 1987. Pharmacol. Biochem. Behav. 27: 685-691.
7. BERRIDGE, C. W. & A. J. DUNN. 1986. Reg. Peptides. 16: 83-93.
8. GUILD, A. L. & A. J. DUNN. 1982. Pharmacol. Biochem. Behav. 17: 31-36.
9. DUNN, A. J. & G. VIGLE. 1985. Neuropharmacology 24: 329-331.
10. GISPEN, W. H., V. M. WIEGANT, H. M. GREVEN & D. DE WIED. 1975. Life Sci. 17: 645-652.
11. REES, H. D., A. J. DUNN & P. M. IUVONE. 1976. Life Sci. 18: 1333-1339.
12. GISPEN, W. H. & R. L. ISAACSON. 1981. Pharmacol. Therap. 12: 209-246.
13. DELANOY, R. L., A. J. DUNN & R. TINTNER. 1978. Horm. Behav. 11: 348-362.
14. DUNN, A. J., J. E. ALPERT & S. D. IVERSEN. 1984. Behav. Brain Res. 12: 307-315.

Location of the Ventricular Site of ACTH Action to Elicit Grooming Behavior[a]

ADRIAN J. DUNN AND RUSSELL W. HURD

Department of Neuroscience
University of Florida College of Medicine
Gainesville, Florida 32610

$ACTH_{1-24}$ elicits grooming when administered into the cerebral ventricles.[1] Injections into substantia nigra, but not striatum, are also effective.[2] Lesions of the septum, preoptic area, mammillary bodies, amygdala, posterior thalamus, and dorsal or ventral hippocampus did not attenuate ACTH-induced grooming,[3] but lesions of the substantia nigra[1] or aspiration of the hippocampus inhibited it.[3]

Following the techniques of others,[4-6] we have used ventricular plugging with cold cream to determine the site of action of ACTH. Stainless steel cannulae (23 gauge) were implanted stereotaxically in a variety of locations. All cannulae were tested by injecting $ACTH_{1-24}$ to elicit grooming. Grooming was scored before and after injection of cold cream (Nivea) through one of the cannulae. The brain was then fixed in formalin and sectioned sagittally on a freezing microtome.

$ACTH_{1-24}$ elicited grooming when injected into the lateral (LV), third (3V), or fourth ventricles (4V), although 4V injection was slightly less effective than LV or 3V. An interventricular foramen block prevented grooming to ACTH in the same, but not the contralateral, LV. Blockade of the cerebral aqueduct by 3V injection of cold cream in seven rats did not prevent grooming induced by LV injection of ACTH.

Effective inhibition of ACTH-induced grooming was obtained only with plugs in the anterior ventral parts of the 3V (AV3V) (10 of 29 rats). Rats with plugs in the dorsal or posterior part of the 3V groomed normally, unless there was also a plug in the AV3V. Significant blockade of grooming occurred only with plugs in the OVLT or median eminence (ME, TABLE 1). All seven rats with a complete plug of the OVLT were substantially inhibited, but two of the six with incomplete plugs in this area and only one of sixteen with no trace of cold cream in this region were also impaired. This one animal had a plug over much of the ME. Seven of nine rats with apparently complete plugs in the rME region showed substantially reduced grooming, but only one of these seven animals was not also blocked in the OVLT. However, two of fourteen rats with no plugs in this region showed decreased grooming. These two rats both had plugs over the OVLT. Plugs in the hypophyseal area (Hyp) did not prevent ACTH-induced grooming.

These results support the AV3V as the site of ACTH action in eliciting grooming, but do not prove it. Cold cream in the AV3V might impair ACTH-induced grooming by another mechanism, although the behavior of the rats was otherwise normal. ACTH might be taken up in the AV3V and subsequentlly be transported to a distant active site. If so, the active site cannot be far away, because the grooming response to i.c.v. ACTH is rapid, reaching maximum in 5–10 min.[7]

[a] This research was supported by the National Institute of Mental Health (grant no. MH25486)

TABLE 1. Effects of Cold Cream Coating of the Third Ventricle Surface on Grooming Induced by Lateral Ventricle Injection of ACTH

	Number of Animals Showing Decreased Grooming Response and Grooming Scores before and after Cold Cream Injection		
Area of the Plug	Area Plugged	Area Partly Plugged	No Plug in Area
Organum vasculoram of the lamina terminalis (OVLT)	7/7	2/6	1/16
Before cold cream	59.6 ± 5.2	61.8 ± 2.4	68.6 ± 3.7
After cold cream	14.8 ± 2.7***	50.7 ± 11.7	58.1 ± 4.4
Rostral median eminence	7/9	1/6	2/14
Before cold cream	55.9 ± 4.1	57.8 ± 4.6	73.9 ± 2.4
After cold cream	26.5 ± 6.8**	58.1 ± 11.6	69.0 ± 3.2*
Caudal median eminence	5/5	1/4	4/20
Before cold cream	62.8 ± 4.0	53.6 ± 4.9	67.8 ± 3.1
After cold cream	14.7 ± 3.9***	44.2 ± 14.5	54.4 ± 4.9*
Hypophysis	2/6	1/2	7/21
Before cold cream	58.0 ± 6.7	63.7 ± 8.0	67.1 ± 2.7
After cold cream	48.7 ± 10.8	29.4 ± 29.4	47.0 ± 5.4***

The behavior of the rats was scored for 45 min starting 10 min after 1.5 μg of $ACTH_{1-24}$ was injected via a lateral ventricle cannula, before and after injection of cold cream through a third ventricle cannula. Animals are grouped into nonexclusive categories according to the location of the cold cream plugs (FIG. 1). The first row of numbers for each location indicates the number of rats, showing a significantly decreased grooming response (>40%) to i.c.v. ACTH after cold cream injection as a fraction of the total number of animals in that category. The second and third rows are the mean ± SEM percentage grooming scores before and after cold cream injection, respectively.

* Significantly different from before cold cream score ($p < 0.05$, **$p < 0.01$, ***$p < 0.001$, paired t-test).

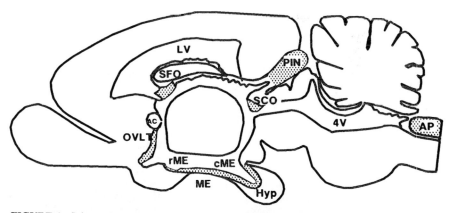

FIGURE 1. Schematic diagram of a sagittal section of the rat brain showing the location of the four areas of the anterior ventral third ventricle (AV3V) used for TABLE 1. The four areas are the organum vasculosum of the lamina terminalis (OVLT), the rostral median eminence (rME), the caudal median eminence (cME), and the hypophyseal area (Hyp). Other abbreviations: AC, anterior commissure; SFO, subfornical organ; LV, lateral ventricle; PIN, pineal gland; SCO, subcommissural organ; 4V, fourth ventricle; AP, area postrema.

Gessa *et al.* previously suggested the hypothalamus as the site of action of ACTH-induced stretching and yawning syndrome (SYS), based on the potency of direct injections, and of the latency from 3V as compared to LV or cisterna magna ACTH injections.[8] It is notable that other studies using ventricular plugging found the AV3V to be crucial for the dipsogenic effects of angiotensin,[5] and the pressor effects of bradykinin.[6]

REFERENCES

1. GISPEN, W. H. & R. L. ISAACSON. 1981. Pharmacol. Ther. **12:** 209–246.
2. WIEGANT, V. M., A. R. COOLS & W. H. GISPEN. 1977. Eur. J. Pharmacol. **41:** 343–345.
3. COLBERN, D., R. L. ISAACSON, B. BOHUS & W. H. GISPEN. 1977. Life Sci. **21:** 393–402.
4. HERZ, A., K. ALBUST, J. MATYS, P. SCHUBART & H. S. TASCHENACHI. 1970. Neuropharmacology **9:** 539–551.
5. HOFFMAN, W. E. & M. I. PHILLIPS. 1976. Brain Res. **110:** 313–330.
6. LEWIS, R. E. & M. I. PHILLIPS. 1984. Am. J. Physiol. **247:** R63–68.
7. DUNN, A. J., J. E. ALPERT & S. D. IVERSEN. 1984. Behav. Brain Res. **12:** 307–315.
8. GESSA, G. L., M. PISANO, L. VARGIU, F. CRABAI & W. FERRARI. 1967. Rev. Can. Biol. **26:** 229–236.

Postcopulatory Grooming in Male Rats Prevents Sexually Transmitted Diseases

BENJAMIN L. HART, ELIZABETH K. KORINEK, AND
PATRICIA L. BRENNAN

Department of Physiological Sciences
School of Veterinary Medicine
University of California, Davis
Davis, California 95616

Following copulatory intromissions male rats always engage in gential grooming. Previous research has revealed that placing collars on male rats to prevent postcopulatory grooming does not alter the temporal parameters of the mating sequence.[1] The fact that human saliva contains a number of antibacterial substances, including lysozyme, lactoferrin, leukocytes, and antibodies,[2] prompted our examination in experiment 1 of the effect of physically preventing genital grooming on the transmission of a marker organism from females to males and, in experiment 2, of the antibacterial effects of saliva for genital pathogens.

In the first experiment male rats were mated to females in which a genital infection had been established with a stock culture of coagulase-positive staphylococcus. This organism was chosen as a marker to represent a genital pathogen. Three groups of males were studied: (1) those wearing long collars with a protruding tongue that prevented grooming, (2) those wearing short collars that allowed grooming, and (3) those wearing no collars. Males were euthanatized by anesthetic overdose 5 days after mating and organs of the reproductive tract were cultured for identification of coagulase-positive staphylococcus. Most male rats with long collars developed an infection in one or more genital organs, whereas none of the subjects with control short collars and few of the rats wearing no collar were infected (TABLE 1).

In experiment 2 male rats were anesthetized and injected with pilocarpine to stimulate saliva production. Saliva from several rats was pooled, sterilized by passage through a Millipore filter, frozen, and then lyophilized. The saliva was later reconstituted to 10% for examination of its antibacterial effect on coagulase-positive staphylococcus as well as on *Pasteurella pneumotropica* and *Mycoplasma pulmonis*, two pathogens implicated in murine genital infection.[3,4] The microorganisms were tested using stock cultures serially diluted from 10^{-1} to 10^{-9} in nutrient broth to which saliva was added and incubated for 24 hours. All three organisms were killed at a dilution that was at least 10^3 more concentrated than found in saline controls (TABLE 2). To test for immunoglobulins an interfacial ring test was performed by drawing reconstituted saliva and killed cells into a Pasteur pipette.[5] This test revealed an antibody-antigen complex at the interface of the saliva and cell suspension for each of the three organisms.

In conclusion, both the behavioral and *in vitro* experiments with saliva indicate

TABLE 1. Number of Subjects Positive for Coagulase-Positive Staphylococcus in One or More Genital Organs Including Urethra and Epididymis

Subjects	n	One or More Genital Organs	Urethra	Epididymis
Long collars	13	10	7	5
Short collars	11	0*	0	0
No collars	9	2**	2	1

* $p < 0.01$, Fisher's exact probability test.
** $p < 0.05$, Fisher's exact probability test.

TABLE 2. Effects of Saliva or Saline on Growth of Three Organisms through a Range of Serial Dilutions[a]

Dilution	Coagulase-Positive Staphylococcus (2.0×10^8 organisms/ml)		P. pneumotropica (3.1×10^8 organisms/ml)		M. pulmonis (4.5×10^6 organisms/ml)	
	Saliva	Saline	Saliva	Saline	Saliva	Saline
10^{-1}	+	+	+	+	−	+
10^{-2}	+	+	+	+	−	+ (563)
10^{-3}	+	+	+	+	−	+ (67)
10^{-4}	+	+	−	+	−	+ (230)
10^{-5}	− (13)	+	−	+	−	−
10^{-6}	−	+	−	+	−	+
10^{-7}	−	+	−	+	−	−
10^{-8}	−	−	−	+	−	−
10^{-9}	−	−	−	−	−	−

Amount of reconstituted (10%) saliva or isotonic saline added to nutrient broth of diluted organism was 0.2 ml, 0.1 ml, and 0.15 ml for coagulase-positive staphylococcus, *P. pneumotropica*, and *M. pulmonis*, respectively.

[a] Key to symbols: +, too many colonies to count except where colony count is given in parentheses; −, no growth except where colony count is given in parentheses.

that genital grooming is an adaptive behavior related to prevention of venereal diseases in males or the transmitting of such diseases to females subsequently mated.

REFERENCES

1. HART, B. L. & C. M. HAUGEN. 1971. Prevention of genital grooming in mating behaviour of male rats (*Rattus norvegicus*). Anim. Behav. **19**: 230–232.
2. BOWEN, W. H. 1974. Defense mechanisms in the mouth and their possible role in the prevention of dental caries: A review. J. Oral Pathol. **3**: 266–278.
3. CORBEIL, L. B. 1980. Criteria for development of animal model of diseases of the reproductive system. Am. J. Pathol. **101**: 5241–5253.
4. HARKNESS, J. E. & J. E. WAGNER. 1983. The biology and medicine of rabbits and rodents. 2d edit. Lea & Febiger. Philadelphia.
5. MAURER, R. H. 1971. Precipitation reactions. *In* Methods in Immunology and Immunochemistry, Vol. 3. C. A. Williams & M. W. Chase, Eds.: 1–54. Academic Press. New York.

Autogrooming and Social Grooming in Impala[a]

LYNETTE A. HART AND BENJAMIN L. HART

Department of Physiological Sciences
School of Veterinary Medicine
University of California, Davis
Davis, California 95616

Grooming is a frequent and prominent behavior of many plains-dwelling antelope in East Africa. Studies of cattle, in which animals are restricted from licking, reveal that grooming is important in removal of ticks.[1,2] The presence of one engorging tick for the equivalent of 1 year can reduce the growth increment in calves by 1.7 pounds.[3] A tick load of even a few engorging ticks could be costly to a slender antelope. Impala typically inhabit deciduous woodland areas, where most species of ticks are found in greater numbers than on open grassland commonly inhabited by some other species of antelope. The purpose of this study, conducted in Kenya, was to analyze features of impala grooming that might differentiate them from other antelope species and would relate to increased tick exposure.

The impala chosen for study were females ($n = 66$) found in groups of 4–80 in a male's territory. For comparison, observations were made also on female Grant's gazelle ($n = 51$) and common wildebeest ($n = 86$), which inhabit open grassland in the same geographic areas as impala. The focal animal technique was used whereby one subject was followed for a 20-min period. Observations were spread evenly throughout the day and subjects in many different herds were observed. Occurrences of social grooming were studied by watching an entire herd, waiting for a social grooming exchange, and recording the exchange from beginning to end. Oral grooming with the tongue or lower incisors and scratching with hind hooves occurred as bouts, each bout comprising a number of episodes. The body parts that antelope are able to reach with their mouths by licking or biting are the shoulder, trunk, and posterior body. With their hooves they can reach the head, neck, shoulder and trunk.

Extrapolating from the 20-min focal animal observations, the approximate daylight daily rate of self oral grooming in impala was 120 bouts, comprising about 1100 episodes. This exceeded that observed in comparison antelope, whereas the impala's rate of hoof scratching was only half that of other antelope (FIG. 1). Impala differed from the other antelope in exhibiting social oral grooming in which two adult females would exchange bouts to the head and neck (FIG. 1). Female impala received (and gave) approximately 27 such bouts (160 episodes) per 12-hour day. These bouts were virtually always completely reciprocal (FIG. 2).

Another difference between impala and the comparison antelope was that impala were often tended by the oxpeckers that are commonly seen on giraffe and buffalo. The behavior of impala towards oxpeckers could be either accommodation or rejection.

[a] The authors acknowledge the University of California, University Research Expeditions Program, and participants for support of this work.

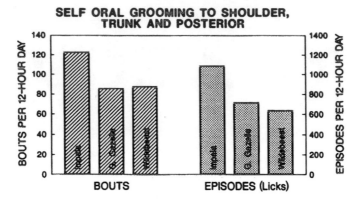

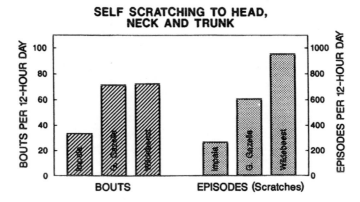

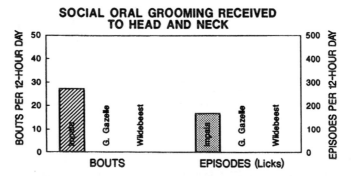

FIGURE 1. Number of bouts and episodes of self oral grooming, scratching, and social oral grooming per 12-hour day, extrapolated from 20-min focal animal observations.

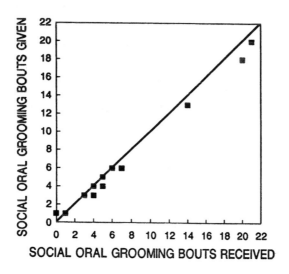

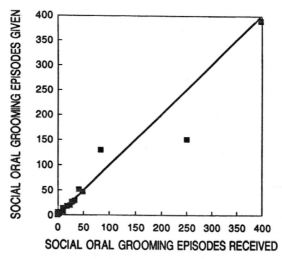

FIGURE 2. Scattergrams of social grooming (**A**) bouts and (**B**) episodes given and received by impala females. The line at 45° represents perfect reciprocity.

Since the impala females that make up a herd are unrelated,[4] the social grooming phenomenon would appear to be one of the few examples of animal behavior that meets the criteria for reciprocal altruism.[5] The presumed increased exposure to ticks could have provided the basis for the evolution of a cost-effective way of obtaining grooming for parts of the body the impala cannot reach with its own mouth.

REFERENCES

1. SNOWBALL, G. J. 1956. The effect of self-licking by cattle on infestations of cattle tick *Boophilus microplus* (Canestrini). Aust. J. Agric. Res. **7:** 227-232.
2. BENNETT, G. F. 1969. *Boophilus microplus* (Acarina: Ixodidae): Experimental infestations on cattle restrained from grooming. Exp. Parasitol. **26:** 323-328.
3. LITTLE, D. A. 1963. The effect of cattle tick infestation on the growth rate of cattle. Aust. Vet. J. **39:** 6-10.
4. JARMAN, M. V. 1979. Impala social behavior: Territory, hierarchy, mating and the use of space. Advances in Ethology, No. 21. Verlag Paul Parey. Berlin.
5. TRIVERS, R. 1985. Social Evolution. Benjamin/Cummings. Menlo Park, CA.

Estrogen Influences on Pregnancy-induced Autogrooming in Mice[a]

PAMELA J. HARVEY AND MARTHA A. MANN

The University of Texas at Arlington
Arlington, Texas 76019

Pregnancy induces high levels of autogrooming in Rockland-Swiss (R-S) albino mice. During the last third of pregnancy, R-S mice allocate significantly more time to nipple self-stimulation and the washing of face, paws, and dorsal areas than do early pregnant and virgin females. The factors responsible for pregnancy-induced autogrooming are unknown. However, it is well known that endogenous changes in estrogen influence a number of sensorimotor processes in rodents (*e.g.*, ref. 2), including the initiation of maternal behaviors.[3] Since circulating estrogen rises near term,[4] at a time when vigorous autogrooming is evident, we sought to examine the contribution of this hormone to late gestational grooming in the mouse.

Adult R-S females were timed-mated and isolated upon detection of a copulatory plug, *i.e.*, on gestation day (GD) 0. Beginning on GD 2, separate groups of gravid mice were treated daily with one of three dosages of estradiol benzoate (*i.e.*, 0.01, 0.02, or 0.04 μg EB) or the oil vehicle alone, or they remained nontreated ($n = 10$ per group). On the evening of GD 18 (less than 24 h prior to parturition) autogrooming was assessed for 30 min by recording the time spent grooming nipple line regions, the ano-genital region, and noncritical regions (*i.e.*, face, paws, and dorsal area[5]). Regardless of treatment condition, all females displayed stereotypical bouts of grooming that proceeded in a cephalocaudal direction. As seen in FIGURE 1, there were no significant differences among the groups for durations of ano-genital or noncritical grooming. However, durations of nipple line grooming varied significantly among these groups. Dams of the EB-treated groups, while not differing significantly from one another, displayed significantly less nipple self-stimulation compared either to dams of the oil- or nontreated control groups.

Estrogen is thought to promote the development of nipple-mammary structures during pregnancy.[6] However, in the previous experiment, supplemental amounts of estrogen may have inhibited nipple growth, thereby preventing oral manipulation and stimulation of the teats. To examine this possibility, additional animals were assigned to the treatment groups described previously ($n = 5-6$ per group). On GD 18, the dams were anesthetized and nipple lengths were determined on all five pairs of teats. TABLE 1 shows that nipple growth was uniform among the five groups and no significant differences were detected.

The grooming of nipple regions would appear to be an estrogen-sensitive component of the autogrooming repertoire of gravid mice. Exogenously administered estrogen decreased the amount of time R-S mice engaged in nipple self-stimulation, yet the administration of this steroid did not alter durations of grooming of other body areas.

[a] This work was supported by the ORF grant no. 15-273 from UTA and MH3977-01 from the NIMH.

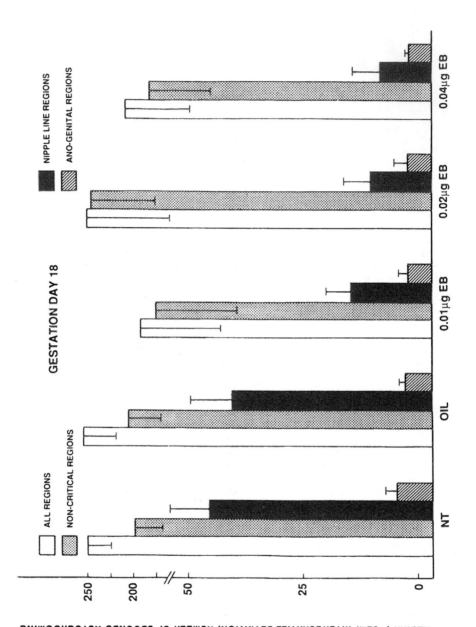

FIGURE 1. At 60 days of age Rockland-Swiss albino female mice were timed-mated, isolated, and randomly assigned to one of five groups. Beginning on GD 2, separate groups of females (n = 10 per group) received a daily subcutaneous injection of either 0.01, 0.02, or 0.04 μg estradiol benzoate (*EB*, Sigma Chemical Co.) suspended in 0.05 ml sesame oil. Ten animals were injected daily with an equal volume of the oil vehicle alone (*OIL*) and ten remained nontreated (*NT*). On the day of the final injection, GD 18, observations of autogrooming were conducted under dim red light between 2100–2300 hours, 2 h following the onset of darkness. Following a 10-min habituation period, autogrooming was assessed for 30 min by recording the time spent grooming nipple line regions, the ano-genital region, and noncritical areas (*i.e.*, face, paws, and dorsal area). Separate grooming scores were then summed to obtain a composite score for all regions. Pregnancy was confirmed by noting the delivery of young 4–12 h following behavioral testing. Median scores computed for each group were subsequently analyzed by separate Kurskal-Wallis ANOVA tests. These analyses revealed significant differences among these groups for durations of nipple self-stimulation ($H(4) = 30.75$, $p < 0.001$) but not for durations of other grooming behaviors. Between-group comparisons of nipple self-stimulation data were made with Mann-Whitney U-tests and showed that dams of the EB-treated groups, while not differing significantly from one another, displayed significantly less nipple self-stimulation compared either to dams of the oil-treated or nontreated control groups ($U \leqslant 6.5$, $p < 0.002$).

TABLE 1. Average ($\pm$ SD) Nipple Length (mm) for R-S Mice on GD 18

	No Treatment	Oil	0.01 µg EB	0.02 µg EB	0.04 µ EB
n	5	6	6	5	5
$\bar{x}$	1.65	1.65	1.52	1.55	1.62
SD	$\pm$ 0.15	$\pm$ 0.12	$\pm$ 0.13	$\pm$ 0.26	$\pm$ 0.15

Adult Rockland-Swiss (R-S) albino mice were timed-mated and randomly assigned to one of the five groups given above. (Housing conditions and methods of hormone administration were identical to those described in the autogrooming study.) On gestation day (GD) 18, the females were anesthetized i.p. with 0.1 ml per mouse Chloropent (Ford Dodge Laboratories), and their nipples were measured under a dissecting microscope. Using dial calipers, all 10 nipples of each female were measured to the nearest 0.001 mm during three independent sets of observations. Observers remained naive to the conditions employed in the experiment and interobserver reliabilities exceeded 0.90 (Spearman's rho) for inclusive data. Nipple lengths for individual animals were averaged and group means were compared by ANOVA ($F(4,23) = 1.03$, $p > 0.05$).

It is unlikely that the observed decrease in nipple autogrooming resulted from a general disturbance of motor behaviors, since EB-treated dams continued to display long durations of noncritical grooming. Moreover, inadequate nipple growth was not a factor associated with the EB-induced suppression of nipple self-stimulation. Estrogen may alter nipple-mammary sensitivity locally or it may change the behavioral threshold for self-stimulation via its spinal or central actions.

REFERENCES

1. HARVEY, P. J. & M. A. MANN. 1987. Dev. Psychobiol. **20:** 593–602.
2. KOMISARUK, B. R., N. T. ADLER & J. HUTCHISON. 1982. Science **178:** 1295–1298.
3. SIEGEL, H. I. & J. S. ROSENBLATT. 1975. Horm. Behav. **6:** 211–222.
4. McCORMACK, J. T. & G. S. GREENWALD. 1974. J. Endocrinol. **62:** 101–107.
5. ROTH, L. L. & J. S. ROSENBLATH. 1967. J. Comp. Physiol. Psychol. **63:** 397–400.
6. TOPPER, Y. J. & C. S. FREEMAN. 1980. Physiol. Rev. **60:** 1049–1106.

Regulation of the Seasonal Changes in Grooming Reflexes by Social Cycles

RALPH JOHNSON[a] AND WALTER RANDALL

Department of Psychology
University of Iowa
Iowa City, Iowa 52242

INTRODUCTION

Cats with pontile lesions develop grooming reflexes and an associated clonictonic seizure.[1] The receptive field for the grooming reflexes undergoes seasonal changes in size thought to be controlled by photoperiod.[2,3] We have found that a concomitant of photoperiod, cycles of human activity that were consistently associated with the light period, is essential for inducing the seasonal changes.[4] In the previous studies food and water were available continuously, and the cycles of human activity consisted of daily care and cleaning and other laboratory routines. In the study reported here, the cycles of human activity and food availablility were carefully controlled and monitored. The data consisted of actograms and measures of the size of the receptive field for the grooming reflexes and were obtained with the cycles of human activity minimized and with the cycles of human activity exaggerated. Exaggeration of the cycles of human activity effected changes in the receptive field in light-dark (LD) cycles, whereas no change was obtained when the human disturbances were minimized. When cycles of human activity were presented in constant light, the seasonal changes in the size of the receptive field for the grooming reflexes were obtained. Manipulations that were effective induced two peaks on the actograms, peaks that were associated with the semidaily presence of humans at the transitions of the daily cycles.

METHODS

Study I

Five cats with pontile lesions and with large receptive fields for grooming reflexes were individually housed in stainless steel cages in a sound-attenuating room. Two LD cycles (LD 9:15, which induces large receptive fields, and LD 15:9, which induces small receptive fields for the grooming reflexes) were used. Both LD cycles were presented with minimal human disturbances and with exaggerated human disturbances. The min-

[a] Present Address: Department of Psychiatry and Behavioral Science, Health Sciences Center, State University of New York at Stony Brook, Stony Brook, New York 11794-8101

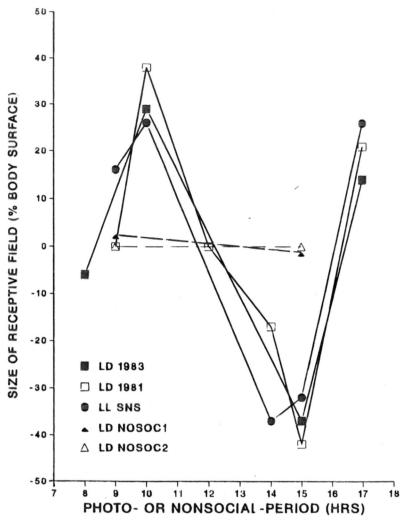

FIGURE 1. The response curves from this and two previous studies.[2,3] The ordinate values are the percentage deviation from the mean for each group and represent stable end points at the indicated periods. The data from study 1 on LD cycles with exaggerated disturbances replicated the previous findings (+18% at LD 9:15 and −17% at LD 15:9) and were not included in this figure. The pattern of changes obtained in study 2 in constant light with social cycles (*LL SNS*) was similar to the pattern from the previous studies in LD. Replicated data (*LD NOSOC1* and *LD NOSOC2*) from study 1 indicate that LD cycles with minimal human disturbances are without effect.

imal disturbance (LD NOSOC) consisted of 15 minutes of caretaking at light onset. The exaggerated disturbances were presented during the light phase of the LD cycles and consisted of a recording of colony sounds that was played throughout the light phase of the LD cycles. In addition, humans were present at all transitions of the cycles and food was available only during the light phase. Thus there were four manipulations, LD 9:15 and LD 15:9, each presented with and without the "social" concomitants of the photoperiod. Replications of the results on the two cycles without the social concomitants were obtained.

Study II

Eight cats with pontile lesions and large receptive fields for the grooming reflexes were housed individually in stainless steel cages in a large room in an LD 10:14 cycle. The exaggerated "social" stimuli were presented concomitantly with the light phase of the LD 10:14 cycle, and food was restricted to the light phase. When the size of the receptive fields had stabilized, LD was changed to LL (constant light), and then five different social cycles were presented in LL (LL SNS). Measures of the size of the receptive field and of activity (photobeam interruptions) were obtained for both studies.

RESULTS

In the study in LD (study 1), a prominent bimodal pattern of activity was found when the "social" factors were systematic concomitants of the photoperiod. The peaks of the bimodal pattern occurred at the concomitant transitions of the LD and social cycles. In the LD cycles with minimal disturbance (LD NOSOC), the bimodal pattern was not apparent. A statistical analysis of these actograms (a treatments-by-treatments-by-subjects design) indicated that no difference in activity pattern between the two LD cycles could be detected when the disturbances were minimized. The analysis of variance did indicate that exaggeration of the disturbance significantly changed the

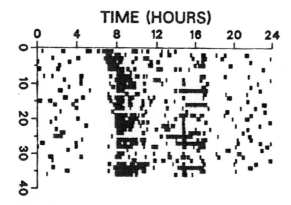

FIGURE 2. An actogram from the main colony showing the separate influences of LD and human cycles in LD 10:14. Onset of light and entry of humans into the room occurred at 0800 local time. On day 7 onset of light and entry of humans occurred 1 hour later because daylight saving time had ended. The blanks in the vertical band of activity at 0800 are days when no one entered the room. Beginning on day 18, a thin vertical line at 1430 is a concomitant of a person entering the colony at this time.

daily activity patterns. The data on size of receptive field and cycles are plotted (FIG. 1) in the typical format of photoperiodic response curves. The predicted change in the size of the receptive field was effected, from 18% to −17%, only when the LD cycles were paired with the social-nonsocial cycles (not shown in FIG. 1).

In study 2, an unexpected decrease in the size of the receptive field occurred when LD 10:14 was changed to LL. The decrease in size suggested that the cats had reversed the phase of the cycle in the absence of LD and were responding as if the remaining social cycle was 14:10 rather than 10:14. In the subsequent changes of the SNS cycles in LL, the cats exhibited orderly changes in the size of the receptive field for grooming reflexes and continued to respond as if they had reversed the phase of the cycles. The response curve is included in FIGURE 1 (LL SNS). The curve is plotted using the length of the nonsocial phase, e.g., the point that is plotted for a period length of 14 is the value obtained from the SNS 10:14 cycle. For comparison, response curves from two previous studies[2,3] are included in FIGURE 1.

An actogram from the main colony (FIG. 2) illustrates the effects of both LD cycles and humans. Light onset and entry to the room occurred concomitantly in the morning at 0800 hour local time. However, humans did not enter the colony at 0800 on Sundays and holidays, but the lights were turned on by an automatic clock. This cat systematically anticipated the LD transitions by inactivity and was active in the presence of humans.

CONCLUSIONS

The results indicate that LD cycles do not regulate the seasonable changes in the size of the receptive field for grooming reflexes. Furthermore, "social" cycles are essential and sufficient. Thus the previous studies on photoperiodic response curves for the size of the receptive field[2,3] were probably studies on socioperiodic response curves; the human activity in these studies was confined to the photoperiod of the LD cycles and thus constitutes an uncontrolled variable. The cat is a human symbiont[5] and apparently more readily influenced by cycles of human presences than by LD cycles. However, both "social" and LD cycles have influences on the activity patterns of the cat (e.g., FIG. 2). Similarly, circadian rhythms in humans may be entrained either by "social" or LD cycles.[6]

REFERENCES

1. RANDALL, W., S. RANDALL & R. JOHNSON. 1985. Behav. Neurosci. **99:** 109–121.
2. RANDALL, W. 1981. J. Comp. Physiol. **141:** 227–235.
3. JOHNSON, R. & W. RANDALL. 1983. Behav. Neurosci. **97:** 195–209.
4. RANDALL, W., R. JOHNSON & S. RANDALL. In preparation.
5. RANDALL, R., R. JOHNSON, S. RANDALL & T. CUNNINGHAM. 1985. Behav. Neurosci. **99:** 1162–1175.
6. WEVER, R. 1985. Ann. N. Y. Acad. Sci. **453:** 282–304.

Kassinin and Xenopsin: Effects upon Grooming in the Mouse[a]

R. J. KATZ

CIBA-GEIGY Corporation
Summit, New Jersey 07901

INTRODUCTION

We previously have demonstrated that grooming syndromes might be elicited in rodents both by a variety of tachykinin neuropeptides, including substance P and eledoisin,[1,2] and that tachykinin-elicited grooming is blocked by neurotensin. Kassinin, a peptide derived from the African frog *Kassina senegalensis*, possess a COOH terminus tripeptide and position 5 Phe residue typical of other tachykinins and has been identified in the mammalian CNS. Xenopsin, an octapeptide derived from *Xenopus laevis*, is structurally similar to neurotensin and may interact with substance P.[3,4] Both peptides were examined for their effects upon grooming based upon their structures.

MATERIALS AND METHODS

Materials and methods of procedure including surgery, habituation, and behavioral quantification procedures were identical with those previously published.[1,2] Doses of Kassinin and xenopsin (product nos. 7111 and 8401, Peninsula Laboratories) were 0–1 µg, were prepared immediately prior to use, and injected via microsyringe in 5 µl Ringer-Locke solution.

RESULTS

Kassinin produced dose-related increases in grooming. Grooming tended to be directed at the head and neck and involved washing. FIGURE 1 provides dose-response curves for the effects of Kassinin. Incidental behavioral observations of these subjects indicated the concomitant presence of circling, forward locomotion, and coarse tremor. xenopsin did not increase grooming. It antagonized substance P and Kassinin-elicited grooming (FIG. 2).

[a] These studies were carried out at Johns Hopkins University and supported in part by a Sloan Research Fellowship in Neurosciences. Findings do not represent positions of the CIBA-GEIGY Corporation.

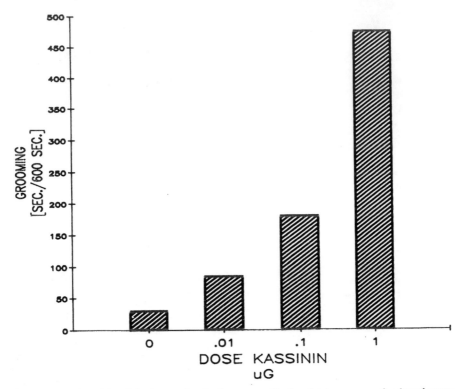

FIGURE 1. Kassinin-elicited grooming in the mouse. Each point is a mean value based upon five to seven determinations. The two higher doses differ significantly from vehicle (as determined by U tests).

DISCUSSION

These findings extend the pharmacological generality of tachykinin-elicited grooming. Given the behavioral structure of the grooming syndrome and previously published data, it is possible to categorize grooming syndromes into head and tail grooming, elicited by opioids, and a facial groom-wash syndrome elicited by tachykinins. Conversely, the class of neurotensin-related neuropeptides may serve as antagonists of tachykinin syndromes.

REFERENCES

1. KATZ, R. J. 1980. Substance P elicited grooming in the mouse. Behavioral and pharmacological characteristics. Int. J. Neurosci. **10:** 187–189.
2. KATZ, R. J. 1979. Grooming elicited by intracerebroventricular bombesin and eledoisin in the mouse. Neuropharmacology **19:** 143–146.

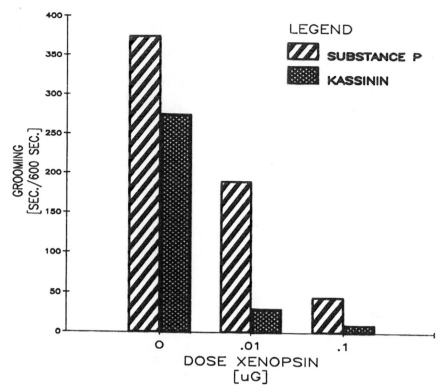

FIGURE 2. Xenospin reduction of grooming in the mouse. Each point is a mean value based upon five to seven determinations. The highest dose of xenopsin produced a significant reduction, determined by U tests, in tachykinin grooming (for both substance P and Kassinin).

3. MAGGIO, J. E. 1985. Kassinin in mammals: The newest tachykinin. Peptides **6**(3): 237–243.
4. FOREMAN, J. C., C. C. JORDAN & V. PIOTROWSKI. 1982. Interaction with substance P receptor mediating histamine release from rat mast cells and the flare in human skin. Br. J. Pharmacol. **77**(3): 531–539.

The Grooming Pattern in Bonnet Macaque, *Macaca radiata* (E. Geoffroy)

G. U. KURUP

Zoological Survey of India
Western Ghat Regional Station
Calicut, India

"Grooming is an important pattern of Primate social response from which have evolved varied and highly significant kinds of social service."[1] In this paper grooming protocol, its strategy, and the main variables affecting its spatial and temporal incidence are analyzed.

This study was conducted for varying periods during a 4-year program. An uninterupted intensive 5-day weekly dusk to dawn study for a period of 18 months formed its core part. The study site was the Indian Institute of Technology campus adjacent to the Guindy National Park in the metropolitan city of Madras, a 1000-acre urban natural site where six groups of bonnet macaque groups lived in free-ranging state. The study group at the beginning of the study consisted of 53 individuals: 10 males, 15 females, 16 subadults, 10 juveniles, and 2 infants. The mean figures for climatic parameters during a 7-year period in which the present study fell were: rainfall, 123.3 mm (0.5 to 483 mm); day temperature, max 32.63°C (28.3°C to 37.3°C) and min 24.5°C (20.6°C to 28°C); relative humidity, 72.7% (61.1%–85.2%). The sampling technique adopted was instantaneous scan sampling at 10-minute intervals with six individuals sampled at a given sampling bout. The tree canopy was classified into three horizontal layers (the lower, middle, and top layers) and three vertical zones (the inner, central, and peripheral zones) for study of placement of individuals. Similarly, a day was divided into four quarters for study of diurnal variations in incidence.

TABLE 1. Grooming Interactions of Age-Sex Class Combination Pairs (%)

Groomer	Groomed						
	M	F	Sm	Sf	J	I	Total
M	5.0	4.0	4.0	0.5	0.5	—	14.0
F	4.0	24.0	4.0	4.0	4.0	8.0	48.0
Sm	6.0	6.0	7.5	0.5	1.0	—	21.0
Sf	2.0	7.0	—	3.0	0.5	1.5	14.0
J	—	1.5	—	1.0	0.5	—	3.0
I	—	—	—	—	—	—	—
Total	17.0	42.5	15.5	9.0	6.5	9.5	

In order to obtain amount of total grooming interaction of a grooming pair, the figure of its grooming the other and that of its being groomed by the other have to be combined. Thus, grooming between female and subadult male is 4 + 6 = 10%. Abbreviations: M, male; F, female; Sm, subadult male; Sf, subadult female; J, juvenile; I, infant.

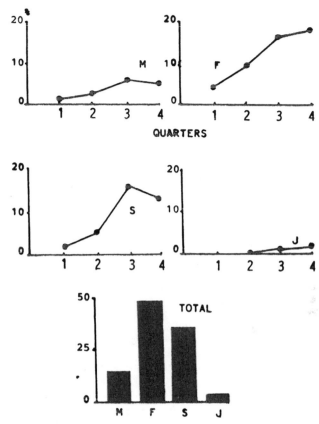

FIGURE 1. Age-sex class differential in grooming variations during different quarters of the day and also as totals for the whole day: *M*, male; *F*, female; *S*, subadult; *J*, juvenile.

The study group spent 13.4% of its time in grooming activity. The share of age-sex classes in this respect is shown in FIGURE 1.

Females and subadults of both sexes perform more grooming than they receive, suggesting that grooming is an urge on the part of these classes. The adult female grooms the most, followed by subadult male (not, as one might expect, subadult female). Adult males, juveniles, and, of course, infants perform grooming much less than they receive it. Juveniles are groomed for double the amount of time they themselves groom. Subadult males are groomed more than subadult females. When grooming activity is considered in proportion with other activities of each class, the relative percentages of time spent are as follows: male, 12%; female, 17.5%; subadult, 16.5%; and juvenile, 8%. The difference between males and females in amount of grooming is highly significant (Z test).

The grooming activity observed was classified into six pairs, with a total of 36 groomer-groomee combinations as shown in TABLE 1. Female-female grooming (24%) is overwhelmingly greater than in all other class combinations. Next, in order, are female grooming infant, the grooming between subadult males, and then the grooming

TABLE 2. Spatial Distribution of Grooming during Different Quarters of the Day (%)

Quarters	L	M	T	I	C	P	Tree	Ground
First	100.0	–	–	20.0	60.0	20.0	17.24	82.76
Second	13.56	86.44	–	11.86	62.71	25.42	74.68	25.32
Third	16.67	81.75	1.57	9.52	73.81	16.67	63.96	36.04
Fourth	9.02	60.66	30.33	8.20	63.11	28.69	64.55	34.45
Day	14.42	73.08	12.50	9.94	67.31	23.08	63.16	36.84

Letters L–P indicate canopy levels: L, lower; M, middle, T, top and canopy zones; I, inner, C, central; and P, peripheral.

of female by subadult female. Male-male grooming is a mere 5%. Homosexual grooming (60.5%) clearly predominates over heterosexual grooming (21%) when all sex-age classes are considered. The rest is autogrooming (18%) by all sex-age classes except infant. Males and subadults appear to autogroom slightly more than females.

Diurnal distribution of grooming shows that maximum grooming takes place in the third quarter of the day and minimum in the first quarter. Afternoon grooming is significantly more than that in forenoon. While the males, subadults, and juveniles groom more in the second half of the day, the females groom significantly more (Z test) in forenoon quarters. Grooming does not vary significantly among seasons.

As to the spatial distribution of grooming, 63% of the grooming in a day takes place on trees. In the canopy, maximum grooming (almost three-fourths) takes place in the middle level and over 67% in the central zone. The top level inner zone is least utilized in this respect. All the grooming in trees in the first quarter (17%) takes place in the lower canopy level, and the grooming in the top level is done in the afternoon only. On the ground, however, first quarter grooming is more than double that of other quarters.

Proximity studies on the basis of interindividual distances showed that for 61% of the time there was near proximity, i.e., nearest individual distance was 1 meter excluding close proximity (physical contact) necessary for grooming.

Correlations of grooming with agonism, mounting (not mating), and resting were tested by the Pearson r correlation coefficient, and it was found that grooming was positively correlated with mounting. Mounting significantly correlated only with grooming. Grooming was often preceded or succeded by mounting or sometimes even interrupted for this purpose. The two therefore form an integral part of the social cohesive mechanism of the group.

REFERENCE

1. YERKES, R. M. 1933. Genetic aspects of grooming, a socially important primate behaviour pattern. J. Soc. Psychol. **4:** 3–25.

Pawgrooming Induced by Intermittent Positive Reinforcement in Rats

CINDY P. LAWLER AND PERRIN S. COHEN

Psychology Department
Northeastern University
Boston, Massachusetts 02115

Past research suggests that excessive drinking (schedule-induced polydipsia[1]) is the prototypical adjunctive behavior in rats exposed to intermittent food reward. Contrary to this, we found that grooming emerged as the dominant adjunctive behavior in five of nine adult male rats (Sprague-Dawley strain) initially exposed to fixed-time (FT) food reinforcer schedules with intermediate or long (120–240 s) intervals between rewards. However, rather than a general increase in all forms of grooming, one particular type of grooming was enhanced; our subjects engaged mainly in repetitive pawgrooming, including rubbing and licking of the forepaws and nail biting. In some conditions, pawgrooming bouts routinely lasted as long as 30 seconds.

Why have other researchers failed to find adjunctive grooming? Perhaps they failed to recognize perseverative pawgrooming as a part of the normal grooming response. We also suspect that the traditional use of short interreinforcer intervals during the initial induction of adjunctive behaviors favors drinking rather than grooming. FIGURE 1 illustrates this point; behaviors of two representative subjects during an FT 30-s schedule are shown. The upper panel shows that drinking is the exclusive adjunctive behavior for a rat initially exposed to a short FT schedule (60 s). The lower panel depicts a subject initially exposed to a longer FT schedule (240 s). For this subject, pawgrooming is the dominant adjunctive behavior.

Although the optimal conditions (long vs. short interreinforcer interval) for inducing pawgrooming and drinking may differ, our comparison of both behaviors within the same experimental context uncovered many similarities. The most obvious is that pawgrooming and drinking share a common motor component (licking). Moreover, the temporal locus of pawgrooming is comparable to that for drinking. Specifically, FIGURE 2 shows that pawgrooming began soon after reinforcement at each interreinforcer interval value and tended to continue further into the interval as the overall interval length increased. Although the total amount of pawgrooming in each condition varied between subjects, the proportion of session time spent pawgrooming generally increased as the interval length decreased; the same was true for drinking. Finally, both drinking and pawgrooming occurred at much lower levels during extinction sessions. These similarities suggest that drinking and pawgrooming may reflect a single underlying process.[2] If so, then pawgrooming, like drinking, may function to reduce physiological stress[3] associated with intermittent reinforcement.

In summary, we report that in addition to drinking, adjunctive grooming in rats is a robust phenomenon. Our research places adjunctive grooming induced by intermittent positive reinforcement in the context of grooming induced by other environ-

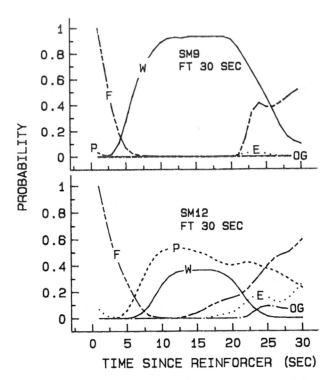

FIGURE 1. Behaviors of two rats during an FT 30-s schedule of food reinforcement: *P*, paw-grooming; *OG*, all other forms of grooming (excluding pawgrooming); *W*, drinking water; *E*, exploration (rearing, walking, and sniffing away from the food site); *F*, food tray entry. Data have been smoothed twice by computing running means of 3 successive seconds. The two rats differed with respect to their prior experimental history. The subject depicted in the *upper panel* (SM9) had been exposed to short (60 s) interreinforcer intervals during the initial condition of the experiment. For this subject, drinking was the exclusive adjunctive activity. The *lower panel* shows data from rat SM12, exposed to longer (240 s) interfeeding intervals during the first experimental condition. This subject engaged in both drinking and pawgrooming. However, pawgrooming occurred more frequently and continued further into the interval. Results from these two subjects and others suggest that the initial inducing schedule can affect the type of adjunctive behavior subsequently displayed in response to other reinforcement schedules.

mental stressors and raises questions concerning the differential control of various grooming components and the relation between grooming, drinking, and other schedule-induced behaviors.

REFERENCES

1. FALK , J. L. 1976. The nature and determinants of adjunctive behavior. Physiol. Behav. **6:** 577–588.
2. COHEN, P. S., T. A. LOONEY, F. R. CAMPAGNONI & C. P. LAWLER. A two-state model of

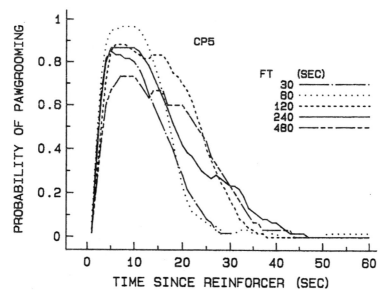

FIGURE 2. Temporal distributions of pawgrooming for subject CP5 for each of several FT food schedules. Data have been smoothed twice by computing the running means of 3 successive seconds. Only the first 60 seconds of the longer interreinforcer interval conditions are shown. In all conditions, pawgrooming peaked in the early part of the interval. At the longer interval lengths, there was a tendency for pawgrooming to spread further into the interval. Distributions of induced drinking for other subjects in this experiment were similar. We also found comparable distributions of induced attack in pigeons in an earlier experiment.[2]

reinforcer-induced motivation. *In* Affect, Conditioning, and Cognition: Essays on the Determinants of Behavior. F. R. Brush & J. B. Overmier, Eds. Lawrence Erlbaum. Hillsdale, NJ.

3. BRETT, L. P. & S. LEVINE. 1979. Schedule-induced polydipsia suppresses pituitary-adrenal activity in rats. J. Comp. Physiol. Psychol. **93:** 946–956.

Selective Enhancement of Grooming Behavior by the D-1 Agonist SKF 38393 in Rats following the Destruction of Serotonin Neurons[a]

IRWIN LUCKI AND ROBERT F. KUCHARIK

Department of Psychiatry
University of Pennsylvania
and
Veterans Administration Medical Center
Philadelphia, Pennsylvania 19104

Receptors for dopamine were classified into two categories, D-1 and D-2 receptors, based on different effects caused by dopamine agonists and antagonists on biochemical responses and in ligand binding studies.[1] Following the availability of the D-1 receptor agonist SKF 38393, grooming behavior and oral dyskinesias were described as behaviors caused by the selective stimulation of D-1 receptors.[2,3] Serotoninergic (5-HT) systems in brain have been suggested to exert an inhibitory role on dopaminergic functions. For example, interference with 5-HT function enhances the activity-increasing effects of dopamine agonists like amphetamine.[4,5] In order to characterize the influence of 5-HT neurons on behaviors caused by the activation of D-1 receptors, grooming behavior and oral dyskinesias induced by the D-1 agonist SKF 38393 were studied following the destruction of 5-HT neurons.

To destroy 5-HT neurons, 5,7-dihydroxytryptamine (5,7-DHT; 200 µg) was injected into the lateral ventricles of eight rats pretreated with desipramine (20 mg/kg, i.p.) under ketamine-xylazine anesthesia. Control rats ($n = 8$) received injections of only the vehicle (1% ascorbic acid). Three weeks were allowed for recovery before behavioral testing. Rats were observed in individual Plexiglas cages (44 × 24 × 20 cm) with a wire mesh floor and their behavior was videotaped for later scoring. Bouts of grooming behavior and episodes of oral dyskinesias were counted 10–60 min following the injection of various doses of SKF 38393 (1.25–20 mg/kg, i.p.). Grooming behaviors included paw licking, face washing, cleaning of the body, licking genitalia, and hindlimb scratching. Oral dyskinesias consisted of tongue protrusions, jaw movements, and perioral muscle fasciculations.

SKF 38393 produced a dose-dependent increase of both grooming behavior and oral dyskinesias in control rats compared with the administration of saline. For both behaviors, drug effects were maximal at 10 mg/kg (see FIG. 1), whereas 20 mg/kg produced less than the maximal response. Rats treated with 5,7-DHT did not differ in baseline grooming behavior or oral dyskinesia episodes from controls. However, 5,7-DHT-treated rats showed an enhanced number of grooming bouts after SKF 38393,

[a] This research was supported by USPHS grant nos. MH 36262 and GM 34781.

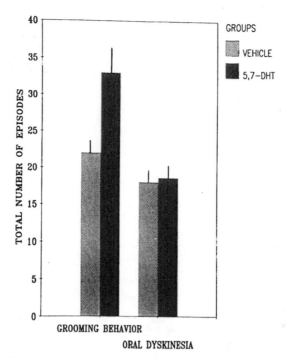

FIGURE 1. The effect of administering the D-1 receptor agonist SKF 38393 (10 mg/kg, i.p.) to rats treated with 5,7-DHT or vehicle controls (n = 8 rats/group) on episodes of grooming behavior and oral dyskinesias. 5-HT depletion caused by 5,7-DHT significantly enhanced the number of grooming behavior episodes caused by SKF 38393 ($t(14)$ = 4.81, $p < 0.01$). In contrast, the frequency of oral dyskinesias caused by SKF 38393 was unchanged in rats treated with 5,7-DHT ($p > 0.05$).

according to analysis of variance ($F(1,14)$ = 9.19, p = 0.008). Although the ED_{50} for increasing grooming behavior was unchanged (controls = 2.8 mg/kg; 5,7-DHT = 2.7 mg/kg), the maximal effect of SKF 38393 on grooming episodes was increased significantly (controls = 20.4 ± 2.2, 5,7-DHT = 32.0 ± 4.9; FIG. 1). More detailed analysis also suggested changes in the quality of grooming behavior caused by SKF 38393 (FIG. 2). Although grooming episodes and total grooming time were increased by 5,7-DHT, the time per grooming episode was significantly decreased ($p < 0.01$). Examining individual behaviors, the 5,7-DHT rats showed fewer sequential behaviors to complete an episode of grooming than did controls. This suggests that 5,7-DHT-treated rats showed a fragmenting of normal grooming behavior into shorter components, or that they were less capable of performing sequenced chains of behavior.

In contrast to the grooming behavior, the number of oral dyskinesia episodes induced by SKF 38393 did not differ from control rats (FIG. 1). Thus, depletion of brain 5-HT content enhanced selectively the increased grooming behavior caused by the D-1 agonist SKF 38393, and may indicate a special role for 5-HT neurons in this behavior. Oral dyskinesias, a second behavior in rats mediated by D-1 receptors, was unaltered by the depletion of 5-HT.

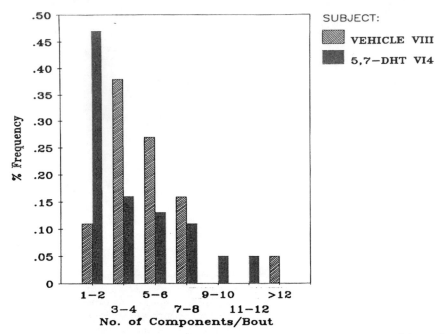

FIGURE 2. Frequency distribution for the number of behavioral components for each bout of grooming behavior caused by the D-1 receptor agonist SKF 38393 for two representative subjects. The control subject emitted 18 bouts of grooming during the 50-min observation period. Only 11% of these episodes involved one or two grooming components, whereas 65% of the grooming episodes were composed of chains of three to six grooming components. The subject treated with 5,7-DHT showed 36 separate bouts of grooming behavior. In contrast to controls, however, 47% of the bouts involved one or two components whereas only 39% of the episodes consisted of chains of three to six grooming components. This analysis suggests that the depletion of 5-HT increases grooming behavior to SKF 38393 by increasing the number of short-chain grooming episodes that are different topographically from the multiple-component chain of behaviors that comprise normal grooming for rats.

ACKNOWLEDGMENTS

We thank Dr. Carl Kaiser for helping us to obtain SKF 38393 from the Smith, Kline & French Laboratories, Philadelphia, PA.

REFERENCES

1. STOOF, J. C. & J. W. KEBABIAN. 1984. Two dopamine receptors: Biochemistry, physiology and pharmacology. Life Sci. **35:** 2281–2286.
2. MOLLOY, A. G. & J. L. WADDINGTON. 1984. Dopaminergic behaviour stereospecifically promoted by the D1 agonist R-SK & F 38393 and selectively blocked by the D1 antagonist SCH 23390. Psychopharmacology **82:** 409–410.
3. ROSENGARTEN, H., J. W. SCHWEITZER & A. J. FRIEDHOFF. 1983. Induction of oral dyskinesias in naive rats by D1 stimulation. Life Sci. **33:** 2479–2482.
4. GERSON, S. C. & R. J. BALDESSARINI. 1980. Motor effects of the serotonin in the central nervous system. Life Sci. **27:** 1435–1451.
5. LUCKI, I. & J. A. HARVEY. 1979. Increased sensitivity to *d*- and *l*-amphetamine action after midbrain raphe lesions as measured by locomotor activity. Neuropharmacology **18:** 243–249.

Role of the Cholinergic System(s) in the Expression of the Behavioral Effects of Bombesin[a]

Z. MERALI,[b] C. KATEB[c] AND P. KATEB[c]

[b] Department of Pharmacology
[c] School of Psychology
University of Ottawa
Ottawa, Canada

INTRODUCTION

Centrally administered bombesin (BN) induces time- and dose-dependent behavioral changes in mammals. The conspicuous pattern of behavioral activation in the rodent includes vastly increased grooming and ambulatory activity. The neural mechanism(s) subserving these potent effects remain unknown. The objective of this study was to investigate whether pharmacological manipulation of the central cholinergic system(s) altered the behavioral response to BN.

METHODS

Male Sprague-Dawley rats (300–350 g) were implanted with cannulae in the third ventricle. Injections of BN and/or cholinergic antagonists were performed in a randomized order. Scopolamine and scopolamine methyl bromide were injected 5 min before BN, and hemicholinium-3 30 min before.

Locomotor and rearing activity were monitored using a computerized infrared beam grid system and grooming elements were recorded by time-sampling procedure by human raters. Each animal was observed for 5 s, every 40 s, for the 1-h observation period. Thus a maximum score of 90 is possible for each grooming element.

RESULTS

Results demonstrate that scopolamine, administered centrally or peripherally, antagonized the effects of BN on grooming activity. The most prominent effects were apparent in facial grooming (facial scratching and licking). Scopolamine alone did not affect grooming behavior, as compared to control condition. It is of interest to note that, in contrast to the above effects, scopolamine potentiated the BN-stimulated

[a] This research was supported by the Natural Sciences and Engineering Research Council of Canada.

locomotor and rearing activity. These results indicate that the scopolamine antagonism of BN-induced grooming is not the result of generalized motoric depression.

To determine whether this antagonism was mediated centrally or peripherally, methylscopolamine (which does not readily cross the blood-brain barrier) was used as a pharmacological tool. Whereas peripherally administered methylscopolamine was relatively ineffective, centrally administered methylscopolamine was extremely effective in blocking the effects of BN on grooming. These data suggest that the site of antagonism is in the CNS.

To further test the role of the cholinergic system in the expression of the behavioral effects of BN, hemicholinium-3 (HC-3), which antagonizes cholinergic functions presynaptically by inhibiting uptake of choline and by inhibiting the synthesis of acetylcholine, was used. Rats pretreated with HC-3 demonstrated a very blunted grooming response to BN. Once again the effects were most prominent for facial grooming.

CONCLUSIONS

These results clearly demonstrate that the central cholinergic neurons must be involved in the expression of the BN-stimulated grooming. It has recently been reported that grooming behavior induced by ACTH also involves cerebral cholinergic neurons.[1] We have previously reported that dopamine receptor blockers also attenuate the behavioral effects of BN.[2] Haloperidol caused a general reduction in BN-induced excessive grooming without changing the composition of grooming behavior,[3] whereas the antagonism of BN-induced grooming by anticholinergic agents were found in the present experiments to be accompanied by a shift in the distribution of grooming elements. The main suppressive effect of these latter drugs appeared to be on the elements of facial grooming, particularly facial scratching. It is of interest to note that naloxone and neurotensin also cause a marked reduction in the scratching component.[3] Thus pharmacological manipulation of the different neurotransmitter systems appears to affect differentially the magnitude and distribution of BN-stimulated grooming elements. The relationship of the various neuronal system(s) in the initiation and expression of BN-induced excessive grooming needs further investigation.

REFERENCES

1. DUNN, A. & G. VIGALE. 1985. Neuropharmacology 4: 329–331.
2. MERALI, Z., S. JOHNSTON & S. ZALCMAN. 1983. Peptides 4: 693–697.
3. VAN WIMERSMA GREIDANUS, TJ. B., D. DONKER, R. WALHOF, J. VAN GRAFHORST, N. DE VRIES, S. VAN SCHAIK, C. MAIGRET, B. SPRUIJT, D. COLBERN. 1985. Peptides 6: 1179–1183.

Maternal and Self-Grooming in Norway Rats

Mechanisms and Consequences of Dissociable Components

CELIA L. MOORE

Department of Psychology
University of Massachusetts
Boston, Massachusetts 02125

Grooming is a behavioral category that includes components dissimilar in various respects. For example, recent work with maternal grooming in rats has identified grooming of the anogenital region as a component with mechanisms of control and consequences for development different from other forms of grooming. Observations of maternal anogenital licking (AGL) and body licking (BL) in 12 Long-Evans dams revealed the two patterns to be relatively independent of one another (FIG. 1). The observed dissociation is consistent with findings that maternal AGL is performed in response to specific pup cues such as water, salt, and odors in urine.[1,2]

In order to determine whether duration of individual AGL bouts was related to amount of available pup urine, pups were weighed before and after a licking episode. Two siblings (one male), varying in age from 2–15 days, were placed with each foster dam. Each pup was weighed to the nearest milligram immediately before and after a 15-min test during which AGL directed to each pup was measured to the nearest second. Weight loss accounted for 21% of the variance in AGL bout duration (FIG. 2, $F(1,151) = 39$, $p < .001$); pup age accounted for another 8%. Older pups produced a greater volume of urine, but their rate of weight loss per second of AGL was significantly greater. No sex differences in rate of elimination were found. These results further support the idea that maternal grooming of the anogenital region is causally distinct from other components of maternal grooming in rats. Unlike other components, it is regulated by salt and water appetite and the salt and water in pup urine.[1] It is also elicited by chemosignals in the urine, including sex-related products of the preputial gland.[3] Chemosignals from male pups are more effective; thus, males receive more AGL. There is no sex difference in other forms of maternal grooming, however. The distinction between genital and other grooming may be extended usefully to analyses of mechanisms for self-grooming. There is a sex difference in genital (but not other) self-grooming, which is related to circulating testosterone in socially housed juvenile rats but independent of circulating testosterone in rats observed alone in unfamiliar cages.[2]

In addition to immediate effects on the organism, grooming may have long-term consequences for development. Because genital grooming provides a qualitatively different pattern of stimulation from other grooming patterns, it may have different developmental consequences. Genital self-grooming has been linked to the pubertal growth of prostate glands and seminal vesicles in juvenile rats, and maternal AGL

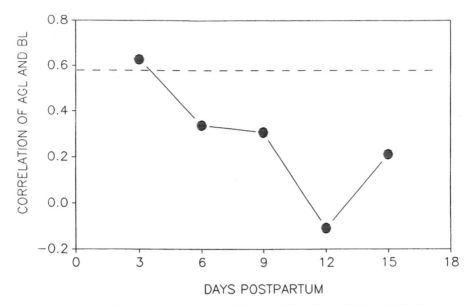

FIGURE 1. Correlation of maternal anogenital licking (AGL) and body licking (BL) for 12 dams each with litters of 8 pups. Data were scored from 4-h time-lapsed (8:1) videotapes recorded in red light during early dark cycle on each of days 3, 6, 9, 12, and 15 postpartum. The correlation was significant (*dotted line* represents $p < 0.05$) only on day 3. Individual dams maintained consistent levels of AGL across days but varied levels of BL. Thus, the mechanism for AGL may remain rather stable whereas that for BL changes with pup age.

has been linked to offspring copulatory behavior.[2] Further, direct comparison of maternal AGL and BL as predictors of offspring behavior has shown that both maternal grooming patterns were related to later offspring behavior but that AGL was more useful for predicting some outcomes and BL was more useful for others.[4] Specifically, BL, but not AGL, was related to juvenile activity level and AGL, but not BL, was related to juvenile self-grooming and copulatory efficiency of adult males. Both BL (both sexes) and AGL (males only) were related to juvenile play.

These results suggest that studies of grooming may profit from separate analyses of independent components of the behavioral category.

REFERENCES

1. GUBERNICK, D. J. & J. R. ALBERTS. 1983. Physiol. Behav. **31:** 593–601.
2. MOORE, C. L. 1986. Ann. N.Y. Acad. Sci. **474:** 108–119.
3. MOORE, C. L. & B. SAMONTE. 1986. J. Comp. Psychol. **100:** 76–80.
4. MOORE, C. L. & K. L. POWER. In preparation.

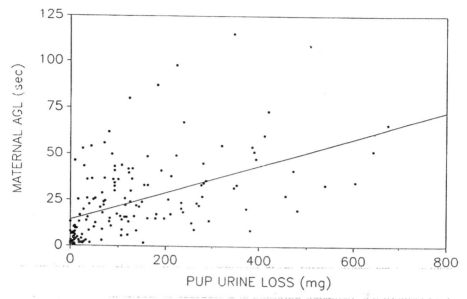

FIGURE 2. Regression of maternal AGL bout length (in seconds) during brief (≤ 15 min) test and pup weight loss (in milligrams) during test. Weight loss reflects amount of urine and/or feces eliminated by pup during test. Gains from milk intake were prevented by terminating test if pup attached to nipple. Because urination has a lower threshold than defecation, all losses probably reflect some urine loss; extreme measures probably include defecation. Sample includes 153 pups that received some AGL during test.

Selective Association of Grooming Acts and Penile Reflexes in Rats

BENJAMIN D. SACHS AND ANTHONY G. MOLLOY

Department of Psychology
University of Connecticut
Storrs, Connecticut 06268

Pharmacologial treatments have been described that increase grooming or penile reflexes or both. In some of these studies only one of these behaviors has been recorded, and for the most part attention has not been paid to the sequential ordering of these behaviors or of their components. It is therefore unknown whether grooming of the genitalia may provoke, or be provoked by, penile reflexes, or whether each behavior may be independently potentiated by pharmacological treatments and the physiological states they provoke. Our study was directed at beginning to answer this question. We examined the frequency, duration, and sequence of grooming acts occurring in association with penile reflexes displayed spontaneously outside the context of copulation.

Male rats ($n = 13$, 2-3 months old) were restrained in supine position with the penile sheath retracted as described previously.[1] The anterior torso and head were enclosed in a clear cylinder with an internal diameter (80 or 85 mm) that allowed the rat to groom his head and forepaws and to flex his neck enough to lick his anterior ventrum. Restraint was provided by paper (masking) tape placed across the midsection and over the base of the hindlegs and tail. After 2 or 3 weekly preliminary tests, the final test was administered. In this 30-min test, the animal's behavior was videotaped for subsequent analysis. A video date-time generator facilitated measurement of intervals and durations.

Grooming was classified according to its target, but in practice only paw licking and ventral licking occurred often enough during the target intervals to be included in the analyses. We assumed that without the restraint most episodes of ventral grooming would have been directed at the penis. Penile responses were classified as erections (engorgement of the corpus spongiosum) or "flips" (engorgement of the corpora cavernosa, resulting in rapid anteroflexion of the glans). These responses occur in clusters of 1 to 6 or more events within 3 to 10 s. Interevent intervals of more than 15 s defined separate penile response clusters (PRCs). For many of the data analyses, we compared grooming during 10 s centered on the midpoint of PRCs with grooming during 10 s centered on the midpoint of intercluster intervals.

As summarized in FIGURE 1, ventral grooming was about three times as likely to occur during PRCs as between them ($p < 0.05$). Licking of the forepaws was also reliably associated with PRCs ($p < 0.01$), but the magnitude of the difference in proportion was not as great as that for ventral grooming. The occurrence of flips within a cluster potentiated both paw and ventral grooming: the mean proportion of PRCs associated with paw licking increased from 0.38 to 0.53 if the PRC contained flips ($p > 0.10$) and the increase for ventral licking was from 0.21 to 0.42 ($p < 0.05$, one-tailed t-test). A major difference between paw licking and ventral licking is that the former preceded

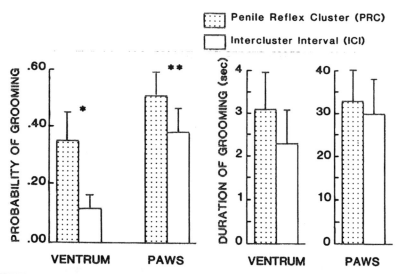

FIGURE 1. Mean probability and duration of grooming of paws and ventrum by supine rats during and between clusters of penile reflexes. Error bars indicate SEM. Significant differences: * $p < 0.05$; ** $p < 0.01$.

PRCs 53% of the time, whereas ventral licking rarely (19%) preceded the first penile reflex of a cluster ($p < 0.05$).

The differential results for paw and ventral grooming underscore the necessity of treating the components of grooming as distinct entitites. The pattern of results in this study suggests that the state of the male that characterizes incipient penile reflexes promotes paw grooming (and presumably other nongenital grooming), whereas afference from the occurrence of penile erections and flips may promote ventral grooming and, in a freely moving rat, genital grooming. In addition, since paw grooming tended to precede penile response clusters, conditions that promote grooming in rats may also promote the display of penile reflexes. Hence, some treatments that increase grooming, particularly of the genitalia, may do so by increasing the rate of spontaneously occurring penile reflexes.

REFERENCE

1. MEISEL, R. L. *et al.* 1986. Physiol. Behav. **37**: 951.

The Functional Significance of Baboon Grooming Behavior[a]

CAROL D. SAUNDERS

Neurobiology and Behavior
Cornell University
Ithaca, New York 14853

GLENN HAUSFATER

Division of Biological Sciences
University of Missouri
Columbia, Missouri 65211

Grooming is a widespread behavior among primates that is often considered an altruistic or cooperative activity. Since grooming has several functions, ranging from skin care and ectoparasite removal to social communication, there have been many selective pressures on this behavior. Previous studies of primate grooming relations have quantified only a few of the parameters necessary for testing sociobiological hypotheses concerning the distribution of grooming.

During 1980–81, an 11-month study of grooming behavior in a group of 45 yellow baboons (*Papio cynocephalus*) was carried out in Amboseli National Park, Kenya. The baboons were individually identifiable, matrilineal relationships were known, and paternal sibships could be reasonably estimated. The amount of grooming between individuals was measured by group scan samples every 30 minutes, and 20 15-minute focal animal samples per day. The focal animals were 12 adult females of different dominance ranks, ages, and kin network complexities.

To determine the major ecological costs and benefits of grooming for this study population, seasonal grooming patterns and ectoparasite-host relations were examined. The time the baboons spent allogrooming decreased slightly as the time spent feeding increased, but the correlation was not significant (Pearson's $r = - 0.538$, $p > 0.1$) suggesting that the cost of grooming with respect to lost feeding time is small. The rate that baboons encounter certain ectoparasites was estimated by weekly samples of ixodid ticks from the habitat using drag methods and carbon dioxide tick traps. Tick populations varied seasonally with respect to rainfall, but changes in tick densities had little effect on the amount of grooming or on the rate of grooming hand or mouth movements.

Two evolutionary theories, kin selection and reciprocal altruism, were considered with respect to grooming partner preferences by using this data base. As predicted by kin selection, grooming bouts occurred frequently between closely related individuals and less often between nonkin. In contrast, there was little support for the reciprocal altruism prediction that preferred grooming partners should be ones likely to reciprocate. As shown in TABLE 1, there was substantial asymmetry in the number of bouts

[a] This research was supported by grants from the NSF, the NIMH, and the Bache Fund of the NAS.

TABLE 1. Reciprocity of the Number of Grooming Bouts, Either Initiated or Given in Response to Solicitation, between Focal Females and Individuals of Different Age-Sex Classes

Age-Sex Class of Partner	No. of Bouts Focal Female[a] Initiates		No. of Bouts Partner Initiates	No. of Bouts Focal Female[a] Gives If Solicited by Partner		No. of Bouts Partner Gives If Solicited by Focal Female[a]
Adult male	174 (41%)	↔	248 (59%)	83 (51%)		81 (49%)
Adult female[b]	862 (67%)	↔	420 (33%)	275 (54%)		235 (46%)
Subadult male	22 (21%)	↔	82 (79%)	40 (55%)		33 (45%)
Juvenile male	116 (78%)	↔	32 (22%)	37 (79%)	↔	10 (21%)
Juvenile female[c]	24 (9%)	↔	253 (91%)	11 (25%)	↔	33 (75%)
Juvenile female[d]	220 (61%)	↔	140 (39%)	118 (89%)	↔	14 (11%)
Infant	585 (85%)	↔	105 (15%)	32 (91%)	↔	3 (46%)
Totals	2003 (61%)		1280 (39%)	596 (59%)		409 (41%)

Two-way arrows denote significant asymmetry in grooming frequency: $X^2 > \chi^2_{.01(1)} = 6.635$.
[a] Focal females or more dominant females.
[b] Lower ranking adult females.
[c] Distant or unrelated ($r < 0.25$) juvenile females.
[d] Closely related ($r \geq 0.25$) juvenile females.

given and received by the focal females with respect to age-sex class, whether the bouts were initiations or responses to solicitations by the partner.

There are two major kin selection models for explaining the length of time adult female baboons groom their selected partners. The Proportional Altruism Model (PAM) predicts that the groomer should vary the amount of grooming in direct proportion to the degree of relatedness of the partners. The Diminishing Returns Model (DRM) predicts that all grooming should be directed to the groomer's most closely related relative until the groomer's inclusive fitness gain reaches a point of diminishing returns. Different individuals will have different investment return curves depending on the demographic structure of the population. As shown in TABLE 2, the observed distributions of grooming by the focal females among their available kin were best explained by a version of the DRM.[1]

In contrast to the kin selection models, the TIT-for-TAT model of reciprocity[2] predicts that regardless of relatedness, baboons will initiate grooming and then closely monitor their bout lengths throughout an episode based on their partner's previous contribution. In general, bout lengths within a grooming episode were found to be similar to previous bout lengths (mean difference = − 0.058, SD = 2.37) but the correlation between bout length and previous bout length was small (Pearson's $r = 0.30, p < .0001$).

In sum, grooming behavior among yellow baboons appears to be a low-cost activity that extends beyond the hygienic function of ectoparasite removal and occurs frequently between closely related individuals. Of the kin selection hypotheses tested concerning the distribution of grooming among available relatives, the best predictions were generated by a version of the DRM. Tests of reciprocal altruism predictions yielded some support for reciprocity of bout length within grooming episodes, but there was little evidence for equal bout frequencies between individuals of different age-sex classes over longer time periods. Additional tests of these evolutionary models are currently underway.

TABLE 2. Distribution of Grooming by Adult Female Baboons among Available Relatives

Types of Kin Networks	No. of Focal Females per Category	No. of Available Kin	Kin Relationship[a]	OBS[b]	DRM[c]	PAM[d]	RANDOM[e]
Many close	2	6	M-O	0.8217	0.8516	0.5455	0.3000
and distant		6	sibs	0.1781	0.1484	0.2727	0.3000
kin		8	A-N	0.0002	0	0.1818	0.4000
Mostly close	2	7	M-O	0.9998	1.0000	0.9333	0.8750
kin		1	sibs	0.0002	0	0.0667	0.1250
		0	A-N	–	–	–	–
Mostly distant	5	5	M-O	0.4645	0.3350	0.2128	0.1042
kin		31	sibs	0.4818	0.6649	0.6596	0.6458
		12	A-N	0.0537	0	0.1277	0.2500
Few distant	3	0	M-O	–	–	–	–
kin		15	sibs	0.9529	1.0000	0.8824	0.7895
		4	A-N	0.0471	0	0.1176	0.2105

[a] M-O, Mother-offspring ($r = 0.50$); Sibs, full or half siblings ($r = 0.25$ to 0.50); A-N, aunt-niece/nephew ($r = 0.125$ to 0.25); r, coefficient of relatedness.

[b] OBS, Observed proportion of grooming =

$$\frac{\text{Min. of grooming given by focal female to a relatedness class}}{\text{Total min. of grooming given by focal female throughout study.}}$$

[c] DRM, Proportions expected by the Diminishing Returns Model, as formulated by Schulman and Rubenstein,[1] with $a_i = 2.5$.

[d] PAM, Proportions expected by the Proportional Altruism Model.

[e] RANDOM, Proportions expected by the number of individuals per relatedness class.
Kendall τ between OBS and DRM = 0.929.
Kendall τ between OBS and PAM = 0.809.
Kendall τ between OBS and RANDOM = 0.523.

REFERENCES

1. Schulman, S. R. & D. I. Rubenstein. 1983. Kinship, need, and the distribution of altruism. Am. Nat. **121:** 776–788.
2. Axelrod, R. & W. D. Hamilton. 1981. The evolution of cooperation. Science **211:** 1390–1396.

Effects of Lidocaine or Capsaicin on Scratching and Grooming Induced in Mice by Centrally or Peripherally Injected Bombesin[a]

H. WHEELER,[b] T. P. BLACKBURN,[c]
N. J. W. RUSSELL,[c] AND A. COWAN[b]

[b] Department of Pharmacology
Temple University School of Medicine
Philadelphia, Pennsylvania 19140

[c] Bioscience Department II
ICI Mereside
Alderley Park, Macclesfield, United Kingdom

Bombesin (BN) is a tetradecapeptide originally isolated from frog skin.[1] Several species respond to small doses of centrally administered BN with behaviors that are outwardly suggestive of changes in skin sensation. In mice, BN acts at spinal, as well as supraspinal, levels to cause dramatic scratching and grooming of the flanks, neck, ears and face. Intraperitoneal injection of BN also causes mice to scratch and groom excessively, but much larger doses are required.[2]

In the present work, the role of skin as a trigger for these BN-induced behaviors was studied following (1) topical application of lidocaine, a local anesthetic, and (2) pretreatment with capsaicin, a sensory neuron blocking agent.

MATERIALS AND METHODS

All mice were obtained from Temple University Skin and Cancer Hospital. Male Skh: HR-1 hairless mice (20–25 g) (n = 6–7) were dipped for 30 seconds in either saline or 4% lidocaine HCl topical solution U.S.P. The head and ears were painted separately. Evidence of local anesthesia was obtained as follows: one ear of each mouse (n = 12) was painted with either water or 4% lidocaine and the number of wiping motions to that ear was recorded over the next 5 minutes. Capsaicin (0.01%, Fluka) was then painted on the same ear and the number of wipes was again noted for 5 minutes. Capsaicin-induced wiping was significantly ($p < 0.001$, Student's t test) reduced in mice pretreated with lidocaine.

Male ICR albino mice (15–20 g) were pretreated with vehicle (2% ethanol : 2% cremaphor : saline) or capsaicin (50 mg/kg, s.c.) under pentobarbital anesthesia (50 mg/kg, i.p.) at 7 and 6 days prior to experimentation. On the day of the experiment,

[a] This study was supported by grant no. DA 03681 from the National Institute on Drug Abuse.

a drop of 0.01% capsaicin was instilled into an eye of each animal. Only vehicle-control mice showing a pronounced wiping response and capsaicin-pretreated mice showing no response were used.

Quantitation of grooming or scratching (hereafter termed "grooming") was as follows. Mice were each placed in an individual Plexiglas box, given BN (Sigma) s.c. (at $t = 0$ min) or i.c.v.[3] (at $t = 15$ min), and observed for grooming (at $t = 15$ min) for 15 min. Each of four mice ($n = 6-7$) was monitored for 5 out of every 20 seconds. Results were graphed as percentage of the maximum (i.e., 45) possible number of grooming episodes (MGE). The A_{50} value of BN (i.e., the dose of BN that is required to give 50% of MGE) was estimated by linear regression analysis.

RESULTS AND CONCLUSIONS

When BN (0.003, 0.03, and 0.10 μg) was given i.c.v., A_{50} values were comparable in saline-treated and lidocaine-treated mice, that is, 0.005 μg and 0.006 μg, respectively. When BN (5, 10, and 20 mg/kg) was given s.c., the lidocaine-BN dose-response line was displaced downward from the saline-BN line to a significant ($p < 0.05$) extent. The potency ratio was 2.10 (1.27–4.80). A_{50} values were 12.99 mg/kg (lidocaine) and 6.75 mg/kg (saline). The cues associated with peripherally injected BN are therefore sensitive to lidocaine. In contrast, lidocaine has no overt effect on centrally injected BN.

Pretreatment of mice with capsaicin (50 mg/kg, s.c., at -7 and -6 days) had no marked effect on behavioral responses to BN (0.003 and 0.10 μg, i.c.v. or 10 and 20 mg/kg, s.c.) Similarly, injection of ICR mice with capsaicin (50 mg/kg, s.c.) on day 2 of life did not attenuate grooming induced by BN (0.003 and 0.10 μg, i.c.v.) when the animals matured (unpublished results).

REFERENCES

1. ANASTASI, A., V. ERSPAMER & M. BUCCI. 1971. Experientia (Basel) 27: 166–167.
2. COWAN, A., P. KHUNAWAT, X. ZU ZHU & D. E. GMEREK. 1985. Life Sci. 37: 135–145.
3. HALEY, T. J. & W. G. M. MCCORMICK. 1957. Br. J. Pharmacol. Chemother. 12: 12–15.

Effect of Adrenocorticotropic Hormone (4–9) Analogue Org 2766 on Preening Behavior of Domestic Chickens

SANDRA WILLIAMS AND MAUREEN BESSETTE

Biology Department
Simmons College
Boston, Massachusetts 02115

Numerous investigators have shown significant increases in grooming behavior in rats,[1,2] in yawning and headshaking in pigeons,[3] and in preening in chickens[4] after intraventricular administration of ACTH and $ACTH_{1-24}$. An $ACTH_{4-9}$ analogue, Org 2766, has been shown to attenuate passive avoidance behavior in rats,[5] delay extinction of avoidance behavior in rats,[6] and normalize social behavior in rats.[7] The present experiments were conducted to test the effect of Org 2766 on the preening behavior of stressed chickens.

In experiment 1, eight female chickens were exposed to four treatments: (1) no handling; (2) 5 µg Org 2766/kg plus 10 minutes of handling; (3) 50 µg of Org 2766/kg plus 10 minutes of handling; and (4) 1 ml of 0.75% saline plus 10 minutes of handling. Injections were intraperitoneal.

Treatments were given in random order using a double-blind procedure. An injection was followed by a 5-minute period to allow uptake of the solution, then 10 minutes of handling. After a 10-minute delay, behavior was observed for 15 minutes in 30-second intervals.

In experiment 2, seven chickens received the same four treatments as in the first, except that Org 2766 was injected subcutaneously at doses of 25 and 50 µg/kg, and the time period for handling was reduced from 10 to 15 minutes.

Preliminary inspection of the data of these experiments showed that the only sizeable difference in behavior involved preening; therefore only this behavior will be discussed. The results of experiment 1 (TABLE 1) indicate that the preening behavior of handled saline-treated birds is greater than that of nonhandled controls, and the preening of handled Org-2766-treated birds is lower than that of either handled or unhandled saline-treated birds. Because the analysis of variance for single factor experiments with repeated measures[8] revealed that the distributions were homogeneous — $F(3,21) = 0.88$, $p > 0.05$ — we did not compare the treatment effects statistically.

In experiment 2, the distributions were found to be nonhomogenous, $F(3,18) = 7.54$, $p < 0.01$. A significant decrease in preening was found in the 50 µg/kg (handled) chickens versus the saline-treated handled birds (TABLE 2). The lack of significant difference in preening of nonhandled versus 25 µg/kg (handled) and 50 µg/kg (handled) birds suggests that birds pretreated with Org 2766 and then stressed exhibit preening behavior similar to nonstressed birds.

The influence of Org 2766 may be due to altered activity of endogenous opiate

TABLE 1. Experiment 1: Effect of Handling and Intraperitoneal Injections of Org 2766 on Incidence of Preening in Chickens

Treatment (n = 8)	Total Preening Events	Preening Events per Bird (Mean ± SD)	% Change[a]
No handling	26	3.25 ± 6.11	—
Saline plus handling	38	4.75 ± 5.06	46.2
5 µg/kg Org 2766 plus handling	21	2.63 ± 3.29	19.2
50 µg/kg Org 2766 plus handling	16	2.00 ± 2.27	38.5

[a] Each treatment compared to no handling.

TABLE 2. Experiment 2: Effect of Handling and Subcutaneous Injections of Org 2766 on Incidence of Preening in Chickens

Treatment (n = 7)	Total Preening Events	Preening Events per Bird (Mean ± SD)	% Change[a]
No handling	6	0.86 ± 1.07	—
Saline plus handling	49	7.00 ± 4.28[b]	716.7
25 µg/kg Org 2766 plus handling	26	3.71 ± 4.03	333.3
50 µg/kg Org 2766 plus handling	20	2.86 ± 2.19[c]	233.3

[a] Each treatment compared to no handling.
[b] Significantly different from no handling, $p < 0.05$, Newman-Keuls method.
[c] Significantly different from handled saline-treated birds, $p < 0.05$, Newman-Keuls method.

systems, a possibility consistent with evidence that ACTH-induced grooming involves high-affinity opiate receptors.[9] This effect may be caused by competitive binding of the analogue, causing it to become an antagonist and thereby preventing endogenous ACTH, produced during stress, from having its usual behavioral effects.

ACKNOWLEDGMENT

We are grateful to Organon International for supplying Org 2766.

REFERENCES

1. COLBERN, D. R., R. L. ISAACSON, E. J. GREEN & W. H. GISPEN. 1978. Behav. Biol. **23:** 381–393.
2. DUNN, A. J., E. J. GREEN & R. L. ISAACSON. 1979. Science **203:** 281–283.
3. DELIUS, J., B. CRAIG & C. CHAUDOIR. 1976. Tierpsychology **40:** 183–193.
4. WILLIAMS, N. S. & D. L. SCAMPOLI. 1984. Pharmacol. Biochem. Behav. **20:** 681–682.
5. FEKETE, M., F. DRAGO, J. M. VAN REE, B. BOHUS, V. M. WIEGANT & D. DE WIED. 1983. Life Sci. **32:** 2193–2204.
6. GREVEN, H. M. & D. DE WIED. 1973. Prog. Brain Res. **39:** 429–442.
7. NIESINK, R. J. M., & J. M. VAN REE. 1983. Science. **221:** 960–962.
8. WINER, B. J. 1971. Statistical Principles in Experimental Design, 2nd edit. McGraw-Hill. New York.
9. DUNN, A. J., S. R. CHILDERS, N. R. KRAMARCY & J. W. VILLIGEIR. 1981. Behav. Neural Biol. **31:** 105–109.

Index of Contributors